Key-Concepts in MIN

Series Editor

Klaus Dieter Maria Resch
Department of Neurosurgery
Nuevo Hospital Civil \Dr. Juan I. Menchaca
Hospital Escuela de la Universidad de Guadalajara
Guadalajara
Mexico

More information about this series at http://www.springer.com/series/16498

Klaus Dieter Maria Resch

Key Concepts in MIN - Intracerebral Hemorrhage Evacuation

Volume 1: Basics

Springer

Klaus Dieter Maria Resch
Department of Neurosurgery
Nuevo Hospital Civil \Dr. Juan I. Menchaca
Hospital Escuela de la Universidad de Guadalajara
Guadalajara
Mexico

ISSN 2662-7205 ISSN 2662-7213 (electronic)
Key-Concepts in MIN
ISBN 978-3-030-46511-7 ISBN 978-3-030-46513-1 (eBook)
https://doi.org/10.1007/978-3-030-46513-1

This Springer imprint is published by the registered company Springer Nature Switzerland AG
The registered company address is: Gewerbestrasse 11, 6330 Cham, Switzerland

Foreword

The main idea and concept of MIN (minimally invasive neurosurgery) is the minimization of the intraoperative trauma. The approach should be as small and atraumatic as possible, but simultaneously as invasive as necessary to perform the operation controllable and safely. MIN is based on the assumption that the postoperative clinical result will be the better, the less invasive and non-destructive the approach will be performed. This statement is valid not only for neurosurgery but obviously for any type of surgical intervention. The question arise why this nearly self-evident idea of MIN was not recognized and applied earlier from the beginning in neurosurgery.

A retrospective view into the history of neurosurgery shows that the idea to make the neurosurgical operation as atraumatic as possible was intended already by the pioneers of neurosurgery at the beginning of the twentieth century in Europe and in the United States. However, at that time, the restricted radiological and intraoperative localization and illumination of the operative field made a greater and more traumatic access to the region of pathology often indispensable. Examples for this statement are the development of the transsphenoidal approach to the pituitary region, which was developed minimal invasively before the World War I by Oscar Hirsch at the ENT unit in Vienna and Harvey Cushing in Baltimore. However, a general safe and less destructive application of this approach required better illumination and zoom which became possible not before the introduction of the microscopes in the 1970s. Edoscopes by Nitze are another example Prototypes that have been tested in neurosurgery already in the third decade of the twentieth century in the USA, but a broad clinical application was possible only after improvement of the optical quality of the endoscopes by HH Hopkins in the 1960s.

Meanwhile, in the last 40 years, the technical developments in imaging by means of CT and MRI made visible also small intracranial lesions. Additionally, computer-aided navigation systems made it possible to localize the craniotomy and the lesions in space with high accuracy during surgery. Programs at workstations became able to simulate and plan the approach of an intracranial procedure in 3D virtual space. Together with highly developed rigid and flexible endoscopes, this advanced technology was in the late 1980s ready for clinical application in MIN. Nevertheless, it has to be admitted that the term "minimally invasive surgery" was originally neither introduced nor applied by neurosurgeons. The British visceral surgeon John EA Wickham coined the term minimally invasive surgery in 1984 and published this philosophy in the British Medical Bulletin in 1986. The main purpose was to distinguish this new type of endoscopic and robot-assisted abdominal surgery from the standard open procedures.

The neurosurgeon who deserves the greatest merits in developing and popularizing the minimally invasive technique for neurosurgery was in the late 1980s and 1990s is Axel Perneczky (1945–2009). After receiving the chair of neurosurgery at the University of Mainz in 1988, he picked up the idea of minimally invasive surgery and applied it consequently to neurosurgery. However, he restricted the term MIN not to endoscopy alone, but he understood under MIN all techniques which make the operation less invasive and simultaneously safer, such as neuronavigation and 3D planning workstations.

Endoscopy became a useful tool during a part of a MIN procedure. Before starting clinical application, basic anatomic work was performed with Manfred Tschabitscher, professor of Anatomy in Vienna to establish the ventricular anatomy from the endoscopic view. Klaus Resch joined in that time the team in Mainz and contributed to the basic work with postmortem inspections of the intracranial pathologies. Later he integrated the ultrasound into the endoscope for application intracranially.

As all new and promising neurosurgical techniques also, the MIN was in the first decade of clinical practice challenged by great parts of the neurosurgical community. Nevertheless, it continued to improve and expand also to the field of spinal neurosurgery. I think at this point it is time to make a comprehensive survey of the concepts and methods of MIN.

Klaus Resch, who was involved in the development of MIN from its very beginning and continued to improve this technique during his whole neurosurgical career, gives in these four volumes a comprehensive overview of the methods and clinical application of minimally invasive neurosurgery. In this first volume, Klaus Resch concentrates on the historical evolution of the MIN and the era of Perneczky. Then he gives an overview about MIN key concepts and key techniques with mouth-tracked high-zoomed microsurgery, neurosonography, and neuroendoscopy.

The volumes I and II give a theoretical and technical background for MIN, and volumes III and IV are dedicated to clinical application in cases of intracerebral bleedings.

Peter Grunert
Senior Consultant Neurosurgery/Stereotaxy
University of Saarland
Homburg, Saar, Germany

Foreword

Congratulation to this outstanding book. I have known the author since his time as medical student in Heidelberg starting in 1989. He attracted my attention during the dissecting course in human anatomy. His enthusiasm and his ability in dissecting even complex regions of the body were outstanding. After his preclinical exam, he asked me if I was willing to become the supervisor of his medical thesis. He explained that his desire to become a neurosurgeon had already come up at his time in high school, and the work in the dissecting course had reinforced his plan. He also proposed a topic for his thesis himself. He wanted to examine trans-oral pathways to the brainstem in human corpses, which was at that time a no man's land. His enthusiasm and his skill in dissecting convinced me that I allocated him cadavers donated for medical research to start with his work.

From then on, when a cadaver was available, he spent days and nights in the basement of the department, checking the possibilities for trans-oral routes to the brainstem. His thesis "Beitrag zum transoralen Zugang zum Hirnstamm" has been highly awarded by the Medical Faculty of the University of Heidelberg.

Thereafter, as a young physician in the Department of Neurosurgery in Heidelberg, he maintained his close relationships to anatomy continuing his work refining several variants of trans-oral pathways to the basal aspect of the brain. By dissecting non-fixed specimens, he demonstrated the intact subarachnoid cisterns in relationship with the arteries on the basal aspect of the brainstem; finally, he extended his work to pathological situations. He always had his full heart in his work. During his subsequent residency at the Department of Neurosurgery in the University of Mainz, we lost contact to each other. It came as a real surprise, when in 1993 he sent me his book: *Perneczky, Tschabitscher, and Resch: Endoscopic Anatomy for Neurosurgery*, showing to me that my former thesis student had succeeded to reach a position in the leading group of neurosurgeons at that time.

During the many subsequent years, he developed together with others the concept of minimally invasive neurosurgery (MIN), the success of which, as he emphasizes, would have not been possible without a thorough knowledge of the anatomy, precise anatomical planning, and accurate performing of an exact designed procedure. At that time, when the imaging and computer-aided navigation came up in neurosurgeries, it has led to the believe that thorough direct knowledge of the anatomy is of less importance. This has turned out to be wrong. No question, imaging-guided surgery represents a great advance, but nevertheless the new techniques required profound anatomical experience to control and understand the images delivered by the new tools. It has become necessary as ever to have a second look into the anatomical training laboratory to get the clinical anatomy knowledge in order to be able to underlay this knowledge to the virtual pictures of imaging techniques.

On this fundament, the further evolution of MIN and the description of the MIN key techniques for neuromicrosurgery, neurosonography, and neuroendoscopy (presented in volume I and II) became a full success. These techniques need an adapted anatomical understanding, which strongly relies on the primary anatomical basis. Each new technique requires the knowledge of specific anatomical features and recognitions permitting the intracranial orientation. This is the basis for the final success of a surgery.

Taking together, the clinical success of MIN, here described in four volumes, will always depend on profound knowledge of the anatomy, which, as Klaus Resch always has emphasized, can be learnt and must be learnt by anybody who pursues the career of a neurosurgeon. I am sure this book will have the resounding success that it deserves.

Wilhelm Kriz
Professor and Chairman Emeritus
Anatomy and Cell-Biology
University of Heidelberg
Heidelberg, Germany

Foreword

In many of the surgical specialties (including Neurosurgery) there is the idea that 'the most important thing is to develop the surgical technique', which is only partly true.

My encounter with Axel Perneczky took place in the 1990s, in a context of frustration with the results I observed in complex neurosurgical cases during my training as a resident in Mexico. That made me question what neurosurgery was for me and prompted me to look elsewhere for an answer. I got that answer in April 1996 in Mainz Germany. In the operating room there, a simple person before a microscope, showed me not only a new and refined technique, but also a new path in neurosurgery. In my time in Mainz I learned from him a new way of seeing and understanding neurosurgery. I began my contact with Professor Axel Perneczky and with Minimally Invasive Neurosurgery, which would change my life and that of many neurosurgeons in Mexico.

I was lucky enough to meet the creator of this new philosophy in neurosurgery and to learn, together with his disciples, at the very center where minimally invasive concepts and techniques were generated and at a key moment in the transformation of neurosurgery. Receiving the teachings of Professor Axel Perneczky generated in me a commitment to transmit them, which began more than 20 years ago in Jalisco and western Mexico. We also established a Germany/Mexico Cooperation MIN-Education-Project since 2005 with Dr. Klaus DM Resch.

'The concept guides the technique', creates it, shapes it and recreates it to develop it to a culminating point in which it becomes obsolete 'through another concept that emerged from that very evolution'. Axel Perneczky took us - through minimal invasion - to a site where there was a better understanding of neurosurgical approaches without neglecting the humanistic aspects by considering less damage to brain structures, but with maximum therapeutic results.

22 years ago, the search for an answer for a neurosurgery that offered better options and results for patients continues to have the same answer, the answer of Axel Perneczky, the answer of the minimally Invasive concept in neurosurgery.

The book by Dr. Klaus DM Resch, one of his most important pupil and collaborator, is a recognition of the work and contribution of Prof. Axel Perneczky to

Neurosurgery and is the perfect tribute to a simple man who played the banjo solo for himself and who revolutionized the lives of many neurosurgeons and neurosurgery itself. Thank you very much Dr. Klaus D. M. Resch for this excellent and well-deserved tribute to our Professor and teacher, Dr. Axel Perneczky.

Hector Velazquez Santana
Chairman of Neurosurgery
Nuevo Hospital Civil JIM
University of Guadalajara
Guadalajara, Mexico

Preface

Volumes 1 and 2 are dealing with the preclinical basics, which are prerequisites to do this kind of surgery that is presented by illustrative cases, and in **volume 4**. This organization of the content is made to show that the **key techniques** and this surgery can be learned. Everybody who learns the key techniques of volume 1/2 will be able to get the results of volume 4.

The Real Da Vinci Code: Viewing <> Experience <> Science

To make the content more attractive and comprehensive, it is illustrated richly, and there are usual figures but also many conceptual symbolizing graphics (**Graph No.**), being part of the text itself. They are along the concept of Leonardo da Vinci: "To see is to understand." They are information carriers and apply for both hemispheres of the readers' brain by a visible message, not just a finding or photography.

Chapter 2 Vol. 1 presents the professional vita of A. Perneczky and the evolution of the keyhole concept. The long way and the fast evolution of techniques, concepts, and scientific events are reviewed, illustrated with the main figures and graphics.

The second part of this chapter is dedicated to the remembrance of A. Perneczky with many obituaries from friends and colleagues around the globe. It gives a multi-perspective insight by contemporary neurosurgeons, also giving Perneczky the place he deserves, and preventing neglecting the idea of MIN in future.

Chapter 3 describes the **key techniques** and **concepts** with review to the roots, which are the microsurgery concept of M.G. Yasargil and the key-hole concept of A Perneczky. The recent evolution is driven towards the **MIN Key Concept**, not asking for key-holes anymore, rather than for the **keys for MIN**. Furthermore, this new concept is pathophysiological based and needs a pathophysiological thinking rather than an imaging guidance. Perfect surgical indications are based on pathophysiology, the way of surgery on imaging. What is going on in this brain? This question cannot be answered sufficiently by slice-imaging (MR, CT, PET, etc.), it needs a real-time imaging on site from the perspective of the surgeon during surgery, and neurobiological knowledge. The following summarized key techniques of MIN are not a technical hype or fun for the surgery but tools to fulfilling the concept of MIN: Minimizing the overall trauma.

The most unknown MIN key is **ergonomics** applied to neurosurgery. This is the very core of the MIN Key Concept and is introduced in detail at the end of the chapter. It is presented in three major application fields: environmental (OR suit design), procedural (work-flow), and mental (neuropsychology of the surgeon) ergonomics.

The key techniques (Vols. 1 and 2) are presented meticulously, as they are not well-established today in neurosurgery: The beneficial use of **mouth-switch tracked microscope** with **high-zooming** leads to an advanced microsurgery with MIN approaches and MIN strategies. The end of the chapter presents the **Ergo-Tool** which is derived from the ergonomics concept and applied for the recent upcoming exoscopes.

High-end neurosonography is indispensable for MIN but is far away from advanced use in neurosurgery still today. Therefore, it is supported by an integrated, short sono-atlas, presenting the major features and rules to start up with.

Neuroendoscopy, once the flagship of the keyhole concept, is regularly in advanced use only in a few departments in Europe. It is in danger to be lost in Europe, getting the place of an exotic discipline beside usual neurosurgery. *The Endoscopic Anatomy Atlas for Neurosurgery* (1993) is transduced into an endo-atlas with life-endoscopy of clinical used aspects (155 figures). Additional endo-anatomy of latero-basal posterior fossa with an approach system is substituted finally.

The new combined imaging technique of **transendoscopic ultrasound** for neurosurgery (ENS) is inserted in a focussed version, an imaging, targeting, and navigating tool for endoscopy (*Transendoscopic Ultrasound for Neurosurgery*; Springer 2005).

LASER (Vol. 2) was running into the fate of an exotic technique at the very beginning after its introduction to neurosurgery. It will be addressed as a key technique due to its nonmechanical use-property and elegant transendoscopic application.

Sealing techniques (Vol. 2) belong to the key techniques in MIN, when dura mater suturing becomes nearly impossible in very small approaches. Sealing is of decisive importance to prevent complications and for very fast closure, a big advantage of MIN.

All these techniques have a common problem, which is the bad education and training status among contemporary neurosurgeons. It is nearly impossible to find a neurosurgeon today who can master all these key techniques, which is the major obstacle of MIN. These key techniques became somehow "unloved children", but they provide the surgeon with all abilities to master most problems of MIN in general. They make MIN evacuation of intracerebral hematomas safe and fast, simple and cheap.

This type of MIN surgery needs a **strong training concept** and practice (Vol. 2). It is present as an anatomical, manual-haptic, and ergonomics **operation simulation environment**. This training setting is shown and explained precisely. The best visual training is given in nonfixed fresh body postmortem, or more practically in **plastinated specimens**, which is therefore added in an own chapter (Vol. 2, Chap. 4). The variety of use and applications of such specimens are highlighted because of their availability, dryness, invulnerability, and brilliance with absolute accuracy and beauty. Moreover, in plastinated specimens, the execution of "Gestalt" Anatomy is possible.

Training is not to fulfill a schedule rather than to inform the brain of the trainee! It is a neuropsychological mind–body training.

As MIN needs a perfect mind and body performance, comparable with a solo artist, a chapter on physical and mental readiness with the concept of **Feldenkrais for MIN** is introduced at the end of neurosurgical basics (Vol. 2, Chap. 5). Last but not least, it provides the MIN surgeon to stay healthy in a high challenging art of work.

This is an **Experience-Based Medicine** (ExBM) contribution. It belongs to a scientific axiomatic field together with EBM, the falsification theory of C. Popper, the paradigm shift concept of Th. Kuhn and many theories and concepts involved in MIN. This **scientific axiomatic web** is the theoretical basics of the book. It is discussed in Chap. 6 "Scientific Conditions for MIN" (Vol. 1) and is indispensable for a scientific aided profession. But, MIN is only perfect with philosophy and a clear concept!

The last Chap. 6 "Scientific Conditions for MIN"(Vol.1) is dedicated to the victims of the STICH trails.

Now I want to thank all those, colleagues and friends but also opponents, who supported or urged this project for MIN, kindly or critically, helpful and challenging. They all have their contributions to keep this growing and strong enough to enable innovations not only by techniques but also by concepts and ideas. An extraordinary recognition I want to give to all the contributors of obituaries for Prof.

DDr. Axel Perneczky, at least it could be managed to publish them late. The neuroophthalmological cases were done in cooperation with Dr. Antal/ Prof. Dr. Mennel. For this interdisciplinary openminded oppotunity I express my gratitude. My thanks also go to the artists who prepared some of the graphic designs, Stefan Kindel and Heide Roesler. Special thanks go to PhD Sylvana Freyberg (Editor, Springer Co) for her supervision of the publication progress. My last thank is given to all my patients and their relatives who courageously pushed through the way against the mainstream to perform a MIN procedure. I thank cordially Dr. H. Velazquez Santana for the International German-Mexican MIN Education Project since 2005.

"Most what you have, you did not get it by yourself, you got it from others, as a present" (Baltasar Gracián, 1647). Finally, I thank for this present.

Guadalajara, Mexico Klaus Dieter Maria Resch

Contents

About the Author

Klaus Dieter Maria Resch had his medical education at the universities of Heidelberg, Zürich, and Vienna. (**Prof. Yasargil/Prof. Perneczky**). He started his work in clinical anatomy and established a **surgical simulation concept** and training environment. In the team of Prof. Perneczky, during his main training period, he elaborated **endoscopic anatomy for neurosurgery** and progressed to **endoscopic surgical simulation technique for aneurysms** (para-endoscopic dissection technique) which became the basis for endoscope-assisted microneurosurgery. From this experience, he derived the **concept of ergonomics in neurosurgery**, by describing the spatial, procedural, and mental conditions of the endo-microsurgical habitat. He used it also as a training environment for MIN. In 1996 he introduced **trans-endoscopic ultrasound** into MIN to navigate endoscopes in complex hydrocephalus and cystic lesions. In 2004 he started the cooperation project **Germany/Mexico in MIN training** for young neurosurgeons and was awarded as Prof. Honorario at the University of Guadalajara. He developed an **ICH evacuation concept and technique for MIN** with 5 MIN key techniques (mouth-tracked max-zoomed microsurgery, neurosonography, neuro-endoscopy, LASER, and sealing). Recently, he improved the **keyhole concept** into the **MIN Key Concept**. In 2012 he became an elected board member of the International Society on Minimally Invasive Neurosurgery (**ISMINS**).

Dedication to My Great Teacher Prof. Dr. Dr. h. c. Axel Perneczky

1

1.1 The Perneczky Era 1988–2009 (Graph 1.1)

1.1.1 Summary

The essence of the first Axel Perneczky Lecture 2014 will be presented below, with regard to the main events and with reference to the memories.

Drawn from 10 years of personal experience of close co-working with Perneczky (including 8 years at the University Hospital of Mainz), a scientific portrait will first be painted, incorporating the structures he created, but also containing his visions, innovations and inventions, and his teaching activities in numerous courses and hands-on events. His work as an author and editor, his love of art and his role as a promoter are documented. Last but not least, his professional and personal unique selling points complete the portrait.

After the biographical section will come an outline of the thematic section of the lecture (s. Chap. 3), followed by a brief outline of the roots of the development of the minimally invasive movement. Next comes an account of the evolution of the keyhole concept and its current further development. Finally, the significance of ergonomics for the conceptual development of MIN, which has thus far been by and large neglected, is summarised in terms of models, broken down into the three sub-dimensions of the spacial-, procedural- and mental-ergonomics of MIN.

1.1.2 History

The **International Society of Minimally Invasive Neurosurgery** was founded in 2011. Yoko Kato/Japan was elected its first president, and Armando Basso/Argentina, its honorary president. This was shortly followed in 2012 by the founding congress, from 20th to 23rd March in Florence under the presidency of Giovanni Broggi/Italy. This urgently needed interconnection of forces on an international level was, after the sudden absence of Axel Perneczky, on whose life's work the

K. D. M. Resch, *Key Concepts in MIN - Intracerebral Hemorrhage Evacuation*, Key-Concepts in MIN 1, https://doi.org/10.1007/978-3-030-46513-1_1

Graph 1.1 **"Korrektheit schützt vor Tadel, Größe aber erzwingt Bewunderung"**
"The flawless protects against blame, but true greatness compels admiration."
Karl Jaspers (Psychiatrist and Philosopher; Heidelberg/Germany)

society relies to a great extent, an expression of the desire to close this sensitive gap and not give up on this innovative development.

Peter Grunert published two important works on the history of endoscopy in 2009, and on the evolution of minimally invasive techniques in 2013. These clearly present the important stages in the development of techniques, schools of thought and events, and the historical roots during the Perneczky era.

For the second ISMINS congress in 2014 in Xian/China, under the presidency of Ling Feng/China, it was decided at the Executive Committee Meeting on 23.2.2014 to establish the institution of the "Axel Perneczky Lecture". The honour of presenting this lecture, in memory of my great teacher, was bestowed upon me by unanimous decision.

From 15th to 19th June 1993, the "First International Congress on Minimally Invasive Techniques in Neurosurgery" was held in Wiesbaden under the presidency of Axel Perneczky, with M.G. Yasargil as honorary president (Fig. 1.1). This has become the historic focal point for a global development that still continues today.

"The Society for Minimally Invasive Neurosurgery" was founded during the congress, and the journal "Minimally Invasive Neurosurgery" (Fig. 1.2) was established together with Thieme publishing house.

Fig. 1.1 First international congress on MIN, Wiesbaden 1993

With the support of the company Aesculap, the "Hands-On Course on Endoscopic Neurosurgery", which had already been developed several years before, became a regular institution (Fig. 1.3). These courses introduced hundreds of international doctors to the basics of these techniques and the concept of MIN. Over time, the concept and the courses were developed further, to cover "endoscopy-assisted key-hole micro-neurosurgery".

Also, in time for the congress, the first atlas on "Endoscopic Anatomy for Neurosurgery" (Fig. 1.4) was published by Thieme Publishing House (Perneczky et al. 1993), and with it the morphological basis that had been elaborated for the entire CNS over many years of work in Vienna and Mainz.

As was appropriate for Perneczky's esteem for art, the opening of the congress was accompanied by the vernissage "Kopf – Gehirn – Geist" (Head-Brain-Mind) (Fig. 1.5) organized by the senior author and the choreographer and dancer Heide Roesler. The scientific part was held in the historic buildings of the Wiesbaden casino house.

The thematic design already represented Perneczky's unmistakable MIN concept; that is to say, it was about not defining the concept according to a single technique, but offering a panorama of techniques with minimally invasive capacities: trauma reduction techniques, keyhole neurosurgery, stereotactic-assisted microsurgery, neuro-endoscopy, endovascular neurosurgery and radiosurgery were already part of the programme.

The first generation of special neurosurgical endoscopes from the companies Wolf, Storz and Aesculap were presented at the industry exhibition, together with the "Perneczky aneurysm clip system" and other keyhole instruments (Fig. 1.6).

The MIN concept had already been discussed at a meeting "surgical planning principle" in Mainz in 1990 (Fig. 1.7). The development from a standard technology to a customised technology adapted to the individual patient was already discussed there. The great resistance of influential representatives of German neurosurgery enabled the learning process of conveying the new concept and its further

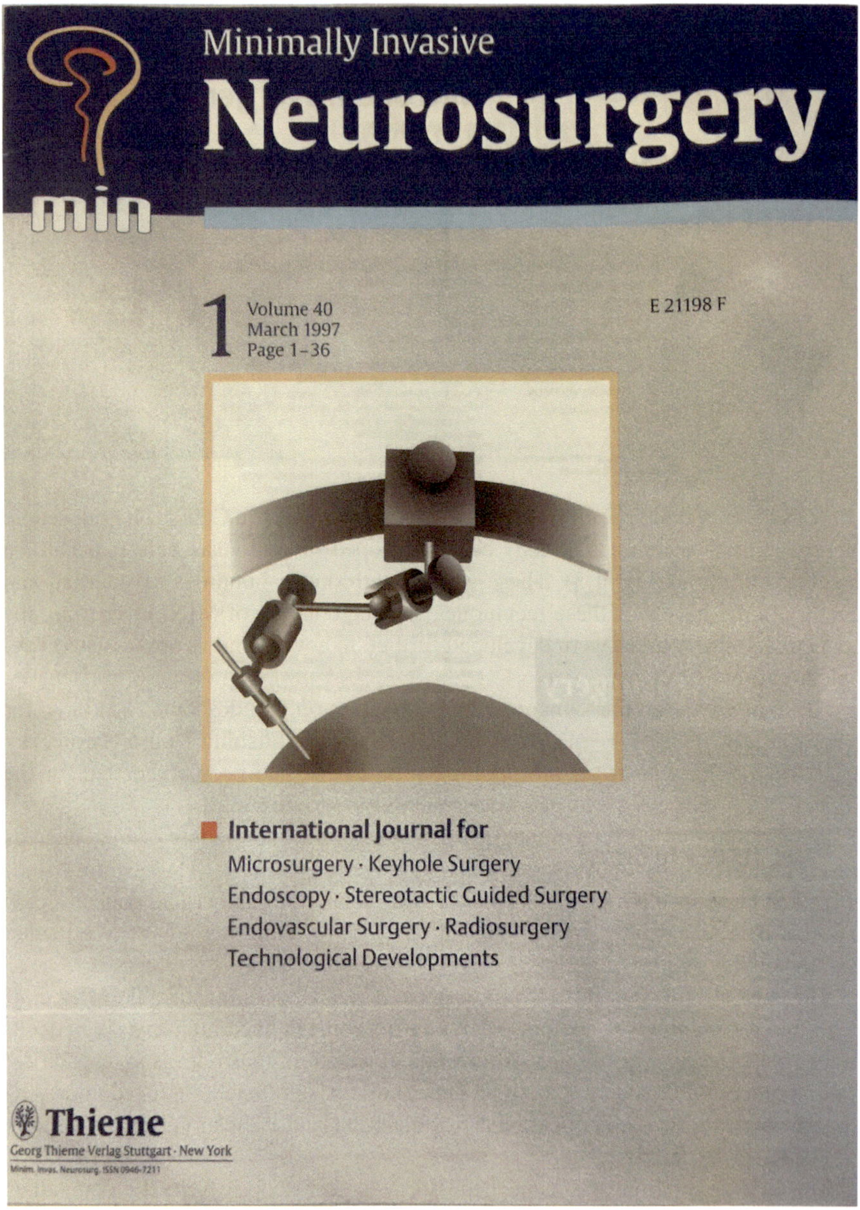

Fig. 1.2 First volume of MIN Journal 1993

Fig. 1.3 MIN Hands-on Cours at Aesculap Academy/Tuttlingen; Germany

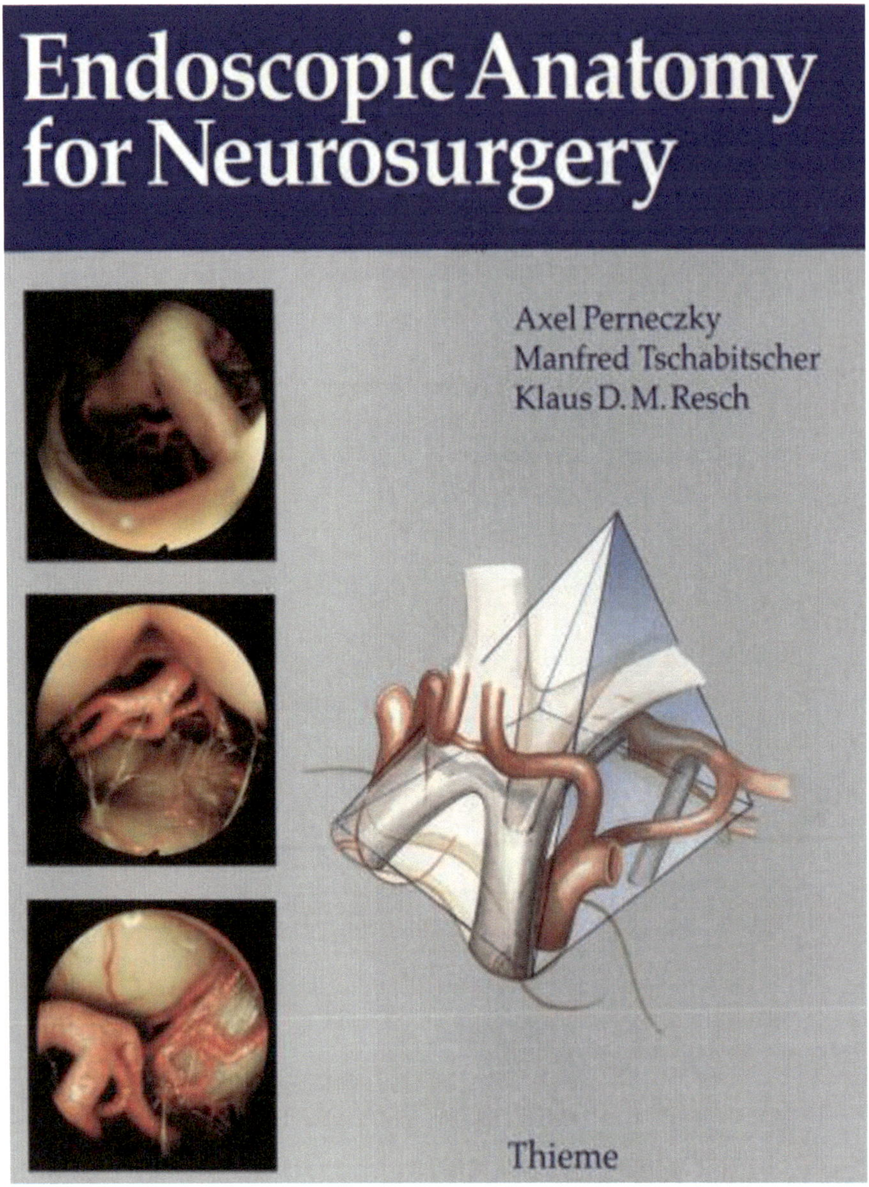

Fig. 1.4 Firs Atlas-Book on Endoscopic Anatomy for Neurosurgery (first major contribution in MIN 1993)

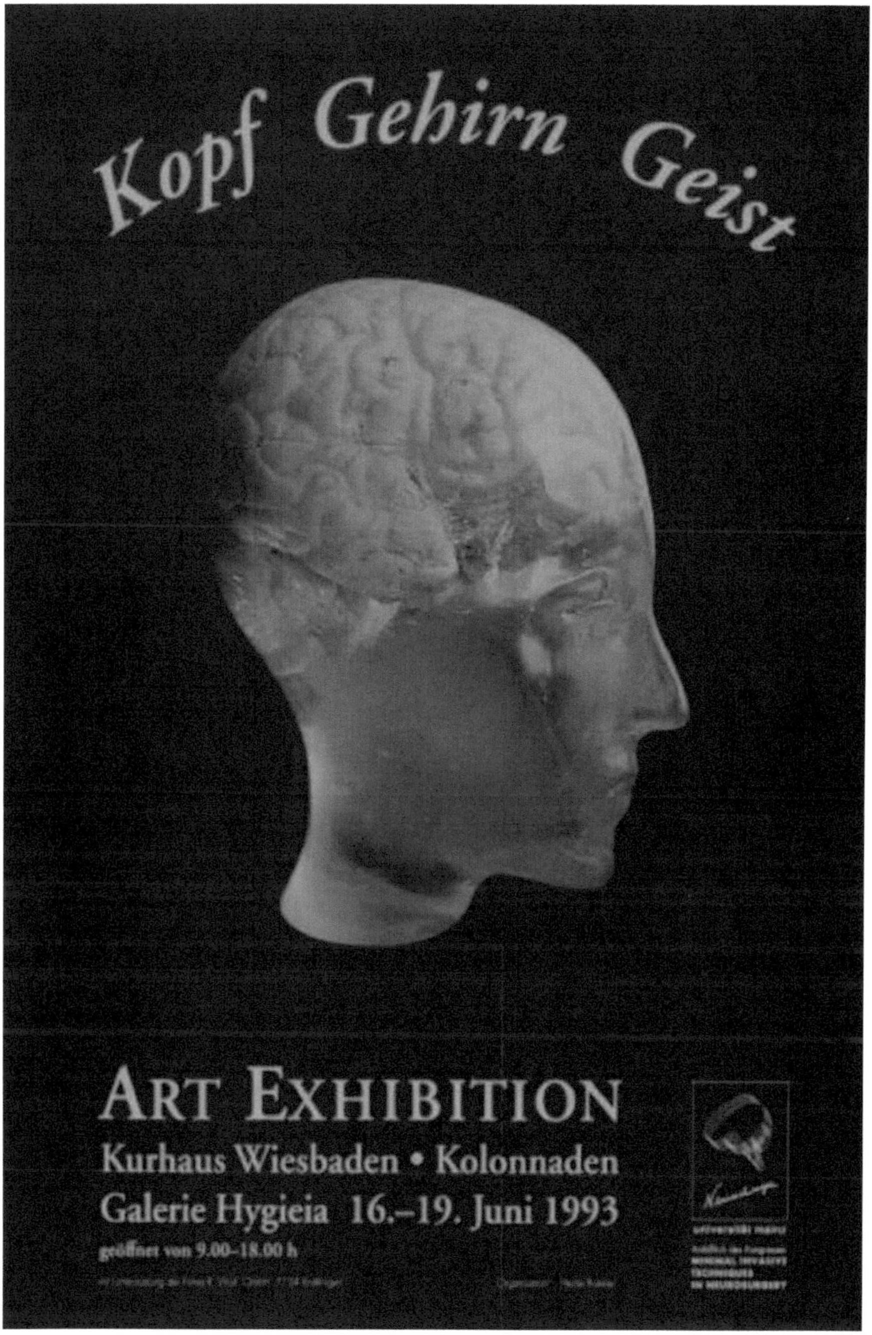

Fig. 1.5 ART-Exhibition/Opening of the first MIN Congress Wiesbaden/Germany. (organized by the author; Heide Roesler/artist and the Academy of Art, Stuttgart/Germany)

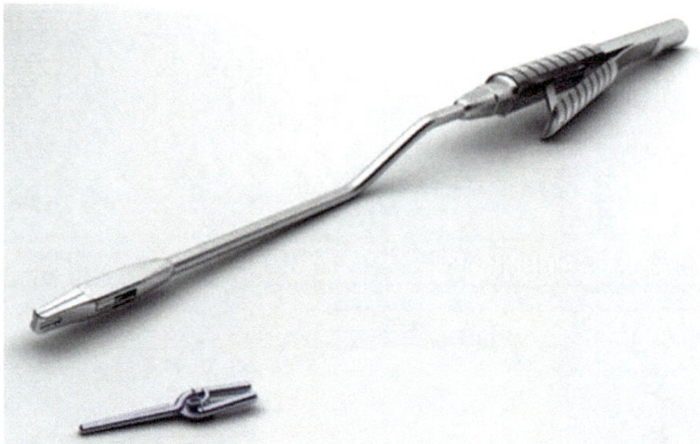

Fig. 1.6 Perneczky clip-set (Zeppeli Co)

Fig. 1.7 Meeting: Surgical Planning Principles; Mainz/Germany 1990

development in a conceptually and technically more transparent way. This development continues to this day. The international success that was sparked in Wiesbaden was not without consequences in Germany and Europe.

The tireless, innovative undertakings by Perneczky were confirmed by the huge worldwide interest at the “Neuro-endoscopy” satellite symposium on 4th–6th May at Frankfurt Airport, on the occasion of the 1995 EANS Congress in Berlin.

The “MINOP I” project to develop an elegant neuro-endoscope system was now presented with the clinical results (Fig. 1.8).

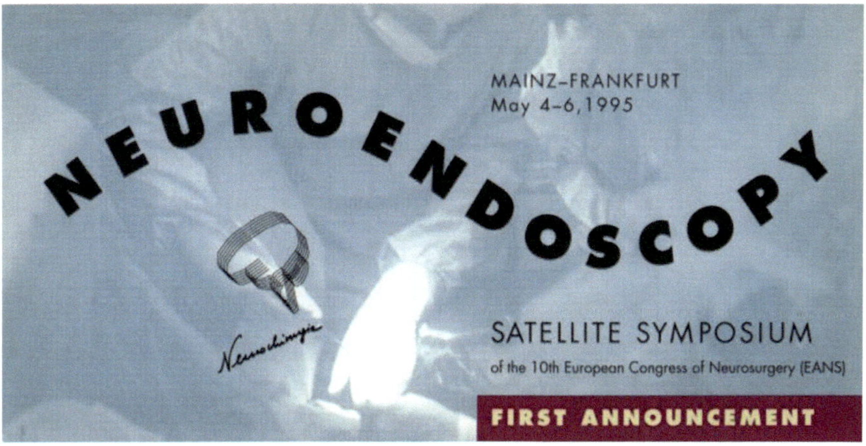

Fig. 1.8 Satellite Symposium of 10th EANS Congress at Frankfurt Airport/Germany

In 1999, Thieme publishing house then published "Keyhole Concept in Neurosurgery". It presented the concept along with 25 case histories from roughly 1200 clinical cases (Fig. 1.9).

Laboratory work from 1990 onwards showed during surgical simulations in 74 deceased aneurysm patients that a paradigm shift from trans-endoscopic to para-endoscopic instrumentation was required. This was leading to the concept of "endoscopy-assisted micro-neurosurgery". In line with Perneczky's MIN concept, the visualization advantages of endoscopy were combined with the instrumental and preparatory advantages of microsurgery. This resulted in a virtually limitless expansion of the use of the endoscope in micro-neurosurgery. This paradigm shift in the concept and technique is at the core of Perneczky's "Mainz School".

In February 1998, Perneczky published the crucial paper "Endoscope-assisted Brain Surgery: Part 1 – Evolution, Basic Concept, and Current Technique" in the journal Neurosurgery. In Part 2, 380 clinical cases were analyzed and presented.

From 10th to 12th June 1998, the "1st Congress on Endoscopically-Assisted Micro-Neurosurgery", an elaboration and evolution of the key-hole concept, then again, took place at Frankfurt Airport (Fig. 1.10). Recent developments, instruments and clinical results were in the focus and discussed within the growing international neurosurgical community.

Throughout this entire development a generation of international physicians was trained in the hands–on courses, taking place up to six times a year, covering a variety of topics and techniques in MIN.

This was the clearest indication of the world-wide appreciation and recognition of this development (Figs. 1.11, 1.12, 1.13, and 1.14).

During that period many different formats of education and training MIN were established and the latest steps of advances were integrated. The courses met a major interest mainly by the young generation of neurosurgeons.

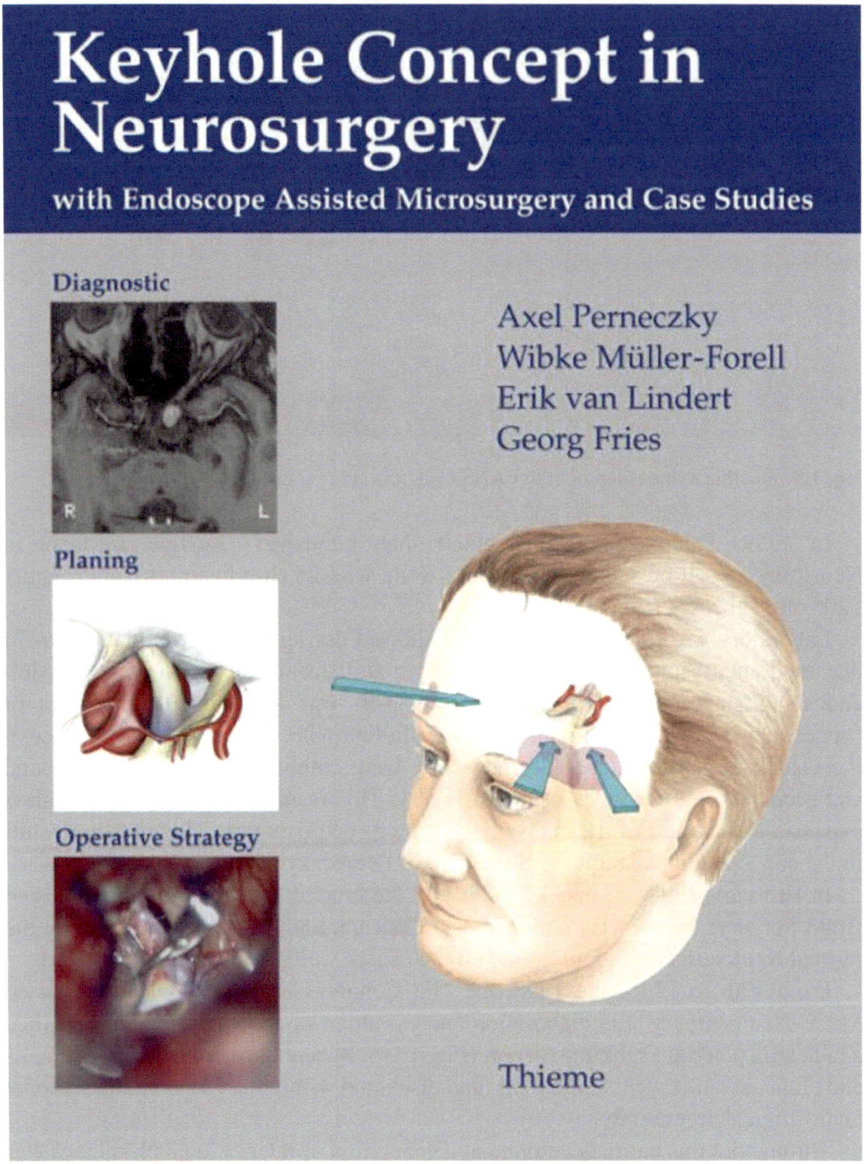

Fig. 1.9 Keyhole Concept in Neurosurgery (2nd major contribution 1999)

Perneczky was a passionate teacher and loved to give his experience personally to the participants of the courses.

The focus was always to connect the techniques with clinical cases and the benefit for the patients.

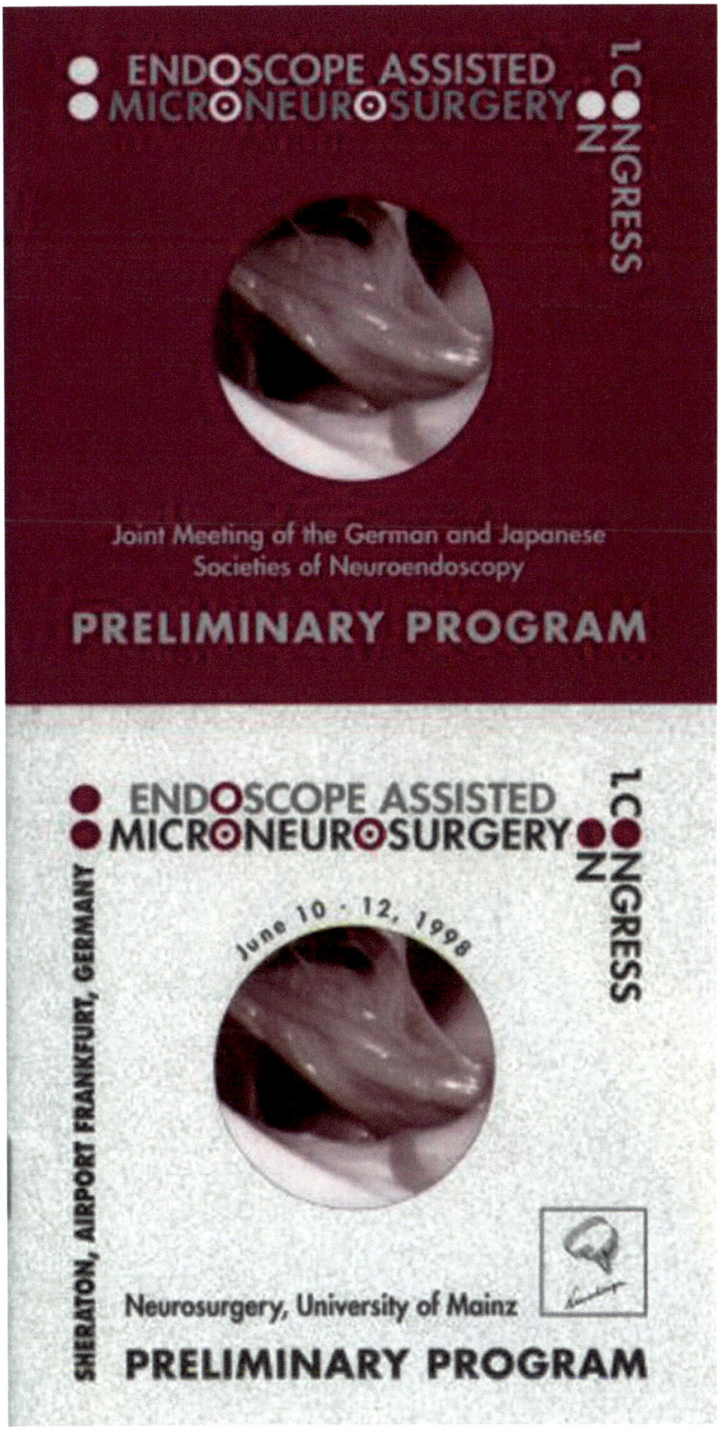

Fig. 1.10 First Congress on MIN (new concept and technique 1999)

Fig. 1.11 Two citiy project on MIN

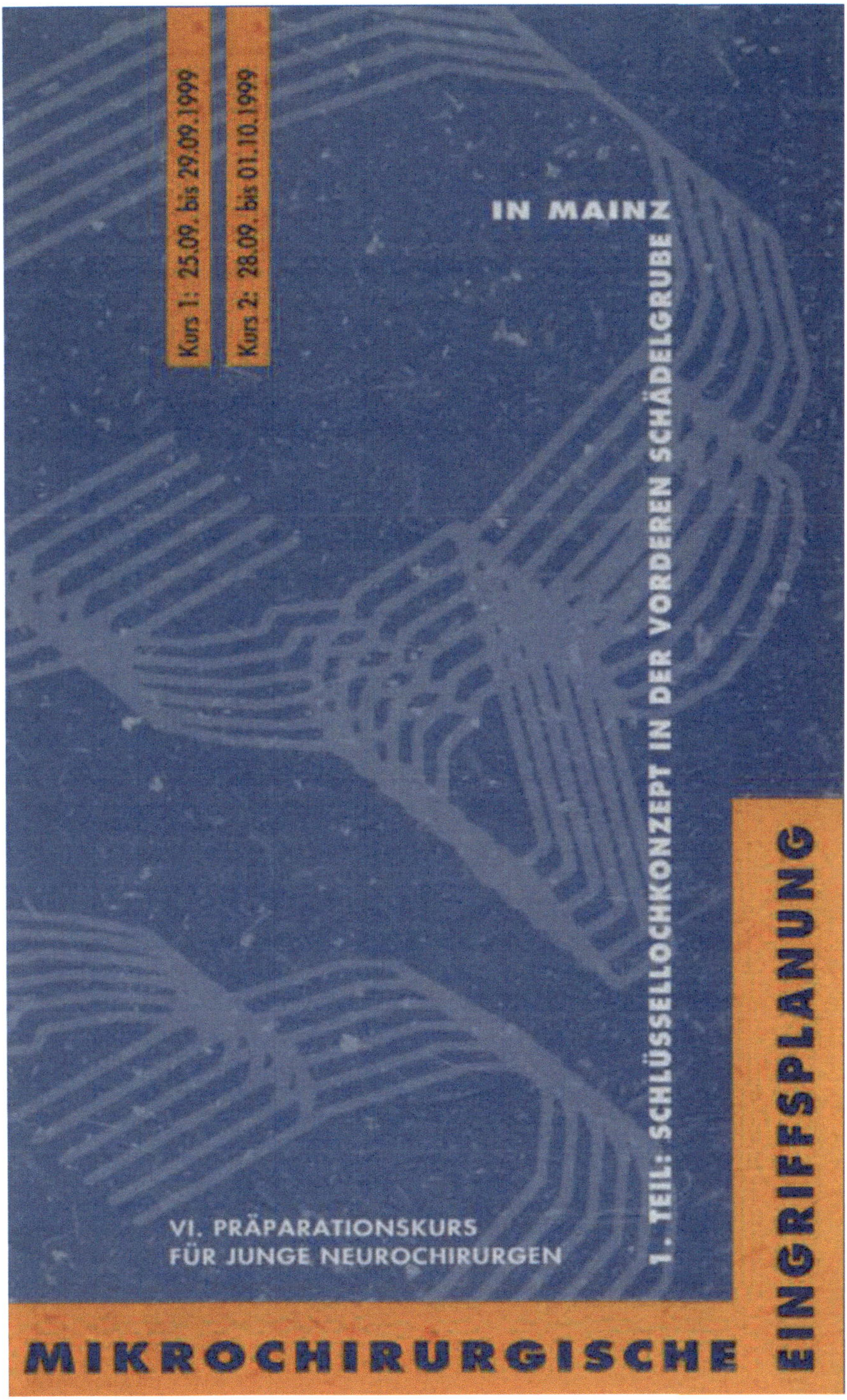

Fig. 1.12 Example of formats of hands-on courses in MIN

Fig. 1.13 MIN 24th hands-on course; Mainz/Germany 2002

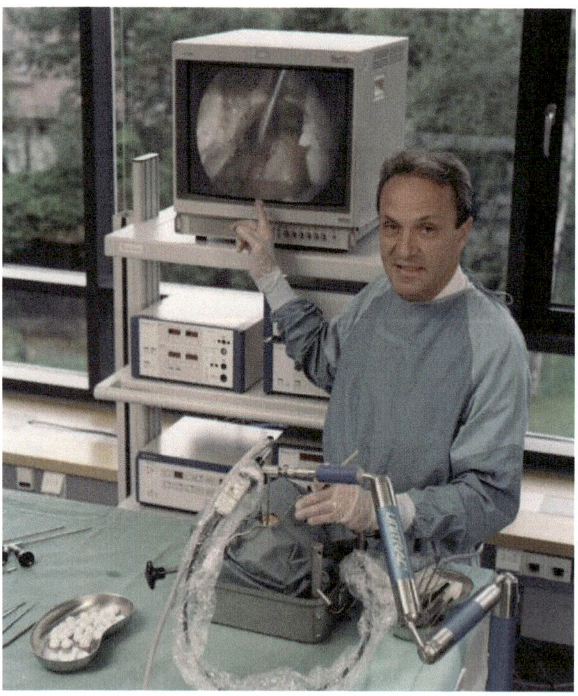

Fig. 1.14 Hands-on course neuro-endoscopy/Tuttlingen Aesculap Academy

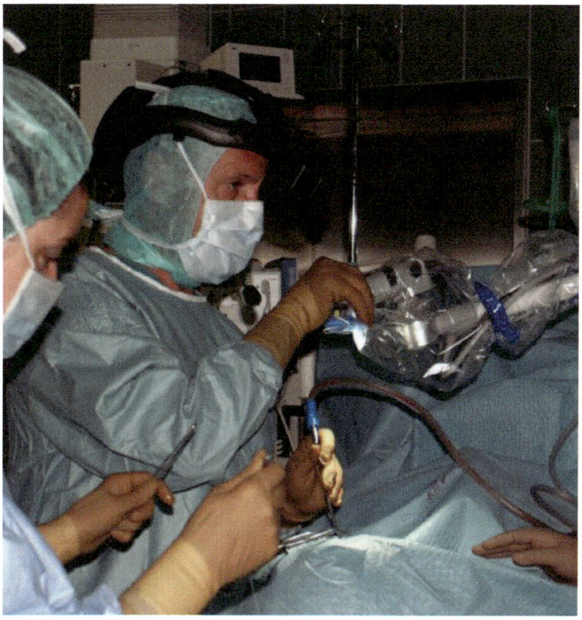

Fig. 1.15 Endoscopy assisted microneurosurgery with HMD system

Problem solving had a high priority as a context of application technical equipment. Techniques were always discussed within the concept of MIN.

The last stage of development was made possible in 1999 by the Olympus endoscopy robotic arm together with the head-mounted display system (HMD). As a result, the bracket problem was solved, which had so far not reached the microsurgery standard of the Contraves. The indispensable bi-manual technique with simultaneous guiding of the endoscope was finally possible for endoscopy-assisted micro-neurosurgery. The HMD system resulted in an ergonomic working method making use of multiple imaging modalities without multiple monitors in the theatre.

For the pre-endoscopic access phase of the operation, the endoscope can be disconnected, and the operation and visualisation performed with HMD using the endo-camera as a microscope substitute. Perneczky called this application "**exoscopy**" (Fig. 1.15). Ten years later, there is a hype with several Exoscope types from different companies delivered to the market (s. Chap. 4).

Only then was the ergonomics, practised in laboratory work as a surgical simulation (s. Vol. 2), completely transferred to clinical application. The ergonomic concept for MIN was described in several publications by the author (Resch 1999; Resch 2002; Resch: Transendoscopic Ultrasound for Neurosurgery, Chap. 5; Springer 2005).

However, it remained a blind spot within even MIN and will be described and discussed later below (s. also Chaps. 3 and 4) (Graph 1.2).

There is no place in MIN for "lets open and see what we can do". Everything has to go through a meticulously planning procedure before being applied in the patient.

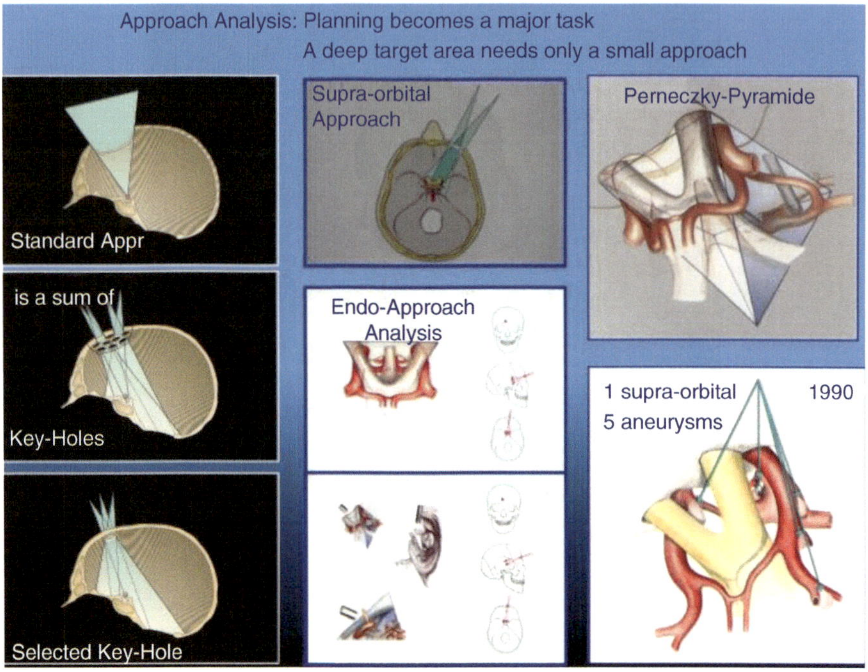

Graph 1.2 Planning concepts in MIN

Not the size but the precision of the approach planning is the goal in MIN. Geometrical figures (Perneczky-Pyramid) may assist to get the planning comprehensive and to simplify all procedures.

Such can lead to be able approaching five aneurysms through one single supra-orbital approach. (This was recognized very controversial, leading to an animosity stigmatisation of MIN by the representatives of the German Society of Neurosurgery/DGNC, at that time: Mainz Meeting 1990.)

It was also possible to publish this Perneczky MIN surgical setting, which differed considerably from the micro-neurosurgical setting according to Yasargil. However, only the first volume of "Keyhole Approaches in Neurosurgery" was published by Springer publishing house, completed by Robert Reisch in 2009 (Fig. 1.16). In addition, Reisch expanded upon Perneczky's MIN spectrum through the addition of the young field of "per-nasal endoscopic surgery on the base of the skull" started by Jho, Kassam, Cappabianca and others.

This account of Perneczky is true to the man himself but characterises only the neurosurgical part.

It is important to emphasise Perneczky's relationship to art, which subtly and powerfully shaped his fine nature. His scientific and clinical environment was accordingly accompanied by artistic activities of his own, and those he performed together with others (Fig. 1.17). This lent a classical and humorous ethos to everything. This was noticeable in his day-to-day work and above all at the bedside, where a pleasant aura was generated around him (Graphs 1.3 and 1.4).

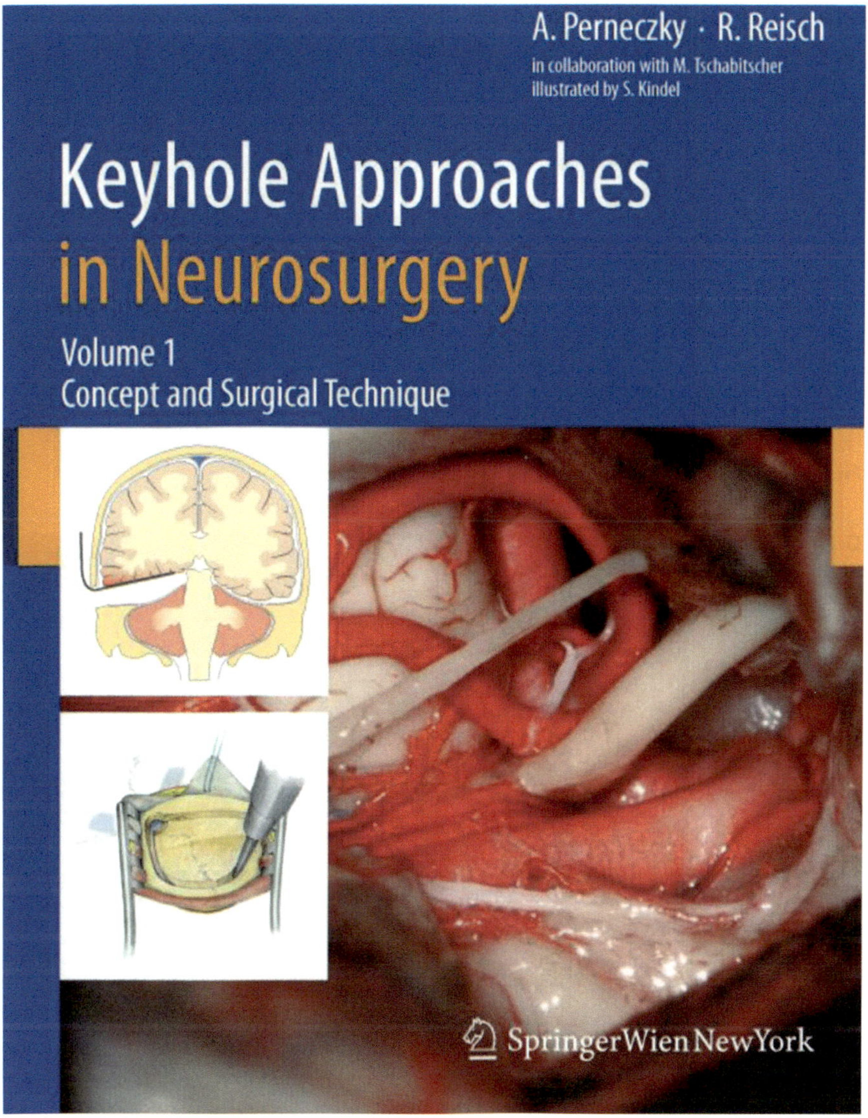

Fig. 1.16 Keyhole approaches in MIN (3rd major contibution in MIN 2000)

Fig. 1.17 "Skull Base Combo" (subgroup of "Skull Base Allstars")

Graph 1.3 The Personality

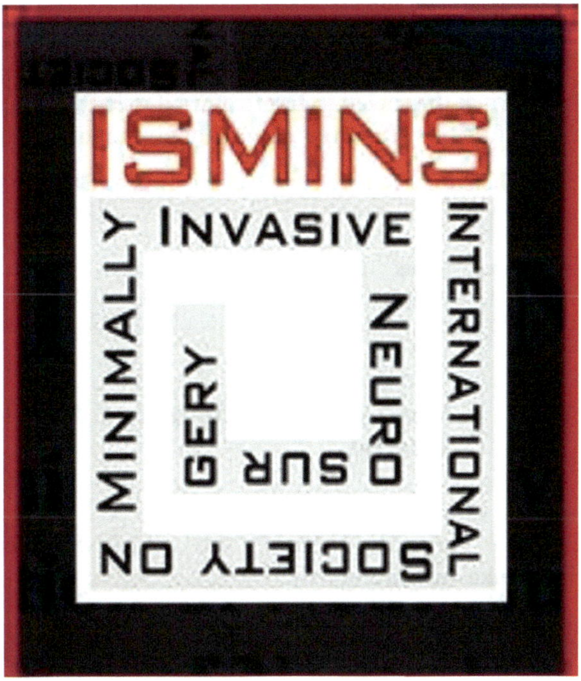

Graph 1.4 Logo of **theismins.com**

From this experience of 10 years of close cooperation with Perneczky during his peak years, the intention here is to summarise conclusively, without commentary, his unique professional and personal characteristics (Table 1.1):

Table 1.1

1. The apparent ease with which he performed neurosurgery
2. His customisation of standard interventions
3. His fight for every opportunity to minimise trauma
4. Online learning, even during ongoing surgery
5. Analytical unbundling with reduced synthesis
6. Compensation for the equipment's ergonomic handicaps
7. Procedural reduction and clear working towards an aim
8. Geometric analysis of the access point and precise design of the approach
• His phenomenally empathetic charisma in everyday life
• The disciplined self-restraint that was typical of him
• His self-detached way of accepting tasks
• His ignoring of social and cultural stigma
• The primacy given to constructive problem-solving
• Consistent, systematic openness to ideas

This urgently needed interconnection of forces on an international level was, after the sudden absence of Axel Perneczky, on whose life's work the society relies to a great extent, an expression of the desire to close this sensitive gap and not give up on this innovative development.

theismins.com

Acknowledgements The "cover picture" was painted by Heide Roesler (long-term collaboration of art and neurosurgery) in acrylic on canvas 60 × 80. Stefan Kindel, longtime graphic artist of Perneczky, provided the Figs. 1.1, 1.4, 1.6, 1.7, 1.8, 1.9, 1.10, 1.11, 1.12, and 1.15.

1.2 In Memoriam Prof. Dr. Dr. h. c. Axel Perneczky/ Obituaries

(s. also: https://youtu.be/NB1gmoenvnc)

2nd Perneczky-Lecture 4th ISMINS Congress/ 2018 Moskau, Burdenko-Institution

Anatomy of the human spirit and **phylogenesis of consciousness** are forming history. All history is a product of the human brain and represents the ability of the brain in its time. Remembering and memory are the basics of awareness, so history is a cultural memory and awareness of culture. According to Jaspers, "history is from where greatness is speaking to us from the past". Future is a vector from the past. For human future it is of tremendous impact to keep the memory of great people high and to recognize their "mythos". Mythos means to tell a meaningful history, creating the fundament of science. Perneczky was a great man and had some unique personal and professional qualities (s. Table 1). During 3 years of being his personal assistant, I discovered his gifted grace, which cannot be acquired, and this guided him to do extraordinary results.

The most memorable characteristic was an aura he developed in the dealing with his patients. This was the core of all his efforts. From this natural ethics and exot heric philosophy and empathy, he created all his projects and his surgical mastery. He presented an enduring gentleness and patience, not accepting any violence of routine and stupidity of useless standards. One could give the patients each promise, knowing that he would keep it reliable.

His boldness made him promote the evolution of MIN. The difference to others on the same way, before or in parallel, was: he had the vision, to understand the new idea: the endoscope was not just a tool but a landmark of MIN, the key-hole was not a burr-hole but a concept of precision and result of planning, imaging was not a diagnosis but an information pool to create the individual approach.

"He, who does not see the world with the eyes of a friend is not worth, that the world takes notice of him." ***J. W. v. Goethe***

And yes, "The flawless protects against blame, but true greatness compels admiration." ***Karl Jaspers***

His candle burned not long, but because it spent generously more light in that period.

The author of the book

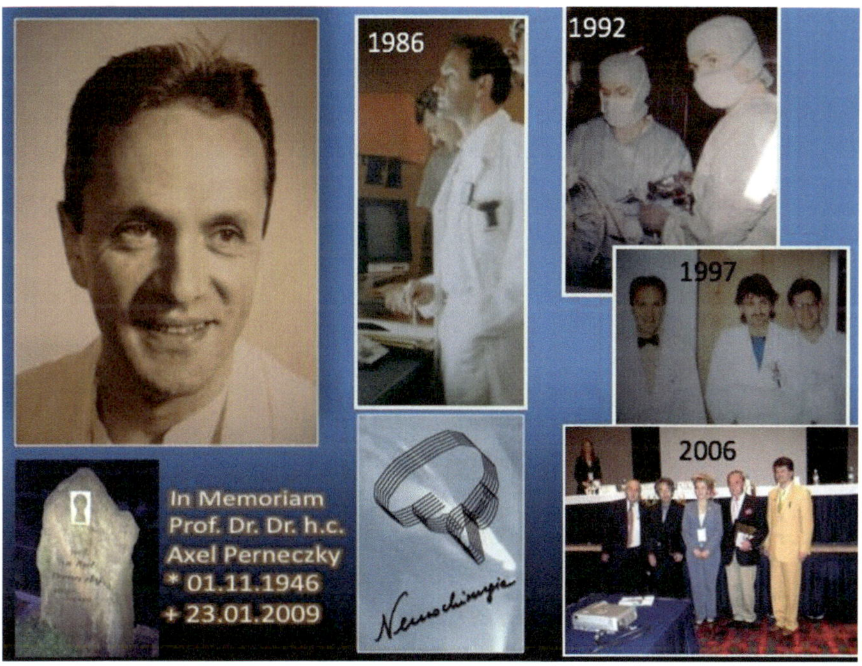

3 Obituary Links

1) Axel Perneczky: A Remembrance
Grotenhuis, J. André MD, PhD; Cohen, Alan R. MD
Neurosurgery:
June 2010 - Volume 66 - Issue 6 - p 1036–1038 doi: 10.1227/01.NEU.0000369185.36188.12 Legacy: Institutions and People
2) Nachruf Univ.-Prof. DDr. Axel Perneczky
Journal für Neurologie Neurochirurgie und Psychiatrie 2009;10(1), 99
Prim. Univ.-Prof. Dr. Karl Ungersböck
Minim Invasive Neurosurg 2009; 52(1): 1-4
DOI: 10.1055/s-0029-1202818
3) Obituary
© Georg Thieme Verlag KG Stuttgart · New York
Axel Perneczky, 1.11.1945–24.1.2009
N. J. Hopf, R. Reisch

[Operative Neurosurgery] 21.07.19

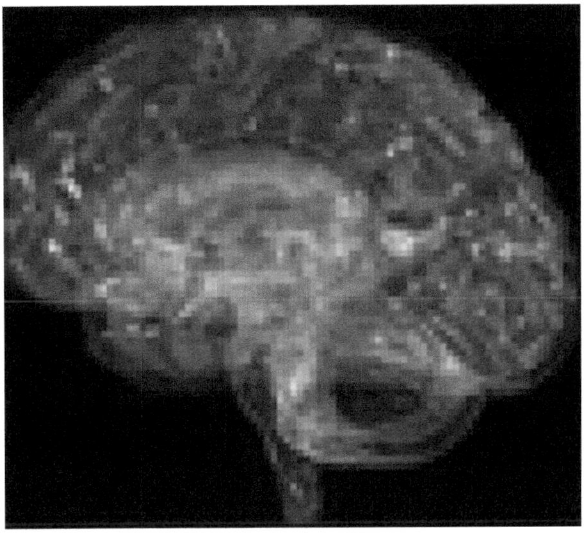

1.2.1 Guest Contributions I

1.2.1.1 Axel Perneczky

Axel Perneczky (1945–2009), a Hungarian stemming consultant of neurosurgery at the University of Vienna.

In Vienna he was at that time a recognized neurosurgeon in the field of vascular microsurgery. The philosophy of minimally invasive surgery suited his general neurosurgical concept very much. His ambition was to minimize as much as possible the intraoperative trauma for the patient respecting also the cosmetic aspects. This could be achieved by careful and thorough preoperative planning of the best approach which was based on a detailed analysis of individual anatomy and topographic relationships of the lesion visible in the radiological images. The lowest functional intraoperative trauma required sometimes new and unusual approaches such as the contralateral approach to the suprasellar region or through the lateral ventricle. The new approaches required detailed knowledge of topographic anatomy, which he acquired while working as demonstrator at the anatomical institute in Vienna and in the late 70th while working at the laboratory of Gazi Yasargil in Zurich.

Perneczky understood neurosurgery as applied neuroanatomy. In 1988, Perneczky became Chairman of the Neurosurgery at the Johannes Gutenberg University in Mainz. There, he organized a team of neurosurgeons which started to realize his idea of minimally invasive neurosurgery. The clinical application preceded work in basic research.

With his friend Manfred Tschabitscher, anatomist at the University of Vienna, Perneczky studied the anatomy of the ventricles and the basal cisterns from an endoscopic view.[1)] The neurosurgeon Klaus Resch completed in Mainz the anatomical study by postmortem endoscopic inspections. He convinced also Perneczky of the superiority of rigid endoscopes over flexible endoscopes for neurosurgery due to the much better image quality. Additionally, a laboratory was established for anatomical studies on cadavers. The first meeting in Mainz in 1989 on the subject of minimally invasive neurosurgery was too early and not successful. The breakthrough of this new technique in neurosurgery was the international meeting in Wiesbaden in 1993 for minimally invasive neurosurgery. This meeting, organized by Perneczky, gathered for the first time all neurosurgeons involved worldwide in neuro-endoscopy and was also attended by many internationally recognized neurosurgeons. Their basic approval of this new method had a stimulating effect for the irradiation of these new techniques and philosophy around the whole world. During this meeting a society for minimally invasive neurosurgery (MIN) was established with international conferences every 2 years and a journal with the same name. The spark of minimally invasive neurosurgery spread also to the industry and with Minop1 and Minop2 projects for development of minimally invasive technology, instruments became supported and sponsored by many companies. For Perneczky minimally invasive neurosurgery comprised not only endoscopy. The endoscope was only a tool during the operation or a part of the operation. Minimally invasive neurosurgery comprised all surgical instruments or devices and operative techniques which helped to diminish the intraoperative trauma including 3-D operation planning workstations, navigation devices, and endoscopes. He demonstrated that by the key-hole effect also huge deep-seated brain tumors can be satisfactorily controlled and removed through a small craniotomy.[2) 3)] Perneczky was not the founder of minimally invasive methods in neurosurgery. These ideas started to be realized even by the pioneers of the neurosurgery. Interestingly, the great historic neurosurgical personalities were usually creative spirits not only in one but also in several neurosurgical fields. However, Perneczky deserves the merit of bundling and further developing all the new technical developments existent in the late 80th under a common neurosurgical concept of minimally invasive neurosurgery. Although Perneczky was one of the protagonists of this new philosophy, he would not be successful if not other neurosurgical centres in Europe and in other continents had at the same time simultaneously similar ideas.

https://operativeneurosurgery.com/doku.php?id=axel_perneczky Seite 1 von 2
axel_perneczky [Operative Neurosurgery] 21.07.19, 21)04

1) Perneczky A, Tschabitscher M, Resch KDM. Endoscopic Anatomy for Neurosurgery. Thieme; 1993.
2) Perneczky A, Müller-Forell W, van Lindert E. Keyhole Concept in Neurosurgery: With Endoscope Assisted Microsurgery and Case Studies. Thieme; 1999.
3) Perneczky A, Reisch R, Kindel S. Keyhole Approaches in Neurosurgery. Vol. 1: Concept and Surgical Technique. Springer; 2008.

4) **Grunert P.** From the Idea to Its Realization: The Evolution of Minimally Invasive Techniques in Neurosurgery. Minim Invasive Surg. 2013;2013:171369. Epub 2013 Dec 17. Review. PubMed PMID: 24455231; PubMed Central PMCID: PMC3877623.

P. Grunert: axel_perneczky.txt · Last modified: 2016/07/07 10:13 (external edit)
In Memoriam A. Perneczky, 27.04.2010 Guadalajara/ Mexico

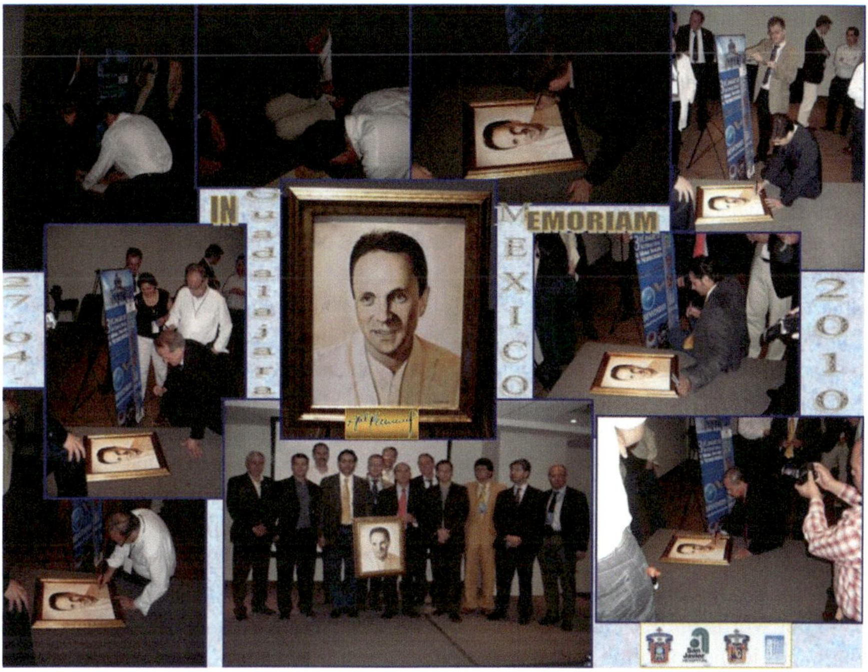

1.2.1.2 In Memoriam, Axel Perneczky

In 1994, when I concluded my specialty in Neurosurgery in Guadalajara, Jalisco (second largest state in Mexico), I did not know that the name 'Axel Perneczky' would end up signifying and determining the most important aspects of my life as a neurosurgeon. When I graduated in 1994 from one of the classic neurosurgery schools in Mexico (Antiguo Hospital Civil de Guadalajara), I was daunted by the scope and results of neurosurgical procedures of those times. The limited functional results in the patients discouraged my desire to stay and practice neurological surgery and the thought of "if that's neurosurgery" made me prefer not to continue it. The search for an answer took me to Mainz, Germany: in April 1996 I arrived at the Johannes Gutenberg University Hospital in Mainz, and without further ado, from day one, I knew in the operating room the answer to the question that had made me travel more than a ten-thousand kilometers ... "Neurosurgery was Axel Perneczky,

Neurosurgery was Minimally Invasive." A couple of months in the Mainz Department of Neurosurgery was enough for me to know that I was experiencing the beginning of a new trend in neurosurgery, and two years it took me, under the mentorship of its main character, Chief "Axel Perneczky", to change and reconstruct the deeper concepts of neurosurgical care…neurosurgery was a good and important thing, which was completely worth it; Minimally Invasive Neurosurgery was completely good for me, for Mexico, and even for the world. So, I said goodbye to that closeness and uninterrupted work in Mainz, hand in hand with someone who had decided, for some reason still unknown to me, to favor myself with having me almost daily next to him, on the operating table ... - Héctor, ¿What's next? - Return to Mexico to operate, operate and operate, applying what I learned from you every day. - Well thought, go ahead. (March 1998, Mainz, Germany). In May 1998 we performed the first endoscopic surgery on a pediatric patient with hydrocephalus at the New Civil Hospital in Guadalajara, starting in Mexico what has been the German School of minimal invasive Neurosurgery (NEUROMIN) in Mexico, Axel Perneczky's school. On January 24, 2009, neurosurgery worldwide and especially those of us who were students, colleagues and friends, suffered the loss of the number one driver of minimal invasive Neurosurgery, Prof. Dr. Axel Perneczky, when he died in the city from Mainz, Germany. In February 2010, in the city of Guadalajara, Jalisco, Mexico, we held a farewell 'In memoriam' meeting with the one who was our teacher (Dr. Martin Bettag, Dr. Andre Grotenhuis, Dr Klaus DM Resch, Dr. Niko Hopf, Dr. Robert Reisch, Dr. Henry Schroeder, Dr. Dieter Hellwig, Dr. Giorgio Frank, Dr. Marco A. Barajas, Dr. Héctor Velazquez-Santana, Dr. Américo Do Santos, Dr. Carlos Gagliardi among others). 2020 marks the 27th anniversary of the First World Conference on Minimally Invasive Neurosurgery (Mainz-Wiesbaden, Germany), 22 years of Minimally Invasive Neurosurgery in Guadalajara, Mexico and 11 years since the death of Prof. Axel Perneczky (24 January, 2009), a loss that we honor with our words in the Neurosurgery services, our actions in the hospital rooms and our hands in the operating rooms of countless hospitals in Mexico. - Héctor, ¿What's next? - Operate, operate and operate. - Well thought, go ahead. – IN MEMORIAM, TEACHER AND MENTOR, DR. AXEL PERNECZKY

Dr. Héctor Velázquez Santana. Head of Neurosurgery Department of Neurosciences Nuevo Hospital Civil "Dr. Juan I. Menchaca" (School Hospital of the University of Guadalajara) hvelazquez@hcg.gob.mx hevesan@yahoo.com.mx Address: Salvador Quevedo y Zubieta 750 Independencia Oriente. 44340 Guadalajara Jalisco, México.

1.2.2 Guest Contributions II

Dr. AhmedAl Ferayan Consultant Neurosurgeon King Fahd National Guard Hospital Riyadh/Saudi Arabia

Dr. Tareq Kanaan, Dr. Ali Ayyad Consultant Neurosurgeons Dept. of Neurosurgery University of Mainz/Germany

Laligam N. Sekhar, MD, FACS

Professor and Vice Chairman, Director of Cerebrovascular and Skull Base Surgery, University of Washington, Seattle, USA

Peter Black, MD, PhD President, World Federation of Neurosurgical Societies

Franc D. Ingraham Professor of Neurosurgery, Harvard Medical School, Boston, MA

Founding Chair, Department of Neurosurgery, Brigham and Women's Hospital/USA

Takanori Fukushima, M.D., D.M.Sc Professor & Director Carolina Neuroscience Institute, Raleigh NC

Professor, Duke University Medical Center, Division of Neurosurgery, Durham NC/USA

Prof. Dr. Robert Reisch Centre for Endoscopic and Minimally Invasive Neurosurgery, Clinic Hirslanden Zurich. Witellikerstrasse, 40CH-8032, Zurich, Switzerland

Dieter Hellwig M.D., Ph.D. Head of the Division: Stereotactic and Functional Neurosurgery; International Neuroscience Institute Hannover (INI), Rudolf-Pichlmayr-Straße 4; 30625 Hannover/Germany

Yoko Kato, M. D. Ph.D. Chair, Education and Training Committee of the WFNS

Secretary, WFNS FoundationProfessor, Department of Neurosurgery

Fujita Health University/Japan

President of ISMINS (International Society on Minimally-Invasive Neurosurgery)

Obituary

Tribute in Memory of Professor Axel Perneczky (1945-2009)

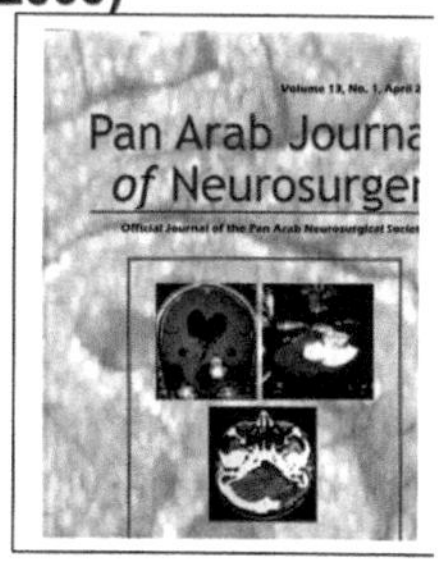

When I received the request to write an obituary for Prof. Axel Perneczky I found it to be one of the most challenging tasks in my career for two reasons, one is because of the close relationship between us and the second is the huge responsibility in writing such an obituary which perfectly describes the man and his many achievements. His life could be divided into many distinctive periods.

The first period was a tough beginning. He was born in Krasnogorsk near Moscow at a prison camp following the abduction of his mother, who was a newspaper reporter, from Berlin by the Russian army. After 7 months his Mother and her new born son were released and allowed to return to her home town in Budapest. His father joined them later from French captivity.

In 1964 Prof. Perneczky obtained his high school degree and joined the Medical School at the University of Budapest.

In 1965 the family escaped the eastern block to Austria. Here, Prof. Perneczky completed medical schooling and joining the neurosurgical team under the guidance of Prof. Kraus and later Prof. Koos. He also spent one year with Prof. Yasargil obtaining microsurgical skills. He received his Board Certification in 1980.

In 1988 he became the Chairman of the Neurosurgical Department at Mainz University and began to apply his vision of minimal invasive neurosurgical techniques, as well as endoscopic and navigation systems.

He organised the 1st International Minimal Invasive Neurosurgical Meeting in 1993 in Wiesbaden, Germany, at which many of the world's pioneers of neurosurgery attended. He was now also the Editor-in-Chief of the Journal of Minimal Invasive Neurosurgery. He also took the mantle of Founder and President of the German Society of Endoscopic Assisted Neurosurgery and Neuronavigation.

Prof. Perneczky was a devoted teacher and many residents, including myself, were privileged to be taught microsurgical techniques under his guidance. He organised 6 endoscopic courses per year. He developed with manufacturers the world renown aneurysm clips and many varieties of endoscopes.

He published approximately 250 articles and his input in the books on Color Atlas of Microsurgery (co-authored), Endoscopic Anatomy for Neurosurgery (co-authored) and Keyhole Concept in Neurosurgery (sole author) sold countless copies and are still used by neurosurgeons all over the world.

On the more personal side, I knew him to be a dedicated family man with amazing communication skills with his trainees.

Obituary

He played the guitar with the 'Skullbase Jazz Band', as well as being an excellent painter.

Prof. Perneczky's death is a great loss to us all, both from a professional standpoint and a personal one. Our deepest condolences are given to his wife Angela and his six children, Alexa, Ines, Paul, Julia, Max and Tom.

As Prof. Perneczky visited us on many occasions, not only to share his knowledge and skills but also to enjoy the desert he loved, we have the honour to include a few of the photos of times we were able to enjoy together here in Saudi Arabia.

Dr. Ahmed Al Ferayan
Consultant Neurosurgeon
King Fahd National Guard Hospital
Riyadh
Saudi Arabia

We were very lucky to know Prof. Perneczky beyond the limits of our professional relationship. We always addressed him as Chief (Boss). This is not a title a man gets or a position that somebody offers you, it is something you have, something that you are born with. It is a characteristic, a charisma, a talent and a vision. Prof. Perneczky had them all. He had the most generous and modest personality combined with all the characteristics of a leader.

As a Boss, a teacher and as a friend there was always something that surrounded him, and that is the enormous charisma which laid in his silence; his few well chosen words as he spoke and his selective meaningful movements as he operated.

His surgical skills were amazing. He was able to introduce a new concept in neurosurgery which at one point nobody thought possible. His vision is now established as a standard in many places all over the world.

His death is a great loss to the Neurosurgical Society and a personal loss to his students, those who loved him and looked up to him. His huge legacy, his precious books and his marvellous paintings reflect his immortality.

Those who wonder about the way he died may rather wonder about the way he lived.

Dr. Tareq Kanaan
Dr. Ali Ayyad
Mainz University, Germany

1.2.3 Guest Obituaries

I first met Prof. Axel Perneczky at Vinko Dolenc's symposium on Cavernous Sinus in 1986. I knew his Chairman in Vienna Professor Wolfgang Koos quite well, and he had very good co-workers such as Englebert Knosp, and Christian Matula. I was impressed with his work on the removal of the anterior clinoid process, and the treatment of para-clinoidal aneurysms. We became good friends. Since then, he moved to Germany to become the Chairman at the University of Mainz and here he started his seminal work on Endoscope Assisted Microsurgery for aneurysms and skull base tumors through keyhole approaches. Although I did not completely agree with the concept of the "keyhole approach", I also had a great interest in Endoscope assisted Microsurgery, and still do. I believe that this is a great contribution to Neurosurgery made by Professor Perneczky.

He and I met several times in many conferences in exotic places such as Rejkavik, Iceland, and Foz de Iguazu, in Brazil, where we were speakers. We always had good discussions, and often played tennis together. He was a great friend, and a man of principles. For example, after the US led invasion of Iraq, he refused to attend any meeting in the USA. He was somewhat vindicated when no weapons of mass destruction were found in Iraq, but more recently, some semblance of democracy has been established in Iraq, and this may have stimulated democratic movements all across the Middle East.

Axel was a great neurosurgeon, and great friend, and I miss seeing his friendly face, and having a nice, frank discussion with him about some topic. I hope, that someday people will remember me with equally fond memories.

Laligam N. Sekhar, MD, FACS
Professor and Vice Chairman
Director of Cerebrovascular and Skull Base Surgery
University of Washington
Seattle, USA

1.2.3.1 Tribute to Axel Perneczky

Professor Axel Perneczky was an academic neurosurgical visionary whose legacy lives on through his students.

Trained initially as an anatomist in Vienna, he spent most of his life as chief of neurosurgery in Mainz, Germany. There he developed endoscopy for ventricular and pituitary lesions and vascular decompression; eyebrow incisions for frontal craniotomy; and minimally invasive spine surgery. He was interested not only in surgery but in science, particularly the engineering science of brain imaging and the application of functional MR and diffusion tensor imaging to minimally invasive cerebral surgery. His extensive lecturing, writing, and teaching in the operating room revolutionized neurosurgery.

He was a great clinician, a wonderful mentor, a dedicated investigator, and an accomplished world traveler. I remember his modest demeanor at many meetings, but always with a conviction that his concepts were right. Although his approaches were often astonishing, they were always backed up by experience, and his pioneering work must have been difficult for many to accept.

There is excitement even today in much of the work he did, and his concepts are now main-stream in neurosurgery. What distinguished him among other world-class neurosurgeons, however, was his dedication to training the next generation and the subsequent commitment of his trainees to his cause. His fervor has been transmitted to a group of neurosurgeons who have continued to develop these techniques: Charalampaki, Ayyad, Grotenhuis, Resch, Grunert, Knosp, Fischer, Stadie, Fries, Conrad, Cohen, Weschehold, and many others who continue to refine techniques and create new applications. This book is a great example of their affection for him and his work.

Professor Perneczky was in many ways the ideal academic neurosurgeon. He emphasized good patient outcomes rather than technical bravado, used his knowledge of anatomy to create innovative approaches, and believed particularly in teaching the next generation. It is an honor to help to celebrate his life in this memorial volume.

Peter Black, MD, PhD

President, World Federation of Neurosurgical Societies

Franc D. Ingraham Professor of Neurosurgery, Harvard Medical School, Boston, Ma Founding Chair, Department of Neurosurgery, Brigham and Women's Hospital

1.2.3.2 Minimally Invasive Keyhole Concept in Spinal Tumor Surgery

Robert Reisch

Centre for Endoscopic and Minimally Invasive Neurosurgery, Clinic Hirslanden Zurich

Corresponding author

Prof. Dr. Robert Reisch

Centre for Endoscopic and Minimally Invasive Neurosurgery Clinic Hirslanden Zurich

Witellikerstrasse, 40CH-8032, Zurich, Switzerland

Tel: +41 44 387 2853, Fax: +41 44 387 2861, Email robert.reisch@hirslanden.ch

1.2.3.3 Abstract

On account of their space occupying effect, intraspinal tumors may become symptomatic causing severe neurological deficits. The gold standard of treatment is the total removal of the lesion. However, extended dorsal and dorsolateral approaches may cause late complications due to iatrogenic destruction of posterolateral elements of the spinal column.

In this article, Axel Perneczky's concept in minimally invasive spinal tumor surgery is described. Two illustrative cases demonstrate the safety and feasibility of keyhole fenestrations, exposing the intraspinal space. The first case is a 67-year-old woman with a one-year history of severe back pain and slight gait disturbances. Neuroimaging revealed a right-sided extradural T11/12 schwannoma with compression of the spinal cord and laterally extension through the intervertebral foramen. The tumor was removed through a contralateral left sided hemilaminectomy. The second case is a 38-years-old man with progressive spinal ataxia according to an intra-medullar C6/8 ependymoma. Here, bi-segmental interlaminar fenestrations were performed, allowing safe and minimally invasive tumor resection. In both cases, postoperative MR imaging demonstrated complete tumor removal, the patients showed no neurological deterioration and no vertebral instability or pain syndromes.

1.2.3.4 Keywords

Contralateral hemilaminectomy, Interlaminar fenestration, Keyhole spinal surgery, Preoperative planning, Spinal schwannoma, Spinal ependymoma

1.2.3.5 Introduction

The main problem is well known: despite the benign biological behavior, natural history of intraspinal tumors is unfavorable on account of their space occupying effect and subsequent neurological deterioration among patients (1). The gold standard is total removal of the lesion without postoperative neurological worsening. In order to achieve optimized visualization of the intraspinal structures, bilateral extended laminectomy has traditionally been recommended (2). However, numerous clinical and biomechanical studies reported late complications after extensive dorsal approaches with postoperative spinal instability, severe kyphoscoliosis or epidural scarring and spinal arachnoiditis; importance of the loss of the posterior bony protection has also been published (3-6). In the case of lateral extension of the tumor, the laminectomy may be enlarged by partial costo-transversectomy; however, this exposure can be complicated by postoperative pneumothorax and may cause denervation of the paraspinal muscles over several segments (7). Furthermore, using extended approaches the incidence of postoperative local pain is higher with subsequent use of analgesics, longer hospitalization and delayed rehabilitation.

1.2.3.6 Axel Perneczky's Concept in Minimally Invasive Spinal Tumor Surgery

The aim of minimally invasive neurosurgery is not to harm the patient by creating a tailor-made limited less traumatic approach based on skilled preoperative planning and detailed neuroanatomical and neurofunctional knowledge.

The interlaminar fenestration approach is par excellence of this concept in spinal tumor surgery. The keyhole exposure offers minimal soft tissue injury and restricted osteo-destruction, thus achieving the clear benefit of keyhole neurosurgery: the minimal approach-related traumatization.

However, limited approaches offer important restrictions in spinal surgery that should also be considered in a critical way.

The major limitations are the 1) predefined surgical corridor, the 2) difficult intraoperative orientation, the 3) insufficiency of available micro-instruments and the 4) decreased illumination in the deep-seated field.

Therefore, the spinal approach should be performed in an exact and individual way, according to the individual patho-anatomical situation. Two preconditions of a precisely tailored surgery are the 1) distinguished preoperative approach planning and 2) personal "self-made" performance of the surgery by the surgeon himself. Note that this personal performance should include positioning of the patient, skin incision, interlaminar fenestration, and surgical exposure of the target region.

The second drawback of keyhole procedures is the difficult intraoperative orientation. Real-time imaging and intraoperative use of navigation systems may increase surgical safety if the limited approach leads to a confusing situation. In addition, meticulous neuromonitoring with continuous SSEP/MEP measurements must be performed in every spinal tumor case, allowing careful protection of neurological functions.

The narrow viewing angle and almost coaxial control of dissection causes an additional problem. Intraoperative use of slim and tube-shaft designed micro-instruments may help to overcome this limitation in spinal keyhole surgery.

The fourth and probably main difficulty of keyhole approaches is the significantly reduced optical control during surgery. For this purpose, surgical microscopes can be effectively supported by endoscopes with the advantages of 1) increased light intensity, 2) extended viewing angle and 3) clear depiction of details in close-up position.

1.2.3.7 Illustrative Cases

Case I: Epidural Schwannoma

Patient's presentation This 67-year-old woman was admitted with a 1-year history of severe back pain with extension to the right side of the chest and slight gait disturbances. Clinical examination revealed radicular pain with par-aesthesia, hyp-aesthesia and hyp-algesia in accordance to the right T11 nerve root, and a slight gait ataxia without paresis of the lower extremities. No disturbance of bladder function was detected. Magnetic resonance image demonstrated a right-sided intra-extraspinal situated extradural hourglass schwannoma at T11/12. The intraspinal part caused severe compression of the spinal cord; lateral extension through the intervertebral foramen showed close connection to the pleura (Fig. 1A).

Approach planning The following approaches were discussed taking into consideration the individual patho-anatomical situation. Bilateral laminectomy with wide exposure of the ipsi- and contralateral side. This approach allows good visualization of the tumor removal, however, with destruction of the vertebral lamina on both sides, spinous process, interspinous ligaments, yellow ligaments and facet joints. Using ipsilateral hemilaminectomy approach the dorsal static structures are less

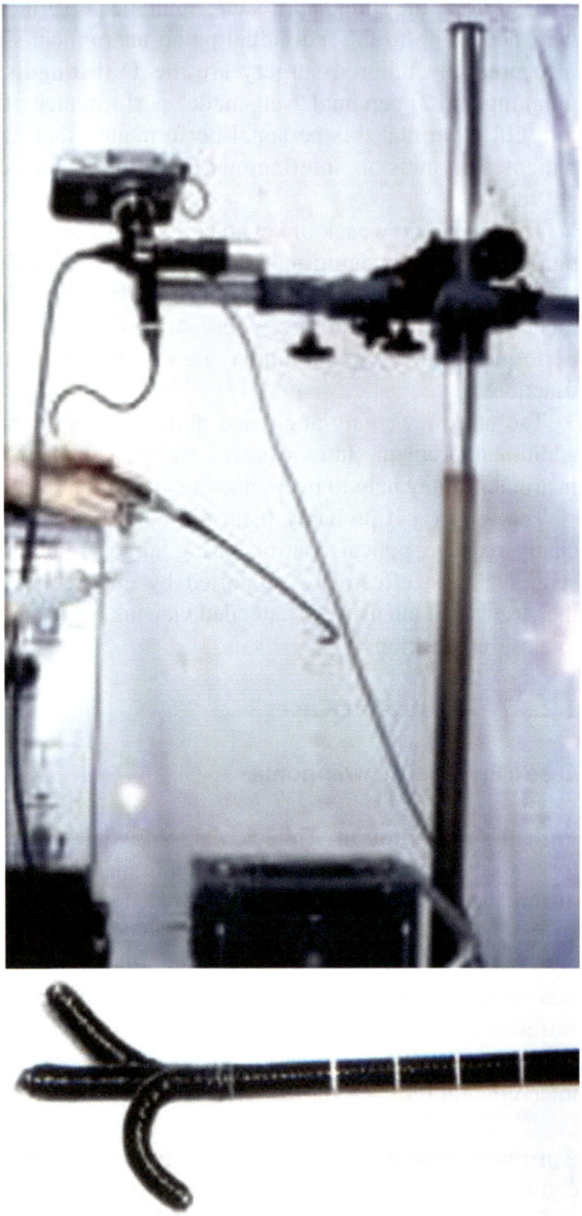

Fig. 1 (**A**) Innovative operating neurofiberscope mounted on the supporting stand with Xenon cool light supply (Olympus model 1969). (**B**) Tip of the bending portion of the neurofiberscope (3 mm), 30 degrees up and 130 degrees down automatic electric control by the foot pedal

likely to be injured than by a conventional laminectomy. However, in this case on account of the lateral extension of the tumor, a partial costo-transversectomy would be necessary. Using contralateral hemilaminectomy approach the exposure of the tumor is equivalent to that after laminectomy, however, without necessity of an ipsilateral costo-transversectomy (Fig. 2). According to the use of keyhole concept

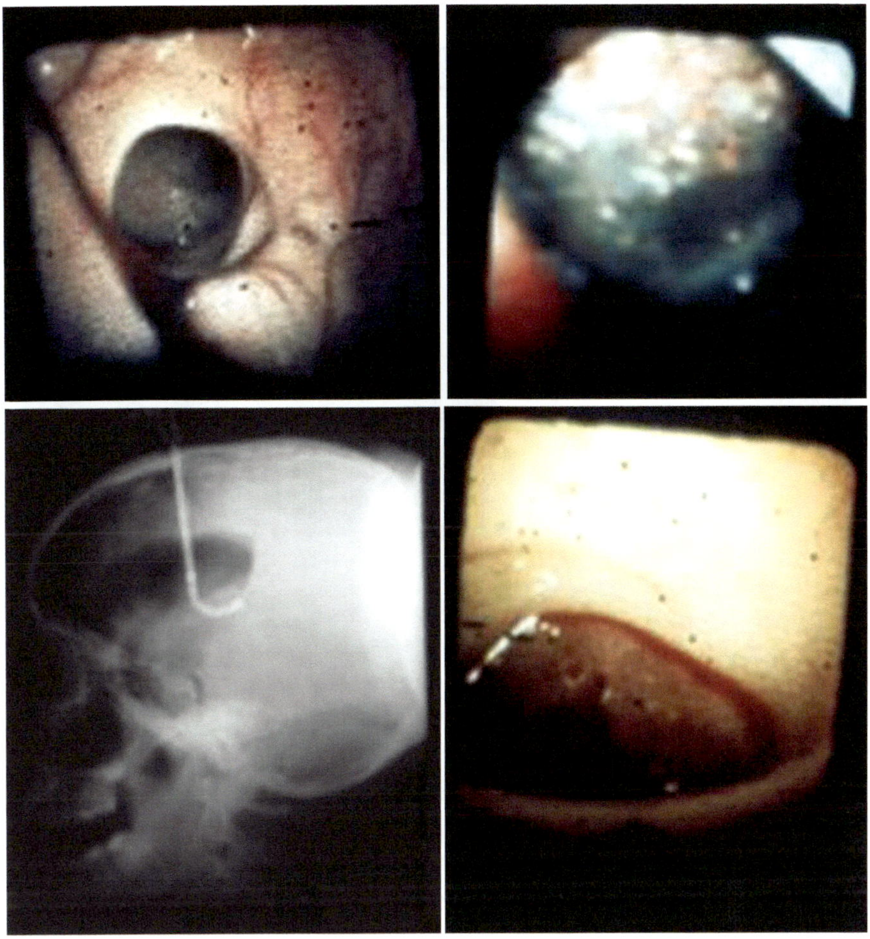

Fig. 2 (Cases from 1968–1969). (**A**) Typical picture of foramen of Monroe with septal and thalamostriate veins and the choroid plexus. (**B**) Typical appearance of cystic craniopharyngioma. (**C**) example of pineoblastoma

in spinal microsurgery, the surgical dissection along the tumor axis allows minimal destruction of eloquent vertebro-muscular and neural structures.

Surgery After longitudinal median skin incision the paravertebral muscles and ligaments were stripped sub-periostally on the left side and reflected laterally. A limited hemilaminectomy was carried out at level T11, the facet joints were not exposed. Using a high-speed drill and small Kerrison punches, the base of the spinous process and the contralateral vertebral arch were carefully undermined to gain optimal access to the opposite side (Fig. 3). The dissection along the tumor axis allowed minimal manipulation of the spinal cord could and neighboring structures.

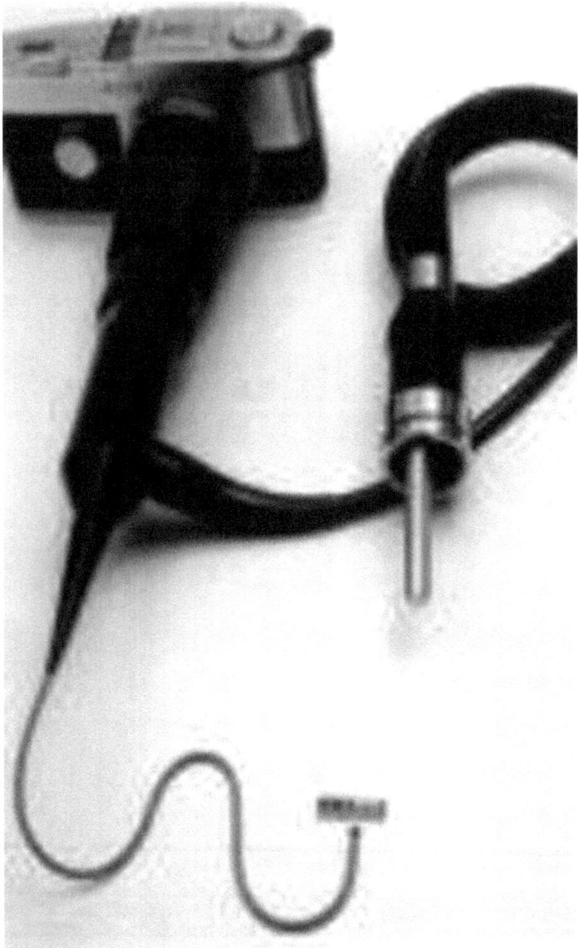

Fig. 3 Development of a 2 mm soft catheter endoscope for cisterna magna and intraspinal investigation

The tumor could be totally removed via extradural approach, complete resection was controlled with endoscope-assisted technique. The histopathological examination showed a benign schwannoma; intraoperative monitoring showed no changes in SSEP/MEP.

Postoperative course On the first postoperative day the patient showed hypaesthesia in accordance to the T11 nerve. The patient did not reveal paresis of the lower extremities, gait or vegetative disturbances. A slight local wound pain could be managed with analgesics. Three weeks postoperatively, after neurophysiological training the patient could subsequently return to her previous employment. Postoperative MR imaging demonstrated that the schwannoma was completely removed (Fig. 1B).

Case II: Intramedullary Situated Ependymoma C6/8

Patient's presentation This 38-years-old man developed progressive gait disturbances without back pain. Neurological evaluation revealed severe medullar ataxia without paresis of the lower extremities and radicular par-aesthesia according to the right C8 nerve root. No disturbance of bladder function was detected. Magnetic resonance image revealed an intraspinal situated tumor C6/8 (Fig. 4).

Approach planning The following approaches were evaluated. Multilevel laminectomy with wide intraspinal exposure: This approach allows wide visualization however, approach-related destruction of the vertebral lamina can cause postoperative late deformity. Multilevel laminoplasty avoid postoperative loss of posterior bony protection, however, the exposure causes wide bilateral muscular injury. A unilateral approach with hemilaminectomy or multi-segmental interlaminar fenestrations offers comparable intraspinal exposure, however with minimal muscular destruction and without injury of the dorsal static structures (Fig. 5).

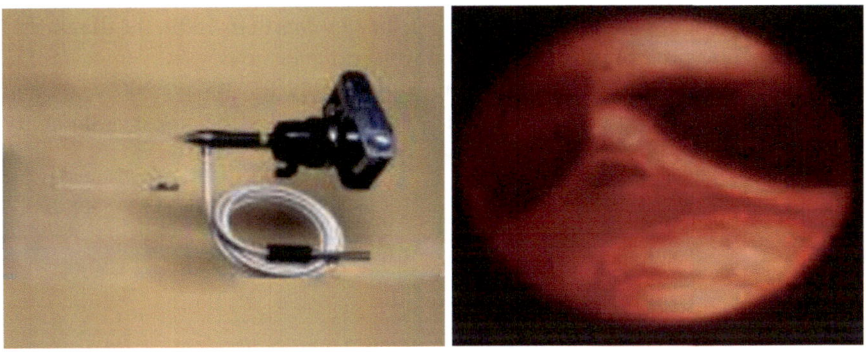

Fig. 4 (**A**) Celfoc glass rod neuroendoscope of needle type (outer diameter 0.8 mm–1.2 mm). (**B**) A picture of C-P angle endoscopy showing the abducens nerve and the Dorello's foramen

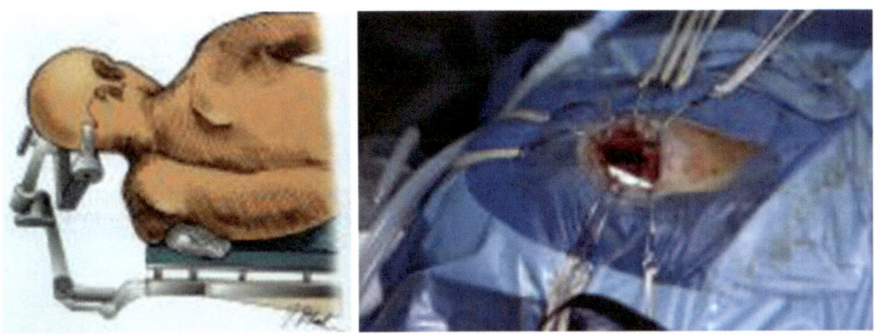

Fig. 5 Lateral position and the location of the trigeminal (upper) and the facial nerve (lower) keyholes (1 inch skin incision and the dime-size bone opening)

Surgery After longitudinal median skin and fascia incision the paravertebral muscle was reflected laterally to the right side. Using the operating microscope, interlaminar fenestrations were performed in the levels C5/6, C6/7. With the high-speed drill and Kerrison rongeurs, the contralateral vertebral arches were gently undermined, achieving optimal access into the spinal canal. After dural opening, the tumor was removed under continuous neuromonitoring. Under the preserved C6 lamina, complete tumor resection was controlled with endoscope-assisted technique (Fig. 6). The histopathological examination showed a benign ependymoma.

Postoperative course Direct postoperative, patient showed minor impairment of gait ataxia without vegetative disturbances. During the 5-days postoperative stay, patient recovered completely and could return to his previous employment four weeks after surgery. Postoperative MR imaging showed complete tumor removal (Fig. 7).

1.2.3.8 Discussion

There is agreement among authors that the treatment of choice of spinal schwannomas is the complete surgical removal. However, concerning operative exposure, opinions vary.

Since the first dorsal laminectomy, described by Smith in 1829, the diagnostic and therapeutic possibilities have been developed enormously (8). However, the operative methods have not been changing markedly for almost 200 years and laminectomy is described in recent publications as the most commonly

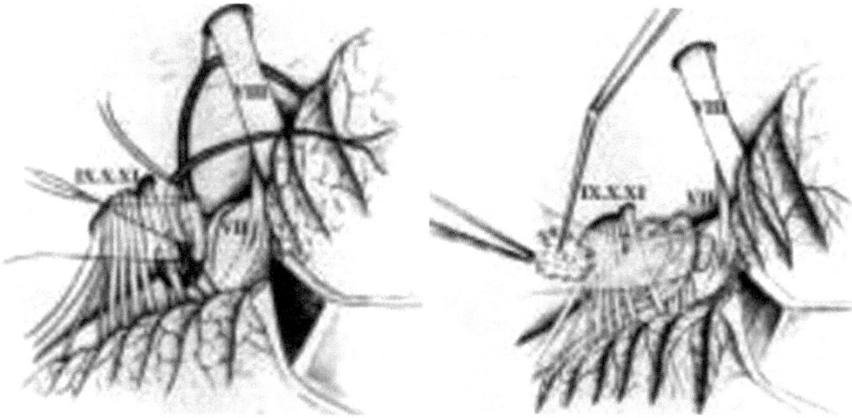

Fig. 6 (**A**) Shredded Teflon tape sling method and fibrin glue fixation for AICA loop transposition. (**B**) Shredded Teflon tape sling and covers for mobilization and repositioning of VA and PICA loops

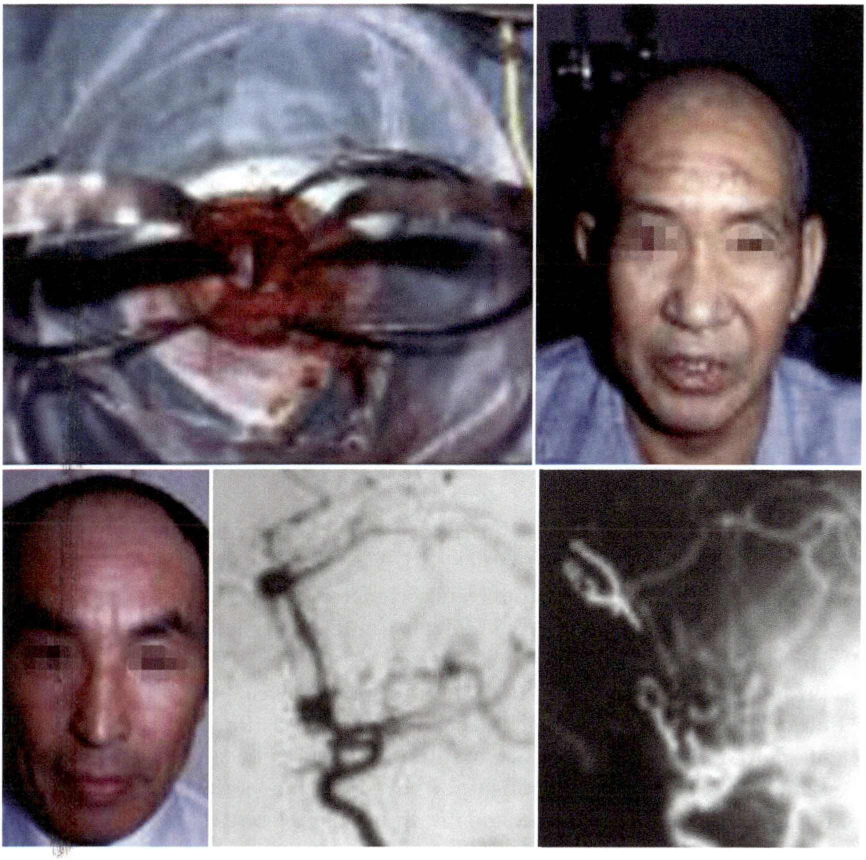

Fig. 7 (**A**) Midline interhemispheric keyhole approach for Acom aneurysm clipping (3 cm trephine opening). (**B**) Midline interhemispheric keyhole approach via forehead crease incision repairing Acom and A2-3 aneurysms

performed neurosurgical procedure (9). Otherwise, the importance of preserving the posterior bone elements with the attached ligaments and muscles is widely recognized (6). In order to reduce these postoperative complications, Raimondi (10) recommended osteoplastic laminotomy, Eggert (4), Yasargil (11) and Sarioglu (12) unilateral hemilaminectomy. Perneczky popularized a more limited approach, using multi-segmental inter-arcuar fenestrations to observe intraspinal lesions (13).

In this article, two cases with intraspinal tumors were demonstrated, operated through minimally invasive spinal keyhole approaches. After preoperative planning of surgical treatment taking into consideration the individual pathoanatomical situation, both lesions were approached by effective limitation of iatrogenic destruction of posterior elements of the spinal column. The contralateral hemilaminectomy approach allowed optimal visualization along the tumor axis and safe surgical preparation in the opposite intraspinal area without extended laminectomy or costotransversectomy. To our knowledge there has been no previous report of a spinal schwannoma treated successfully by contralateral exposure.

Both operations were performed from the author in 2001 as neurosurgical resident under the fruitful assistance and intra-operative support of Prof. Axel Perneczky. This paper is addressed to him with grateful thanks in infinite respect.

Acknowledgments: We express our gratitude to Stefan Kindel for his artistic assistance.

1.2.3.9 Minimally Invasive Neurosurgical Treatments (MINT)

Takanori Fukushima, M.D., D.M.Sc

Professor & Director Carolina Neuroscience Institute, Raleigh NC

Professor, Duke University Medical Center, Division of Neurosurgery, Durham NC

Professor Axel Perneczky and the author were cooperating to promote Minimally Invasive Neurosurgical Techniques and Treatments in the 1980s and 1990s. For over two decades, we maintained good colleagueship as well as a personal and professional friendship. I have been pioneering the development of innovative neuro-endoscopes and of minimally invasive keyhole microsurgical techniques. Professor Perneczky possessed an extremely agreeable and warm personality and had excellent microsurgical skills and neurosurgical knowledge. Professor Perneczky was a man of science and academy, publishing a number of scientific articles and text-books. In commemoration of Professor Perneczky's great clinical neurosurgical contribution and promotion of minimally invasive neurosurgical concept, I will discuss the idea of minimally invasive neurosurgical techniques and the development of keyhole micro-operative procedures.

I started neurosurgery in 1968 and my first specialization was stereotactic treatment of parkinsonism. I performed stereotactic thalamotomy in over 300 cases with Parkinsonism and at the same time I developed the neuro-fibro-endoscope as well as transsphenoidal pituitary microsurgical technique and keyhole supermicroneurosurgery. Over the past four decades of my entire career, I devoted myself for the development of Minimally Invasive Neurosurgical Technique and concept. In this article, I present the development of an innovative neuro-fibro-endoscope, establishment of a keyhole microsurgical approach for hemifacial spasm, trigeminal neuralgia and glossopharyngeal neuralgia as well as the development of a mini craniotomy and lateral supra-orbital minimally invasive approach.

Development of the Neuro-Fibro-Endoscope

Since the first description of a ventriculoscope by Walter Dandy in 1922, various kinds of ventriculoscopes have been reported by Fay, Putnam, Guiot and Scarff. Most of these ventricular endoscopes were used for the endoscopic cauterization of the choroid plexus in the treatment of hydrocephalus. All of the previous neuroendoscopes were rigid endoscopes with large caliber and were restricted for the visual diagnosis and management of hydrocephalus. In 1968, I developed a 3 mm and 4 mm ventriculo-fiberscope with a 68-degree wide viewing angle lens. The tip of the fiberscope could be bent 30 degrees up and 130 degrees down angle with electric bending control[1]. The fiberscope has a channel for irrigation, aspiration and insertion of an electrocautery probe. The results of the clinical application of this ventriculo-fiberscope was published in *Journal of Neurosurgery* in February 1973 with the operative results in 37 clinical cases[2]. This neuro-fibro-endoscope was mounted on a supporting stand and the entire scope was sterilized. The insertion of the biopsy probe and coagulating probe were inserted with a special Y connector in the middle of the fiberscope tube with irrigation and suction control. Photograph and motion pictures could be taken for recording of the operative event. **Figure 1** illustrates the supporting system and the fiberscope (Figure 1A) and the bending section of the tip controlled electronically with the foot lever (Figure 1B). **Figure 2** demonstrates various pictures of the intraventricular lesions. **Figure 3** illustrates a special 2 mm catheter endoscope for spinal endoscopy and **Figure 4** illustrates the needle endoscope (Fig. 4A), which was used for spinal endoscopy and particularly C1-C2, lateral endoscopy[3]. This Celfoc needle endoscope was also used for CP angle endoscopy[4] through the retromastoid burr hole (Fig. 4B). The clinical results of Celfoc needle endoscopy and spinal endoscopy were reported in *Neurochirugica* in 1975[3]. Results of endoscopic biopsy for intraventricular tumors were reported in *Neurosurgery* in 1978[5]. Because of the development of a keyhole microneurosurgical technique and minimally invasive craniotomy with innovative supermicrosurgical instruments, I stopped using these neuro-endoscopy techniques by the end of the 1970s. I never imagined that the revival and revitalization of neuro-endoscopy could occur in the middle of the 1980s by Perneczky and others.

Dime-Size Keyhole Microsurgery: Microvascular Transposition (MVT) for Hemifacial Spasm, Trigeminal Neuralgia and Glossopharyngeal Neuralgia

It should be highly commended of the effort and research of Professor Perneczky for the popularization of the neuro-endoscopy, endoscope assisted microsurgery and keyhole microsurgical approaches. Since 1978, Fukushima developed a keyhole microneurosurgical approach and microvascular transposition (MVT) procedure in the surgical management of hemifacial spasm, trigeminal neuralgia and glossopharyngeal neuralgia. Hemifacial spasm, trigeminal neuralgia and glossopharyngeal neuralgia are distinctive clinical entities and are extremely annoying and distressing disorders that cannot be cured by any other medical management. With the clinical neurosurgical work by Peter Jannetta since 1970, the pathophysiology and etiological mechanism have become clear that hemifacial spasm, trigeminal neuralgia and glossopharyngeal neuralgia are caused by vascular loop compression onto the exit-entry zone of the cranial nerves. I started to perform vascular loop transposition surgery in September 1978. I performed 400 operative procedures from 1978 to 1980 and I established a dime-opening keyhole retrosigmoid approach. A superior dime size opening is used for trigeminal neuralgia and an inferior retrosigmoid keyhole for hemifacial spasm and glossopharyngeal neuralgia. A number of the previous literature make an inappropriate or wrong surgical approach to the cisternal segment of the seventh and eighth cranial nerves where either the PICA or AICA loop were manipulated with high risk of hearing complications. I noted that vascular loop compression is not around the cisternal portion of the seventh nerve. The compression site for hemifacial spasm exists at the nerve root exit zone on the lower pons over the proximal ninth nerve. Previous microvascular decompression procedures of inserting prosthesis material such as Ivalon sponge, Dacron or Teflon pledget are totally inappropriate procedures. Surgery is not the insertion of prostheses in between the facial nerve and the vessel, but mobilization and transposition of the compressing vascular loop towards the jugular foramen and fixation with shredded Teflon tape and fibrin glue. Nothing should touch the nerve root. This is the correct microsurgical procedure for hemifacial spasm. This keyhole microsurgical curative treatment should not be called a Micro Vascular Decompression (MVD) but should be designated as Micro Vascular Transposition (MVT). Figure 5 illustrates the patient positioning and location of the one inch skin incision and a dime size keyhole bone opening. Operative techniques of vascular loop transposition and fixation with shredded Teflon tape sling and fibrin glue is shown in Figure 6A for

AICA-MVT and in Figure 6B for VA PICA-MVT. I performed the MVT procedure in over 3300 cases from 1978 until 2008, over 30 years with a one-time surgical cure of 96%[6]. Throughout my 30 year surgical experience, my keyhole microvascular transposition operation carries 0.1% major risks and 1 or 2% of minor risks. The offending vascular loop was the AICA branch in 35% of cases, PICA branch in 30% of cases and vertebral artery compression in 20% of cases.

For trigeminal neuralgia, it is noteworthy that in my series of 2300, 10% of typical class 1 genuine trigeminal neuralgia showed the presence of a brain tumor, mostly epidermoid tumor (9%) and acoustic neuroma and meningioma (1%). In addition, I had 12 cases of cryptic angiomatous malformation in and around the trigeminal nerve root. The other 80% of patients had typical vascular loop compression mostly by the superior cerebellar artery, 7% by the vertebro-basilar artery trunk and an additional 8% by an AICA loop compression. Throughout these vascular loop transposition procedures, an initial one-time surgical cure was obtained in 96% of cases. In typical hemifacial spasm cases, there was no negative exploration and all patients had vascular loop compression. In typical trigeminal neuralgia cases, despite typical tic douloureux syndrome, 10% had no vascular loop compression. For these negative cases, the cause was unknown. I performed glossopharyngeal neuralgia surgery in 60 patients and the results were excellent. Long term follow-up data over 20 years shows my vascular loop transposition in glossopharyngeal neuralgia was excellent with over a 95% cure rate[7].

Development of a Mini Craniotomy and a Lateral Supra-Orbital Keyhole Approach

In 1980, I developed a midline keyhole interhemispheric approach for the clipping repair of ACOM aneurysms (Figure 7A–B). I developed the lateral supra-orbital keyhole approach (Figure 8) for anterior circulation aneurysms such as internal carotid PCOM aneurysms and MCA aneurysms. In 1981, I made the first report of the lateral supra-orbital keyhole pterional approach for aneurysms (Figure 9). For this type of keyhole procedure, I developed keyhole aneurysm clip appliers (Figure 10) which are about half size of a standard Yasargil clip applier. I have been continuously using these keyhole aneurysm appliers for over a 30 year period. In 1982, I switched from the U-shaped or horseshoe craniotomy flap to the straight incision and trephine minimally invasive craniotomy technique for calvarial lesions and subcortical tumors (Figure 11).

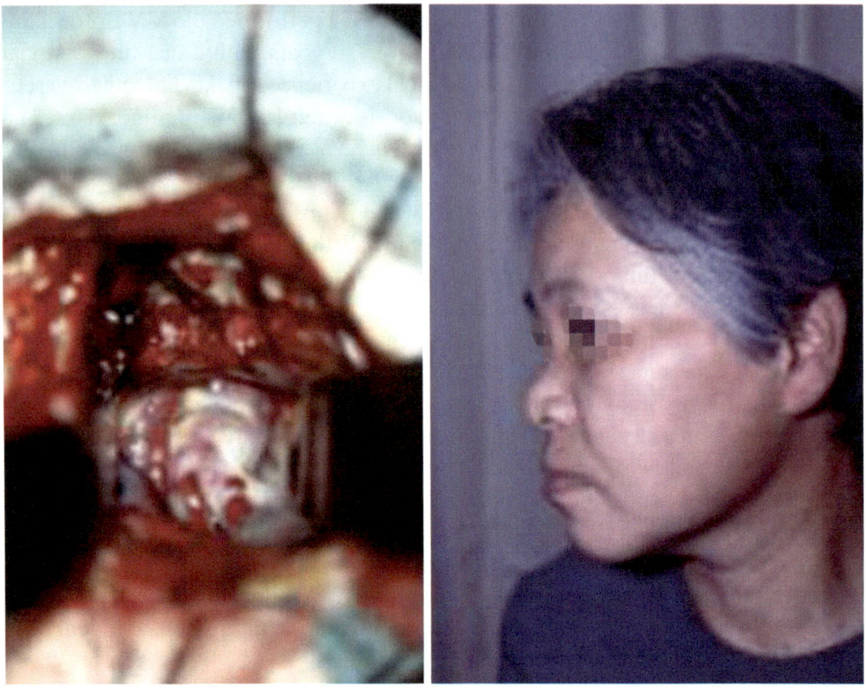

Fig. 8 Lateral supra-orbital 3 cm incision and keyhole bone opening for clipping of Pcom aneurysm (1981)

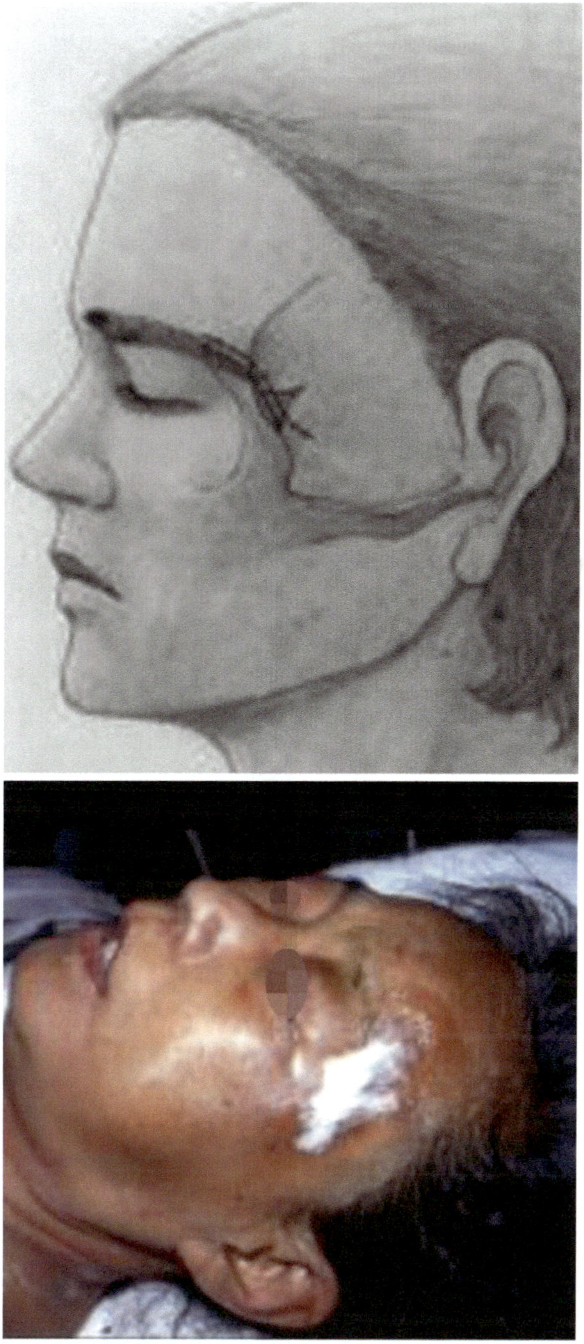

Fig. 9 Illustration of lateral supra-orbital skin incision and a case operated for ruptured MCA aneurysm (1981)

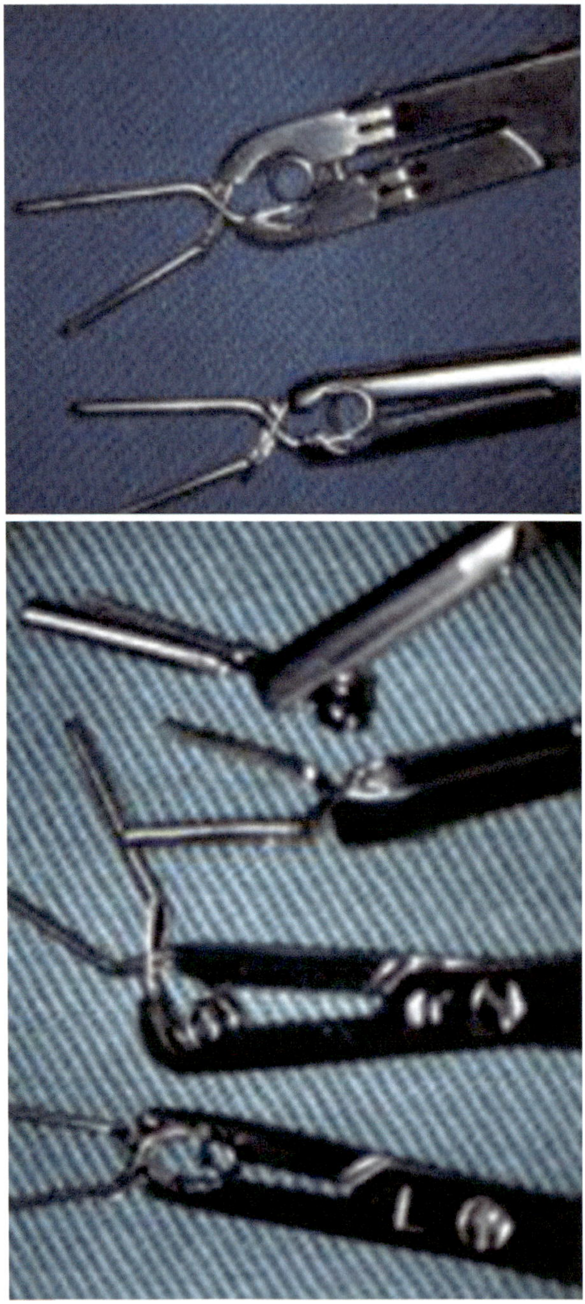

Fig. 10 Fukushima designed keyhole aneurysm clip appliers; double joint model and angled groove models (13 with various angles)

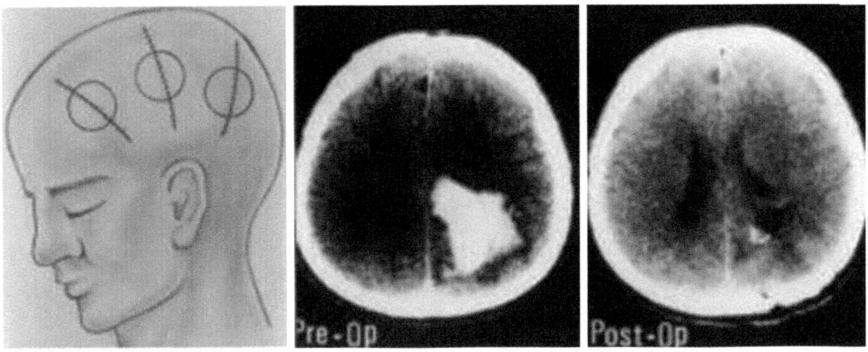

Fig. 11 Straight incision and mini-craniotomy for calvarial subcortical lesions

In 1982, I developed an ultrasound B-scope neurosurgical imaging device with a one inch mini-probe[9] for visualization of subcortical lesions such as brain tumors and cavernous hemangiomas (Figure 12). In 1995, I developed a keyhole ultrasonic aspirator which is less than half the size of the American CUSA hand piece (Figure 13)[8]. I also developed a needle ultrasonic aspirator for percutaneous aspiration and removal of gliomas[8]. All of these ultrasound developments were published in the 1980s and 1990s as a scientific article[9].

Although my current specialty of skull base surgery for treating large and extensive cranial base tumors may require wide surgical exposure, I still maintain a minimally invasive or less invasive philosophy. My previous technical developments, microsurgical Fukushima instruments and special Fukushima Retraction Holder System[10, 11] facilitate accurate performance of Minimally Invasive Neurosurgical Treatment (MINT). In conclusion, I hope the next generation of neurosurgeons in the world will follow and maintain our effort of minimally invasive neurosurgical concept and technique for the benefit of patients.

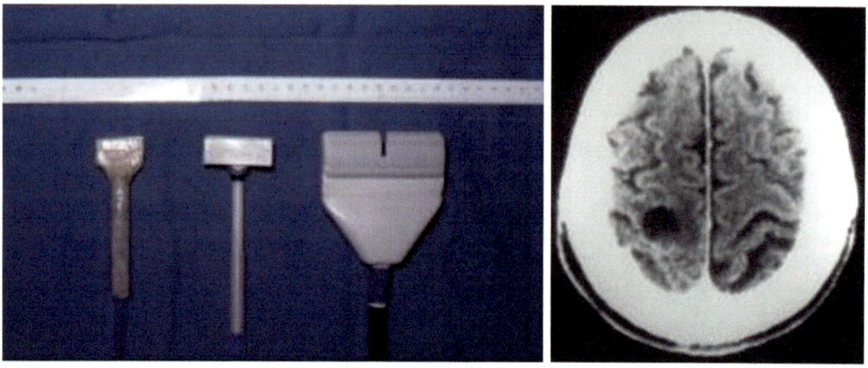

Fig. 12 One inch mini-probe for linear electronic ultrasound imaging (TOSHIBA model, 1982)

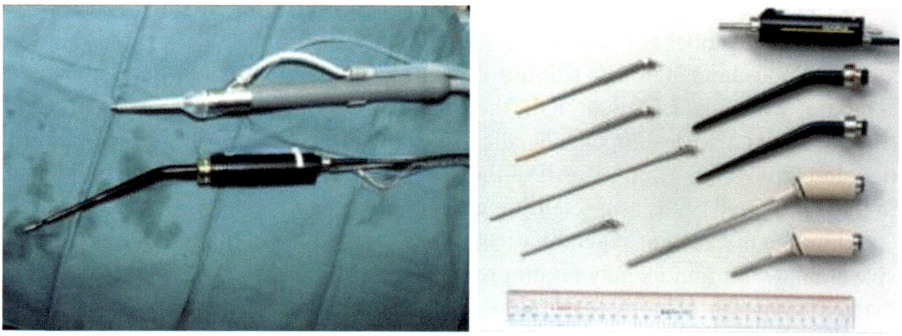

Fig. 13 Development of a keyhole ultrasonic aspirator with a small light-weight hand piece (Olympus 1995) and needle type tips, utilizing a special technology of electrostriction transducer and not electromagnetic type (American CUSA)

References

1. Fukushima T. Endoscopic diagnosis of ventricular lesions by newly designed ventriculo fiberscope. ***Neurologia Medico-Chirurgia*** 11: 1971

2. Fukushima T et al. Ventriculo fiberscope, A new techniques for endoscopic diagnosis and operation, Technical note. ***J Neurosurg*** 38:251-256,1973

3. Fukushima T. Schramm J. Klinischer versuch der endoskopie des spinalkanals: kurzmitteilung. ***Neurochirugia*** 18:199-203,1975

4. Fukushima T. Endoscopy of Meckel's cave, cisterna magna and cerebellopontine angle. ***J Neurosurg*** 48: 302-306,1978

5. Fukushima T. Endoscopic biopsy of intraventricular tumors with the use of a ventriculofiberscope. ***Neurosurgery*** 2:110-113,1978

6. Fukushima T. Microvascular decompression for hemifacial spasm, results in 2890 cases in Neurovascular surgery. ed. L.P. Carter, R.F. Spetzler, 1133-1145,1995, McGraw-Hill.

7. Sampson J, Fukushima T et al. Microvascular decompression for glossopharyngeal neuralgia; long term effectiveness and complication avoidance. ***Neurosurgery*** 54:884-890,2004

8. Sawamura Y, Fukushima T. Development of a hand piece and probes for a microsurgical ultrasonic aspirator: Instrumentation and application. ***Neurosurgery*** 45:1192-1197,1999

9. Fukushima T. Intraoperative ultrasound localization of small subcortical lesions with a newly-developed 2.5 cm linear mini probe. ***Recent advances in neurosonology***, ed. M. Oka, 487-491,1992, Elsevier

10. Fukushima T, Sameshima T. Mastronardi L, Friedman A. ***Fukushima's Microanatomy and Dissection of the Temporal Bone Second Edition***. AF-Neuro Video Inc. 2007

11. Fukushima T, Nonaka Y, Day J, Friedman, A et al. ***Fukushima Manual of Skull Base Dissection Third Edition***. AF-Neuro Video Inc. 2010

1.2.3.10 The "Marburg" Concept of Minimally Invasive Endoscopic Neurosurgery (MIEN) (1988-2008) A Historical Reflections

Dieter Hellwig M.D., Ph.D.; Head of the Division:
Stereotactic and Functional Neurosurgery; International Neuroscience Institute Hannover
Rudolf-Pichlmayr-Straße 4; 30625 Hannover
Phone: 0049/511/27092456; Fax: 0049/511/27092457; eMail: hellwig@ini-hannover.de

First congress on "Endoneurosurgery and Endoscopic Stereotaxie" Marburg 1990

I always admired this friend, gentleman, teacher, phantastic neurosurgeon, extraordinary scientist, musician, bon vivant, charismatic man of human kind. We first met at the meeting in Marburg in 1990 and after this so many times over the years. The common idea was to establish minimally invasive methods into the neurosurgery using endoscopes as a new tool. He from the microsurgical approach, I was coming from stereotactic neurosurgery. Finally both of us were successful and we met in the middle. Endoscopy-assisted microsurgery is now totally accepted in skull base surgery and vascular neurosurgery. On the other hand, pure endoscopy guided by stereotaxy and neuronavigation has its focus on treatment of hydrocephalus, intracranial cystic lesions and tumour biopsies. Of course there had been a lot of criticism that neurosurgery should be "per se" minimally invasive, however Axel Perneczky introduced many innovations to reduce surgical trauma and to improve patient's outcome. As I knew him when confronted with new technologies he never says "we don't need", but "let's try, if we can use and improve it". He was a very open-minded character in a sometimes unflexible and rigid neurosurgical world. We miss him.

In 1989 we introduced "endoscopic stereotaxy" as a new operative technique into neurosurgery. This method was first scheduled to optimize stereotactic biopsy. In its further development it proved to be effective for other indications. In analogy to the term minimally invasive surgery, coined by Wickham and Fitzpatrick in 1990, we defined these procedures as Minimally Invasive (Endoscopic) Neurosurgery (MIEN). MIEN refers to neurological interventions in which by use of endoscopes larger openings of intracranial or intraspinal spaces can be avoided.

Our preliminary indications had been:

- Endoscopic – stereotactic biopsy of intracranial space occupying lesions
- Ventriculoscopy and endoscopic third ventriculostomy (ETV) in diagnosis and treatment of hydrocephalus
- Endoscopic evacuation of intracranial cystic space-occupying lesions

- Endoscopic evacuation of intracerebral haematoma
- Endoscopic evacuation of multiloculated, septated chronic subdural hematoma
- Endoscopic evacuation of subacute and chronic brain abscesses
- Endocavitary syringostomy

A lot of instruments have been developed and operative technologies have been applied over the years to make neuroendoscopy safe, precise and effective.

Some of the preliminary indications have been established, some have been abandoned. Surprisingly in the last decade a lot of new applications have been added into the daily routine of the neurosurgical OR especially for skull base surgery.

In conclusion after years of doubts, discussions and criticism, now there are well-defined clearcut indications and standards, which make endoscopy an accepted technique like microsurgery in modern neurosurgery.

1.2.3.11 "The True Perfection of Man Lies not in What Man has, but in What Man is."

- Oscar Wilde

Prof Axel Perneczky, founder of minimally invasive neurosurgery was a man with a vision. He was not only highly talented as a neurosurgeon but was a great educationist too. He traveled around the world to educate young neurosurgeons with passion.

I was impressed by his graceful attitude in the OR when I visited his institute. A few months before he passed away, he looked just fine after recovering from stroke and he had even started to educate YNS again. I shall always remember Prof. Axel Perneczky as a very talented, outstanding neurosurgeon and a committed teacher who influenced my professional life.

The late Prof. Axel Perneczky in Jordan, 2009

Yoko Kato, M.D. Ph.D.
Chair, Education and Training Committee of the WFNS
Secretary, WFNS Foundation
Professor, Department of Neurosurgery
Fujita Health University
President of ISMINS

Evolution of the Key-Hole Concept: The MIN-Key Concept

2

2.1 Recent Roots of MIN

2.1.1 Origins of the Keyhole Concept: M.G. Yasargil and A Perneczky

2.1.1.1 M.G. Yasargil

In "A Legacy of Microneurosurgery: Memoirs, Lessons and Axioms", in 1999, **Yasargil** summarised the development of micro-neurosurgery. It was not the technical innovation of the microscope alone that brought the breakthrough, but new concepts: neuro-anatomy (subarachnoid approach), new surgical strategy (intra-tumoral approach), new instrument families (bipolar, clip system, Leyla), mouth-guided microscope (real-time focus and field-of-view adaptation during double-handed operation) and a new ergonomic setting. This development has matured over decades to a high standard, reaching a globally recognised gold standard. In Volume IV A of "Microneurosurgery", the basic subjects of neuroanatomy, neurophysiology, neuropathology and neuroradiology were presented in a sustainable conceptual way, thus forming a paradigm methodology for innovation that is currently threatened with extinction. As early as 1975, Yasargil and Fox make use of the keyhole concept, making reference to the principle of dissecting the brain in as gentle a way as possible, namely trans-cisternal. For Yasargil, keyhole corresponds to the concept of trans-cisternal dissection.

The microscope gave the new direction of neurosurgery the name but the overall environment and workflow as well as the new concept was the secret of success. The name of these overall conditions is ergonomics and can be seen in (Graph 2.1).

The ergonomics might be very personal however, it must be perfect, and no single detail should be unaware or neglected.

Together with the corner-stones seen in (Graph 2.1) Yasargil preferred absolutely silence in the OR, and a dark room, only the light of the microscope and some monitors, and only the whispering noise of the sucker ruled the atmosphere throughout the operation. Speaking was strictly forbidden and not necessary in his very

K. D. M. Resch, *Key Concepts in MIN - Intracerebral Hemorrhage Evacuation*, Key-Concepts in MIN 1, https://doi.org/10.1007/978-3-030-46513-1_2

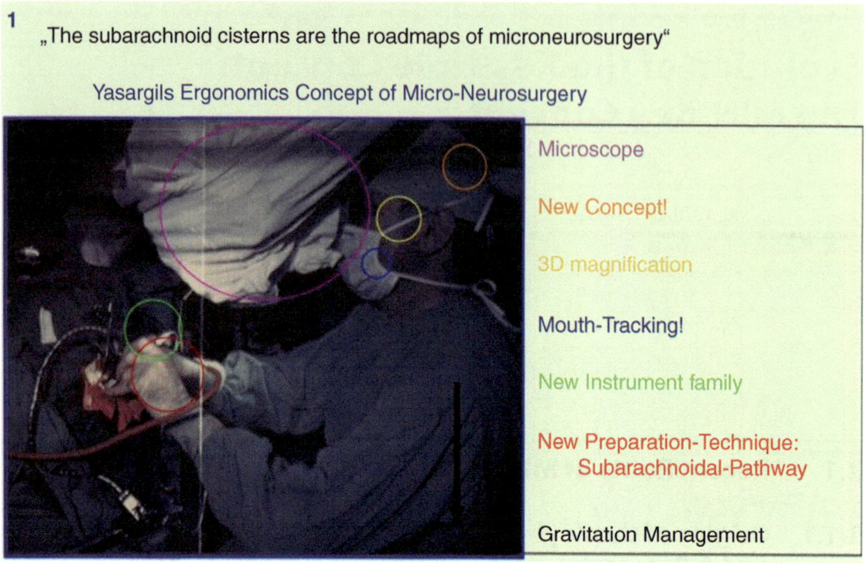

Graph 2.1 Yasargils ergonomics setting: gold-standard of microneurosurgery

effective setting. The complete team workflow was directed to the outcome of the patient through the perfect needs of the surgeons' brain. He created a flow status with maximum of over-sensitivity and concentration like a solo pianist during the concert. The complete surgical suite, work-flow and mental-flow, the ergonomics environment functioned smoothly and like an unique organism.

2.1.1.2 Perneczky

Perneczky further developed this principle and also incorporated the surgical approach. The now improved imaging allowed the personalised anatomy of the individual patient to be shown. This means an individualised access approach can be made through precise analysis and planning. If one analyses the surgical site at the end of an operation through a standard access point, one regularly recognises that a maximum of one third of the standard access was used for the intradural part of the operation. The question then was whether this third and its precise localisation could have been analysed and planned before the operation. Current imaging actually allows for this if appropriate analysis and planning is carried out according to this principle. In most cases, this planning and the greater precision then allow for a smaller access point, thus minimising the trauma.

The further development of the endoscopy video link increased the need for planning and the demand for precision during implementation yet further. The problems of microscopic visualisation through a keyhole access point were largely compensated by endoscopy. This required the paradigm shifts towards "endoscopy-assisted micro-neurosurgery", described above. The access requirement for non-superficial processes was now basically reduced to just a 3 mm endoscope and the requirements

Graph 2.2 The key-hole question (a 20-years history!)

for manipulating the instrument. This conceptual development had considerable consequences for clinical thinking, indications and surgical ergonomics.

The development in the Perneczky school was so rapid, that communication with the professional community became increasingly difficult, and many misunderstandings arose (Graph 2.2).

The need for training also grew significantly, especially since the results attracted more and more interest. However, "Keyhole Concept in Neurosurgery" from Thieme in 1999 did not sufficiently meet the demand. The structures, concepts and techniques created by Perneczky remain the benchmark for further development for many decades to come. On the other hand, this development has not reached the level of a gold standard, not least because the margin for error is still too low (Graph 2.3).

The principles of Perneczky's working in the final level, made a complete new ergonomics necessary. The corner-stones of this environment changed markedly, best discovered in the direct visual comparison (Graph 2.4). The new conditions made a new training in the laboratory indispensable. The essence of evolution was a change from micro-technique to micro-system-technique. The technical environment became small, modular, intuitively and emergent, and finally by the way potentially cheap and simplified. The recent evolution of the main-stream is directly opposite: large, loud, expensive, time-consuming, in one word: un-ergonomic.

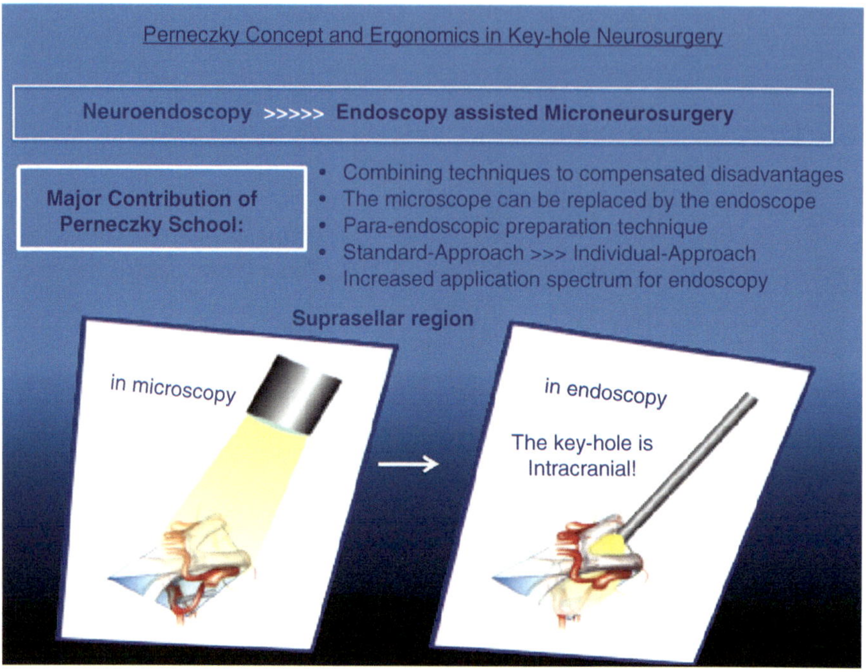

Graph 2.3 Change from micro-/endo-neurosurgery to endoscopy assisted microneurosurgery (major contribution of the perneczky-school)

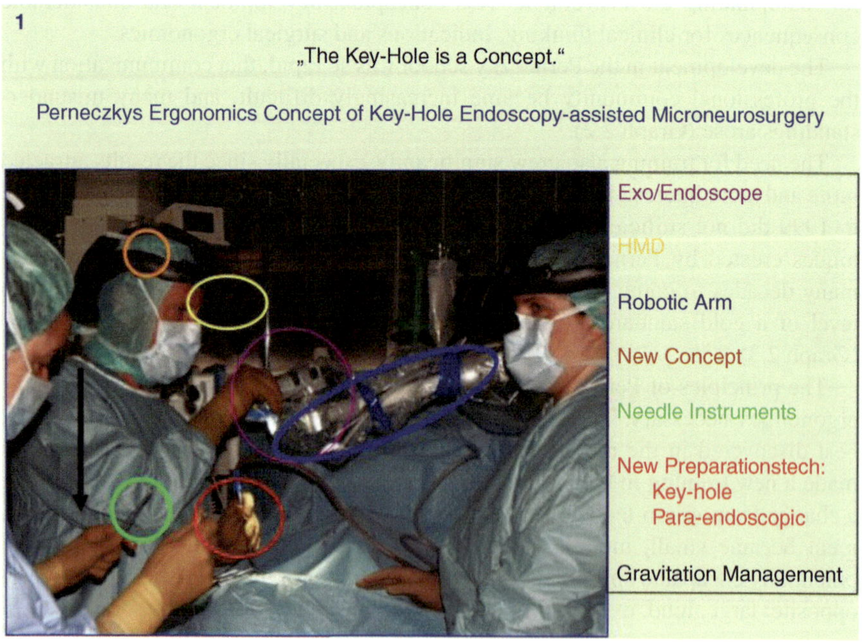

Graph 2.4 Perneczky ergonomics setting (gold-standard of the future)

When studying the ergonomics of Yasargil and Perneczky, each of them was perfect ergonomically. Everything else was very different, however, we see, that the results depend more on ergonomics perfection rather than on personal behaviour.

2.1.2 Further Development of the Keyhole Concept to the "MIN-Key Concept"

2.1.2.1 The MIN-Key Concept

In order to overcome these stated misunderstandings about the key-hole concept, a **further development of the concept** was proposed by the author during the Perneczky-Lecture at ISMINS 2014 in China.

During the evolution of the key-hole concept there was a long phase of surgical improvement and feasibility. It was understandable that the focus was on surgical approach positioning, size and planning. The aim anyway, has always been to minimize the overall trauma. To make the concept more comprehensive it was further developed also during all the steps of MIN evolution (Graph 2.5).

The **keyhole concept** essentially consists of two notions. After 20 years of discussing the "hole", the focus should now be on the "key". **What are the conceptual and technical "keys"** of minimally invasive neurosurgery? (Graph 2.6).

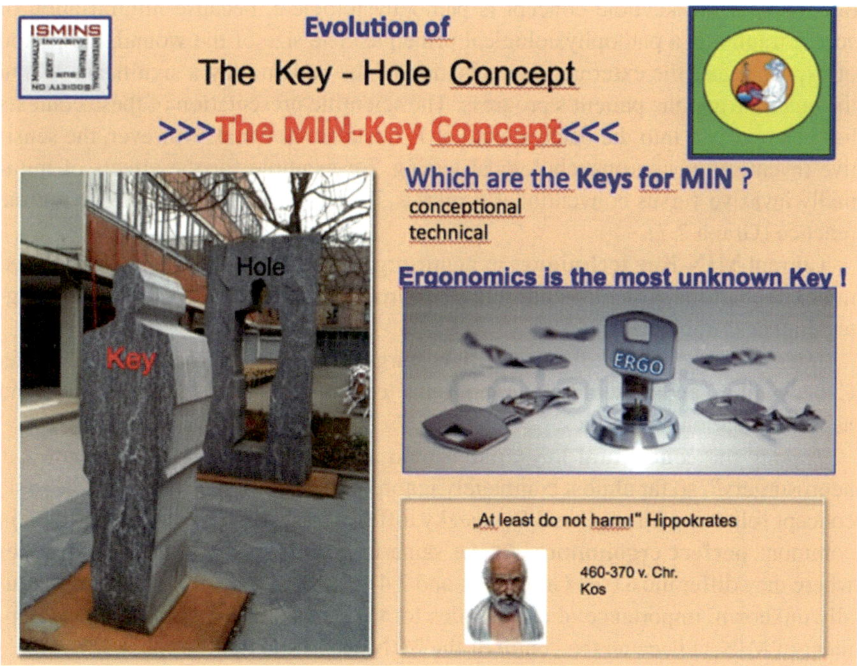

Graph 2.5 The MIN-key concept

Graph 2.6 The hole-discussion

The "keyhole" concept is not anatomically justified, but only its execution, in connection with the planning and precision of the individual access point. The justification for the keyhole concept is pathophysiological, because minimisation of access trauma is a pathophysiological principle. The size of the wound, the period of exposure and the external stigmatisation of the patient play a significant role in the outcome and the patient's progress. The scientific presentation of these contexts has been pushed into the back-ground or is mistaken as banal. However, the sensitive inventory of neuropsychological testing, for example for the effects of minimally invasive versus conventional surgeries, is still too time-consuming in routine practice (Graph 2.7).

Current **MIN-Key techniques** in neurosurgery include mouth-switch-controlled micro-neurosurgery, neuro-endoscopy, neuro-sonography, LASER and sealing-techniques (Graph 2.8).

Interdisciplinary techniques include targeted radiation therapies (Y-Knife, Cyber-Knife), endovascular interventional techniques, some bio-techniques and neuromodulation techniques.

Among the **conceptual keys**, one should above all emphasise "ergonomics in neurosurgery", so far almost completely ignored. The two main roots of the keyhole concept following Yasargil and Perneczky differ greatly. But they have one thing in common: **perfect ergonomics**. At the same time, it is in the ergonomic features where they differ most (see Graphs 2.1 and 2.4). Due to the extraordinary, and virtually unknown, importance of ergonomics for the conceptual and technical development of MIN, it became the centre of the **MIN-Key Concept** (Graph 2.9).

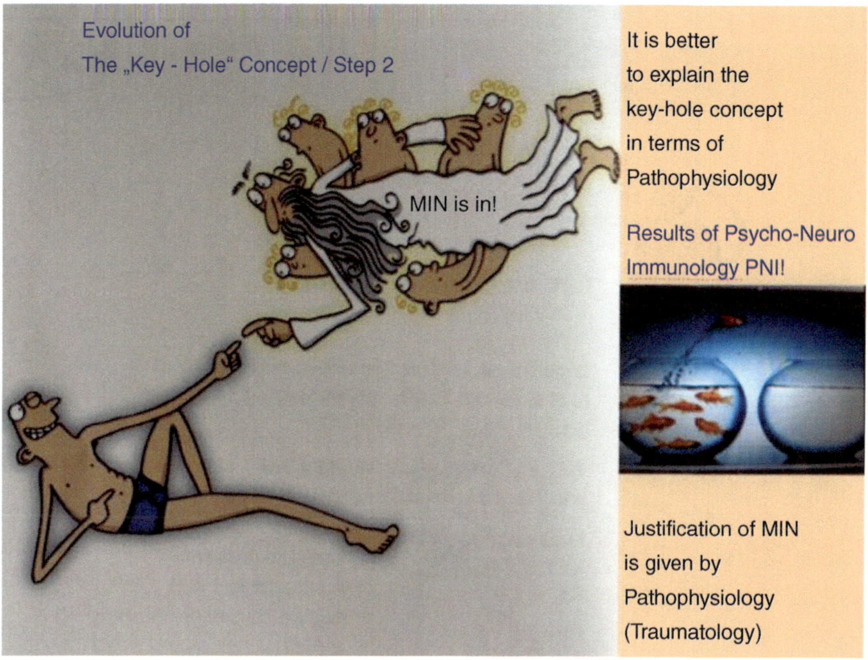

Graph 2.7 Pathophysiology as basics for MIN-key concept

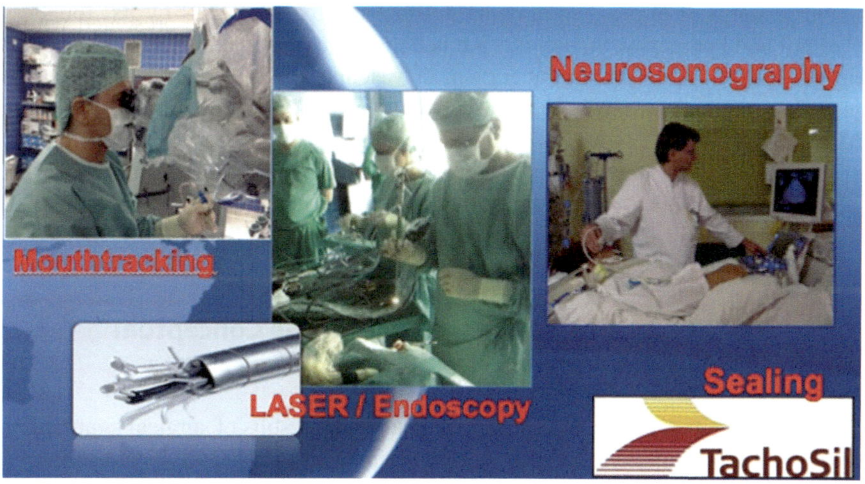

Graph 2.8 The 5 MIN-key techniques

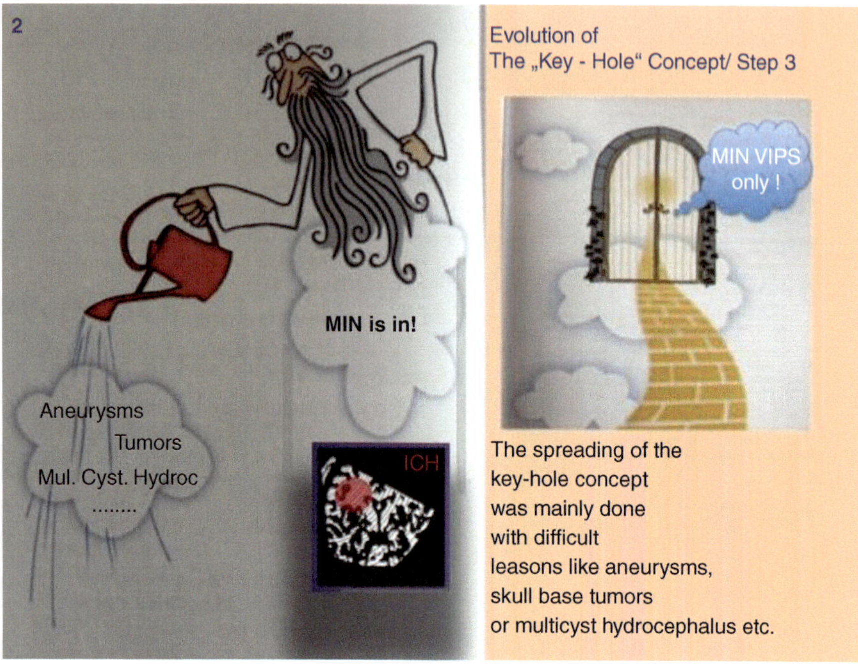

Graph 2.9 MIN-Key concept only for VIPs?

Surprisingly ICH-evacuation came not into the focus of MIN by its' founders Yasargil and Perneczky, though it is a much more frequent pathology than the prestigious lesions like a tumour, aneurysm or vascular lesion. No "golden" trial can make it comprehensive, why ICH-cases do not have the same importance and same MIN indication as a tumour case or a vascular case. This counts even more, as ICHs are mostly easier to be evacuated and only neurosurgery can provide this. This **MIN-Lack for ICH evacuation** is simply inacceptable. Do not leave them back in the stroke units!

2.2 The Importance of Ergonomics for the Conceptual Development of MIN

On 30th July 1986, the senior author presented some results of surgical simulations at the Department of Neurosurgery Medical University of Vienna: Plastination and Dissection in the Area of the Clinical Anatomy of the Head ("Transnasal and Transoral Access to the Vertebro-basilar System"). From 1988 onwards, this surgical simulation method was applied to 74 aneurysmal cases post-mortem. As of

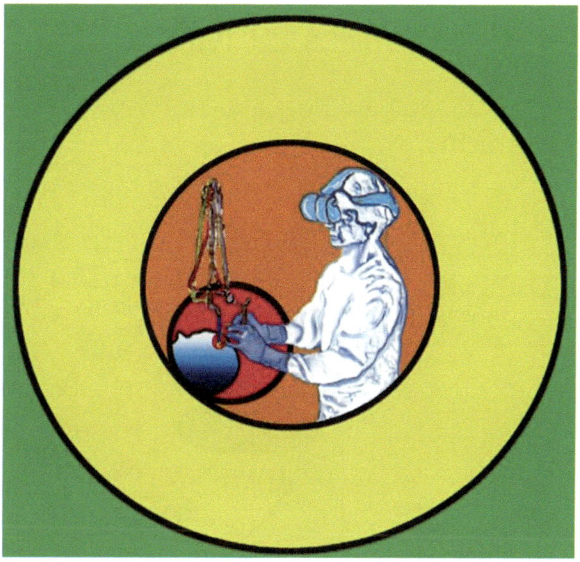

Graph 2.10 MIN-environment. Best ergonomics model: HMD visualization holding-devise robotic arm. Awareness for different sensitivity of ergonomic emergent segmentation of the MIN suite

1990, endoscopy was added to this model in Mainz, and even in the first case of aneurysm, it was shown that a post-mortem trans-endoscopic view of an aneurysm was impossible using trans-endoscopic dissection.

The answer was a **paradigm shift from a trans-endoscopic to a para-endoscopic** technique. During this work, however, it became above all clear that the success of a complex para-endoscopic view through a "key-hole" access point depended absolutely on **ergonomics**.

In the current clinical situation, there is a kind of **"ergonomics chaos"** in the operating room that disturbs the operative process in various ways (Graph 2.11). The work in the surgical simulation setting allowed for an investigation into ergonomic criteria. This happened during hundreds of endoscopy-controlled surgical simulations via keyhole access, but in a technically still un-advanced prototype setting.

In summary, **preclinical** findings from lab work revealed that para-endoscopic dissection results were only possible by changing the ergonomics of the overall process and setting. In addition, it became obvious, as a future subject of clinical investigation, that the techniques available in the clinic, as used here, appeared not to be safe enough for clinical use with a para-endoscopic mode of operation, not so much because of the lack of detailed developments, but because the setting of the ergonomics did not take sufficient account of such clinical surgical processes. It has been experienced that most instruments and devices in their current form disturb the ergonomics of the system: **dissector/dissection object** (head) (see Graph 2.10) and

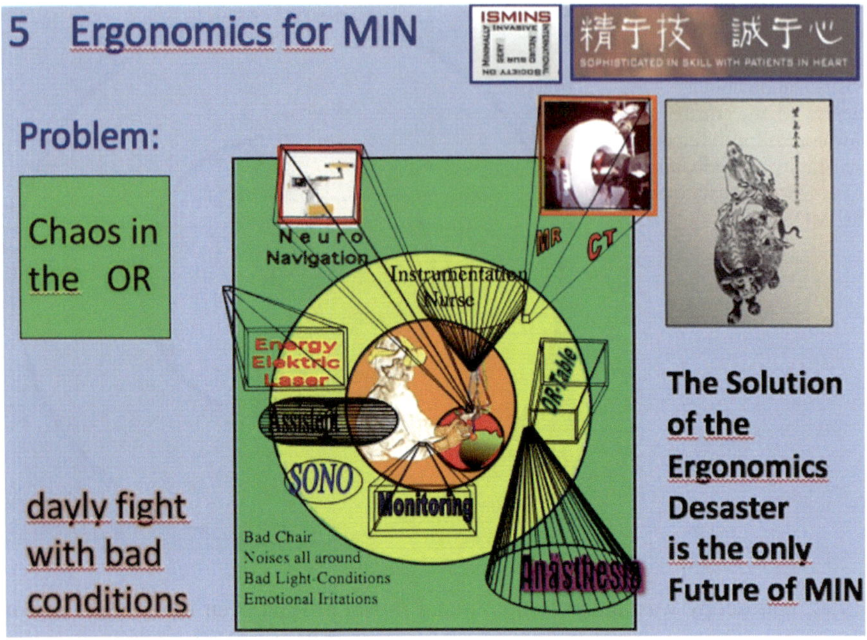

Graph 2.11 The ergonomics chaos in the OR

cannot be smoothly integrated into the existing process, unless, on a case by case basis, at the expense of safety and efficiency.

Ergonomic considerations are a rarity in neurosurgery and, when they occur, they are usually related to a detailed problem. The current development of MIN aims at minimising total trauma, including access trauma, through greater use of general state-of-the-art technology (see Day et al. 1997). At the same time, the technical environment as a whole is developing and is influencing this process in a partly uncoordinated way. As a result, the neurosurgeon faces a large number of devices, instruments, and computers that are rarely in harmony with one another and are often un-advanced or not adapted for neurosurgery (see Graph 2.12).

A crucial factor for the effectiveness of the setting is that the entire development is subordinate to one goal, namely, not to disturb the ergonomic relationship between the surgeon and the patient. The generation of neuro-navigation and intra-operative MRI and CT devices today, for example, cause time-consuming-, spatial- and ergonomic-impairment of the operation. Indeed, it seems necessary to introduce the notion of "**Ergonomics Trauma**" (see Graphs 2.11 and 2.12).

Clinically, the specified ergonomic criteria have been able to be observed in parallel during thousands of operations over the course of 30 years. All ergonomic errors are paid for by a loss in quality or by wear and tear. The **instruments** must allow the hands to work in a relaxed way without access being impeded. Each handpiece must be designed accordingly so as to be extra-axial to the thin, tubular functional pathway (family of knitting needle instruments). The **monitor** should be

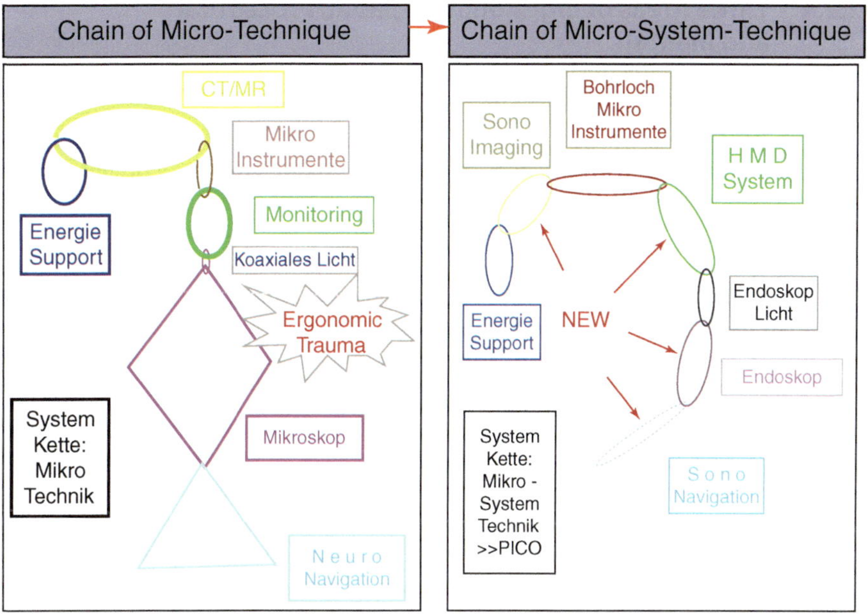

Graph 2.12 Microtechnique >>> Micro-System Technique

directly level with the surgeon's head and be positioned so as to align with a straight, forward-directed look, that is, generally over the patient's thorax. The better alternative is the **MHD** system (Graph 2.10). The **endoscope holder** must be able to be operated precisely and effortlessly, comparable to guidance of the Contraves system, and every position must be able to be maintained without vacillation. The endoscope's **camera** must be able to be adjusted and varied to satisfy the needs of the instruments with regard to dissection, with zoom and focus compensating the instrument-endoscope competition effect.

The introduction of "**neuro-imaging**" and **computer-assisted planning systems**, based thereon, led to increasingly precise anatomical diagnoses and surgical designs (Perneczky 1992). Correspondingly, systems followed that also helped to implement such precise specifications: so-called "**neuro-navigation**". This was followed by systems that compensated for the navigation systems' lack of real-time characteristics: "**intra-operative imaging**" (CT, MRI), which was ultimately accompanied by "**functional imaging**" and "**intra-operative monitoring**". The consequences of this development apparently went unnoticed: the increasing **deterioration of surgical ergonomics** due to un-advanced overall design. Titles such as "Interactive Image-guided Neurosurgery" (Maciunas 1994) or "Neurosurgery for the Millennium" (Appuzzo et al. 1999) have not only provided valuable stimulation for the technical development potential of the future, but have also created a cuphoria around the technology, where the art of neurosurgery seems to have mutated into technology, while ergonomics has fallen by the wayside (see Graph 2.11).

2.2.1 Three Areas of Ergonomics: Spatial, Procedural and Mental

2.2.1.1 Spatial Ergonomics: The Ergonomic Zone Model (Gestalt-Theory) (Graph 2.13)

The **ergonomic zone model** divides the operating room into virtual orbitals of differing ergonomic sensitivity. These start from the centre, the surgeon's hands and patient's head. This is the "red zone" where ergonomic errors have the greatest impact. From here, another three zones (orange, yellow, green) are defined, which are decreasingly susceptible to ergonomic disturbance. Some functional units in the operating theatre cover several zones and in principle carry a high ergonomic risk. They must be "fit for the red zone", for example, although they are mostly located within the green zone.

The ergonomic zone model shows the developer how much the product must adapt to the given setting and be able to be integrated. New technical achievements should be measured by the ergonomic disturbance they cause. For example, a robot that performs a simple procedure such as a biopsy at ten times greater a cost and taking up five times more time and space than a surgeon is an ergonomic disaster, even if, from a mono-dimensional point of view, it just as precise.

With procedural ergonomics, the surgical process, the work-flow, is described from the perspective of **chaos theory (fuzzy logic)**. Chaotic systems are precisely

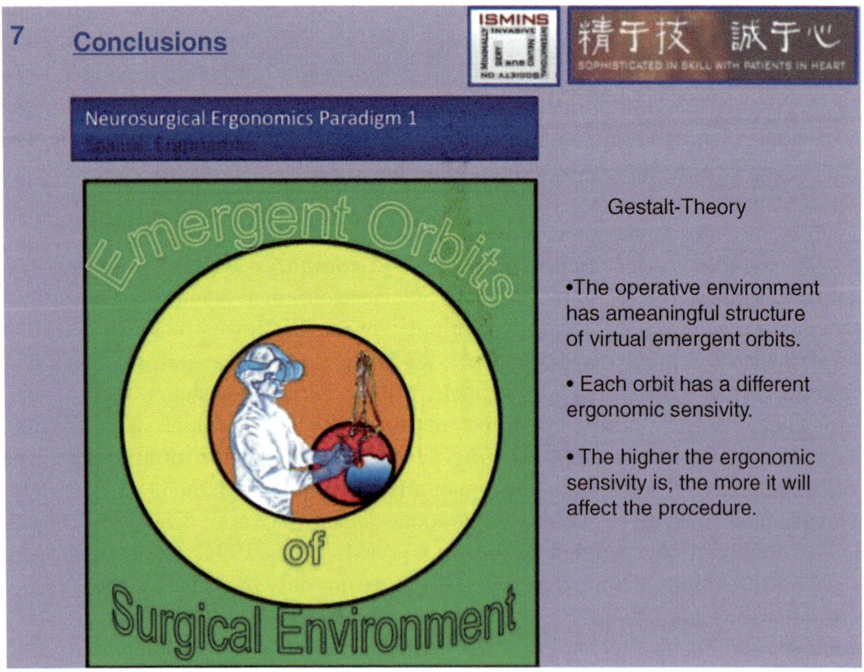

Graph 2.13 MIN ergonomics paradigm I

definable in terms of their details, but their behaviour is unpredictable. It can be influenced, however, by the **peripheral parameters** of the system. Peripheral parameters of the surgical procedure system are: equipment and technology (design/surgical environment), physical and mental fitness of the surgeon (readiness), management of light, climate, noise and gravity, and, last but not least, the work-flow in the team.

Each minor factor that influence the procedure for a long period or even throughout the whole procedure, like positioning of the patient, must be seen as a major factor regarding ergonomics.

This model focuses on the optimisable and trainable elements of the surgical procedure rather than on system components that are difficult or impossible to influence.

2.2.1.2 Procedural Ergonomics: The Chaos Model (Graph 2.14)

2.2.1.3 Mental Ergonomics: The Neuropsychological Model (Graphs 2.15, 2.16, 2.17, and 2.18)

Ergonomics is a discipline of systems-analysis. It therefore not only considers the patient and neurosurgical therapy, but also factors in the **neurosurgeon as part of the overall system**. Neurosurgery's blind spot is the **brain of the neurosurgeon** him-/herself. Every action performed on the patient's brain, however, requires a

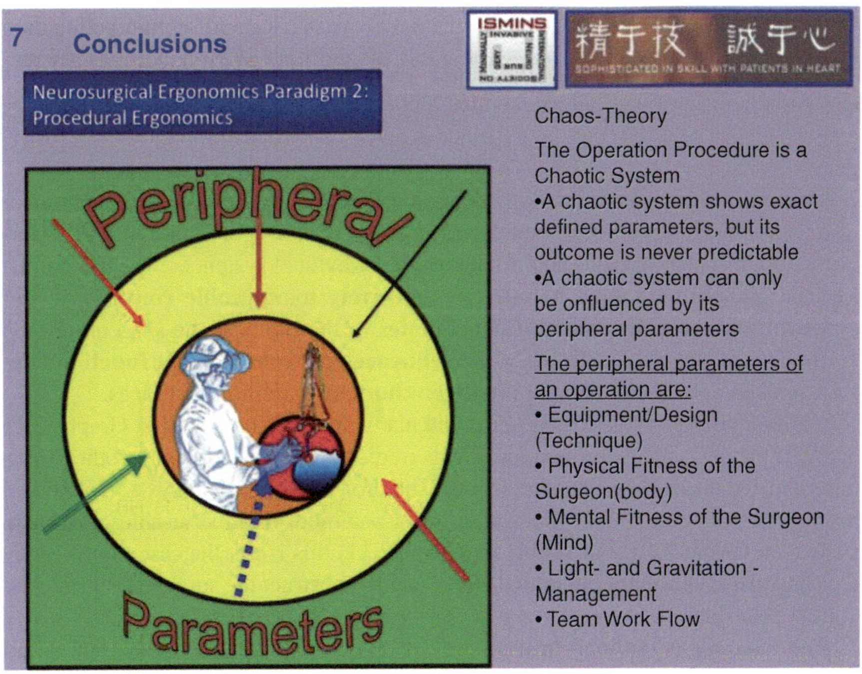

Graph 2.14 MIN ergonomics paradigm II

Neuropsychological Handicaps

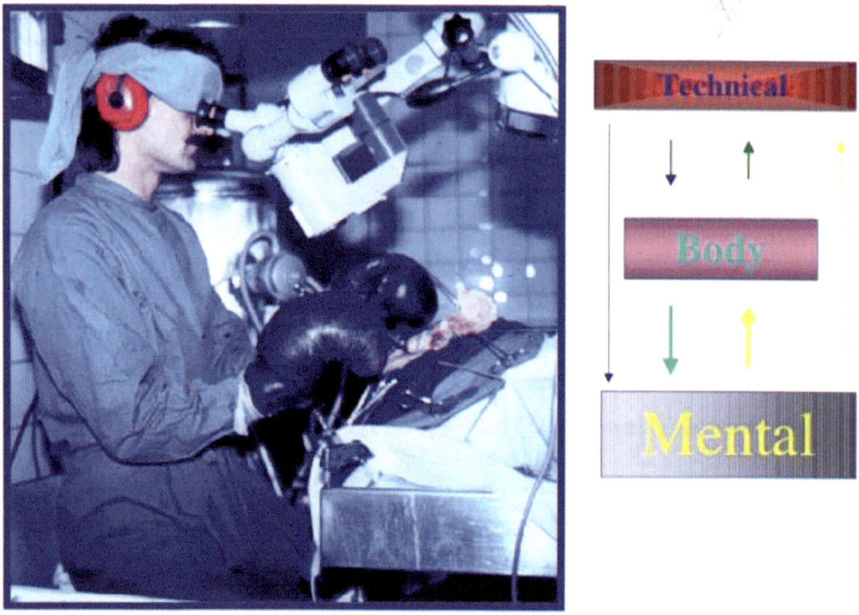

Graph 2.15 Neuro-Psychologigal handicaps for the neurosurgeon

neuronal process in the brain of the neurosurgeon. For this reason, significant results and stimuli to improve the ergonomics and further development of neurosurgery, and especially of MIN, can be expected from a "**neuropsychology of neurosurgery**". Every development has to adapt to the needs of the neurosurgeon.

The multi-tasking in terms of perception and action, the complexity and number of devices and tasks all find their limits in the processes going on within the surgeon's brain. The numerous coordinates of the individual system components of the surgical environment must become an **intuitively manageable convergent and synchronised overall system of coordinates** in the brain of the surgeon (Graph 2.16). This is only possible when the **sequences of technological functions are harmonised** and when there are **intelligent human-machine interfaces.**

Current Neuro-navigation Systems and many systems in the OR (s. Graphs 2.11 and 2.12) show typical ergonomics effects of disturbing the surgeon and the procedure and the work-flow or mental flow. Therefor all coordinate systems involved need to be converged and synchronised to a common, intuitive usable coordinate system (s. Graph 2.16). As visible in (Graph 2.11) this is not the case today. Many data and displays are not integrated and even may obstruct the surgeon to decide the correct priorities.

The three areas outlined in these models are in reality inseparable, and so all three of them are indispensable for the future of MIN.

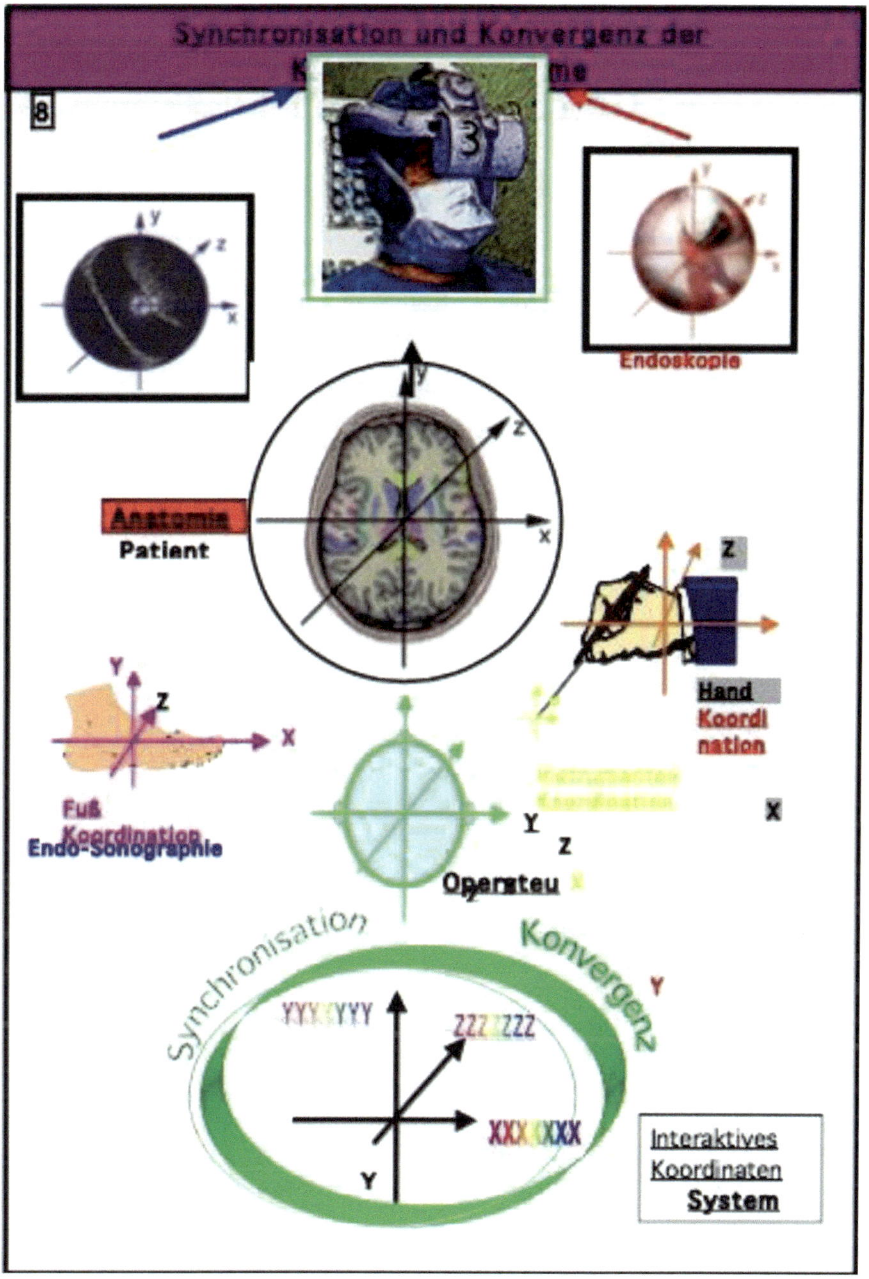

Graph 2.16 Ergonomic coordinate-systems: convergence and synchronisation

Graph 2.17 Ergonomics trauma to the surgeon

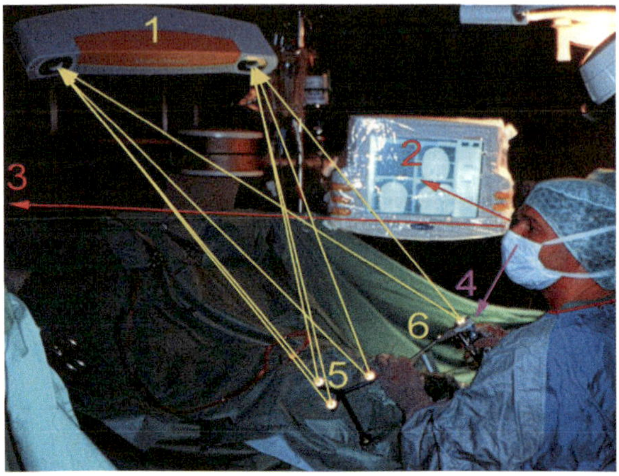

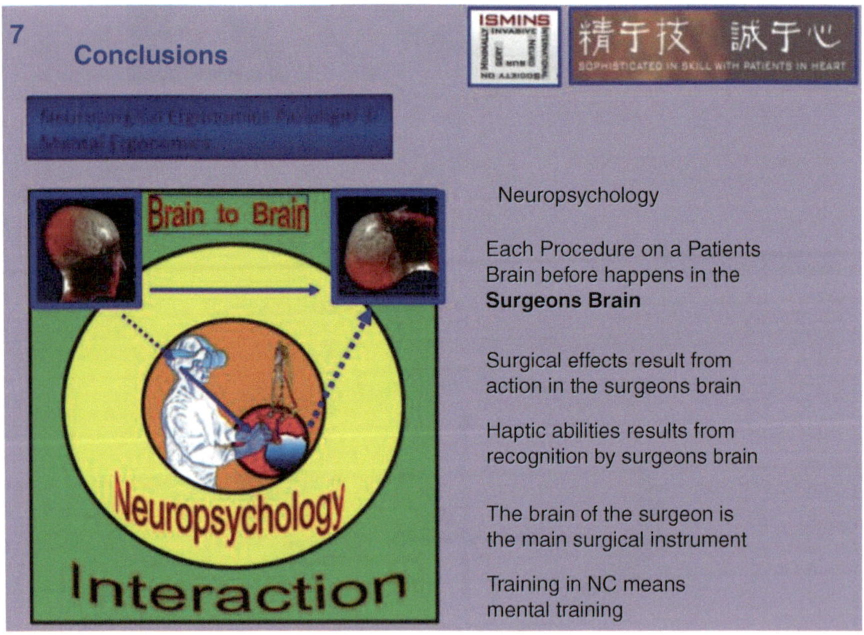

Graph 2.18 Ergonomics paradigm III

Operative Suit Design and Future Neurosurgery

Primate of Ergonomics

The 2014 Perneczky-Lecture has closed with a glimpse into the future of MIN. It is proposed to include other techniques and fields of science outside neurosurgery into the programme: the development of new HMD systems, integrated endo-neuro-sonography (see: **multi-laser technology**, **design of surgical suits** and mental training methods).

A critical review on the reliability of theories (EBM, big data medicine, etc.; s. Chap. 7) is also mentioned, as is the study of psychological resistance to innovation and neuropsychology's perception of the neurosurgeon. This programme is considered a future prerequisite for MIN.

All new fields of neurosurgery will need new effector system (Graph 2.19) additional to MIN there will be **electronic neurosurgery (e-NS)** and **biological neurosurgery (b-NS)**. Ergonomics must be the guiding principle and priority in evolution of the future.

Ergonomics Future of MIN

Future is a vector from the past presenting the direction and speed into present time and further on into future. The analysis of the evolution course and steps from the past until to date (Graph 2.20) give an idea about the history and the direction of

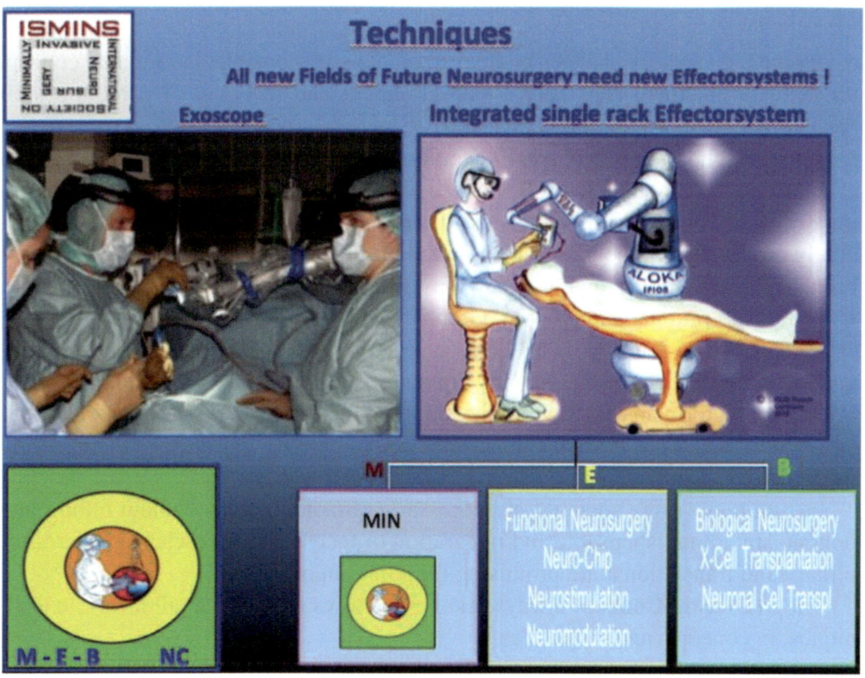

Graph 2.19 **M**IN-/**E**lectronic-/**B**io-/Neurosurgery

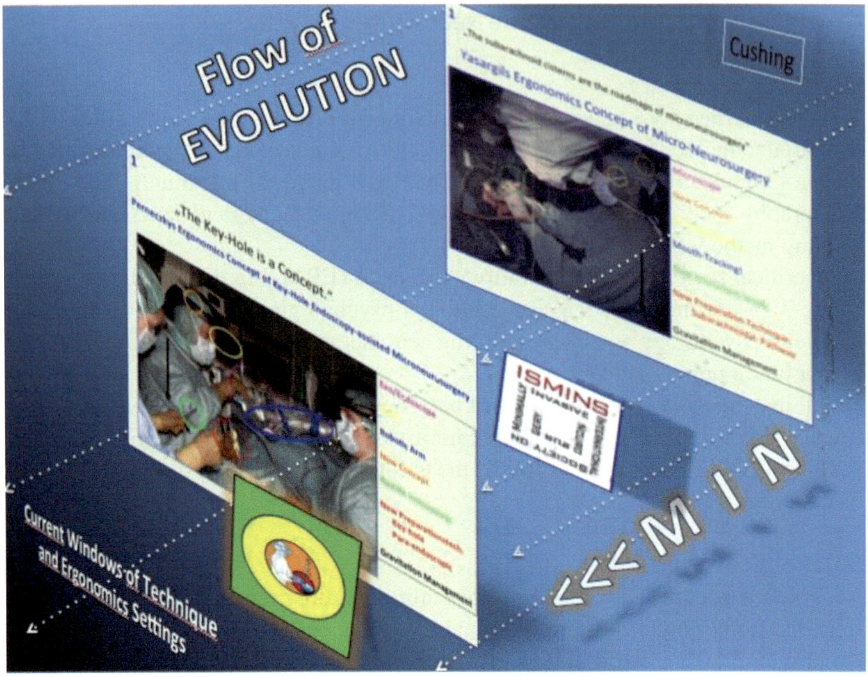

Graph 2.20 Evolution flow of MIN

development. Only its analysis enables to create a responsible and reliable concept for decision-making. The speed of development and the kind of paradigm shift became faster, more complex and occur in parallel. We see a variety of existing paradigms, schools and technical generations, which are in competition. It is not even clear, which will be the follower school of micro-neurosurgery. It will be very important, according to this analysis, if we let the course of the vector run accidently or if we would better influence the evolution. One can see, after all that have been described above, that without the guidelines of ergonomics the evolution would be less rational and reliable (Graph 2.21).

Education and training of the next generation need to be designed on the basic understanding about neuropsychology of neurosurgery (MIN). Education and training means, to form and to program the brains of the trainees. This is a sensible challenge for which we should use the same efforts as we do in training Olympic gold medalists.

Institutional strong societies are necessary, reliable theories without falling into dogmatism are indispensable and integration of techniques and sciences interdisciplinary and translational are promising on-going innovations.

The **MIN-Key-Concept** always asks for the **keys of future** at all levels, in techniques, in concepts and in theories.

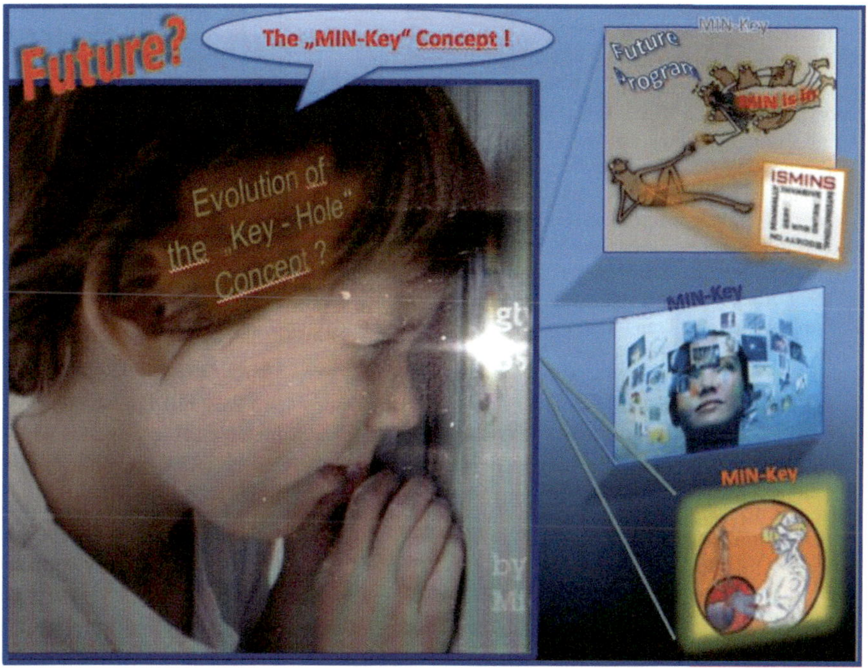

Graph 2.21 Future perspectives of MIN

Suggested Reading

Al-Mefty O. Surgery of the cranial base. Boston, MA: Kluwer Academic; 1989. p. 3–11.

Apuzzo MLJ. New dimension of neurosurgery in the realm of high technology: possibilities, practicalities, realities. Neurosurgery. 1996;38:625–39.

Arnold P, Farrel MJ. Can virtual reality be used to measure and train surgical skills? Ergonomics. 2002;45:362–79.

Barnes RW. Surgical handicraft, teaching and learning surgical skills. Am J Surg. 1987;153:422–7.

Beltoft I, Pedersen JP, Trangeled M, Vase JR. Decision support tool for craniotomy based on analysis of possible complications. MI8 Group 826-2003, HST. Aalborg: Aalborg University Press; 2003.

Dahlerup B, Perneczky A. The Standard Flex - EndoMultiUnit (EMU): a surgical working station for endoscopic neurosurgical procedures. In: Satellite Symposium of the 10th European Congress of Neurosurgery EANS, Mainz Germany, May 1995; 1995.

Dörr W. Gestalt theory and morbid anatomy. Virchow Arch (Pathol Anat). 1984;403:103–15.

Feng L. 3rd Announcement: 2nd ISMINS Congress 2014, Xi Ang, China. 2014. http://www.wfns.org/filebin/member_uploads/ISMINS2014_3rd_Announcement.pdf.

Gordon D. Trends in surgery-suite design. Health Care Design. 2007. http://www.healthcaredesignmagazine.com/article/trends-surgery-suite-design.

Gross R. Chaos und Ordnung. Dtsch Ärztebl. 1991;88(25–26):A-2273.

Haase J. Image guided surgery/neuronavigation/surgiscope - a reflexion on a theme. Minim Invasive Neurosurg. 1999;42:53–9.

Haase J, Museaus P, Boisen E. Virtual reality and habitats for learning microsurgical skills. In: Andersen P, Qvortrup L, editors. Virtual applications. New York, NY ISBN 1-85233-658-7: Springer; 2004.

Haase J. Control and structure of a training programme: the view of a non- academic hospital. In: Training in neurosurgery: Proceedings of the Conference on Neurosurgical Training and Research, Munich, Germany, October 6-9, 1996. Reulen, H.-J. and Steiger, H.-J. (eds.). Wien: Springer. Acta Neurochir. 1997;69(Suppl):79–82. ISSN/ISBN: 3211830022

Jendrysiak U, Resch KDM. Ergebnisse der klinischen Erprobung der Operationszugangsplanung mit NeurOPS. In: Bildverarbeitung für die Medizin. Algorithmen-Systeme-Anwender. Proceedings-Band S. Heidelberg: Springer; 1999. p. 187–91.

Kaufmann HH, Wiegand RL, Tunick RH. Teaching surgeons to operate - principles of psychomotor skills training. Acta Neurochir. 1987;87:1–7.

Kockro RA, Serra L, Tseng-Tsai Y, et al. Neurosurgical planning and training in a virtual reality environment. In: 11th European Congress of Neurological Surgery, Copenhagen 19–24 Sep 1999, Abstractbook; 1999. p. 75.

Kockro RA, Serra L, Tseng-Tsai Y, et al. Planning and simulation of neurosurgery in a virtual reality environment. Neurosurgery. 2000;46:118–37.

Kikinis R, Gleason L, Moriarty TM, Moore MR, Alexanader E III, Stieg PE, Matsumae M, Lorensen WE, Cline HE, Black PML, Jolesz FA. Computer- assisted interactive three-dimensional planning for neurosurgical procedures. Neurosurgery. 1996;38:640–51.

Kikinis R, Black PM, Jolez FA. Image guided surgery. In: Aachener Workshop on Navigierte Hirnchirurgie, 4–5 Sep 1998, Abstractbook; 1998. p. 4.

Linke DB. Cognitive neuroscience foundations for a theory of neuronavigation. In: Computer-aided surgery: abstracts from CIS; 1997. p. 3–16.

Larsen OV, Haase J, Østergaard LR, Hansen KV, Nielsen H. The virtual brain project - development of a neurosurgical simulator. Stud Health Tech Inform. 2001;81:256–62.

Long D. Competency based training in neurosurgery; the next revolution in medical education. Surg Neurol. 2004;61:5.

Lurija AR. Das Gehirn in Aktion, Einführung in die Neuropsychologie. Hamburg: Rowohlt; 1993. p. S230–58.

Matern U, Koneczny S. Safety, hazards and ergonomics in the operating room. Surg Endosc. 2007;21(11):1965–9.

McDonald JM. Mental readiness and its link to performance excellence in surgery. Ottawa, ON: Kinetek; 1993.

McBride DK. Individual differences in the performance of highly learned skill. Percept Mot Skills. 1998;86:985–6.

Nunez G, Kaufman H. Ergonomic considerations in the design of neurosurgery instruments. J Neurosurg. 1988;69:436–41.

Patrick J. Training: research and practice. London: Academic Press; 1992.

Patkin M. Ergonomics applied to the practice of microsurgery. Aust N Z J Surg. 1977;47:320–8.

Perneczky A, Fries G. Endoscope-assisted brain surgery: part 1-evolution, basic concept, and current technique. Neurosurgery. 1998;42(2):219–25.

Perneczky A, Müller-Forell W, van Lindert E, et al. Keyhole concept in neurosurgery. Stuttgart; New York, NY: Thieme; 1999.

Perneczky A. Planning strategies for the suprasellar region. Philosophy of approaches. Neurosurgeons. 1992;11:343–8.

Resch KDM. Beitrag zur Zugangsanalyse und zum Zugangsdesign des transoral-transpharyngealen Weges zum Hirnstamm. 1991. Dissertation, Universität Heidelberg.

Resch KDM. Chapter 5. Ergonomics. In: Transendoscopic ultrasound for neurosurgery. Heidelberg: Springer; 2005.

Resch KDM. Ergonomie in der Neurochirurgie (Abstract). J Neurol Neurochir Psychiatr. 2004;5(Sonderheft 1):19.

Resch KDM. MIN: transoral transpharyngeal approach to the brain. Neurosurg Rev. 1999;22:2–25.

Resch KDM. Postmortal inspection (PMI) for neurosurgery: a training model for endoscopic dissection technique. Neurosurg Rev. 2002;25:79–88.

Sampath P, Long DM, Brem H. The Hunterian neurosurgical laboratory: the first 100 years of neurosurgical research. Neurosurgery. 2000;46:184–95.

Schueneman A, Pickleman J, Hesslein R, Freeark RJ. Neuropsychologic predictors of operative skill among general surgery residents. Surgery. 1984;96:288–95.

Shah J, Darzi A. Surgical skills assessment: an ongoing debate. B J U Int. 2001;88:655–60.

Suh HA. Leonardo da Vinci – Skizzenbücher; LIBERO 2014. New York, NY: Black Dog; Leventhal Publisher; 2005.

Urban V, Wapler M, Neugebauer J, Hiller A, Stallkamp J, Weisener T. Robot- assisted surgery system with kinesthetic feedback. Comput Aid Surg. 1998;3:205–9.

Walker AE, editor. History of neurological surgery. Baltimore, MD: Williams & Wilkins; 1951.

Webster RW, Zimmerman DI, Mohler BJ, Melkonian MG, Haluck RS. In: Westwood JD, et al., editors. A prototype haptic suturing simulator. Medicine Meets Virtual Reality 2001. Amsterdam: IOS Press; 2001. p. 567–9.

van Weizäcker V. Der Gestaltkreis. Stuttgart: Thieme; 1950.

Witt H, Kozianka J, Waleczek H, et al. Das Erlernen und Optimieren minimal-invasiver Operationsverfahren am menschlichen Leichnam. Chirurg. 1999;70:923–8.

Witzke DB, Hoskins JD, Mastrangelo MJ, Witzke WO, Chu UB, Pande S, Park AE. In: Westwood JD, et al., editors. Immersive virtual reality used as a platform for perioperative training for surgical Residents. Medicine Meets Virtual Reality 2001. Amsterdam: IOS Press; 2001. p. 577–58.

Yasargil MG. A legacy of microneurosurgery: memoirs, lessons, and axioms. Neurosurgery. 1999;45(5):1025–92.

Yasargil MG. Microneurosurgery, vol. IV A. Stuttgart: Thieme; 1993.

Yasargil MG, Fox JL. The microsurgical approach to intracranial aneurysms. Surg Neurol. 1975;3(1):7.

Yasargil MG. Microsurgery applied to neurosurgery. Stuttgart: Thieme; 1969.

Key Techniques of MIN: Mouth-Tracked High-Zoomed Microneurosurgery and The "Ergo-Tool"

3

3.1 Evolution of Visualization

We see the legendary working setup of Yasargil using the mouth-tracked microscope with the Contraves joint-automatism which revolutionized the world of neurosurgery.

The actual version of mouth-traced microscope and its' adjustment are described below.

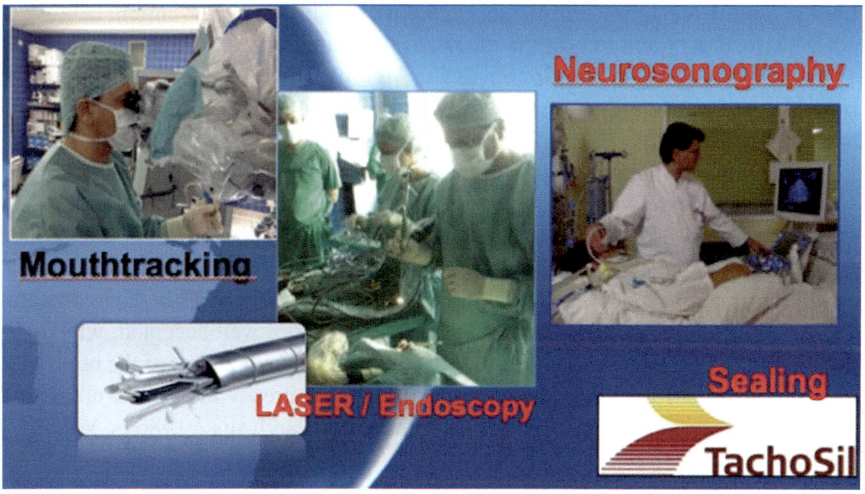

Graph 3.1 Mouth-tracked microneurosurgery and "Ergo-Tool"

K. D. M. Resch, *Key Concepts in MIN - Intracerebral Hemorrhage Evacuation*, Key-Concepts in MIN 1, https://doi.org/10.1007/978-3-030-46513-1_3

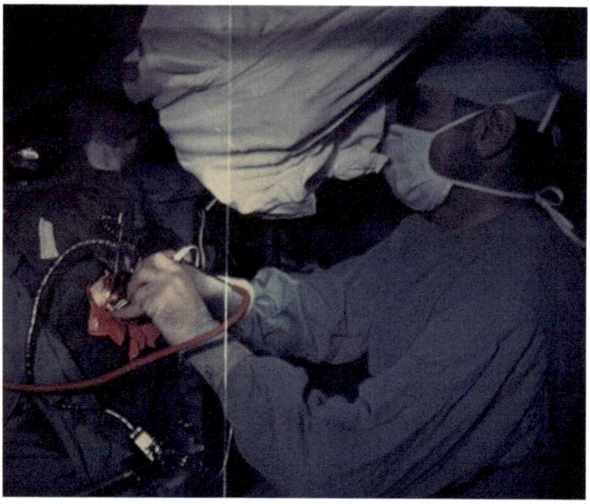

Fig. 3.1 Ergonomics setup of yasargil

Scientific principles were described in the previous chapter.

3.1.1 History

Precisely guided magnification tools began in the late 1960s of twentieth century to enter neurosurgery and to become the next gold-standard: microneurosurgery.

The manufacturing of the Contraves Co. Mouth-Switch system in 1975 was published in use by Yasargil et al. 1977 (Yaşargil MG, Vise WM, Bader DC: Technical adjuncts in neurosurgery. **Surg Neurol 8**:331–336, 1977).

This equipment allows to guide the microscope mostly without the use of the surgeon's hands while carrying on with surgical manipulation at the same time. A 40%-time spare was experienced in trained hands. Special designed surgical mask to take the switch bladders without contact to the mouth makes hygienic use easier, however, strong or double layer masks can serve as well. During 30 years in neurosurgery, gave the impression that this equipment was not well accepted, which is a big pitfall to the evolution of micro-neurosurgery. The senior author of the book used this equipment during all his career and has trained it already in the anatomical and pathological laboratory during surgical simulation research since 1986 (s. Vol. 2).

3.2 Visualization and Ergonomics

The blades of the mouth-switch are fitting between the teeth, the oculars are exactly adapted to the eyes with their rubber-tube-adapter, so the floating microscope is fixed by a 3-point system: two oculars and mainly the mouth-switch blades. Biting on the blades releases the blocked joints, so the microscope is sweeping if correct balanced before. For adjusting the mouth-witch and the microscope see below.

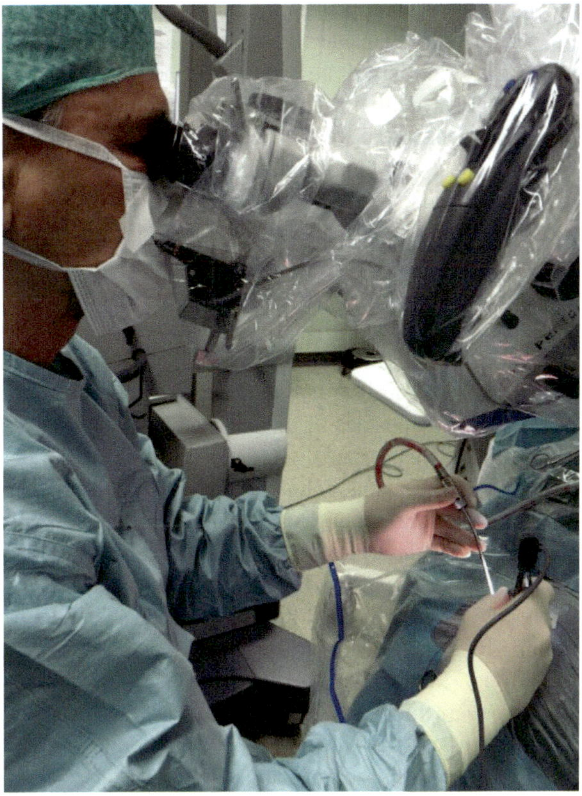

Fig. 3.2 Actual mouth-switch (zeiss; pentera)

The characterization of the mouth-switch as a tool to spare time and to guide the microscope mostly without needing hands of the surgeon therefor is a vast underestimation of this equipment regarding MIN. For advanced MIN, working close to the limits of microsurgical technique, the mouth-switch became one of the **key-techniques of MIN**.

In MIN very small approaches can be used, as a **result of precise planning**, and are sufficient if the analysis and approach geometry is well defined. But under these conditions, it is essential to zoom-up to maximum. Ideally the visual field is congruent with the working field which might be just a burr-hole. But if one zooms up, having a visual field that fits into the burr-hole, one will have a **focus range of** not more than **1 mm**. Therefore, you need to adapt the focus "online" which under such conditions will not work by hands. Such a busy focus control must and can be only realized by the mouth switch, during working through the narrow approach. Once you try to operate microsurgical through the burr-hole with low zoom, using the microscope as a kind of too expensive magnification glasses, you will not have enough visual control, not taking true advantage of the magnification capacity.

Only if a small approach, like a burr-hole, is presenting exclusively for the central visual field near and to the macula, the area of sharp seeing, the optimum of visual information will be present for the surgeons' brain.

Our neuro-ophthalmological and neuro-psychological knowledge describes the immense meaning of this visual information conditions to the brain of the surgeon and the consequences for the quality of surgical manipulations. To present all information exclusively from the working field of interest (ROI) only to the area near and to the macula, causes a maximum of concentration by information in 0.x mm-dimension optical solution. At the same time the peripheral retinal information, which is not needed but disturbs the information flow during operation, will minimally enter the brain of the surgeon. **Zooming selects the most important information to the brain of the surgeon**.

The motor control of the tracking by the mouth-switch (focus and visual field) is realized by the muscles of the cervical spine. Therefore, cervical physiology of the surgeon is involved, causing no side effects in mean healthy people, as encountered in some other tracking techniques. This is due to the special neurological and neuropsychological integration of the cervical spine, being extremely connected and supervised by the brain. The brain has an exclusive information-pathway for the cervical spine, because it balances the head, and is always optimal informed about the heads' position related to the body, but exclusive from the body. Some physical trainings and exercises should be known and practised, therefore (s. Vol. 2, Feldenkrais for MIN).

Moreover, this conditions of informative work flow, which can be obtained by using the mouth-tracking safely during operation, will produce the "mental state of flow" (Mihály Csíkszentmihályi 1975).

Only mouth-tracked microsurgery can obtain this level of advanced MIN. During the challenge of risky dissections through small gaps and channels, only in this maximum of concentration by mental-flow, advanced MIN can be responsible and optimal realized. This working condition belongs to the field of "mental ergonomics", analysing and describing working in an environment at the limits of techniques (s. Chaps. III and V in "Trans-endoscopic Ultrasound for Neurosurgery" 2005; Springer).

Mouth-tracked microsurgery is a unique league but does only work if the tool is precisely adjusted and pre-surgical trained. It can be learned! Once one has mastered the steep learning curve you will never like to operate without it. Once you have not learned to use it correctly, you will not understand the outstanding advantages; there are two different worlds of working ergonomics: with, or without mouth-tracking.

3.2.1 Mental Navigation

We will learn in Chap. 6 (endoscopy): **Once you focus your vision, you need to widen your mind!** (Graph 3.5, s. below).

Graph 3.2 Mental navigation

It was Wolfgang Seeger (Leonardo of German Neurosurgery/some 20 Vols. of neurosurgical drawings), chairman of Neurosurgery Univ. Freiburg/Germany, who taught us, that you must not see everything in micro-neurosurgery like in macro-neurosurgery, but you must know and imagine, what you do not see in the microscope. This is the mental condition, corresponding to the technical and procedural condition of micro-neurosurgery. Mental navigation (Graph 3.2) means that you know how to deal with your own brain during surgery (the white spot of neurosurgery/s. this chapter, mental ergonomics).

In neuro-endoscopy and MIN, this is the rule and daily reality, otherwise you are doing something else but not MIN.

3.2.2 Micro Zooming of the ROI (Region of Interest)

In the example of a microsurgical setting, zooming rules out **97%** of that information which is not necessary to see for the procedure (Graph 3.2). During preserving the oculomotor nerve at the last part of tumour-resection (Graph 3.3), it is necessary to zoom the ROI, providing best information to the eye and brain of the surgeon. If you don't zoom, a deeper focus range will enable to adapt the focus by hands. But then you keep your brain visually busy with about 97% visual field that presents no relevant visual information at that moment. (The only information would be orientation, which should be mastered mentally and not visually in MIN!) One does not take full advantage of the microscope without using the full range of magnification. However, it can be done, but it will always be compensated by the surgeon and the patient.

Without mouth-tracking and full use of magnification, the work-flow and mental-flow will always have a break to re-adjust the microscope by hands. Moreover, one cannot risk manoeuvres with space or time challenges, leaving them undone, or exposing the patient to higher risks of enlarged approaches and increased operation-times.

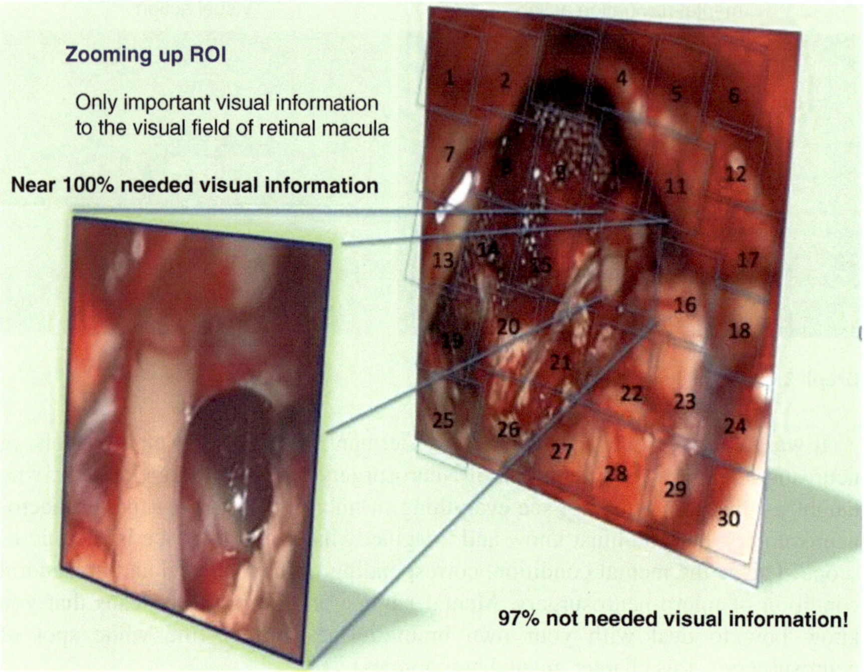

Graph 3.3 High-Zooming

Mouth-tracking and full zooming pushes micro-neurosurgery into the visual advantages of neuro-endoscopy, in the modus of endoscopy-assisted use, however, without the major advantage of endoscopy: wide focus range and strong illumination. But anyway, a major advantage is the possibility for fast acting and the core-abilities of micro-dissection. Mouth-tracking, zooming and mental-navigation must be trained to an integrated intuitive ability, needing 10 years to maser-level!

3.2.3 Zoom and Focus Range

During surgery, the focus-level of 1 mm focus-range is taken into the working canal advancing to the depth together with the tip of the instruments.

This enables very safe visual control "online" and also a precise adjustment of the working window.

Neuro-psychological, it provides the surgeon with the comfortable feeling to face each challenge fast and precise.

During maximal zooming through a small approach with a 10 cm depth of working canal and a focus range of 1 mm, there will be 100 focus-levels and the focus has to follow "online" with the instruments into the depth. This can only be done by mouth-tracking safe and fast enough. The ROI is "brought to the surface", providing

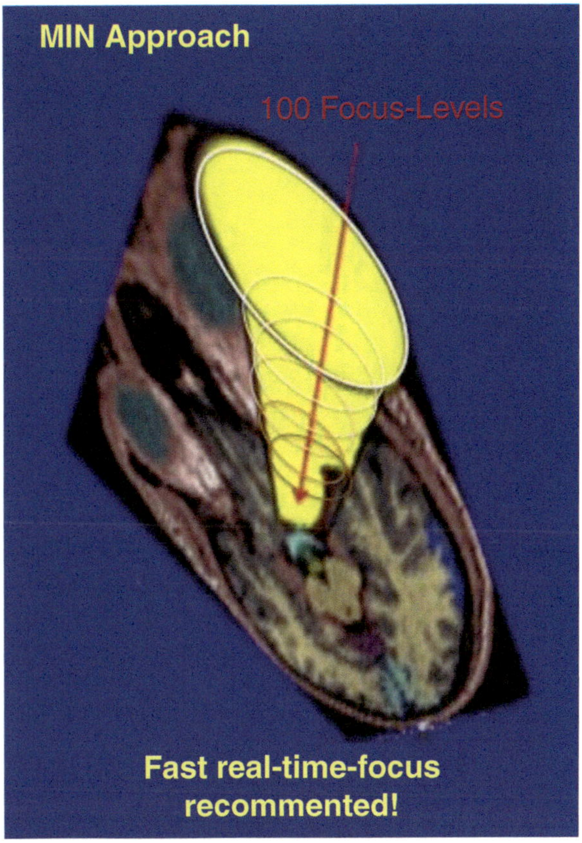

Graph 3.4 Focus-Levels in the working-canal

the surgeon with the best and only relevant information, taking the focus taken into each depth.

3.2.4 Burr-Hole Focus-Levels in MIN

In the setting of a burr-hole approach in MIN, there is no alternative to work safely but with maximal zoom. This enlarges image of the burr-hole to nearly the whole visual field providing the surgeon with the ROI in each focus level. This must be done "online" throughout the working canal, which cannot be done by hands and can only be realized fast and safe by mouth-switch.

In this level of advanced MIN, only optimal focusses by tracking the microscope into the depth, is safe enough for this kind of surgery. It is indispensable to use the full capacity of the magnification.

At the ground of the working canal (example level 4) a fast and safe coagulation is possible by bipolar forceps. Also, along the working canal, in each level, which can be 100 focus levels during 10 cm, the parenchyma and bleeding vessels can be

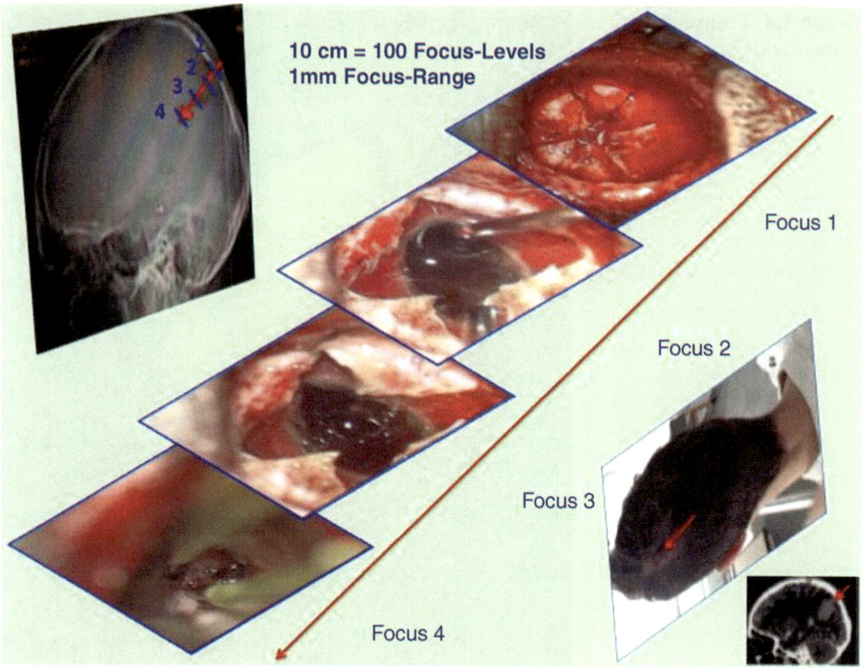

Graph 3.5 Focus-Levels

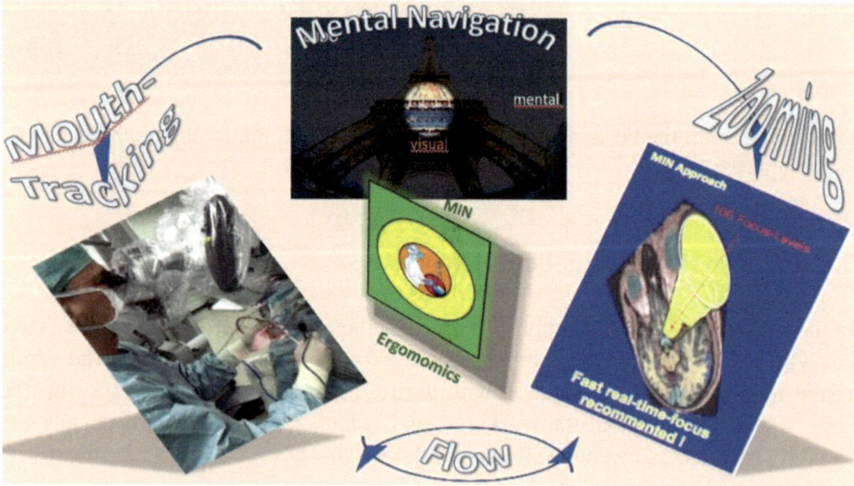

Graph 3.6 Flow-effect in MIN by mouth-tracking and high-zoom

seen sharp for fast coagulation. After perfection of this working, viewing and dissecting will become a natural intuitive feeling, high precision and safety, convincing yourself to face each challenge (Graph 3.6).

Mental navigation, **mouth-tracking** the microscope and **high-zooming** are causing a mental ergonomics that gives the neurosurgeon the ability to come into "flow", enabling extraordinary concentration and standing for challenging tasks to accomplish. All components of this condition must be trained, and only when interfering one another, this master-class of psycho-manual level will be reached. This is not the same level of average microsurgery but causing a work-flow that allows for using a MIN approach (like a burr-hole) as a full competent microsurgical approach, for example. All this can be learned, and it must be learned and trained to be able doing this kind of advanced micro-neurosurgery.

This is especially interesting in ICH-evacuation by MIN, because a liquid to soft consistency of tissue with a well-defined, mostly moderate risk, has to be approached. However, ICH-evacuation is, in low risk cases (epidural ultrasound!), an excellent model to learn and train MIN strategies and techniques. Of course, this is also a basic technique for all indications in MIN.

3.2.5 Adjustment of the Balance System

The above described abilities of the mouth-tracked microscope do work only if it is perfectly adjusted and balanced! (Fig. 3.3).

To bring the holding device of the microscope into balance, after having the observer ocular arms and the surgeons' oculars in the needed position, the cursor spot is placed in the focus area (blue) by moving the holding arm until this is accomplished. Then the auto-balance button start is used to begin the auto-balancing procedure.

This balancing of the system must be as precise as possible to enable an **easy floating microscope** and easy to be moved by the mouth-switch without force. This balancing procedure is mandatory before adjusting the mouth-switch itself.

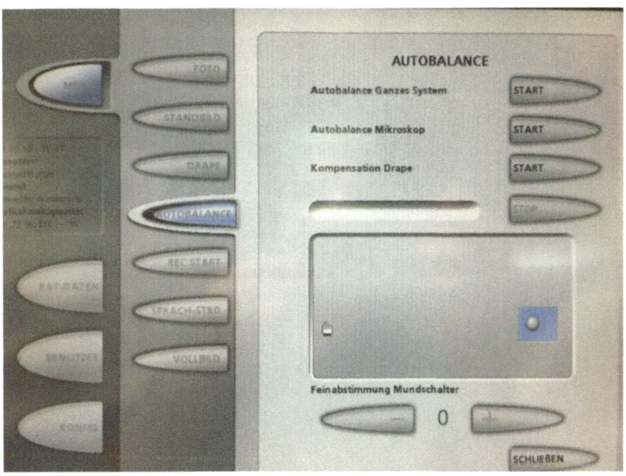

Fig. 3.3 Monitor surface of balance-adjustment

As this level of working environment is applicable at the limits of technique and conditions, it will work only in high precision of all components. At this point, at the limits, all systems can reach the conditions of chaos (s. this chapter, procedural ergonomics). Chaotic systems can only be controlled by preparing the peripheral parameters. Here the peripheral parameter is the perfect preparation and adjustment of the microscope and the mouth-switch.

3.2.6 Adjustment of the Oculars

Before the mouth-switch should be adjusted, the oculars need to be adapted to the surgeons' condition or pre-set, and the ocular-piece angle should be in a comfortable position for relaxed viewing by the surgeon. Distance and ocular-tube must fit perfectly to the surgeon. One should take patiently time to get a good result for sufficient focus tracking.

Only then, the mouth-switch should be adapted to the mouth of the surgeon: all three screws must be open and joints of the switch completely easy going. While looking comfortably through the oculars the switch-blades are placed only with their distal ends between inferior and superior teeth line. Once the switch is in a position with comfortable feeling, all screws can be fixed.

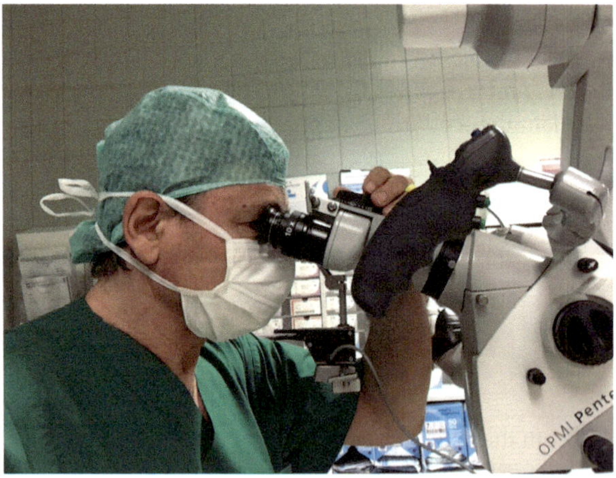

Fig. 3.4 Step 1: occular adjustment

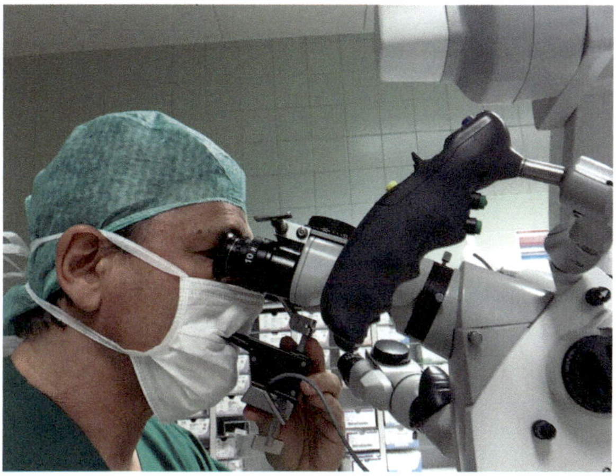

Fig. 3.5 Step 2: mouth-switch adjustment

3.2.7 Adjustment of Mouth-Switch

While looking in relaxed head/neck position and eyes, the mouth-switch can be placed in the perfect relation to the mouth and the oculars. The mouth-switch and the oculars serve as a three-point contact to the head of the surgeon enabling a tight control of the movements that are transduced from the head to the microscope.

To enable the correct position of the switch-blades without perforating the surgical mask, the mask had to be tightened during maximal opened mouth. The thickest surgical mask type should be used, in case, the mask can be prepared with a plastic drape inside to keep it dry. The superior line of the surgical mask (nose-eye-line) should be fixed by drapes to the skin to avoid moistening oculars. Everything should be prepared patiently and end in a comfortable feeling for the surgeon.

3.2.8 Testing the Microscope: Focus and Field of View and Floating

Once the system is adjusted correctly and causing a good feeling, enabling a safe and easy guiding of the microscope, it should be patiently tested. Not perfect working (feeling) must be corrected. The time needed for this is well invested for a perfect working system, which is indispensable in advanced MIN.

Normally this takes only a few minutes. This **key-technique of mouth-tracked micro-neurosurgery** works only in perfection and causes the way to perfection.

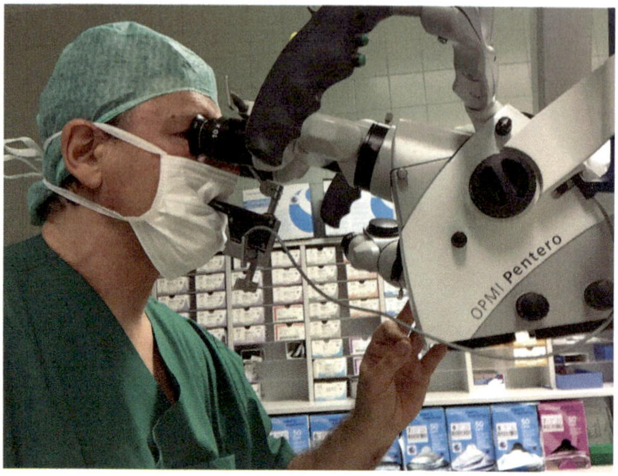

Fig. 3.6 Step 3: floating test for focus visual field and easy-going

3.2.9 Evolution of the Mouth-Piece

An interesting evolution is the introduction of a new device into neurosurgery by D. Pitskhelausri (**J Neurosurg 121**:161–164, 2014) The author has no experience with this equipment.

Principally, MIN in general works only in a high precision environment of all techniques. The further key-techniques will be described in following Chaps. 5 and 6.

3.3 Features on MIN Evolution in the Past and Near Future

Some presentations of steps from microneurosurgery to MIN in the past and present may provide us with an information and a feeling about the ongoing evolution of techniques and concepts. We can imagine a **flow of evolution** to better understand reality (Graph 3.7).

3.3.1 Evolution of Visualization

From the very beginning of MIN, visualization was a key issue. We have progressed through three major stages of visualization in neurosurgery as far as technical tools are concerned: about 1900 macro-techniques, 1970 micro-techniques, and 1980 endo-neurosurgery. Each visualization has its own specific implications for ergonomics, which has impact on the entire procedure and all manual and mental abilities necessary for success (Graph 3.8).

The latest generation of visualization is imaging and postprocessing and means a completely different dimension. It is used in **virtual-reality augmented**

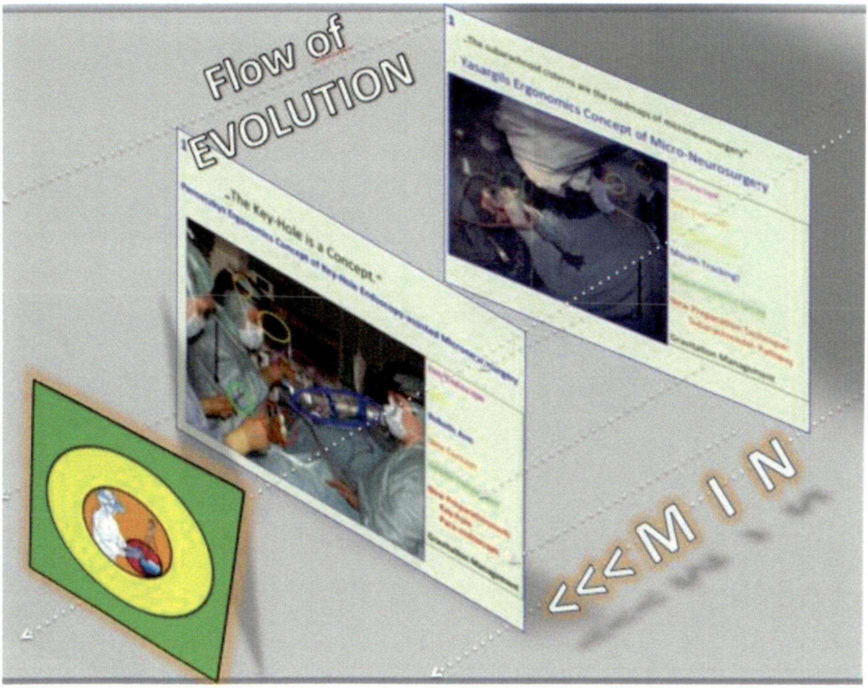

Graph 3.7 Evolution of MIN

neurosurgery and is mainly a matter of research. The ergonomics, neuropsychological- and mental-implications of this type of visualization are not sufficient tested regarding impact on the surgeon as well as on the procedure. However, it became the basics for future visualization techniques and new tools.

Every component that comes into the environment will influence the complete system (s. this chapter, spacial ergonomics). The closer the new component involves the red area of the OR-suit (micro-habitat), which is the head of the patient and the hands of the surgeon (operative field), the more it will have an impact on the procedure and may also disturb it. Therefore, all new devices should be analysed how they fit into the environment and how it interacts there with other components.

3.3.2 Different Visualization Settings and Tools

Actually, the medical industry has finally picked up the concept of "exoscopy", evolving, more or less, the system of Olympus-robotic-arm to a new generation of visualization tools. It is microsurgery through a video-chain, looking on big 3D-displays with high optical solution named high definition or 4K. It is derived from endoscopy assisted micro neurosurgery concept combined with new computer

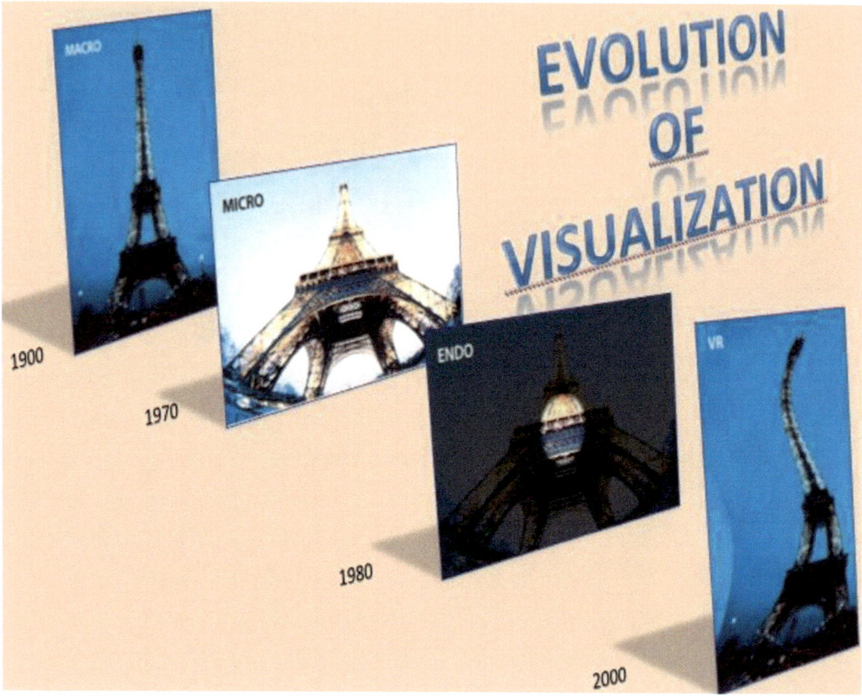

Graph 3.8 Visualization in MIN

power and digital processing of the optical signal. The 3d display came in favourite before the head-mounted-display (HMD)-system.

The display solution has been already used for decades in endoscopy; however, the HMD system would be the better solution regarding ergonomics. But more users are trained of course in display watching. Side-effects of this reality for neurosurgery are not examined. Overall ergonomics of the new systems, especially regarding work-flow and behaviour of the system in critical surgical situations, is completely unclear. It must be compared with the gold-standard of advanced mouth-tracked microsurgical technique (Yasargil-Standard).

3.4 Chaos in the OR-Environment

In the **laboratory setting of surgical simulation** with different generations of visualisation- and ergonomics-settings (Fig. 3.7), the possibilities of future-neurosurgery were narratively tested in prototypes **since 1982** (s. Vol. 2). However, at that time, the industry and professional community was not open for this evolution, while being into big machines and the non-real-time navigation concepts, so called "neuro-navigation systems". The result was the "**ergonomic chaos in the OR**" (s. Fig. 3.10 and Graph 3.10), and the meaning of ergonomics as the red line for future

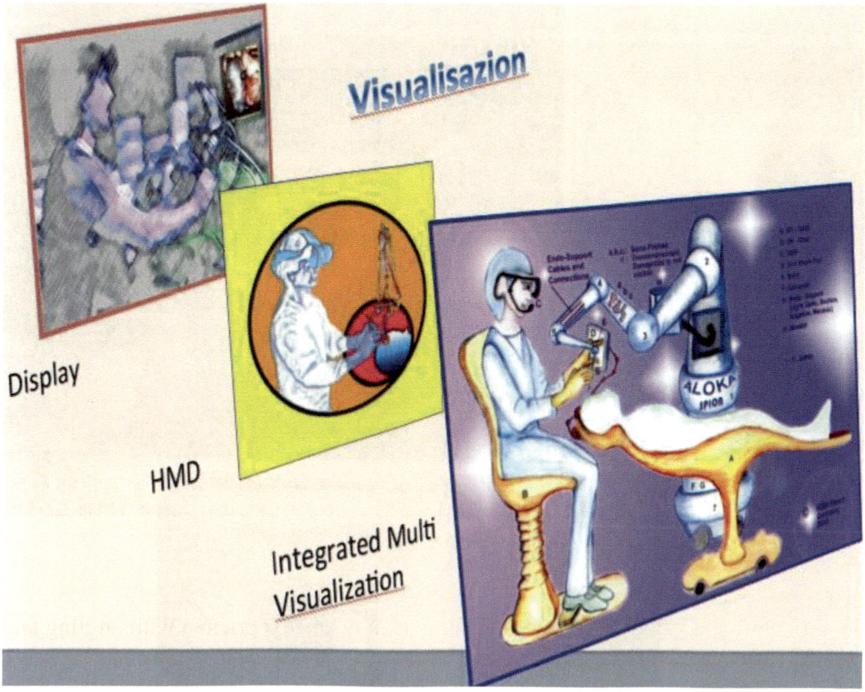

Graph 3.9 Visualization techniques in MIN

development became clear (s. Chap. 5 in: "Trans-endoscopic Ultrasound for Neurosurgery", Springer 2005).

The **meaning of ergonomics** came gradually in the focus of the operating-simulation concept during the lab-work. The contemporary concept at that time preferred techniques which progressively disturbed the workflow of surgery.

A model of this circumstances was necessary and the insight of the **priority of ergonomics** in future development. This was finally formulated for three fields: space-, procedure- and mental-ergonomics (this chapter).

This concept drew a red line into future evolution of MIN, provoking changing of the organization of OR, changing of the work-flow of the procedure and changing of the mental preparation of the surgeon himself.

3.5 Laboratory Settings

In the laboratory setting (1990), the "exoscopic"- and "endoscopic"-concepts and prototype-systems already had been examined. The final question was, which concept enables to interact with the complete equipment to realize a perfect ergonomics. In Fig. 3.7 the principles are visible. The best solution seemed to be an endo-/exo-scope combination (Graph 3.12), which has been done in the PICO-Prototype

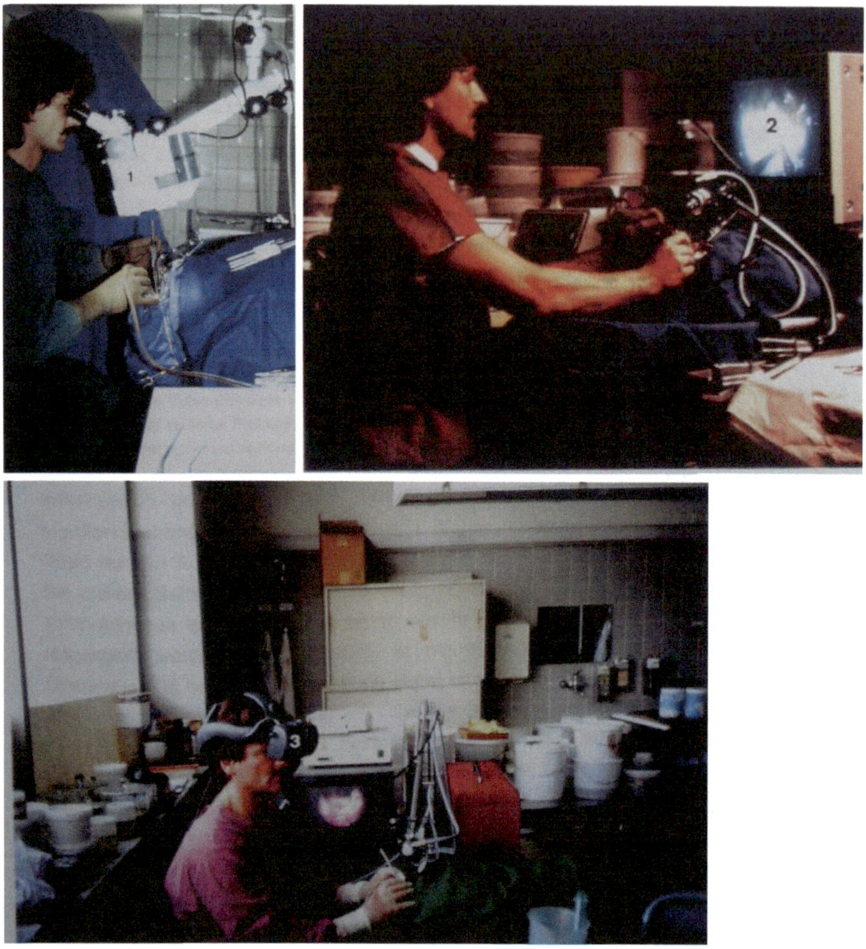

Fig. 3.7 Laboratory setup (1982–1999)

(Graph 3.13). Ergonomics and optical-solution however, were better in the Endoscope-Concept at that time (Fig. 3.8).

In the operation simulation lab-model, the visualization was applied in the exoscopic- and endoscopic-use. The prototype holding-device carried also the cables for camera, light and irrigation. In this setting para-endoscopic, starting exoscopic, strategy was evolved in over 70 cases (s. Vol. 2) (Resch, Neurosurg Rev 25;2002; accepted 12-2000).

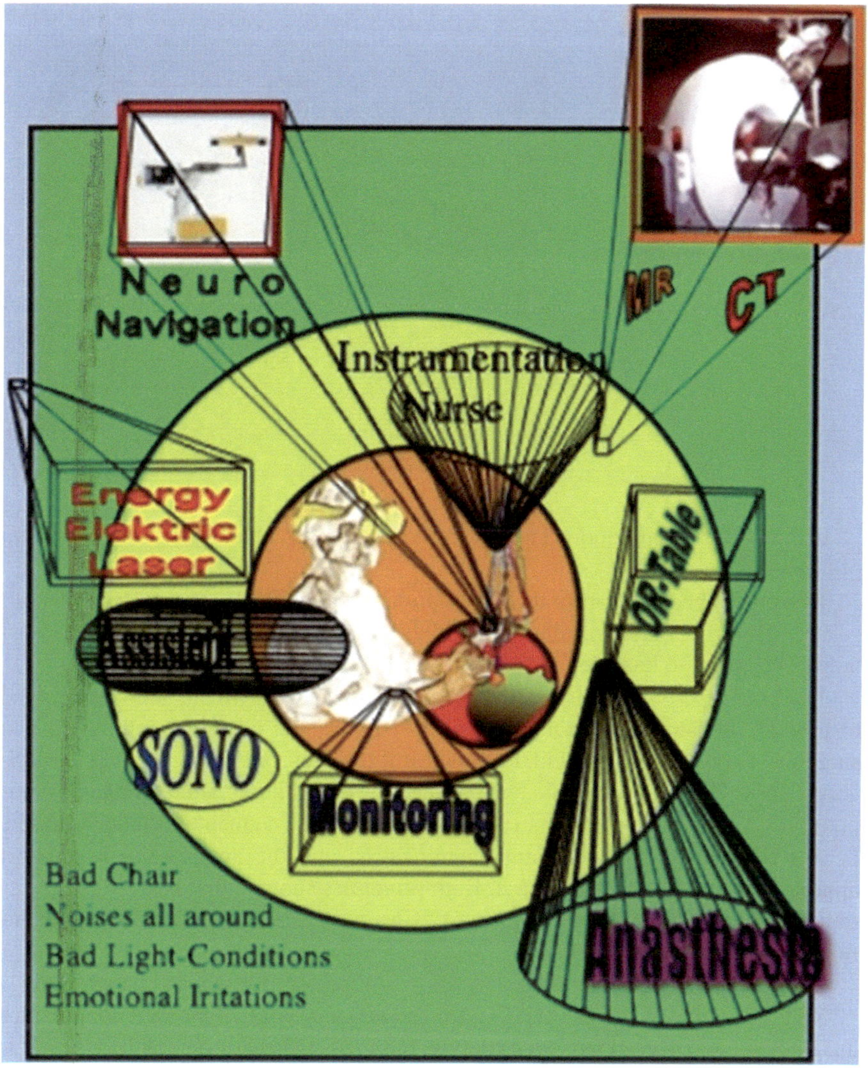

Graph 3.10 Ergonomics chaos in the OR

3.5.1 The PICO EU-Project (2004–2007)

In the program of the **PICO-Project** (**P**ara-endoscopic **I**ntuitive **C**omputer-Assisted **O**peration-system/EU Draft-Project 2004–2007) two arms of use were integrated: Endoscope (Graph 3.11) and Exoscope (Graph 3.12). Actually, both arms are present by some new systems in clinical use already.

This is the major graphic for the concept of para-endoscopic dissecting tech nique, from which clinically endoscopy-assisted microneurosurgery was derived (Graph 3.12).

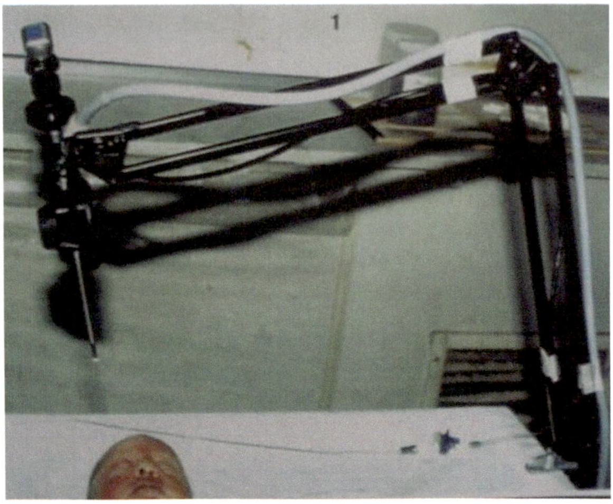

Fig. 3.8 Endoscope-/Exoscope-Prototype

In the operation simulation lab this was tested in over 74 aneurysms post-mortem (47 cases).

This model became part of the PICO-project (Graph 3.13).

For MIN, the optical resolution and the light is superior to the actual new systems. They need to be tested in MIN approaches.

The Exoscope-Systems have the advantage staying extracranial but with the need to have an undisturbed trajectory area for vision. There is a competition between surgeons' hands and visual field. It is unclear if the exoscope can be applied in MIN-approaches. The Endoscope-System has the problem of intracranial safety and competition between endoscope and surgical instruments. The main advantages are the perfect illumination, the brilliant magnification and the application in smallest approach of just a burr-hole in case. Both principles should be judged by two questions: **Is there a benefit for the patient? Can the system realize sufficient ergonomics in MIN?** (s. below).

The PICO project, primarily starting as a research effort was changed into a "DRAFT-Project" of the EU funding for the industry with 13 partners (companies and universities). It resulted in a prototype (Graph 3.13) but never reached the clinical application. However, it contained the two arms of endoscope and exoscope in one system.

The main difficulty anyway turned out to be making all partners comprehensive about the really innovative concepts, which could not be understood and realized with conservative strategies. Especially the new concept of ergonomics was not acquired. It seemed impossible to convey an imagination on the challenge of the mission, to guide an endoscope intra-cranially in the brain with more safety than a perfect Contraves-system holds a surgical microscope. To have the equipment intracranial and at the same time to control a difficult procedure fast in a minimal volume, was the extraordinary difference to the extra-cranial microscope. The system was steered by navigation of the surgical instruments.

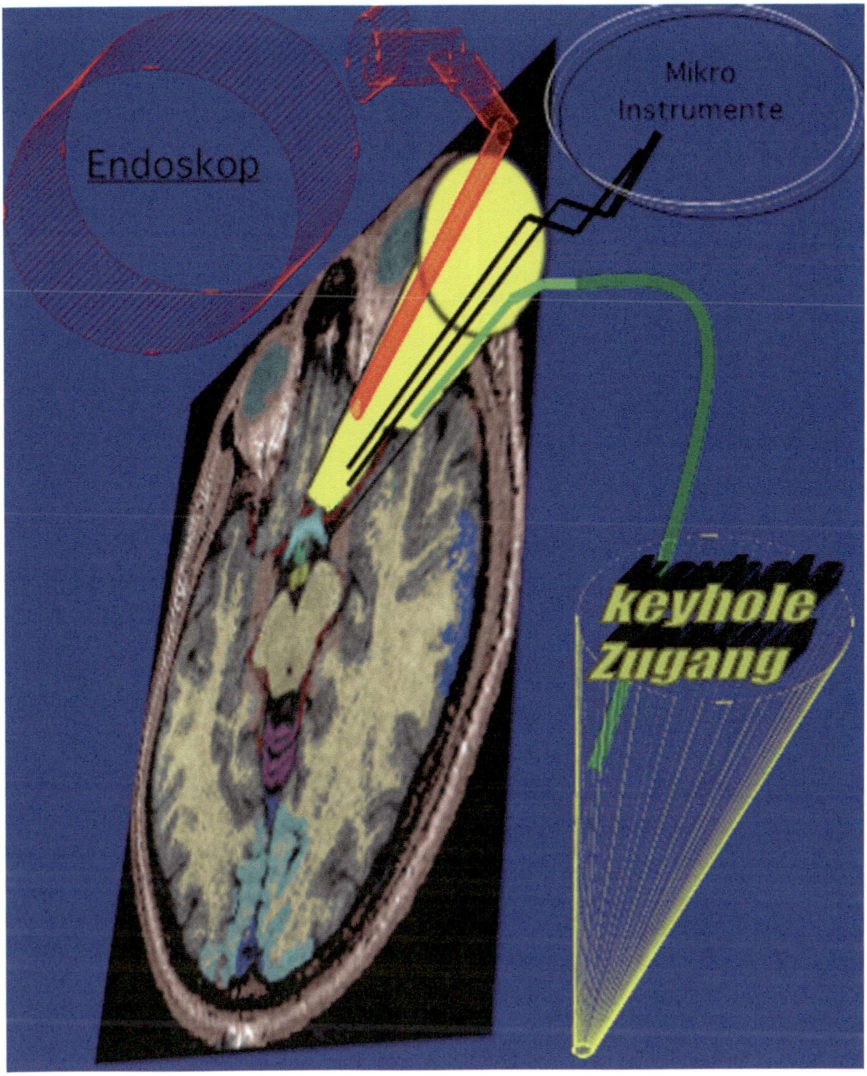

Graph 3.11 PICO-Book 2001

3.5.2 First Exoscope

This **prototype** from Olympus Co. was final used by Perneczky under the term "**Exoscope**" (Fig. 3.15, Chap. 2) (Fig. 3.9).

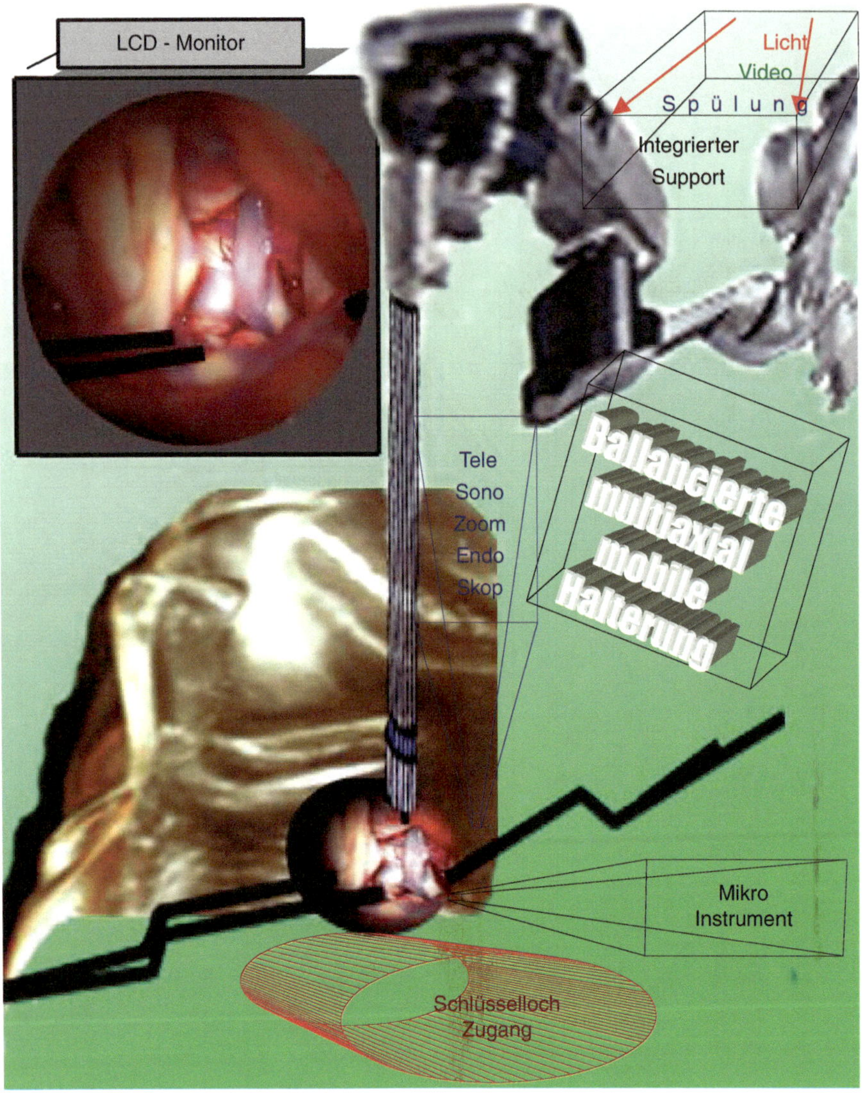

Graph 3.12 PICO-Book 2001

The senior author could experience the system at the Japanese national congress of neurosurgery in Osaka in 10/2000. It was within one testing clear, that the new equipment solved perfectly the major problem of indispensable holding device for endoscopy-assisted micro-neurosurgery. It could have been the brake-through for this new application and indications of MIN. However, it was economically not successful. The first paper in 2004 could not change this pitfall, though being ergonomically perfect (Fig. 3.10).

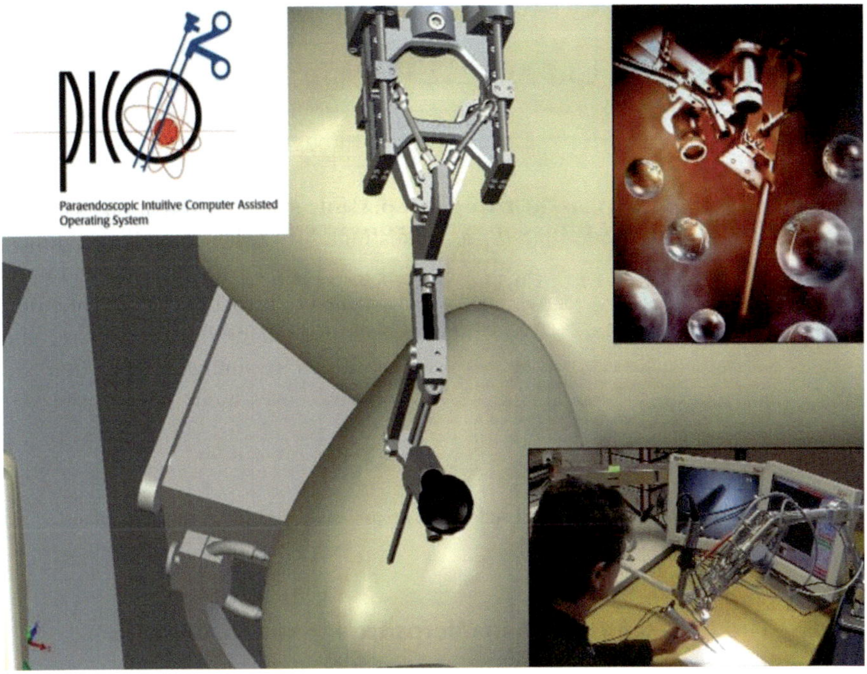

Graph 3.13 PICO prototype 2

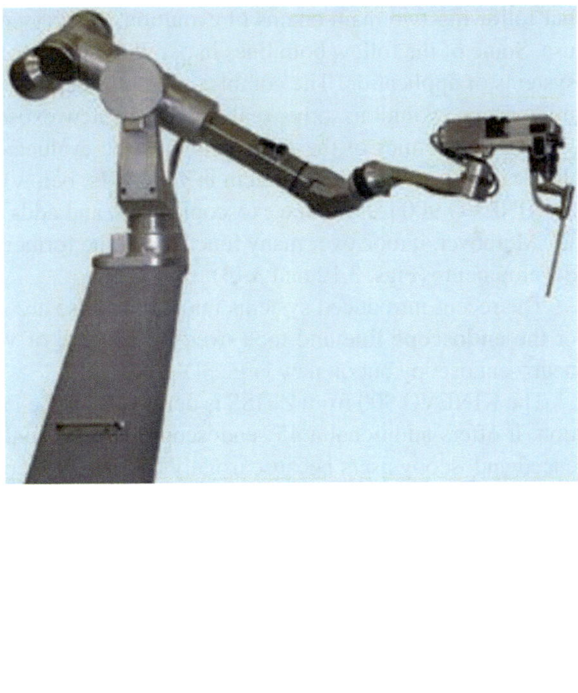

Fig. 3.9 Olympus prototype A3 in 2000

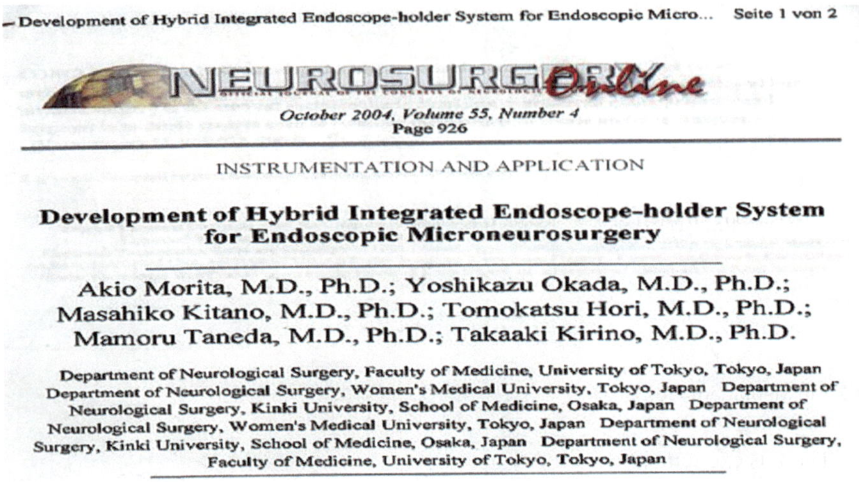

Development of Hybrid Integrated Endoscope-holder System for Endoscopic Micro... Seite 1 von 2

NEUROSURGERY Online

October 2004, Volume 55, Number 4
Page 926

INSTRUMENTATION AND APPLICATION

Development of Hybrid Integrated Endoscope-holder System for Endoscopic Microneurosurgery

Akio Morita, M.D., Ph.D.; Yoshikazu Okada, M.D., Ph.D.; Masahiko Kitano, M.D., Ph.D.; Tomokatsu Hori, M.D., Ph.D.; Mamoru Taneda, M.D., Ph.D.; Takaaki Kirino, M.D., Ph.D.

Department of Neurological Surgery, Faculty of Medicine, University of Tokyo, Tokyo, Japan Department of Neurological Surgery, Women's Medical University, Tokyo, Japan Department of Neurological Surgery, Kinki University, School of Medicine, Osaka, Japan Department of Neurological Surgery, Women's Medical University, Tokyo, Japan Department of Neurological Surgery, Kinki University, School of Medicine, Osaka, Japan Department of Neurological Surgery, Faculty of Medicine, University of Tokyo, Tokyo, Japan

Fig. 3.10 First international papaer on endoscope-/exoscope-device (by a. Morita)

3.5.3 The Actual Situation in Microsurgery and Visualization

Actually, new optics and camera-systems, and large LCD-screens in 3D combined with robot-arms, optimized the exoscope-chain strongly, leading to a new generation of application systems.

There are many systems entering the market actually, differing in several aspects, but following two main chains of evolution, the exoscopic use and the endoscopic use. Some of the follow both lines in parallel, realizing a single system or separated systems of application. The complex systems keep integrating all abilities from the microscope evolutions, other realize only the new exoscope function.

The ergonomics of these systems must be evaluated because it is unclear how they will influence the OR-system in general (s. below) (Graph 3.14) (Fig. 3.11).

KINEVO 900 follows the exoscopic chain and adds an endoscopic tool in parallel. Moreover, it took over many functions of the former microscopes and additional developments (Figs. 3.12 and 3.13).

The recent introduced systems into clinical use are following the **exoscope-line** or the **endoscope-line** and took over the concept of video-surgery from classical neuro-endoscopy but on new large 3D-displays.

The KINEVO 900 from ZEISS is derived from the Penthero Microscope generation. It offers additional a 45°-endoscope, which is only recommended for experienced endoscopy users because usually 0°-, and 30°-endoscope are applied, higher degrees are rising the danger to damage neural structures.

The Modus V (Synaptive) is more along the Olympus Prototype A3 in the exoscope use. It is designed without an endoscopic component and without the microscope functions (Figs. 3.14 and 3.15).

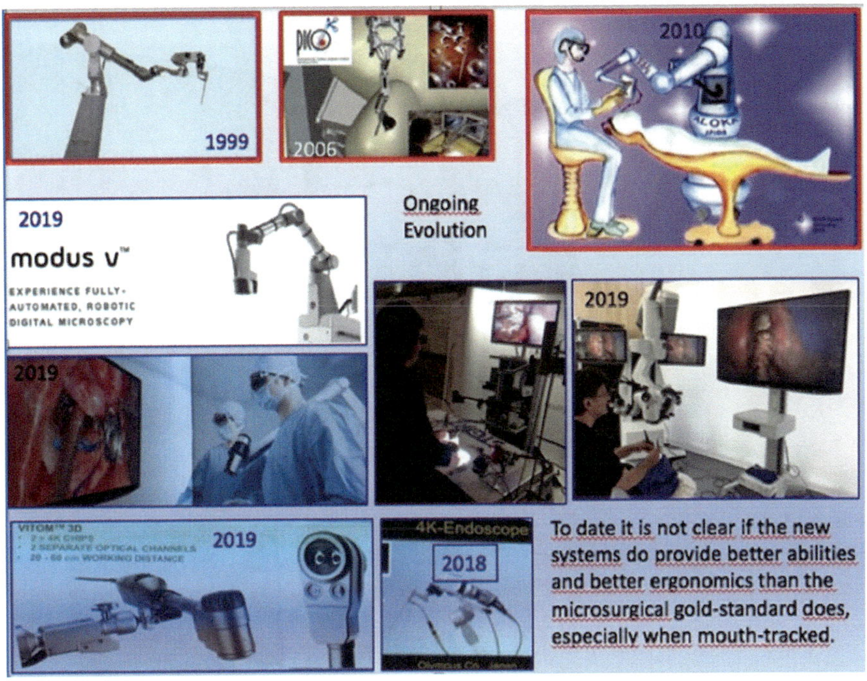

Graph 3.14 Roots and evolutions of exoscopes

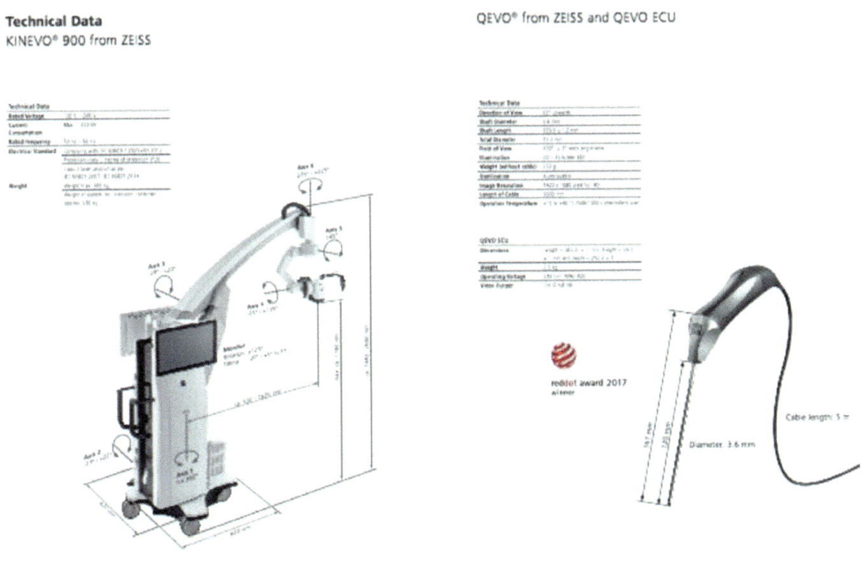

Fig. 3.11 Zeiss KINEVO 900 system

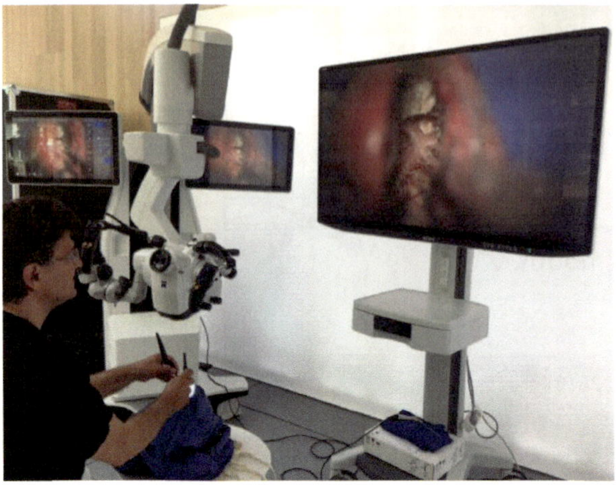

Fig. 3.12 Zeiss KINEVO 900 system test (see vol. 2; Chap. 4) (Pernasal view to pituitary gland and basilar head)

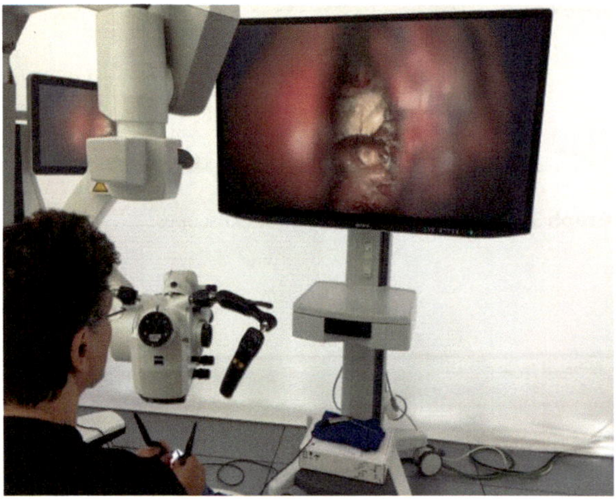

Fig. 3.13 Closeup of Fig. 3.12

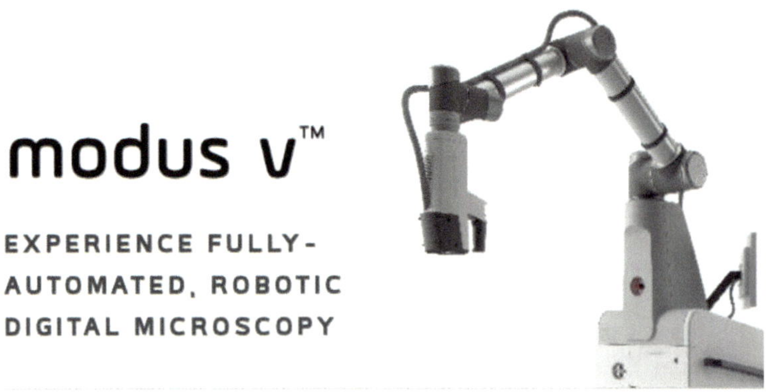

Fig. 3.14 Modus V Exoscope (Synaptive Co)

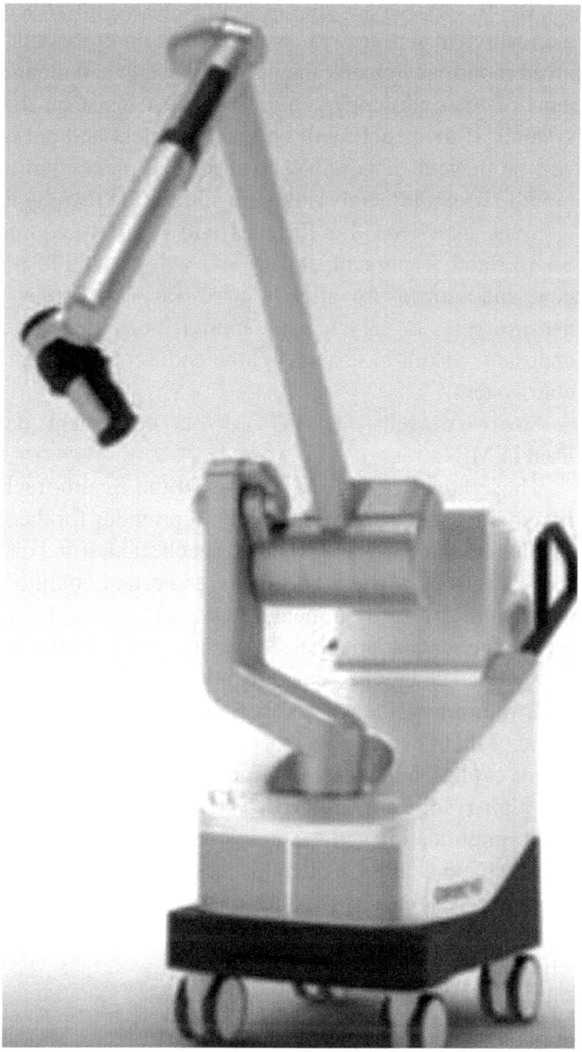

Fig. 3.15 Orbeye exoscope (Olympus Co)

Olympus offers a new Exoscope, "Orbeye", as well as a 4K 3D Endoscope, dividing the **Exoscope-line** from the **Endoscope-line** and so representing the actual course of evolution of equipment for endo-/exo-scope assisted micro-neurosurgery.

However, it seems that the ergonomics problems of tracking and zooming are solved by the exoscope concept rather than with the endoscopic concept. It can also be seen as a hybrid concept and a compromise between classical microsurgery and classical endoscopy to make the assisted micro-neurosurgery more attractive for the surgeons of the changing generations. But the MIN concept and consecutive the ergonomics will decide which of the systems is a benefit for the patient!

All three exoscope types need to be tested towards mouth-switch tracked optical microsurgical setting with special focus on ergonomics. They do not need the unloved mouth-switch and they do not have the challenging intra-cranial safety problems of the endoscopy types. One must insist on the trash-hold, choosing these systems, if the benefit will be for the patient and not only for the surgeon alone. It will be difficult if possible, to reach the precession and speed of tracking of the mouth-switch, remembering with admiration thereby the visionary of Yasargil.

Some experienced colleagues will use both equipment and concepts (Reisch, Switzerland; Taniguchi, Japan) and will be able to present data to compare. Two new endoscopes are already used for 3D display endoscopy assisted microneurosurgery. In very narrow approach-channels (pituitary surgery and per-nasal endoscopic skull-base surgery) the endoscope concept will be probably the preferable system.

Storz Co. sells the VITOM 3D equipment developed for ENT, already used in MIN.

Those however, he who ever used the mouth-tracked microsurgery will never like to miss it because of the safety it provides for the surgeon, giving him the feeling that he can control and manage each sudden and danger event within seconds by optimal visual window and focus, if needed "online". This enables a body-mind continuum driven by a mental-flow and an intuitive integrated motor responds to each experienced situation. It is comparable to the flow of a solo music artist in concert and cannot be copied by automatisms.

In **summary**, the new exoscope-systems follow mainly the exoscopic chain, but some of them have the endoscopic chain in parallel with different realizations. The major evolution addresses the visualisation by **big 3D high-resolution LCDs** and **K4 high-resolution cameras**. The non-combined pure exoscope systems present a little simplification and decrease of size, compared with microscopes.

The main challenge of endoscopy, entering the brain, is not given, but the tracking of the system is still disturbing the work-flow, being far away from the brilliance of mouth-tracking.

3.6 Structure of MIN Evolution

3.6.1 Concepts and Schools (Graph 3.15)

In non-industrialized low-income countries, "Cushing-level" and macro-neurosurgery in a basic standard are given in the most parts of the world!

Micro-neurosurgery is the gold-standard in the industrialized middle/high-income countries. However, there exist markedly differences in the level microneurosurgery is applied. The need of minimizing overall-trauma was discussed early (Rossolimo 1907), but the systematic developments and visionary innovations towards MIN was driven by very few exceptional persons (Cushing, Dandy, Yasargil, Malis, Perneczky, Fukushima and others). Of course, there were exceptional interpreters and also some excellent conductors, but only few composers! All the

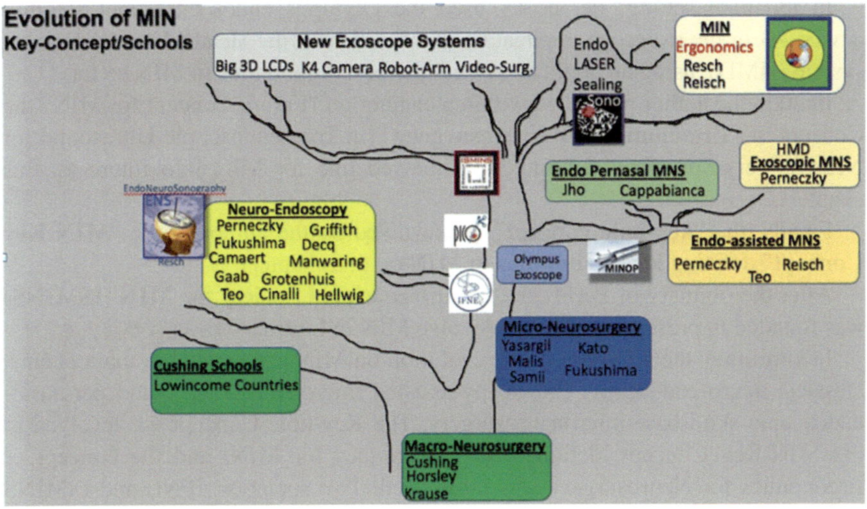

Graph 3.15 Evolution tree of concepts and schools

composers had to struggle lifelong against the main-stream of the profession. The main difference in exoteric philosophy was, taking the "Hippocratic imperative 1" seriously by demanding, that progress must be reaching the patient and functioning in the patient. Success for professionals, for the industry or for statistics did not count for them. Growing of the MIN-tree became a fight for integration of each technical innovation and care to avoid un-necessary overall-trauma.

This motivation became support by the video-chain revolution in endoscopy, and the first mission was to avoid un-necessary shunt-implantation in hydrocephalus. It was Perneczky who identified neuro-endoscopy as a corner-stone of MIN and not just as a tool. However, as a reaction on Perneczky's pioneering world-wide success, a network for neuro-endoscopy was founded (ISGNE >**IFNE**), becoming the arm of classical neuro-endoscopy, mainly driven by paediatric-neurosurgeons.

After the **first international congress on minimally invasive techniques in neurosurgery 1993**, the **MINOP-project** was founded, creating two generations of new neuro-endoscopy systems.

In 1996, the author introduced **"Endo-Neuro-Sonography"** showing navigation-characteristic for neuro-endoscopy (s. Chap. 5).

In parallel, in the operation simulation lab the para-endoscopic preparation technique led to **"Endoscopy-assisted Microneurosurgery"**-concept, becoming the major clinical contribution of the Perneczky-school and increasing the indication of neuro-endoscopy markedly.

Also, in parallel, HD Jho started with systematic **Pernasal Endoscopic Microneurosurgery**, creating a new field of MIN derived from endoscopic pituitary MNS.

The PICO-project produced only a prototype of an exoscope/endoscopy hybrid system with a robot-arm and tracking by navigated instruments, nevertheless serving as a learning step.

In its' final setting, Perneczky used the Olympus endoscopy-robot-holding-device A3 in the exoscope application combined with the Head-Mounted Display system (HMD), using an application with the best ergonomics in MIN so far.

In 2005 the author published within a chapter on "Future Concept for MIN" the Concept of "**Ergonomics for Neurosurgery**" (in Transendoscopic Ultrasound for Neurosurgery/Springer), defining it as the red line for MIN into future (s. this chapter).

Finally, the "**key-hole concept**" was evolved by the author to the "**MIN Key Concept**" defining **key-techniques in MIN** (s. this chapter).

After the obituary of Perneczky the **International Society for MIN (ISMINS)** was founded to preserve and further evolve MIN in future.

In **summary** the structure of the evolution on MIN presents three main chains: classical neuro-endoscopy, endoscopy-assisted microneurosurgery and per-nasal-endoscopic skull-base-microneurosurgery. The Key-hole Concept was evolved to the MIN-Key Concept, defining Key-Techniques for MIN, and the concept of Ergonomics for Neurosurgery was introduced. Two societies, IFNE and ISMINS have been founded to promote MIN.

The new exoscope systems are the actually last step in this short history of MIN Evolution.

3.6.2 Key-Techniques (Graph 3.16)

The different concepts and organized schools in MIN can be characterized by the key-techniques that were used:

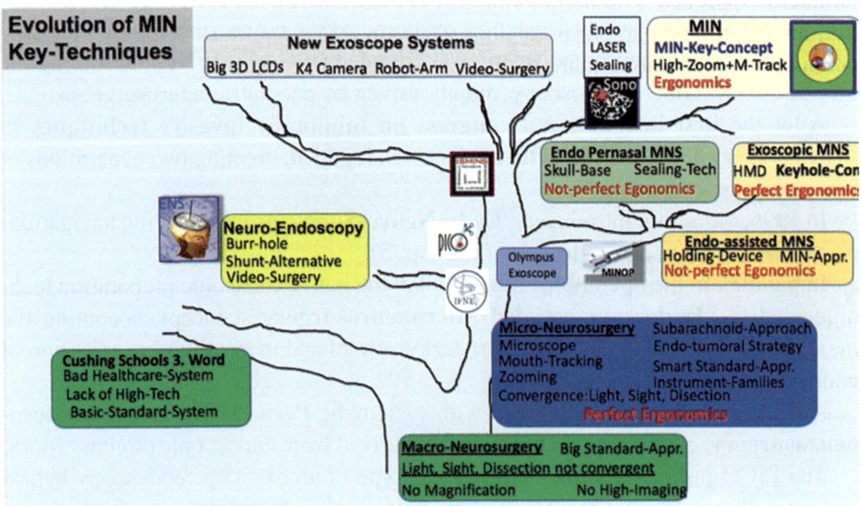

Graph 3.16 Evolution tree of MIN key-techniques

Macro-Neurosurgery had to use large approaches because light, sight and dissection could not have the same corridor, leading to approaches in the form of a pyramid with the base at the surface of the head. Even target of 1 cm^2 needed an approach that had the size of a craniectomy. The absents of imaging did also not allow to perform precise localisation of approaches, and there was no magnification available during surgery. However, it must be remembered, that world-wide in low-income countries the **Cushing-standard** still exists.

It was **Yasargil** who understood, and consequently evolved the introduction of the microscope as the new generation of **micro-neurosurgery**, changing everything: new anatomical and physiological concepts, new instrument-families, new surgical strategies like subarachnoid approach or endo-tumoral resection, new surgical smart-approaches as light, sight and dissection took place now in one single corridor and finally a perfect ergonomics organization of all procedures, of the OR-environment and the behaviour of the surgeon as a part of the system. It grew fast to the new gold-standard in industrialized countries world-wide. The evolution around micro-neurosurgery supported this strongly with neuro-imaging, neuro-anaesthesia and neuro-ICUs, neuro-monitoring and neuro-rehabilitation. The next step is an ongoing **evolution towards MIN**, being started mainly by Perneczky and Fukushima. It started with application of micro-neurosurgery at its' limits by consequently using all available techniques and planning-data to tailor the surgical approach as precise as possible. The discovery, that in standard-approaches regularly only $\frac{1}{3}$ of the approach really was needed and used, led to the question if this really used part of the standard-approach can be analysed and predicted by the individual data of the patient pre-op, mainly from neuro-imaging. This led to the change from standard-approaches to **individual-approaches**, followed by further developments to realize this goal. It was named the **key-hole concept**, however, leading to misunderstandings, as the focus seemed to be on the size of the approach. The size, however, was not the goal but the result of precision and new techniques and concepts.

Into this phase, the revolution of the **video-chain technique in endoscopy**, strongly supported the key-hole concept of Perneczky, accelerating the evolution towards MIN. Endoscopy became the leading surgical technique to realize "key-hole surgery", however, still in a limited indication. During that time, the author not only was working on endoscopic anatomy but also on endoscopic aneurysm surgery simulation in 74 aneurysm cases. It had become clear previous, that aneurysms cannot be dissected by usual trans-endoscopic technique, but only in a para-endoscopic dissection fashion. The advantage of visualization by endoscopy was combined with the advantage of microsurgical dissection-techniques, and disadvantages of both techniques were compensated. This kind of combining different techniques with minimal-invasive characteristics became a principle of MIN. This was soon taken over into clinical application, founding **Endoscopy-assisted Microneurosurgery**. The main problem was the holding-device comparable to the microscope-technique. But Olympus Co. had already the solution by the Robot-A3-System. Perneczky used it finally, and when disconnecting the endoscope from the camera fixed to the robot-arm, he applied the A3-system as a visualisation tool for the opening, naming it **"exoscope"**. With this step, the complete primarily lab model had been translated into clinical use. This setting, together with HMD had a perfect ergonomics (s. Vol. 2).

In parallel, the application on classical neuro-endoscopy was spread world-wide, and ETV became the first well established indication, avoiding shunt-implantation in obstructive hydrocephalus. However, our first six cases, in 1991 were successfully, but having all a clinical history with contra-indications like bleedings or infections. Also, the critical cut-off for ETV in children was decreased from 24 to 6 months. It became by experience clear, that pathophysiology of hydrocephalus was obviously not understood when believing in a mechanical reason (but s. Chap. 6!). The introduction of **trans-endoscopic ultrasound** (ENS; s. Chap. 6) by the author, was applied in a series of 70 complex hydrocephalus-cases in children, convincing that endoscopy will not advance without ultrasound in the future.

The best model for the perspective of MIN, was developed by application of endoscopy-assisted microneurosurgery techniques trans-nasally and per-nasally in pituitary surgery, and then in skull-base surgery. Especially the sealing-techniques came into the level of a key-technique. However, the ergonomics problems are not solved, causing a limited application by few neurosurgeons only (s. Vol. 2).

Finally, in a consecutive series of ICH-evacuation (s. Vols. 3 and 4), the author used five techniques as **Key-Techniques for MIN**, according to the new concept of **MIN Key-Techniques** (s. this chapter): high-zoom mouth-tracking MNS, neuro-sonography, neuro-endoscopy, neuro-LASER and neuro-sealing, not asking anymore for the holes but for the keys of MIN in future. These techniques were raised to "key-techniques" because without mastering them this kind of high-level MIN is impossible.

In **summary**, evolving the concept for MIN and new combinations of techniques made new applications in new indications possible within above mentioned three arms of MIN, and now advancing as a new arm with a combination of five key-techniques. Especially in the neglected fields of ICH-evacuation and endoscopic procedures, this advanced MIN has much better results than ever expected. The principles how to apply which MIN-technique and combination of MIN-techniques individually in each case are the key of success. There is an education- and training-deficit for all these MIN-techniques.

The upcoming exoscope-systems must be evaluated according to the structures and according to the ergonomics of MIN evolution, to get a prediction for their impact on the OR-system behaviour and the future of MIN.

3.7 Evaluation of Exoscope-Systems According to Ergonomics for MIN: The "Ergo-Tool" (s. Chap. 2.2)

To understand the standard of high-zoom mouth-tracked micro-neurosurgery within evolution of MIN, it is most effective to view it in the perspective of the three ergonomics fields OR environment, procedural flow and mental flow (s. this chapter). This gives a perspective for the future of MIN. From the analysis given above, we can start to apply this for the new exoscope principles:

- Which impact will the new exoscopes have regarding the ergonomics parameters OR environment, work-flow and mental flow?

- Can necessary and following changes be made, if needed, in time?
- What happens if not?
- Is there a benefit for MIN evolution?
- Is there a benefit for the patients?

These questions and prove, using the criterions of ergonomics, must be answered to evaluate the exoscope before buying, or to evaluate its' impact after some experience in daily use.

A simple and usual analysis with some statistics of results, naming it "evidence" will not be able to distinguish, which really is the factor that enables the results. The ergonomics system as an evaluation tool (Ergo-Tool) will be much more reliable (s. this chapter).

To apply this tool without the back-ground of experience would be an interesting test in theory but is not well grounded for a reliable result. However, it may make sense to apply the **"Ergo-Tool"** before experience with the exoscope and after a period of using it. All three fields should be tested: the change for the OR-environment, the change of the work-flow and handling and the changes for the mental-flow. What does an exoscope provide in this fields of ergonomics?

The exoscope, as any other equipment may be placed in the OR environment, from different directions and transgressing several virtual orbitals of different ergonomics sensitivity. All transgressed orbits can be disturbed, and a quantitative system for evaluation can be designed, but it is more a qualitative causality. Be aware, that in this ergonomics field, in this system of therapy, the surgeon is integrated! The usual system of patient and therapy is substituted by the surgeon himself! This is missing in all usual research and publications (s. below).

In Graph 3.17, the majority of equipment and all other people of the team are not present and not symbolized. Completion of the environment must be shown and described and symbolized (Graph 3.18).

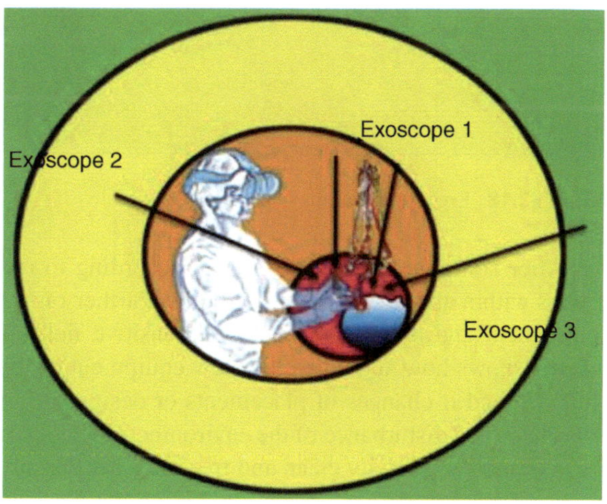

Graph 3.17 Positions of the exoscope

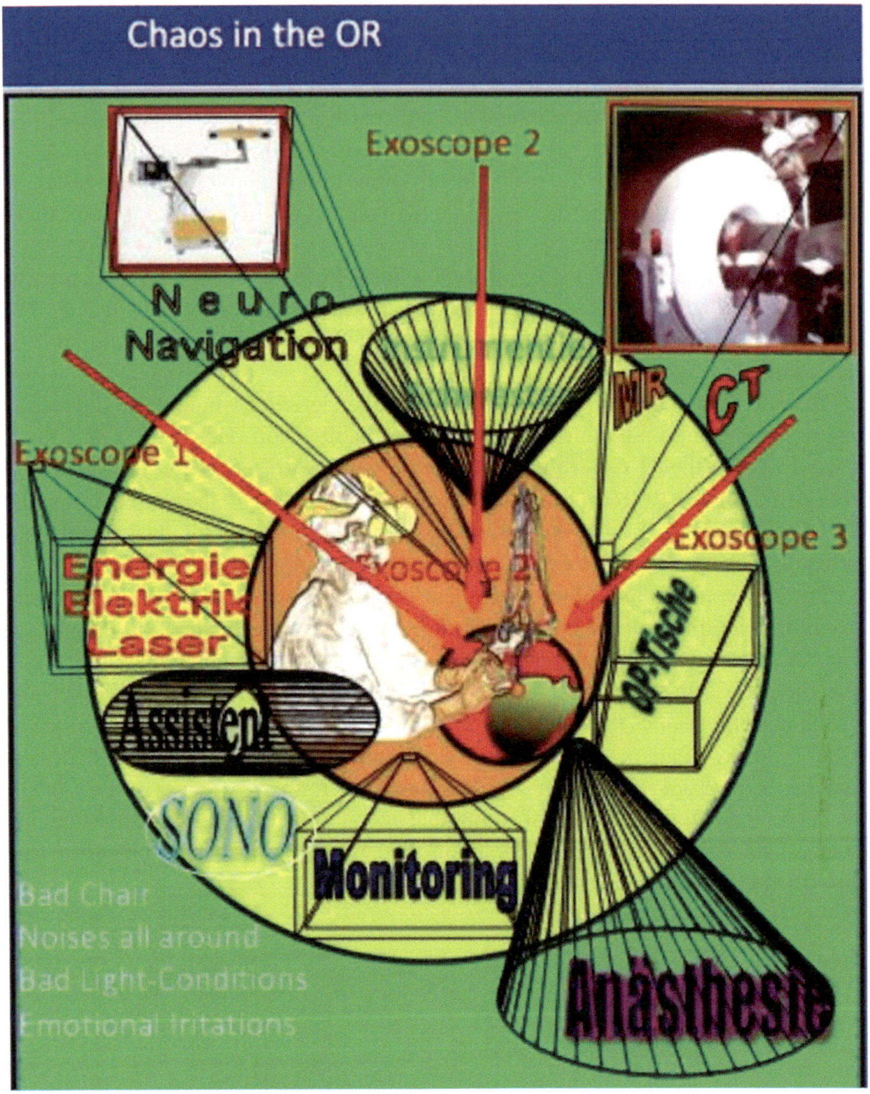

Graph 3.18 Ergonomics chaos in the OR

After completion the environment according to reality, we see the ergonomics chaos within all the ergonomics orbitals. Further-on we recognize, that many of the equipment and team reach the most sensitive field of red and orange area! This model shows how and where the new equipment changes the system and its' ergonomics, and if changes of placements or design or functional evolution may relax the chaos and disturbance of the environment. The goal is to keep the red and orange area of most sensitivity clean and free of chaos (Graph 3.19).

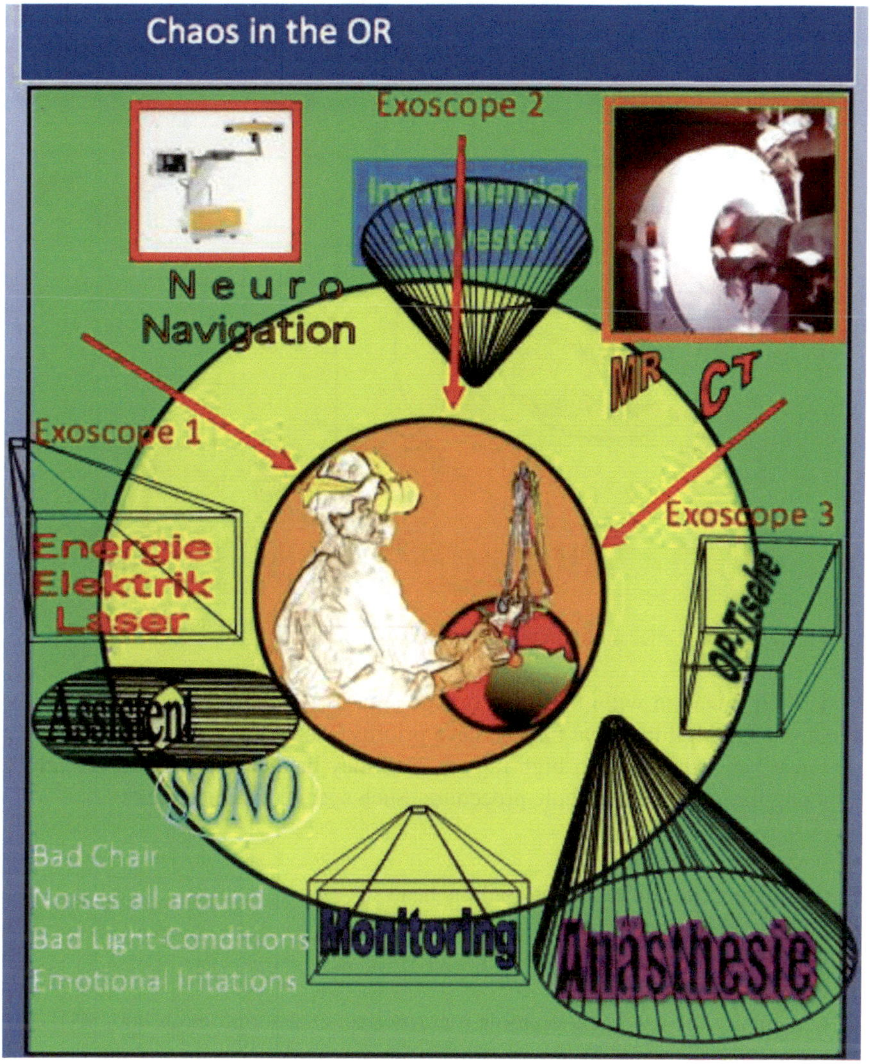

Graph 3.19 Most ergonomically sensitive area red and orange cleared from chaos

After reorganization of the environment, we see the cleared and relaxed red and orange area. This is the meaning and functional strength of the Yasargils' intuitively realized ergonomics, however, caused by leaving everything out of the OR which is not absolutely necessary (Graph 3.20).

The spacial ergonomics model can be used to organize OR-placing the equipment, and by functional analysis and design, but see also the other two models below, application examples see above. In Graphs 3.18 and 3.19 we see how the most sensitive red and orange area are cleaned from ergonomics impairment.

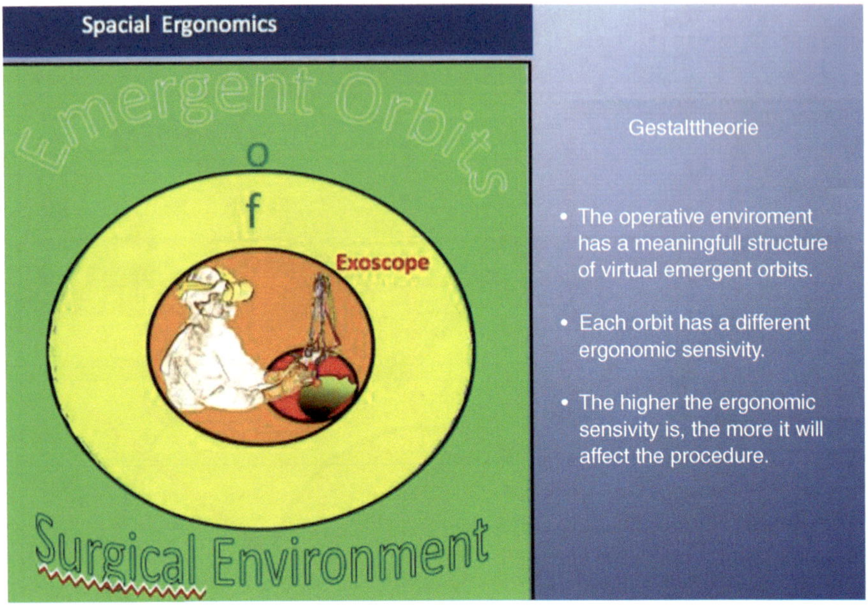

Graph 3.20 Spatial effect of the exoscop in the Ergo-Tool; 1

The organization within red and orange area is a matter of training and art of surgery. Here all things that happen have priority, there is nothing to neglect, even the most banal things get a high impact once they happen there and especially if effecting through-out the whole procedure. Such spacial parameters are:

- Positioning of the patient
- Organizing the equipment around the head
- Draping of the surgical area
- Positioning the light
- Function of the chair
- Organizing the visualization tools (microscope, endoscope, exoscope, HMD)
- Organizing the space for the surgeon to move with the chair or without
- Organizing the pathway of instrumentation (position, direction, smart contact)
- Laminar air-flow in the surgical field (Graph 3.21)

If the operation procedure is a **"chaotic system"**, then the exoscope and all the equipment and even the team are peripheral parameters. But this chaotic process can only be influenced by these parameters and they can be prepared. It is hard for the surgeon to struggle with bad ergonomics and to believe that each disturbance can be compensated. Neglecting preparation of the peripheral parameters, especially if lasting through-out the whole procedure, will have a major impact in the outcome for the patient. Also, it will have an impact on the surgeon and his stress-level and his health. The organizing of the peripheral parameters will have a major impact on

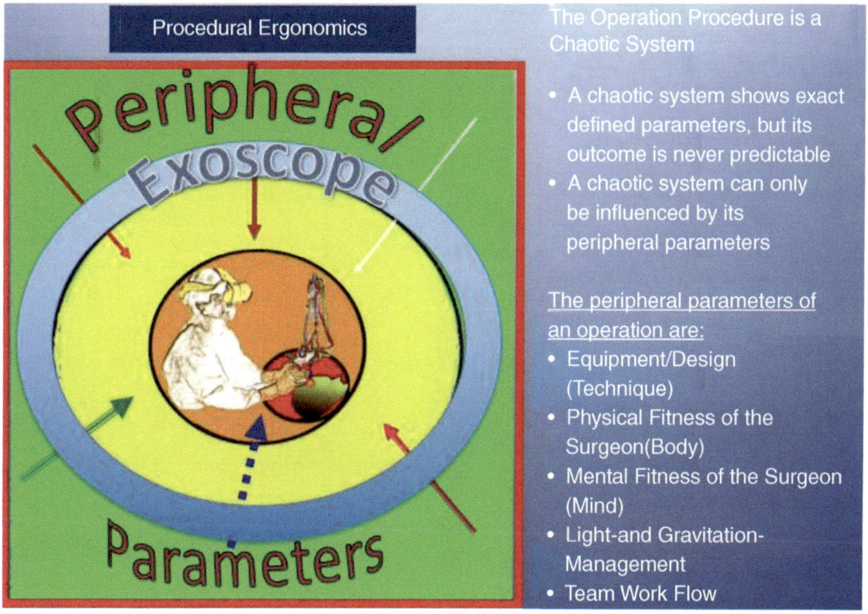

Graph 3.21 Work-Flow effect of the exoscope in the Ergo Tool; 2

time needed and by this on the financial and logistical costs! In nearly all hospitals the OR is the most expensive room and miss-organization due to ignoring ergonomics leads to irrational raising costs, leaving the professionals in helplessness.

Missing an ergonomics concept, more over leads to an educational deficit in having a feeling for time-flow and work-flow.

It must be watched how the exoscope will influence all these parameters in the complex system of the operative procedure. The loss of high-zoom mouth-tracking competence (Yasargil) may be not substitutable by the exoscopes (Graph 3.22).

The most complex, most affecting and most ignored ergonomics field is the mental ergonomics. It may be that the viewing conditions of exoscope will address the visual habits of the young generation. There is a lot of knowledge regarding mental and neuro-ophthalmological consequences of viewing on displays of computers, tablets and smartphones. However, the big 4K 3D-displays in a distance, more comfortable than other digital media, may allow more relaxed viewing. But the concentration of information by high-zooming to a congruence of view and visual field will get lost. The excluding effect of visual disturbances in the retinal periphery will get lost. It is not clear how this will affect the mental-flow. The functional circle of mental-navigation, high-zooming and fast mouth-tracking for focusing and precise visual field (Graph 3.6) will get lost. The maximum of key-hole surgery, like per-nasal endoscopic microneurosurgery might be impossible and the ultimate test for the exoscopes, so far.

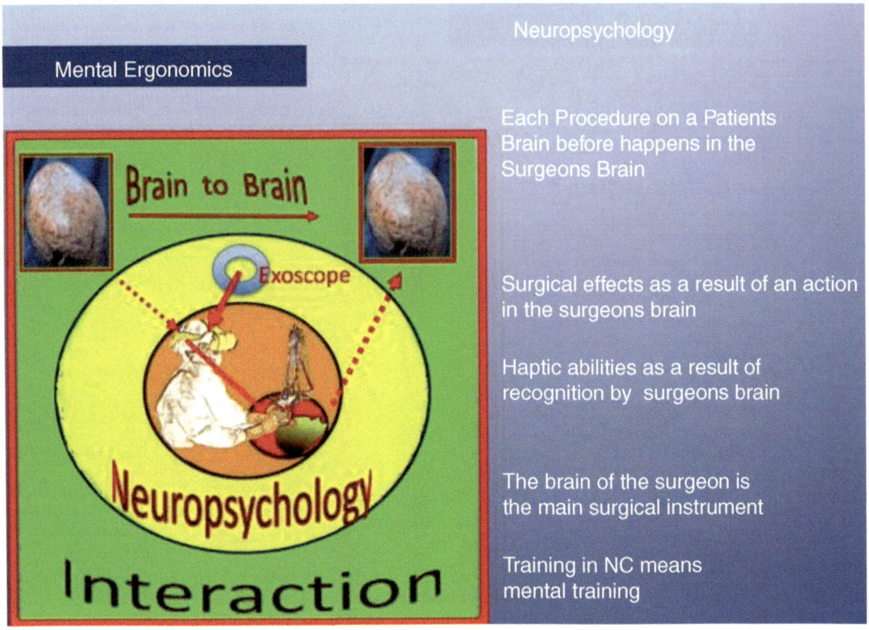

Graph 3.22 Mental effect of exoscope in the Ergo Tool; 3

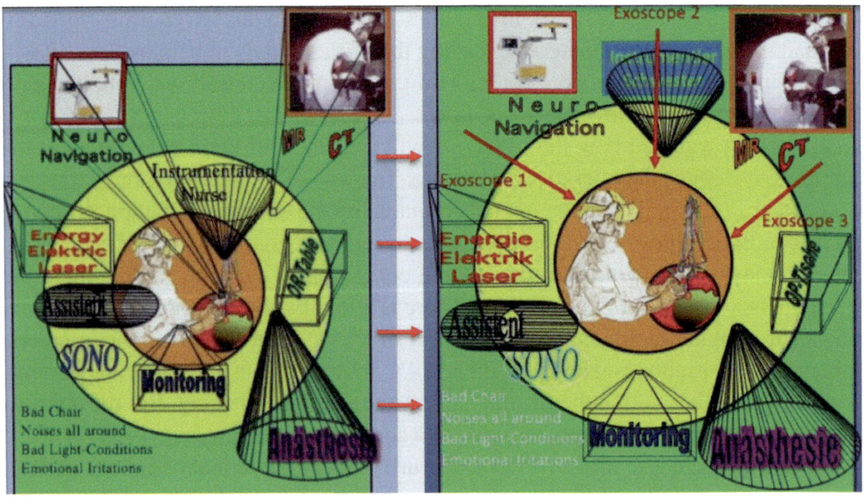

Graph 3.23 Planning ergonomics: overcome ergonomics chaos

May be this will be compensated with more comfort and even fun for the surgeons, but it is not clear if this will arrive the outcomes of the patients. However, pure statistics of outcomes will not address this problem regarding the way, this functions and regarding the mental ergonomics (Graph 3.23).

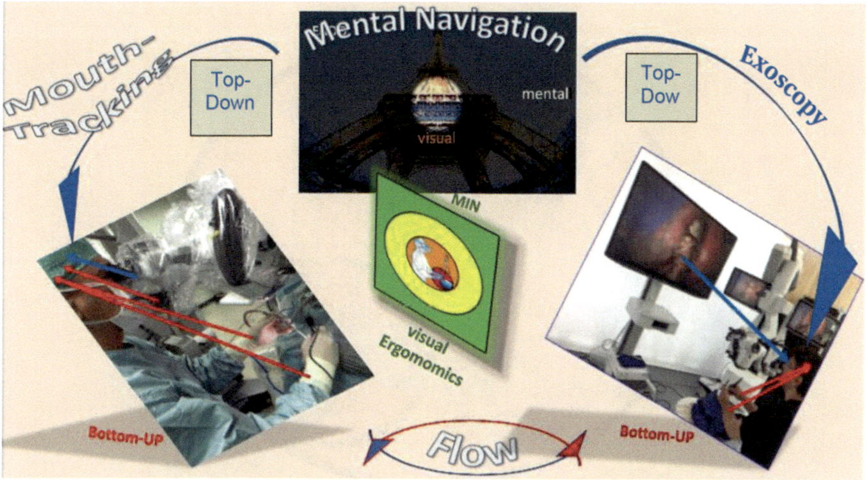

Graph 3.24 Ergonomics flow change microscope >> exoscope

Each new equipment should diminish the OR-chaos and respect the priority of the red and orange area, the field of surgical art, minimizing surgical trauma and saving time and space and logistic efforts.

The **ergonomics system** can more-over be used as an **Ergo-Tool** to evolve the present environment to a more sophisticated place in future.

For a visionary evolution, the most attention must be given to the red and orange field, keeping it free of disturbances by evolution of all equipment, which demands from all orbits major changes.

The benefit in the red and orange orbit may be reached by changing in more peripheral orbits yellow and green.

However, the most important parameters, integrating a new visualization tool, like Exoscopes, are those of viewing with all the partial functions of seeing. The visual functions have strong interaction with cognition, creating imaginations and sensitive hypotheses about the visualized information, influencing the motor action (top-down).

In contrast, the motor action causes a concentration on the present, directing visual action to a target and task (bottom-up). The functional relation between viewing and motor-action is a very crucial one, like eye-hand-coordination. This may undergo a mayor change with exoscopes, changing the mental ergonomics markedly (Graph 3.24).

The visual input in exoscopy is very different from that in microscopy. The complex multifactorial visual function, trying to provide an illusion with a competent correlation to reality, and in strong interaction with cognition and recognition, is in an easy destroyable balance. Visual psychology and visual neuropsychology, as we know from paediatrics, and developing monitoring of impaired children, have a major impact on mental ergonomics. Changing a visualization tool means a new world for the mental flow and its' relation to motor-action.

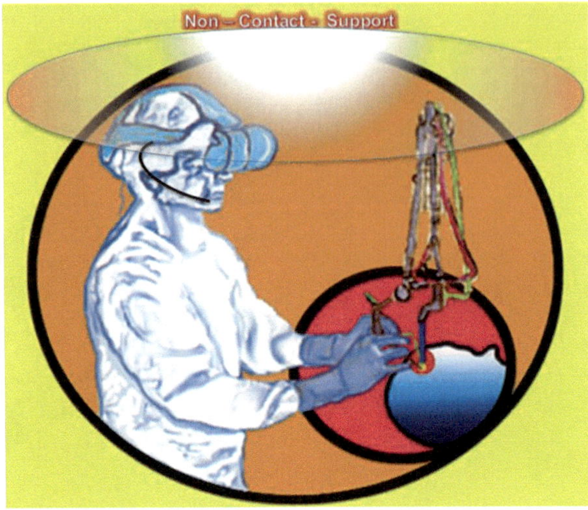

Graph 3.25 Future: bluetooth-assisted MIN

In microscopy the vision is diving into the depth of the operative approach pathway, getting feedback by mouth-tracking of focus and positioning of visual field, and concentration with high-zooming to the only necessary input.

In exoscopy there is no extinction of disturbing input from around the screen. The visual binding is less involving than in microscopy. To realize the body-ergonomics might be easier. Comfort without binding and involving into the target and task might need a strong mental discipline and challenge for the surgeon.

The Ergo-Tool may give some inspiration and assistance for test-settings to examine this functions and changes in future (Graph 3.25).

In a visionary future the ergonomics space will only need the core-environment of the red and orange area, having ruled out all disturbing and immature techniques and equipment. This will be the case in a perfect information-flow environment, leaving the surgeon and the patients head in peace.

It's the goal of the Ergo-Tool to reach this situation and setting. Meanwhile the steps of optimizing can be also assisted by the Ergo-Tool.

The **Ergo-Tool** can support **MIN-Evolution**, reducing financial and logistic costs, concentrating in direction all efforts towards the benefit for the patient. It is a system-analytical concept, regarding everything in the surgical environment.

Suggested Reading

Al-Mefty O. Surgery of the cranial base. Boston, MA: Kluwer Academic Publishers; 1989. p. 3–11.

Apuzzo MLJ. Neurosurgery for the third millennium. Park Ridge, IL: AANS Publication Committee; 1992. p. 11–23.

Apuzzo MLJ. New dimension of neurosurgery in the realm of high technology: possibilities, practicalities, realities. Neurosurgery. 1996;38:625–39.

Auer LM. Robots for neurosurgery? In: Hellwig D, Bauer BL, editors. Minimally invasive techniques in neurosurgery. New York, NY: Springer; 1998. p. 243–50.

Awad IA. History of neurovascular surgery II: occlusive disease, cerebral hemorrhage, and vascular malformation. In: Greenblatt SH, editor. A history of neurosurgery. Park Ridge, IL: AANS Publication; 1997. p. 271–88.

Black PML, Moriarty T, Alexander E III, Stieg P, Woodard EJ, Gleason L, Martin CH, Kikinis R, Schwarz RB, Jolesz FA. Development and implementation of intraoperative magnetic resonance imaging and its neurosurgical application. Neurosurgery. 1997;41:831–45.

Brihaye J, Pertuiset B. Fourteen years to achieve the laying of the foundation of the European Association of Neurosurgical Societies (EANS): Brussels (1957) - Madrid (1967) - Praque (1971). Acta Neurochir. 1994;130:3–7.

Brock M. Proposal of the UEMS for a European harmonization of the training of residents. 50. Jahrestagung DGNC München. Zentralbl Neurochir. 1999;(Suppl):S90.

Davies RA. The Brigham Diary of Loyal Davis: a portrait of Harvey Cushing and a neurosurgical acolyte. J Neurosurg. 1995;82:683–92.

Day DJ, Tschabitscher M. Microsurgical dissection of the cranial base, 1996. New York, NY: Churchill Livingstone; 1996.

Day JD, Koos WT, Matula C, et al. Color atlas of microneurosurgical approaches. Stuttgart: Thieme; 1997.

EC/IC Bypass Study Group. Failure of extra-intracranial arterial bypass to reduce the risk of ischemic stroke: result of an international randomized trial. N Engl J Med. 1985;313:1191–200.

Frerichs KU. Experiences with the operative training in the USA. 50. Jahrestagung DGNC München. Zentralbl Neurochir. 1999;(Suppl):S90.

Hassler WE, Meyer B, Rohde V, Unsöld R. Pterional approach to the contralateral orbit. Neurosurgery. 1994;34:552–4.

Jendrysiak U, Resch KDM. Ergebnisse der klinischen Erprobung der Operationszugangsplanung mit NeurOPS. In: Bildverarbeitung für die Medizin. Algorithmen-Systeme-Anwender. Proceedings-Band S. Heidelberg: Springer; 1999. p. 187–91.

Key A, Retzius G. Studien in der Anatomie des Nervensystems und des Bindegewebes. Bd. 1, S. 1-155, Tafel III, Fig, 1 und 2; Tafel VI, Fig. 1; Tafel VII, Fig. 1 und 3; Tafel XXVII Fig. 2. Stockholm: Samson & Wallin; 1875.

Knosp E, Perneczky A, Resch KDM, Wild A. Endoscopic assisted microneurosurgery. 1. In: International Congress on Minimally Invasive Techniques in Neurosurgery. Abstractbook; 1993. p. 22.

Kockro RA, Serra L, Tseng-Tsai Y, et al. Neurosurgical planning and training in a virtual reality environment. In: 11th European Congress of Neurological Surgery, Abstractbook; 1999. p. 75.

Kriz W. Die Bedeutung der Anatomie in Forschung und Lehre. In: Bauer A, editor. Theorie der Medizin. Dialog zwischen Grundlagenfächern und Klinik, vol. 6. Heidelberg: J. Ambrosius Barth; 1995. p. 60–9.

Lanz T, Wachsmuth W, Lang J, editors. Praktische Anatomie, 1. Teil Kopf, Band B. Berlin: Springer; 1979.

Lanz T, Wachsmuth W, Lang J. Praktische Anatomie, 1. Teil Kopf, Band A. Berlin: Springer; 1985.

Liliequist B. The anatomy of the subarachnoid cisterns. Acta Radiol. 1956;46:61.

Long DM. Competency-based neurosurgical training. 50. Jahrestagung DGNC München. Zentralbl Neurochir. 1999;(Suppl):S90.

Malis LI. New trends in microsurgery and applied technology. In: Pluchino F, Broggi G, editors. Advanced technology in neurosurgery. New York, NY: Springer; 1988.

Mohsenipour I, Goldhahn WE, Fischer J, et al. Approaches in neurosurgery. Stuttgart: Thieme; 1994.

Moskopp D. Experiences and lists of wishes for the operative training of a former assistant. 50. Jahrestagung DGNC München. Zentralbl Neurochir. 1999;(Suppl):S90.

Nagata S, Rhoton AL Jr, Barry MA. Microsurgical anatomy of the choroidal fissure. Surg Neurol. 1988;30:3–59.

de Olivera E, Rhoton AL Jr, Peace D. Microsurgical anatomy of the region of the foramen magnum. Surg Neurol. 1985;24:293–352.

Ono M, Rhoton AL Jr, Peace D, Rodriguez J. Microsurgical anatomy of the deep venous system of the brain. Neurosurgery. 1984;15(5):621–57.

Penzholz H, Piscol K. Experiences with the operating microscope in the treatment of intracranial aneurysms and angiomas. In: Koos WT, Böck FW, Spetzler RF, editors. Clinical microneurosurgery. Stuttgart: Thieme; 1976.

Perneczky A. Planning strategies for the suprasellar region - philosophy of approaches. Neurosurgeons. 1992;11:343–8.

Perneczky A, Fries G. Use of a new aneurysm clip with an inverted-spring mechanism to facilitate visual control during clip application. J Neurosurg. 1995;82:898–9.

Perneczky A, Müller-Forell W, van Lindert E, et al. Keyhole concept in neurosurgery. Stuttgart: Thieme; 1999.

Regelsberger J, Heese O, Horn P, Kirsch M, Eicker S, Sabel M, Westphal M. Training microneurosurgery – four years experiences with an in vivo model. Zentralbl Neurochir. 2011;72(4):192–5.

Resch KDM. Use of plastinated specimens in the demonstration of microsurgical approaches to the cranial base. J Int Soc Plast. 1989;3:29–33.

Resch KDM. Beitrag zur Zugangsanalyse und zum Zugangsdesign des transoral-transpharyngealen Weges zum Hirnstamm Dissertation; 1991. Universität Heidelberg.

Resch KDM, Bohl J, Perneczky A. Postmortal inspection, a new pathoanatomical method. Clin Neuropathol. 1992;11:191.

Resch KDM, Perneczky A. Transorale, intradurale Zugänge für Endo- und Microneurochirurgie. In: Steudel WI, editor. Transfaciale Zugänge zur Schädelbasis. Reinbek: Einhorn Presse; 1995. p. 225–30.

Resch KDM. MIN: Transoral transpharyngeal approach to the brain. Neurosurg Rev. 1999;22:2–25.

Reulen HJ, Steiger HJ. Training in neurosurgery. In: Acta Neurochir Suppl, vol. 69. New York, NY: Springer; 1997. p. 58–82.

Rhoton AL, Saeki N, Perlmutter D, Zeal A. Microsurgical anatomy of common aneurysm sites. Clin Neurosurg. 1979;26:248–306.

Rhoton AL. Microsurgical anatomy of the posterior fossa cranial nerves. Clin Neurosurg. 1979;26:398–462.

Rhoton AL. Microsurgery of the third ventricle: Part 2. Neurosurgery. 1981;8(3):357–73.

Sampath P, Long DM, Brem H. The Hunterian neurosurgical laboratory: the first 100 years of neurosurgical research. Neurosurgery. 2000;46:184–95.

Seeger W. Atlas of topographical anatomy of the brain and surrounding structures. New York, NY: Springer; 1978.

Seeger W. Differential approaches in microsurgery of the brain. New York, NY: Springer; 1985.

Seeger W. Planning strategies of intracranial microsurgery. New York, NY: Springer; 1986.

Sugita K. Microneurosurgical atlas. Berlin: Springer; 1985.

Tulleken CAF. The profile of a manual skilled vascular neurosurgeon. In: 11th European Congress of Neurological Surgery, Abstractbook; 1999. p. 1.

Ulrich P, Perneczky A, Muacevic A. Operative Strategie in Fällen von multiplen Aneurysmen. Zentralbl Neurochir. 1997;58:163–70.

Velazquez-Santana H, Resch KDM. 15 years of Mexico - Germany education project for MIN. In: EANS 2019, Poster. Abstractbook; 2019.

Weinberg R. Future direction in neurosurgery visualization. In: Apuzzo MLJ, editor. Neurosurgery for the third millennium. Park Ridge, IL: AANS Publication; 1992. p. 47–64.

Wen HT, Rhoton A, Katsuta T, de Oliveira E. Microsurgical anatomy of the transcondylar, supracondylar, and paracondylar extensions of the far-lateral approach. J Neurosurg. 1997;87:555–85.

Yasargil MG. Microsurgery applied to neurosurgery. Stuttgart: Thieme; 1969.

Yasargil MG, Fox JL. The microsurgical approach to intracranial aneurysms. Surg Neurol. 1975;3:7–14.

Yasargil MG, Fox JL, Ray MW. The operative approach to aneurysms of the anterior communicating artery. In: Krayenbühl H, editor. Advances and technical standards in neurosurgery, vol. 2. Wien: Springer; 1975. p. 11–128.

Yasargil MG, Antic J, Laciga L, Jain KK, Hodosh RM, Smith RD. Microsurgical pterional approach to aneurysms of the basilar bifurcation. Surg Neurol. 1976a;6:83–91.
Yasargil MG, Kasdaglis K, Jain KK, Weber HP. Anatomical observations of the subarachnoid cisterns of the brain during surgery. J Neurosurg. 1976b;44:298–302.
Yasargil MG. Microneurosurgery I. Stuttgart: Thieme; 1984a.
Yasargil MG. Microneurosurgery II. Stuttgart: Thieme; 1984b. p. 283–97, 297–349, 341.
Yasargil MG. Microneurosurgery II. Stuttgart: Thieme; 1984c. p. 272–5.
Yasargil MG. Microneurosurgery II. Stuttgart: Thieme; 1984d. p. 46–57.
Yasargil MG. Hugo Krayenbühl. Acta Neurosurg. 1985;76:1.
Yasargil MG, Reichmann MV, Kubik S. Preservation of frontotemporal branch of facial nerve using the interfascial temporalis flap for pterional craniotomy. Neurosurgery. 1987;67:463–6.
Yasargil MG. Microneurosurgery IV. Stuttgart: Thieme; 1994a. p. 194–6.
Yasargil MG. Microneurosurgery IVA. Neuroradiology conclusion. Stuttgart: Thieme; 1994b. p. 209.
Yasargil MG. Microneurosurgery IVB, Microneurosurgery of CNS tumors. Stuttgart: Thieme; 1996a. p. 2.
Yasargil MG. Microneurosurgery IVB. Microneurosurgery of CNS tumors. Open MRI system. Stuttgart: Thieme; 1996b. p. 67.
Yasargil MG. Microneurosurgery IVB. Microneurosurgery of CNS tumors. Stuttgart: Thieme; 1996c. p. 1–25.
Yasargil MG. Microneurosurgery IVB. Microneurosurgery of CNS tumors. Stuttgart: Thieme; 1996d. p. 26–8.
Yasargil MG. Microneurosurgery IVB. Microneurosurgery of CNS tumors. Stuttgart: Thieme; 1996e. p. 116–91.
Yasargil MG. Microneurosurgery IVB. Microneurosurgery of CNS tumors. Stuttgart: Thieme; 1996f. p. 194–245.
Yasargil MG. A legacy of microneurosurgery: memoirs, lessons, and axioms. Neurosurgery. 1999;45:1025–91.

Key Techniques of MIN: Ultrasound for Neurosurgery

4

The brain is an ideal ultrasound organ having a complex and fine microstructure causing different sono-intensity. Ultrasound in neurosurgery is a peri-operative concept. It can be used at bed-side (normal unit and ICU), pre- and post-operatively, and intra-operatively. High-end ultrasound technique provides new options for neurosurgery. The complete spectrum of neurosurgical diagnoses presented applicable assistance for therapy in our experience. In 376 intra-operative applications, ultrasound proved to be an excellent neuro-navigation system providing the surgeon with real-time imaging and targeting capabilities. Resection control in 221 tumor cases with targeting in 29 small lesions was very satisfying and in four cases craniotomy correction was possible before opening of dura mater. Compensation of computer-navigation failures was possible in 14 cases preventing possible disasters. The 584 cases of application at the ICU showed a bed-side

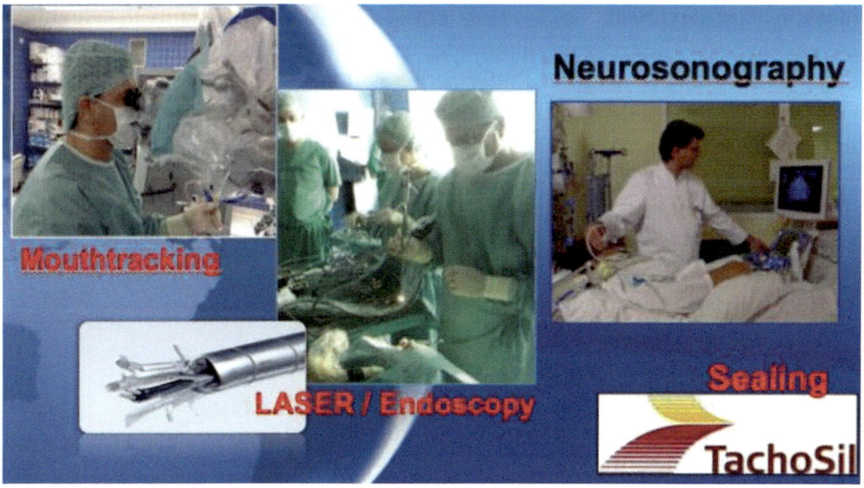

Graph 4.1 MIN key techniques

K. D. M. Resch, *Key Concepts in MIN - Intracerebral Hemorrhage Evacuation*, Key-Concepts in MIN 1, https://doi.org/10.1007/978-3-030-46513-1_4

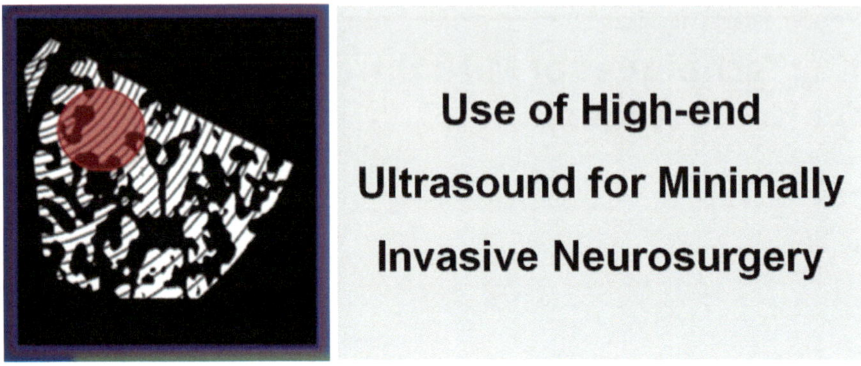

Graph 4.2 ICH sono logo

use, resulting in decrease of risky out-door examination reduce stress for our patients and logistic efforts for the professionals. In summary routine use of high-end ultrasound in neurosurgery enables a broad variety of applications and indications intra- and peri-operatively. Ultrasound contributes to minimally invasive techniques with realisation of good ergonomics in neurosurgery. Most limitations can be avoided by correct usage and sufficient experience and training.

4.1 General Information

Neurosurgery has the privilege to benefit from long time experience and evolution of techniques of many neighbour disciplines using ultrasound since several decades. The purpose of this chapter is to analyse routine use of high-end ultrasound technique in neurosurgery. It does not make any sense to use other quality than high-end in neurosurgery. The reason is, that we have already several high-end imaging present in most cases and additional information will be expected only in the high-end equipment. But then, there will be a lot of results which is not provided by all other imaging techniques.

The main advantages of ultrasound in neurosurgery are:
- Excellent ergonomics!
- Mobile use everywhere in the hospital
- Relative cheap equipment
- Bed-side examination
- No out-door risks for the ICU-patient
- Real-time imaging
- Seeing the pathophysiology within the morphology!
- Minimal logistic efforts
- Minimal personal efforts (one-person-show)
- Intra-operative real-time:
  - navigation (orientation)
  - targeting: angle, depth, approach

  - imaging: acute changes
  - imaging from surgical perspective
- No radiation
- One hand information flow: imaging > diagnosis > therapy!
- Compensation of visualization problems in MIN!
- Compensation of computer-navigation pitfalls!
- Vascular “finger-print” of lesions
- Vascular origin of bleedings (malformation, tumors)
- Resection/evacuation control
- Clipping control
- Adaptation of craniotomy (size and position)
- Time spare in emergency cases!
- First evaluation of TBI (vascular reaction)
- Fast brain-death diagnosis
- High innovation capacity

Ultrasound in neurosurgery is a **peri-operative concept**. It can be used at bed-side (normal unit and ICU), pre- and post-operatively, and intra-operatively. It can be also used at each place you need, like CT-room, ambulance and emergency-room. The equipment comes to the patient, where-ever he is in the hospital, avoiding risky out-door examinations, especially in ICU-patients, softly, easy and repeatingly whenever needed.

Todays ultrasound-equipment is rather small and mobile with excellent ergonomics qualities compared to usual imaging techniques. The equipment fits easily into the OR environment and procedure as well neither disturbing the procedure nor the surgeon. However, first clinical training should be done at the ICU, best in craniectomy cases using a maximal sized window. Intra-operative use success will depend largely from this ICU localised training. The daily application at ICU should be a simulation of the operative situation regarding position of the user to the patient and mentally to be well prepared for the stress situation during surgery (Fig. 4.1 and Graph 4.3).

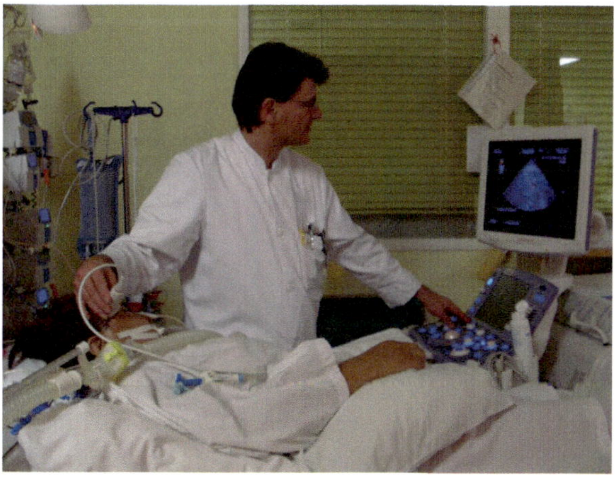

Fig. 4.1 Bed-side examination at ICU

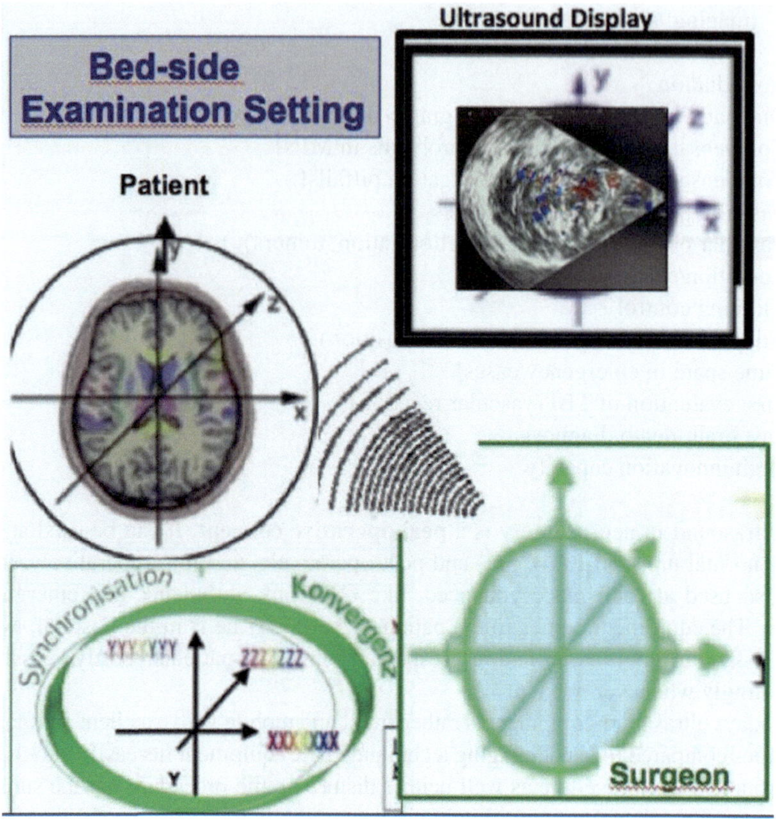

Graph 4.3 Bed-side setup: convergence and orientation

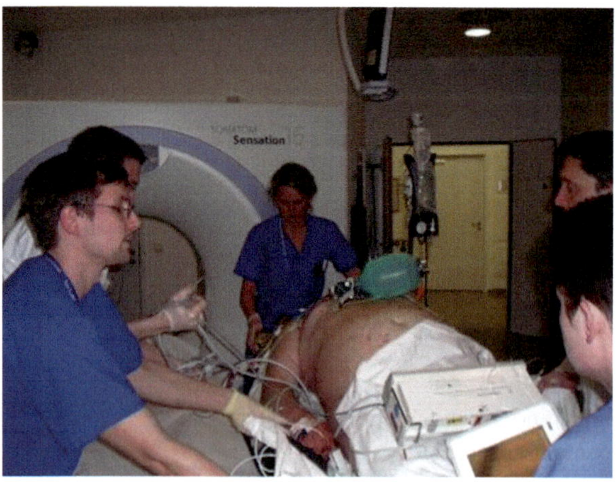

Fig. 4.2 Logistic and personal effort: ICU-patient at CT room

The typical scenario during an CT exam of an ICU-patient (Fig. 4.2). It is danger for the patient, it's expensive, time-consuming and needs a lot of professionals. About half of the CTs might be spared by bedside ultrasound. Many clinical courses and important changes can be monitored and visualized (Fig. 4.1).

The majority of sono examinations are done at ICU in bed-side setting. This is the best training for the intra-operative application. This training will work only if the setting is organized close to that during surgery. The position of the patient and that of the sono user, as well that of the sono image on the monitor need to be in convergence. All coordinate systems (patient, surgeon, image) have to be converged and synchrony. This enables to use the ultrasound as a navigation system and neuropsychologically as training effect for the surgeon (Graph 4.3). The navigation will function only if it can be realized intuitively, using natural neuropsychological functions without the need of complex cognitive control. During the stressful intra-operative situation cognition should be directed towards surgery and ultrasound use should not disturb this flow but being integrated easily.

4.1.1 Bedside Sono-CT

4.1.1.1 Reduction of CT/MR Examinations by Highend-Neurosonography

High-end ultrasound technique may provide new options for neurosurgery. To critically determine indication for CT/MR examinations, especially in the ICU, it has been studied in a primary series, what can be done by actual sonography equipment. Can CT/MR examinations be reduced or partially substituted with the advantage to bringing the technique to the patient and diminish logistic problems and costs in future?

In a series of 1022 consecutive high-end ultrasound investigations there were 584 examinations at ICU of which 92 cases done with the aim that CT or MR examinations might not be necessary thereafter. The ALOKA 5000 equipment was brought to the patient at bedside. A trans-cranial sono-probe 5–7.5 MHz was used in cases with a bone defect of at least 2 × 2 cm.

The sonography images showed an excellent slice-anatomy comparable to that of CT and MR. In this small series CT or MR examinations could be substituted by sonography, at least, the indication for CT/MR could be determined more precisely and diminished in number. Moreover, the logistic problems of CT/MR for ICU patients were reduced. Patients with edema and with critical ICP conditions could be monitored by "bedside sono-CT" with less danger. Additionally, the monitoring of physiological and patho-physiological parameters was always done in the same examination, presenting sonography superior to CT and MR in this selected group of patients.

For reduction of logistic- and financial problems as well as stress for critical care patients, it should be considered to monitoring them by high-end sonography imaging at bedside reducing CT/MR examinations. We bring the equipment to the patient and repeat the examination any time. Quality of imaging is comparable to CT/MR and physiological parameters can be monitored in the same investigation.

In a case of a cerebral infarction (Graph 4.4) with a craniectomy to reduce ICP, a change of the bulging skin flap of the craniectomy (B + C) can be diagnosed safely by bed-side ultrasound using the large ultrasound window.

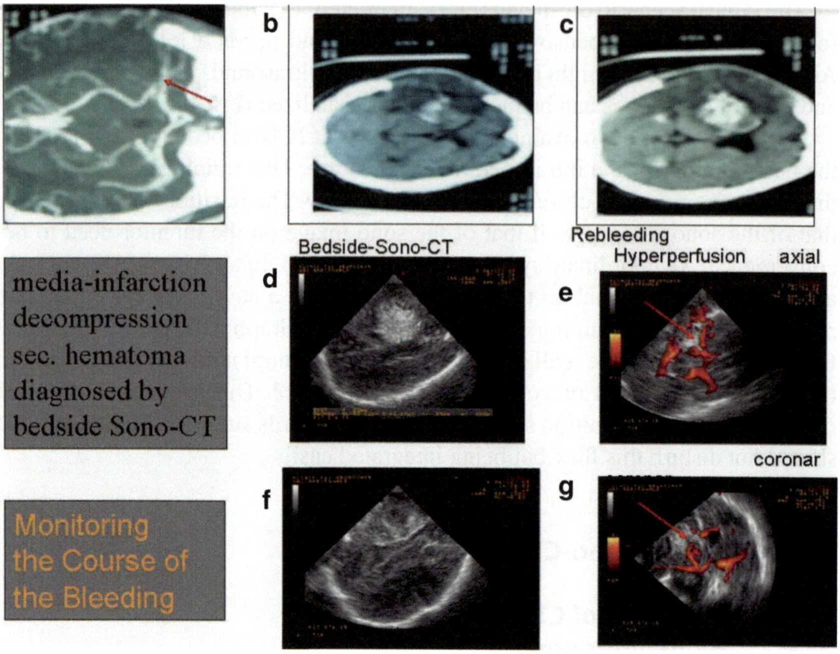

Graph 4.4 Emergency ultrasound-CT at ICU

In such a case, we have additional the best learning model to start clinical neurosonography. In this case the reason of the bulging was a bleeding (B–D) into the infarction. We see also the shift compressioning the lateral ventricle (D). But moreover, in the C-mode a hyper-representation of strial vessels (E + G) is present indicating a control imaging of the vessels by repeated CTA or even DSA.

All morphological changes and moreover many pathophysiological data can be imaged by ultrasound fastly and repeatedly whenever needed with a minimum of effort and without any risk for the patient. The danger of an obstruction of foramina Monroi can be monitored frequently (C) and the course of resorption of the hematoma can by followed as well as the edema resorption (D + F). All this information cannot be acquired better with any other techniques, regarding efforts and costs. Rapidly measures can follow without time-delay and logistic efforts (Graph 4.4).

With this concept and usage of ultrasound at the ICU, half of the out-door CTs can be spared, and at least an indication for out-door CT can be evaluated. The therapy becomes cheaper and more flexible, fast and effective.

4.2 Special Imaging Conditions of Neuro-Sonography

The brain is an ideal ultrasound organ having a complex and fine microstructure causing different sono-intensity. We have 3 standard sono windows (Graph 4.5) fronto-laterobasal (yellow), temporo-basal (blue) and sub-occipido-median

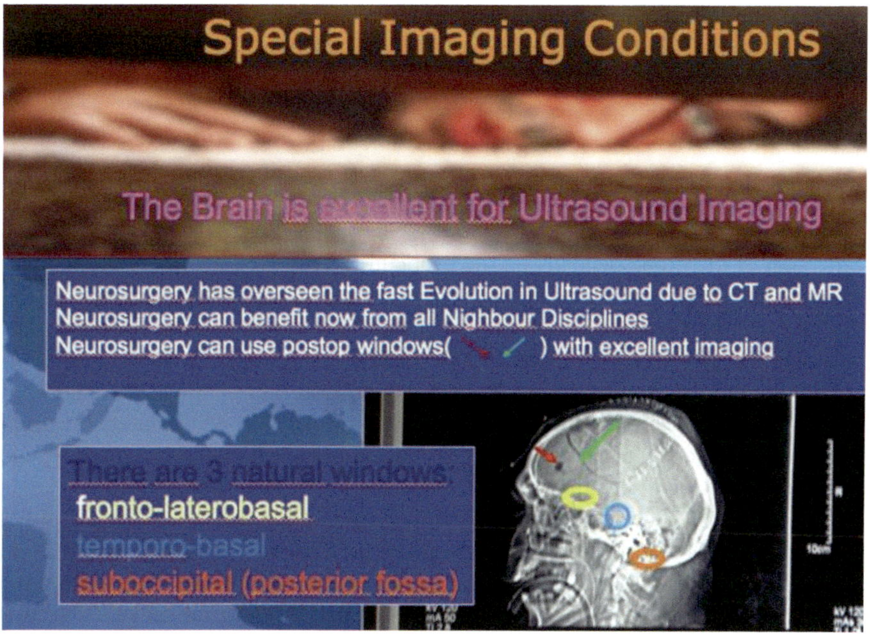

Graph 4.5 Transcranial windows for ultrasound

(orange). But we have moreover post-op burr-holes (red) or craniotomy gaps (green) and even a big craniectomy window (Graph 4.5) enabling imaging quality close to intra-op ultrasound.

The indications and to make use of post-op artificial sono-windows make the difference to the use of ultrasound by neurosurgery compared to neurological use. Imaging usually can be far better due to the advantage of bone-free windows.

4.3 Equipment

4.3.1 Starting the Machine

The practical and technical basics and moreover the physical understanding cannot be presented here in short and should be acquired during a sono-course or by viewing the literature. However, a quick-start guide can be given anyway:

It's helpful to use only one or two machines and to know them perfectly. Different equipment from most companies differ markedly, but some principles are always given. It is essential to know and control if your machine has the **presets** for the type of examinations and the organ you want to explore. Without the **presets for head and brain** one must not start, it will lead to frustration. Also, the probes elected for the problems to image need to be prepared in the machine and by the technicians of the company.

After starting the machine and its computer, one should give the patients data to the computer otherwise the imaging is worthless, documentation is of great importance for clinical work as well as for scientific use in the endless amount of data one will face soon.

4.3.2 Sono-Probes

Selection of the probe depends on the problem to be solved and the kind of sono-window. The Burr-hole probe and the sector-probe are used intra-operatively, TCD-probe extra-operatively and the Endo-probe intra-operatively through the endoscope (Graph 4.6).

The probes of an ultrasound equipment contain half of the knowledge that is applied! The election of the probes must be done carefully according to the problems you want to solve. The use of unpropriate probes will end in bad imaging and frustration. Our own selection of probes is made to support MIN.

The ALOKA 5000 and the Alpha 7 (Hitachi/ALOKA) with four small probes offered a basis to improve minimally invasiveness in our discipline: in this first series TCD probe (2.14–3.75 MHz) was used in 613 cases mainly at ICU; the small part sector probe (3.8–7.5 MHz) was mainly applied intra-operatively in 209 cases and the burr hole probe (3.75–7.5 MHz) was also mainly used intra-operatively in 94 cases. A trans-endoscopic mini-probe (360°, 6F + 8F, 10–15–20 MHz) was used with strictly indication in 75 cases. To assist, MIN linear probes are not able to show enough landmarks in small approaches, whereas sector-probes enable a wide-angled area with sufficient landmarks and anatomical overview and orientation.

4.4 Modes

There are **3 main imaging modes** being the core of use in most applications. They should form the basics of clinical use also in neurosurgery.

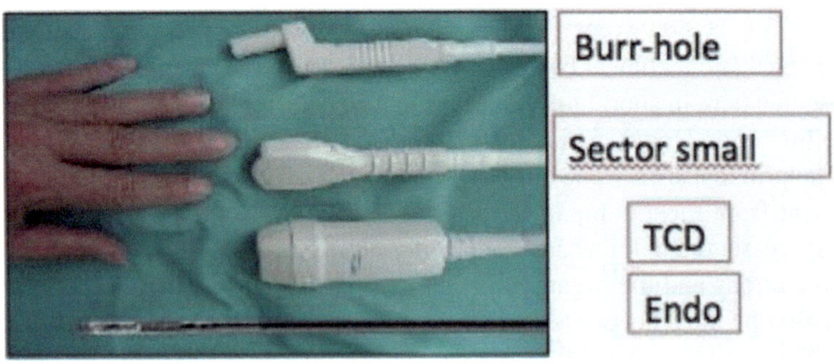

Graph 4.6 Neuro-Sonoprobes for MIN

Typical axial scan of midbrain-level in B-mode showing only a gray scale image. Neural tissue presents dark and connective arachnoid tissue as well as bone areas are visible bright due to high echogenity (Fig. 4.3).

Typical axial scan of midbrain–level in C-mode (colour-mode, duplex-mode) presenting the main vessel of anterior and posterior circulation. Vessels are in colour, in red with flow directed to the sono-probe, in blue with flow directed away from the sono-probe. The direction of flow is visible (Fig. 4.4).

Typical axial scan of thalamus-level in triplex-mode. A cursor is placed in the M1-segment of middle cerebral artery, angle of the cursor is adapted to the vessel. The velocity is shown (yellow) and several flow parameters are visible quantitative (Fig. 4.5).

Typical axial scan of midbrain-level in PW- (semi-3D) mode sowing basal vessels in high-sensitive near immobile flow appearance giving a 3D impression of the vessels and presenting the velocity by brightness of red to yellow scale (Fig. 4.6).

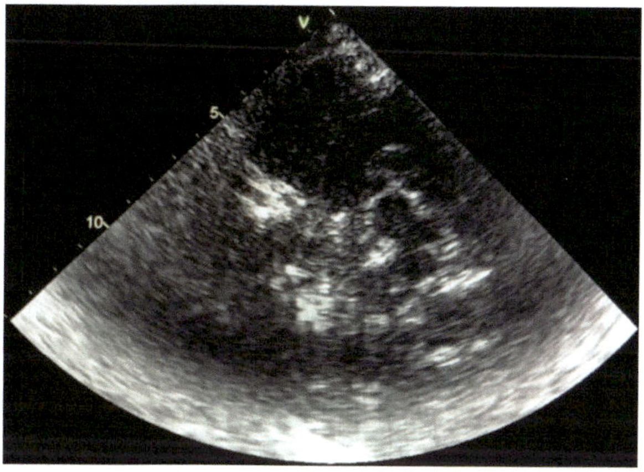

Fig. 4.3 B-mode (Brightness)

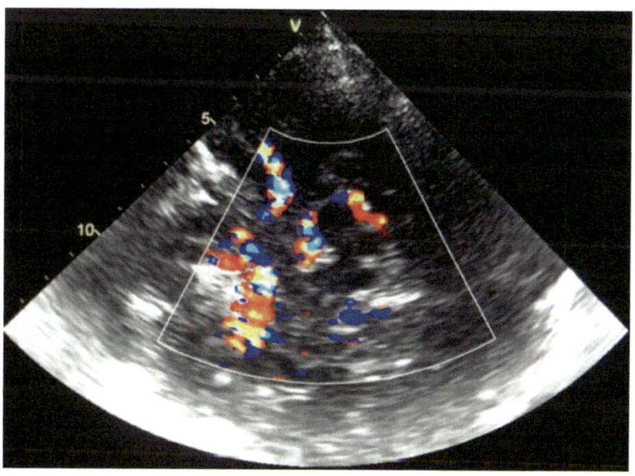

Fig. 4.4 C-mode (Duplex)

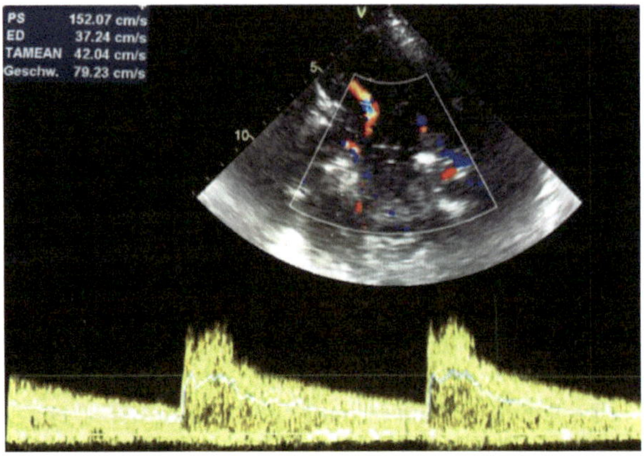

Fig. 4.5 D-mode (Triplex)

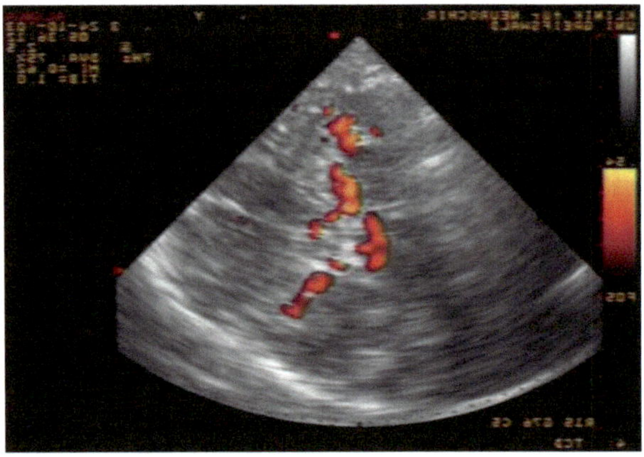

Fig. 4.6 PW-(semi-3D) mode (high-sensitive variant of duplex)

4.4.1 Functions: (Optimation of Image)

The image can be optimized by several parameters which should not be neglected because the clinical information of the image may depend in some cases on them: Focus, depth (Zoom), angle correction, gain and duplex-window can be learned in each course of national and international organisations (EFUMB; WFUMB...).

4.4.1.1 Advanced Functions

Combination with navigation; vibrography; contrast sono; sono-assisted LASER surgery. All these complex and combined techniques cannoet be presented within this volume. Only trans-endoscopic ultrasound will shortly be mentioned due to its meaning in navigation of neuro-endoscopes (s. **E**ndo-**N**euro-**S**onography, below).

4.5 Scan Geometry

The understanding of scan geometry is essential to understand the series of images, which are generated by moving the probe at the sono-window location (s. Graph 4.5). The sono slices can be enabled by angling the probe in small sono windows or by moving the probe parallel along the skin in large windows (craniectomies). Angling on the frontal fontanel (window) are usual in pediatric neuro-sonography, using a perfect trans-cutan window. Scans may be close to CT and MR (orthogonal triplanar) and so far easier to compare with them and to be interpreted. But often they are oblique, becoming difficult in interpretation (Graphs 4.7 and 4.8).

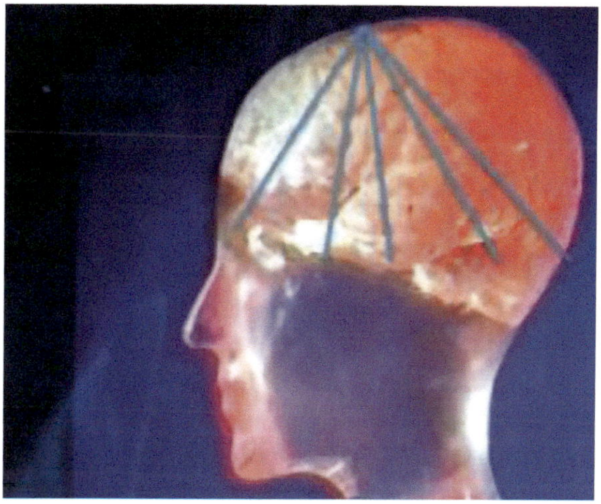

Graph 4.7 Coronar series of sono slices

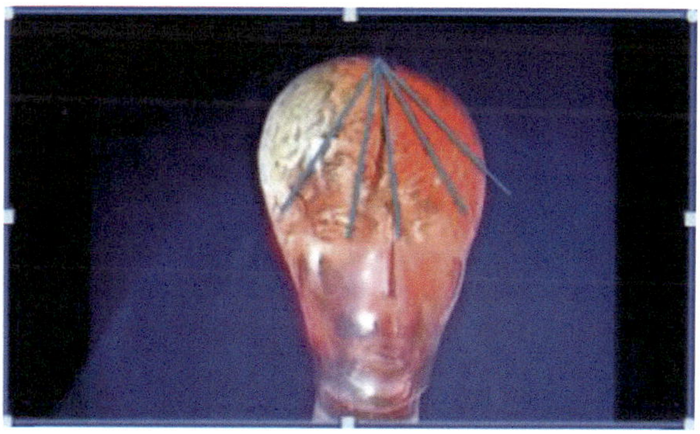

Graph 4.8 Sagittal series of sono slices

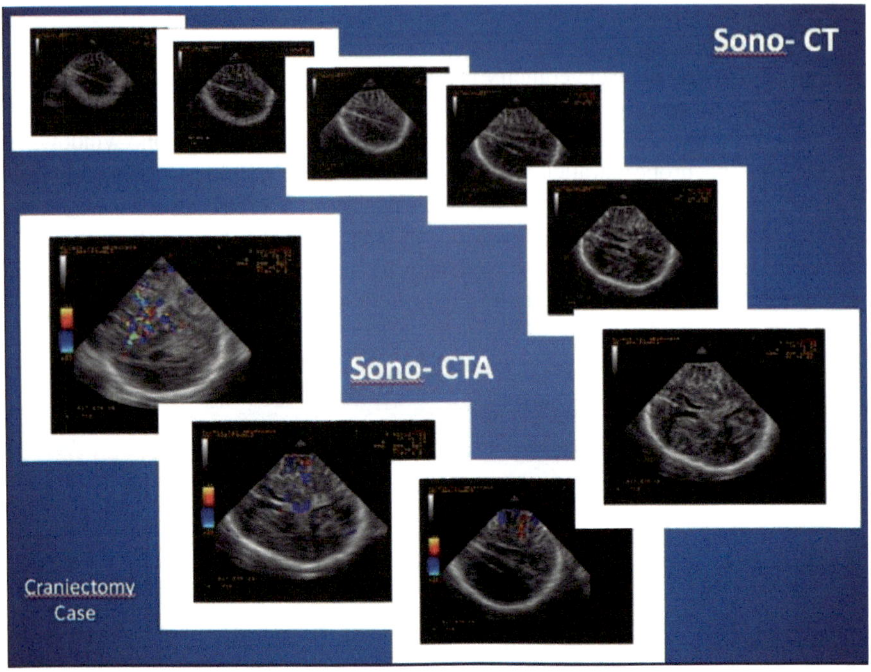

Graph 4.9 Sono-CT and sono-CTA

Technical correct scans will present slices and images close to the information of CT and CTA (s. Graph 4.9). But more over they can show also many patho-/physiological parameters. One needs to know first the triplanar anatomy of some standard levels to be able reading the scans and to have a clear orientation. Convergence of coordinates between patients' anatomy and image-anatomy must be adjusted at the beginning of examination (Graph 4.3). As mentioned above, it is wise to use always a similar examination-setting regarding this convergence of coordinates. The coordinates and orientation of the sono scan should be adapted to the patients' position. Patient and surgeon should "look" to the same direction, also at bed-side, which makes understanding and orientation much easier, intra-operatively it enables a safe navigation ability.

Neuro-psychologicaly these rules are essential to become well and fast trained and to become your own surgical neuro-radiologist for safe and fast imaging, targeting and navigation. Some practical, preliminary examples of triplanar "Sono-CTs", and "Sono-CTAs", shall not replace a neuro-sono-atlas, which however does not exist actually! (Graph 4.10).

This level model is just one system, that can be applied, however it has been helpful in the daily work. The importance is that one should have a system to know how the sono scan is running through the head/brain and which anatomical key structures we have to expect.

The parallel sono scans are usual possible in craniectomy patients at ICU and in surgical standard-approaches intra-operatively. Transcranially there are typical

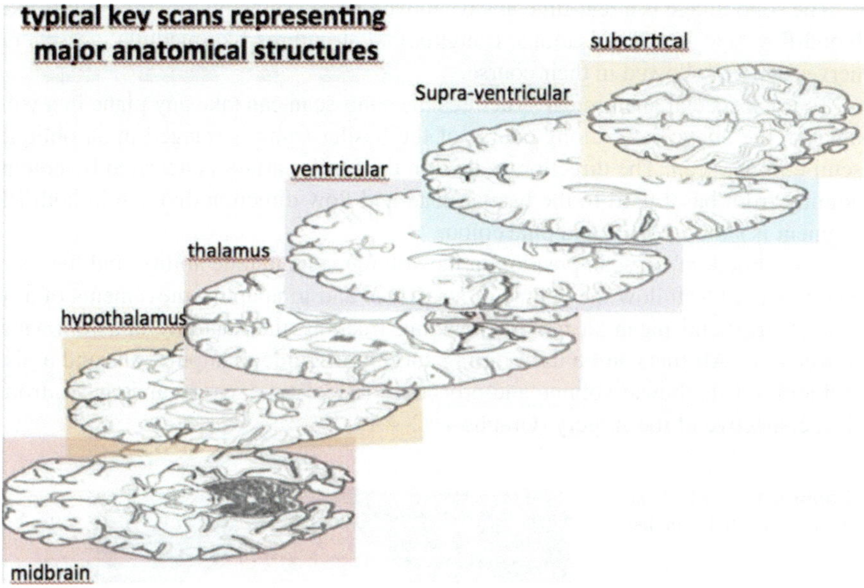

Graph 4.10 Sono-Levels Model

Graph 4.11 Sono-Tracking of Basilar Artery

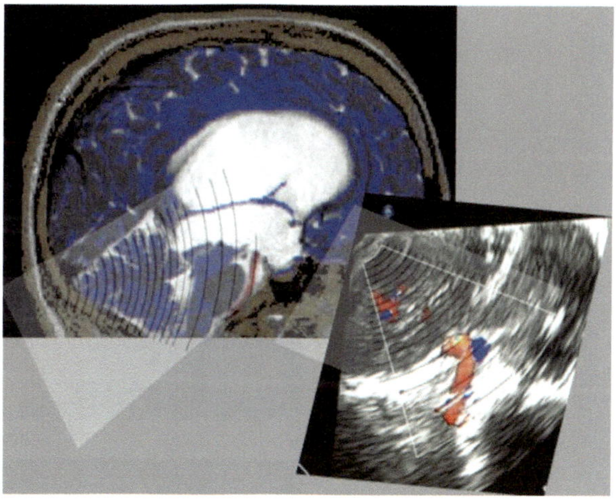

bone-windows (Graph 4.5) or the fontanells in babies (usual until 24 months) making an angling scanning necessary (Graphs 4.7 and 4.8).

In contrast to CT and MR the sono levels can be chosen freely, however it makes sense to do them along CT or MR for easier comparison and for better orientation and identification of anatomical key structures.

There is moreover the possibility to capture a structure by free adaptation and angling to have it best imaged within the scan (Graph 4.11). By this each structure, especially vessels can be tracked.

The sono image is a real-time always moved image showing moving liquids like blood flow and liquor pulsations. Longitudinal structures like medulla, vessels or nerves can be followed in their course.

To track special anatomical structures the sono-scan can take any plane in a volume. In the above case a long course of the basilar trunk is imaged in an oblique semi-coronar scan. The direction of flow in the basilar artery is detected by colour coming from basal until to the basilar head and flow direction division in both P1 segment is shown by red and blue colour.

This "freedom" in scan position in the volume is an unique ability, and the scan can be varied to follow the course of structures and to adapt to movements of any kind. Ultrasound means a moving imaging tracking the anatomy and following movements. All this can be used during surgery to guide manipulations and assist and monitor the howle volume and procedure in real-time, easily and cheap, from the perspective of the surgery (Graphs 4.12–4.14).

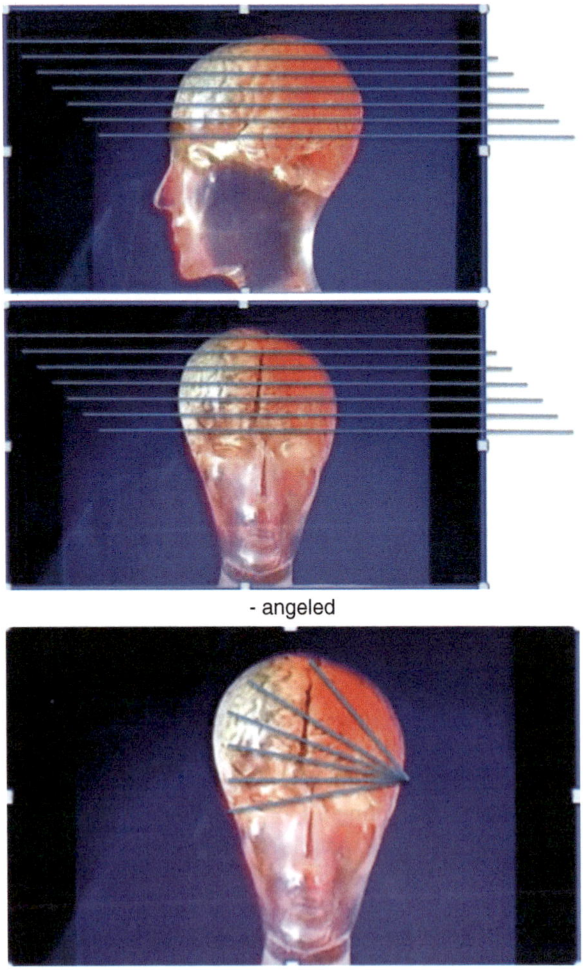

Graphs 4.12–4.14 Axial Slicing—parallel—angled

4.6 Clinical Neuro-Sono Anatomy

4.6.1 Axial

4.6.1.1 Midbrain Level

The midbrain-level is the first approached level, providing a basis for further orientation. Once the midline and the midbrain with its typical butterfly-shape is identified, and the coordinates in the space (anterior, posterior, left and right) are recognized, the orientation of the image in correlation to the patients' head is made, the sono-image can serve for navigation (Figs. 4.7 and 4.8).

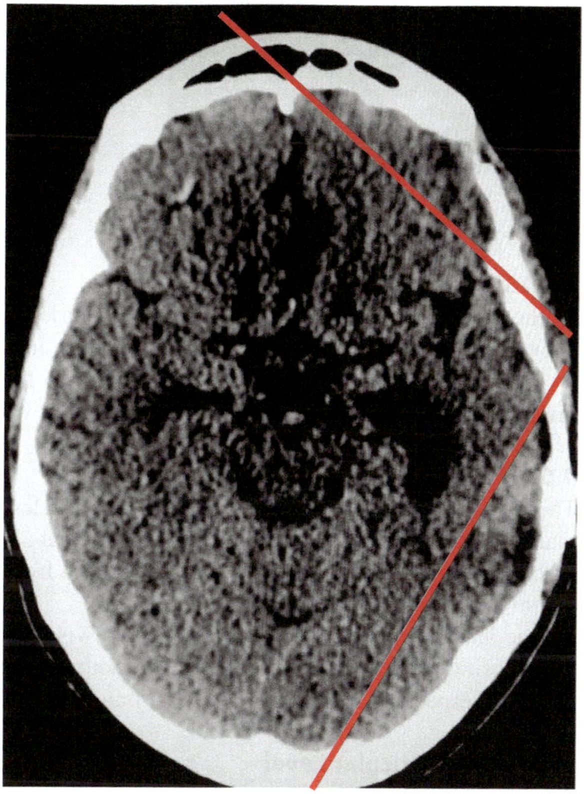

Fig. 4.7 Midbrain-Level CT

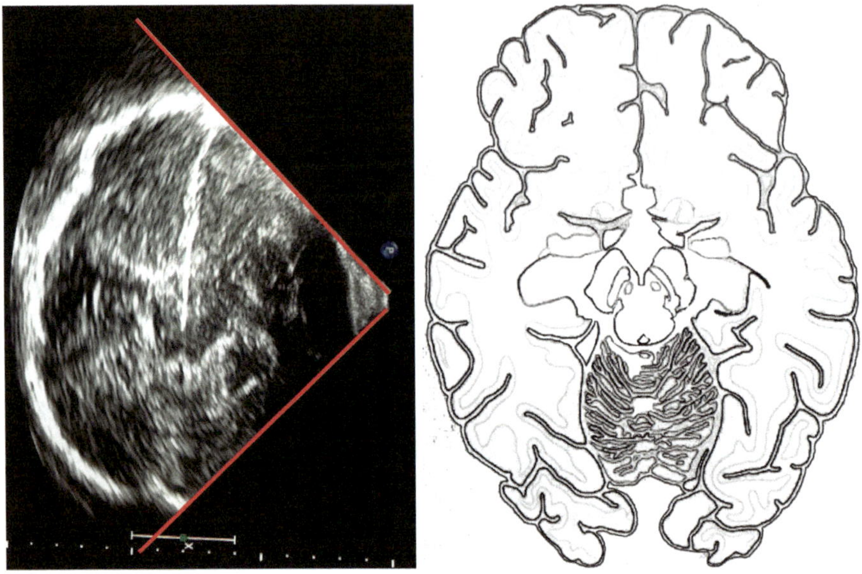

Fig. 4.8 Midbrain-Level Sono

4.6.1.2 Hypothalamic Level

At the hypothalamic level all supratentorial ventricles are present. Sylvian fissure comes into sight and again, the midline is the most important guideline.

Many sublevels can be realized and rotation of the probe will vary the image markedly, enabling targeting of certain structures or volumes (Figs. 4.9 and 4.10).

4.6.1.3 Thalamus Level

The thalamus-level shows all supratentorial ventricles and the insular cistern. The diameter of the third ventricle can be measured, but also frontal horns. Shifting are precisely recognized and measured by comparing frontal falx versus septum pellucidum position.

The internal capsule and pyramide area are visible. Note the proximity of pyramide position and frontal horns, reflecting motoric-symptomatic and severe hydrocephalus, especially in children (Figs. 4.11 and 4.12).

4.6.1.4 Ventricular Level

The ventricular level is dominated by the lateral ventricles and septum pellucidum forms together with frontal and occipital falx the midline. Shifting will be easy recognized. Frontal, parietal and occipital lobes form the shape of the brain at this level. In this case an surgical approach frontal left disturbs the symmetry in the CT, but is not well present on the ultrasound scan, whereas the parenchyma defect of fronto-parietal opercula is well visible (Figs. 4.13 and 4.14).

Fig. 4.9
Hypothalamus-Level CT

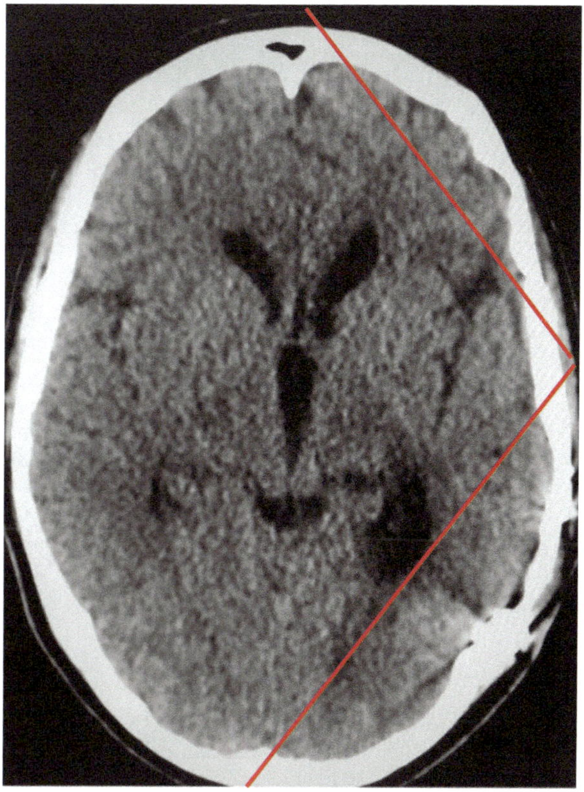

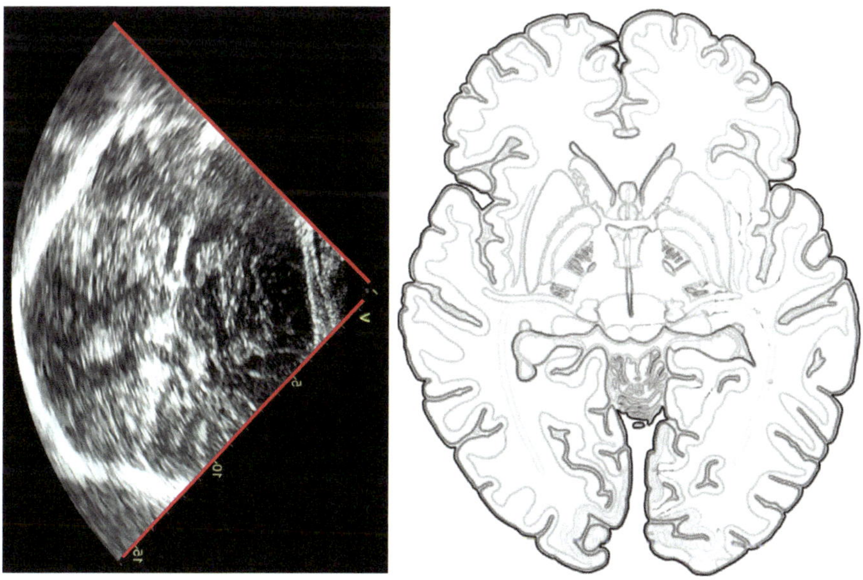

Fig. 4.10 Hypothalamus-Level Sono

Fig. 4.11 Thalamus-Level CT

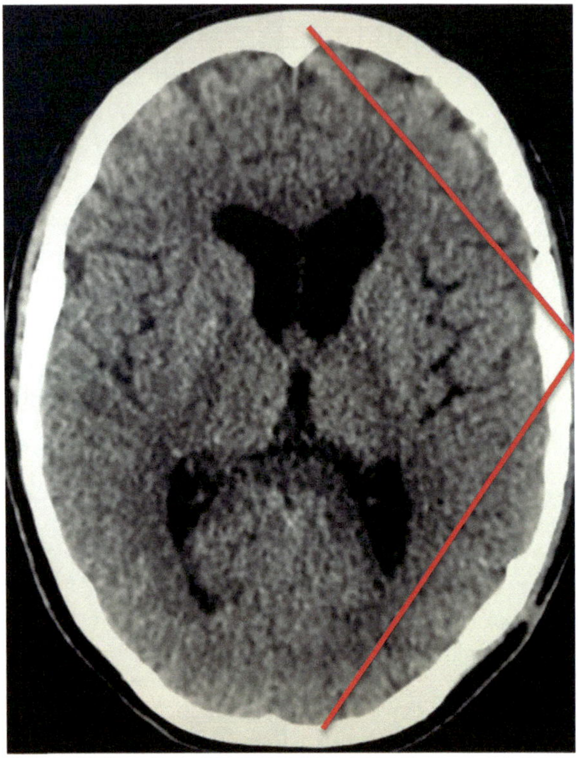

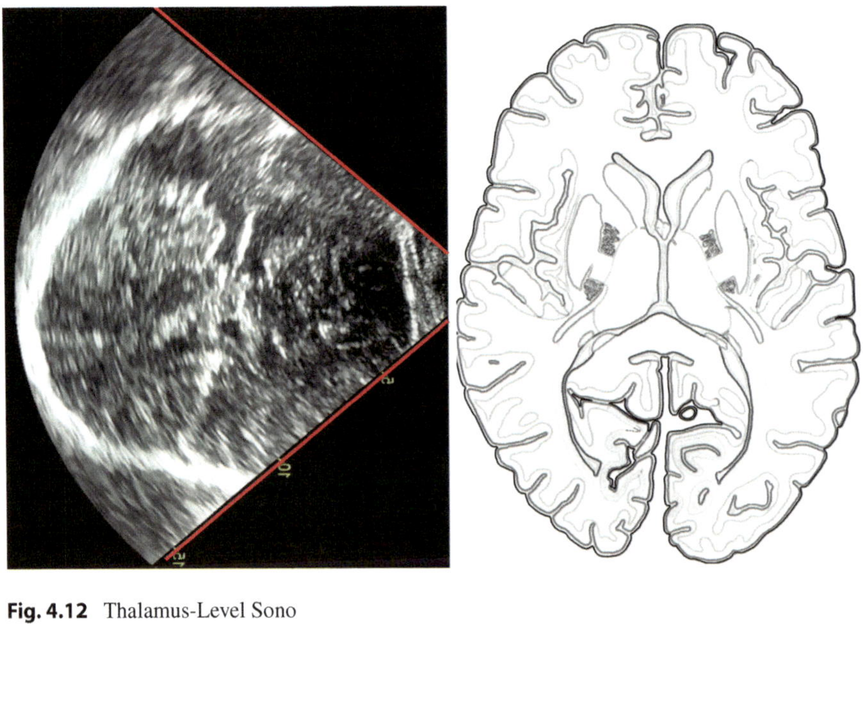

Fig. 4.12 Thalamus-Level Sono

Fig. 4.13 Ventricle-Level CT

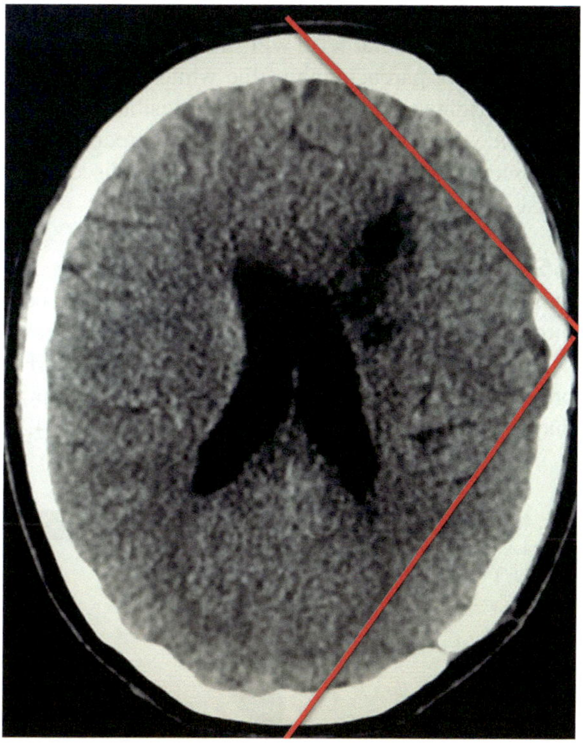

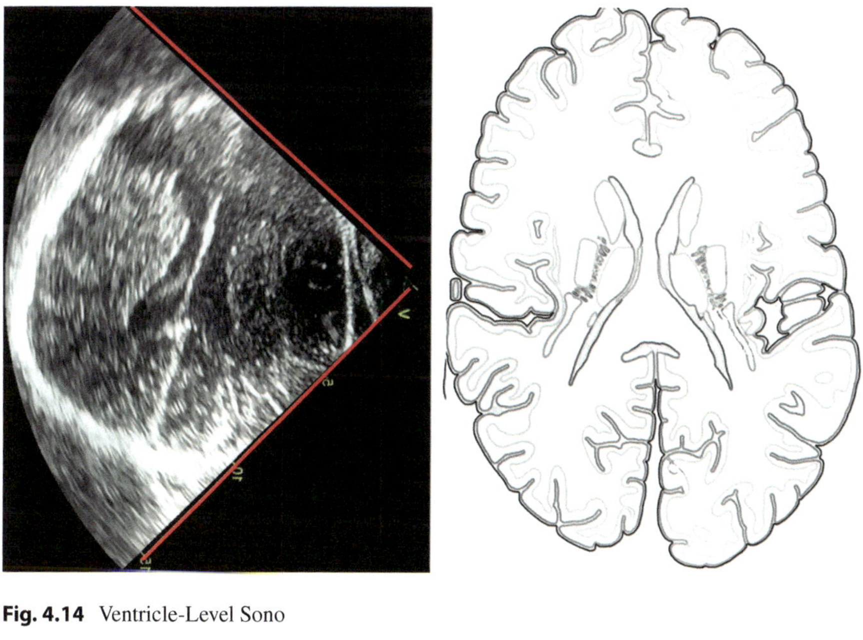

Fig. 4.14 Ventricle-Level Sono

4.6.1.5 Supraventricular Level

The supra-ventricular level will show only superior remnants or no ventricles. Here the semi oval parenchyma of the white matter will be best visible and compared between right and left side. Midline is visible in one line through the head sagittally. The sulco-gyral pattern can be evaluated precisely and asymmetrical effect are easily discovered. Parenchymal defects can be discovered better than in CT (Figs. 4.15 and 4.16).

4.6.1.6 Subcortical Level

Subcortical level will best present the sulco-gyral pattern and the midline is going through all the way sagittally. Asymmetries are easy be visible and parenchyma defects that reach into the cortex like infarctions. Here a slight difference in the presentation of sulcal liquor-space can be seen, representing remnants of swelling during the craniectomy time on the left side (Figs. 4.17 and 4.18).

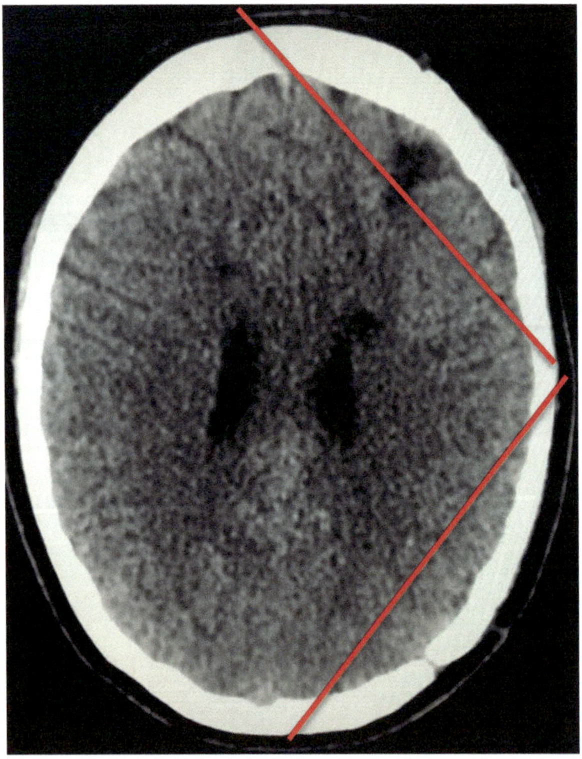

Fig. 4.15 Supraventricular-Level CT

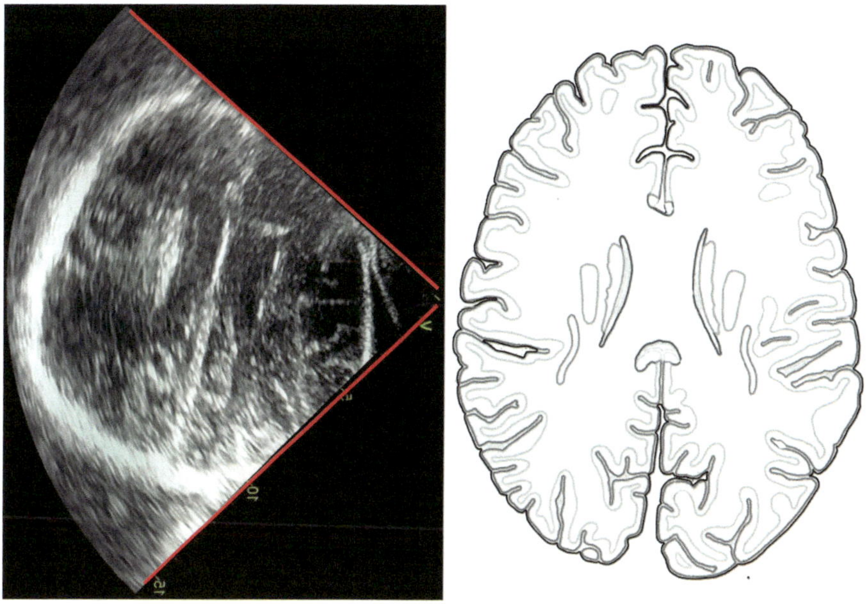

Fig. 4.16 Supraventicular-Level Sono

Fig. 4.17 Subcortical-Level CT

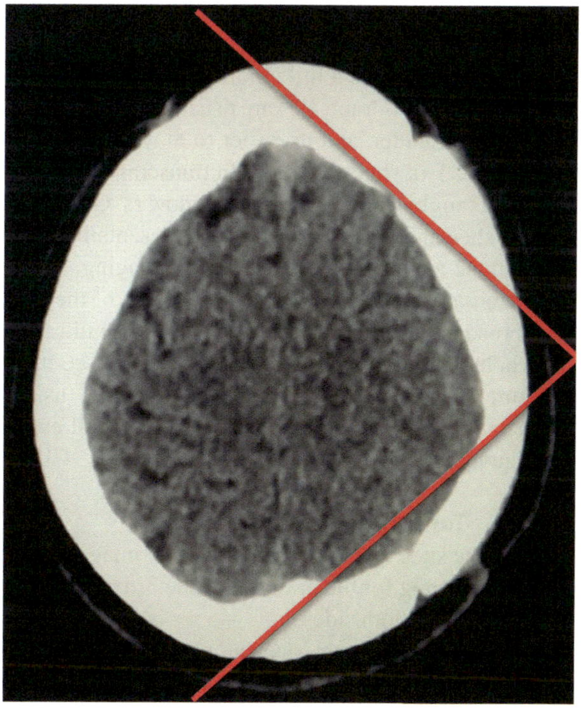

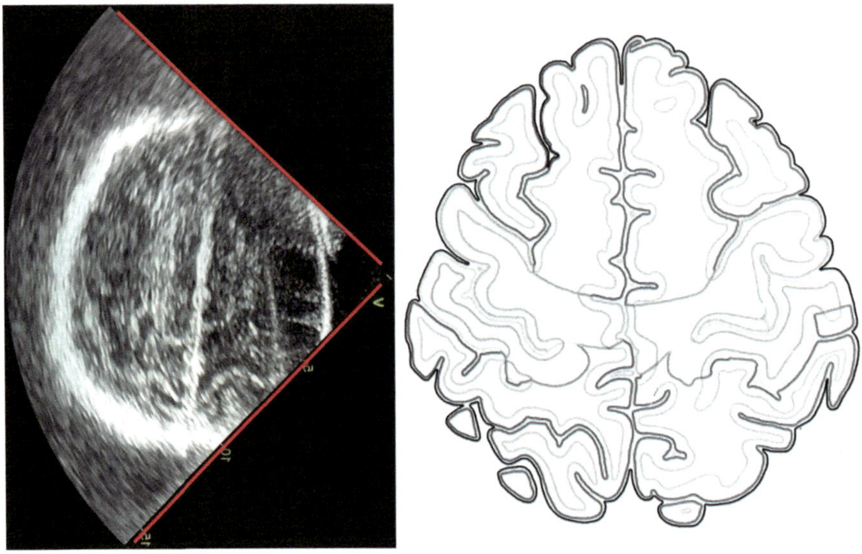

Fig. 4.18 Subcortical-Level Sono

4.6.2 Coronar

4.6.2.1 Frontal Level

The coronar slices are a little more difficult than the axial once. One reason is, that bone-windows (s. Graph 4.5) are usually transversal extended. In craniectomy cases this does not matter, and its easier to move parallel (s. Graph 4.15) for scans comparable to CT in slice-direction. In transcranial ultrasound through a bone-window one has to angle within the small window (s. Graphs 4.16 and 4.17). This causes an oblique slice and change of geometric presentation but can be compensated by playing with the angle and having a mobile imaging.

The broad-angled image causes a torsion of the imaging geometry. However, by such broad angled view one can enable a large ultrasound view with an overview of the whole cranium and brain. This enables to see many landmarks and anatomical structure an provides a navigation capacity, using the slice as a land-map. Parenchyma, midline and ventricles can be well evaluated. Pulsation of the tissue and liquor are visible during examination (Figs. 4.19 and 4.20).

4.6.2.2 Tentorial Path Level

Angling the slice a little dorsally gives an image corresponding to the middle of the tentorial opening for the brainstem. The 3rd ventricle comes into sight and the foramen Monroi on both sides.

In the ultrasound scan, one can play with the angling enabling to see many sublevels and following anatomical structures through the intra-cranial space. This

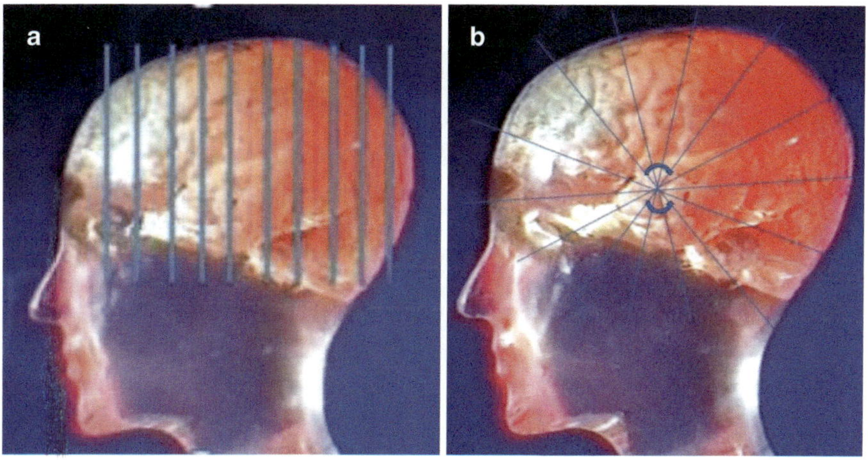

Graph 4.15 (**a**, **b**) Parallel/rotation coronar scans (only after craniectomy)

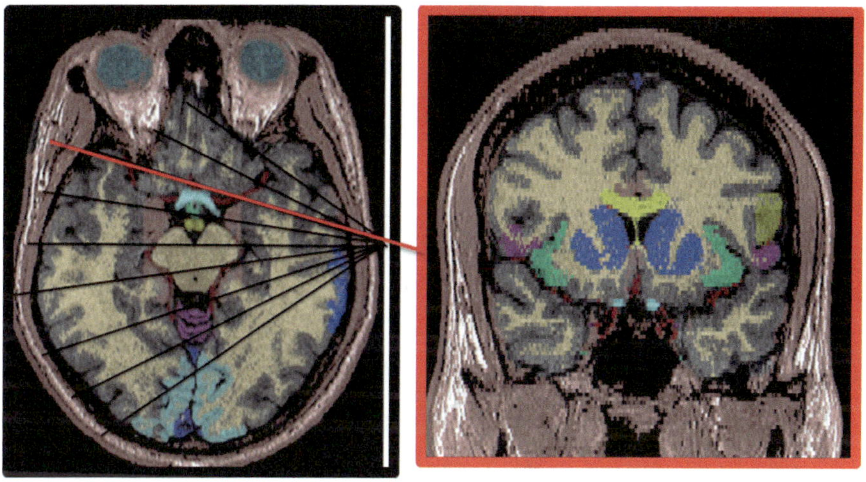

Graphs 4.16 and 4.17 Temporo-basal window and angled coronar scans

moving imaging, comparable with a video, giving this imaging an unique evaluation property.

The transgression of the brainstem with the peduncles and the pyramide tracts are present. The cerebellum and the plexus of the 4th ventricle is imaged, while the fine-tuning present the 3rd ventricle barely, but view degrees angling will change this. The strong signal of the falx presents always the midline, but also shiftings. Angling will give parallel slices frontal to occipital (Graphs 4.16 and 4.17), and rotation (Graph 4.15b) of the probe cause a predominance supratentorial inopposite to infratentorial between coronar and axial slicing (Figs. 4.21 and 4.22).

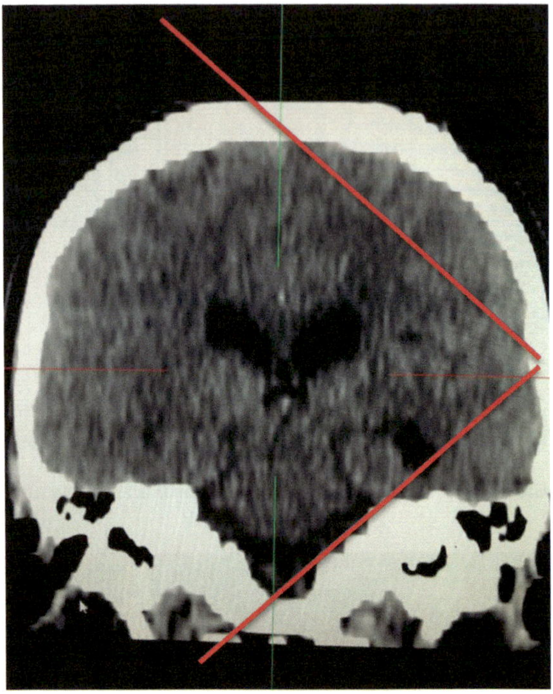

Fig. 4.19 Frontal-Level CT

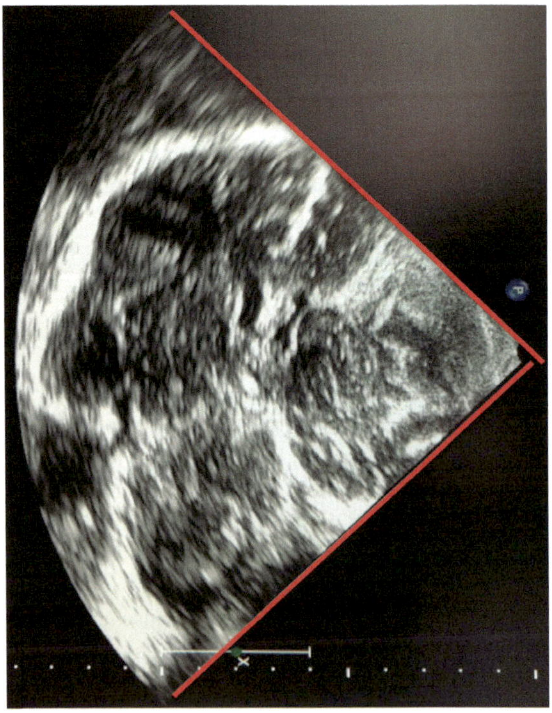

Fig. 4.20 Frontal-Level Sono

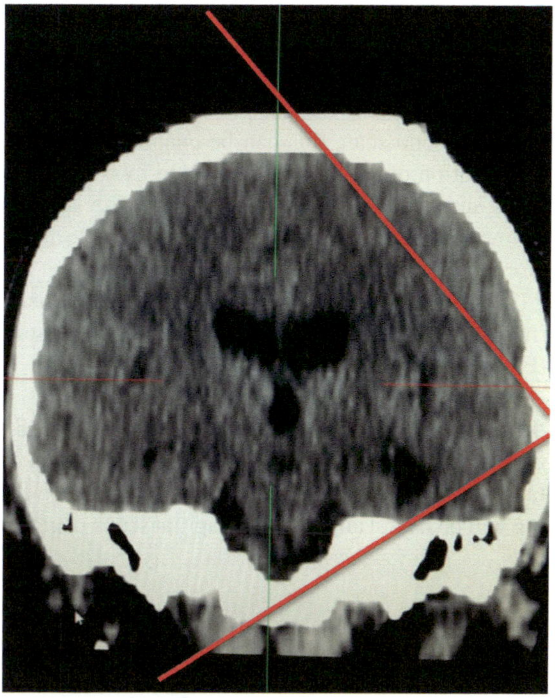

Fig. 4.21 Tentorial-Path-Level CT

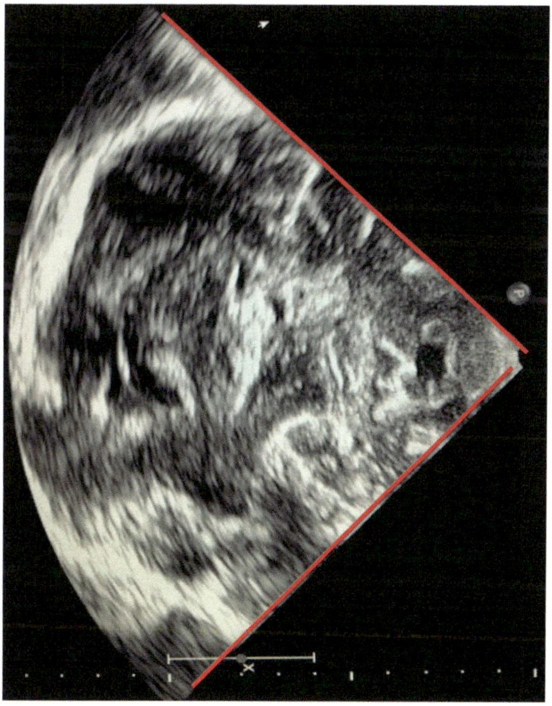

Fig. 4.22 Tentorial-Path-Level Sono

4.6.2.3 Parietal Level

The parietal level shows ventricular trigone and 4th ventricle on one plane, however, it depends on the rotation of the TCD probe (Graph 4.15b).

The pineal gland may be present hyperdens if it is within the scan.

In the sono-scan, this can be enabled by playing with the angling online. The combination of angling vertically and by rotation (Graphs 4.15, 4.16 and 4.17) can focus and target a certain structure or some structures within a chosen plane. Best example is the circle of Willis. This allows to image a structure in a functional or morphologic context.

In this sono-slice the 4th ventricle was one focus together with the topical tentorium shape, to have landmarks on one plane, presenting topography and orientation (Figs. 4.23 and 4.24).

4.6.2.4 Sagittal Paramed. Level (Graph 4.18)

Only in babies and craniectomy patients, rarely in other circumstances, some sagittal scans can be obtained. When the fontanels have closed, by TCD probe the imaging will be useful anyway. The sagittal scans in grownups will always be paramedian

Fig. 4.23 Parietal-Level CT

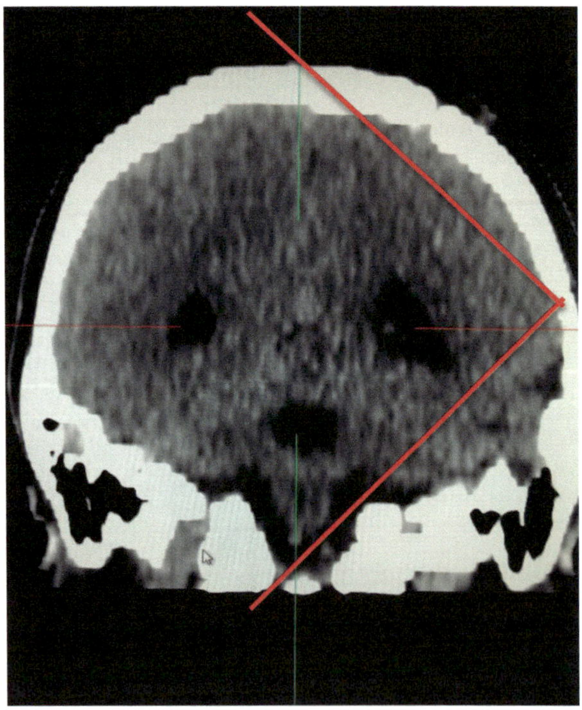

or slightly oblique. This example shows a median-latero-oblique plane, getting the course of the lateral ventricle in this slice.

Playing with the angling will enable to have several focuses in one scan.

However, sagittal scanning is more difficult than coronar and axial once (Figs. 4.25 and 4.26).

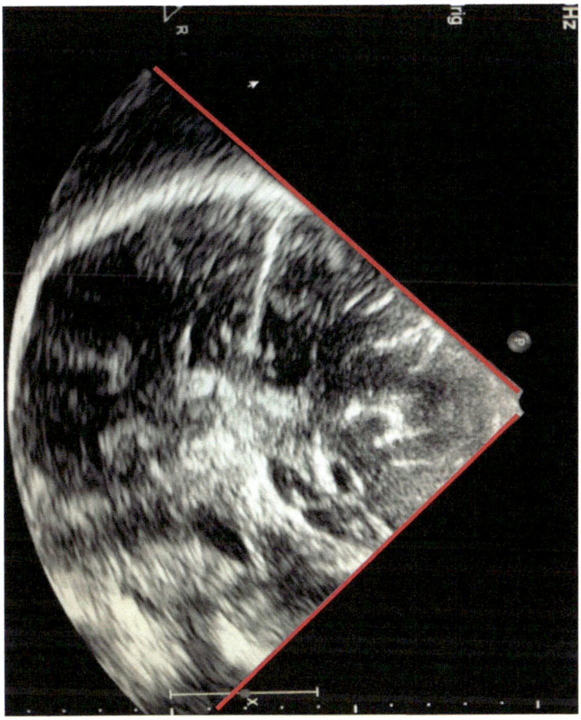

Fig. 4.24 Parietal-Level Sono

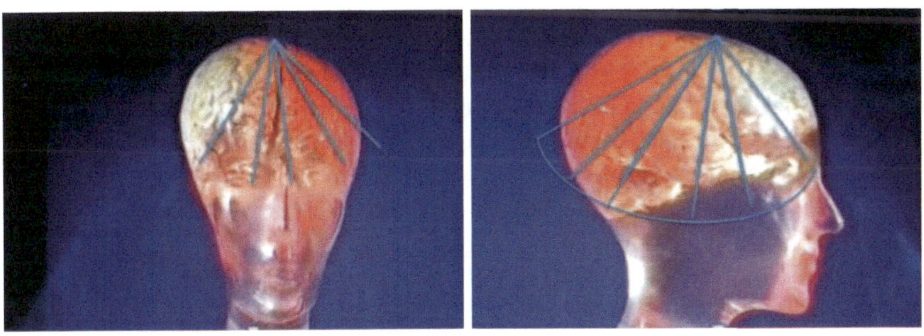

Graph 4.18 Sagittal Levels Model

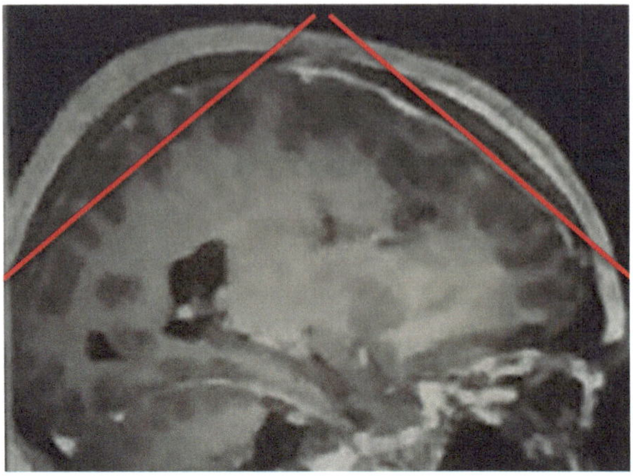

Fig. 4.25 Sagittal-Level MR

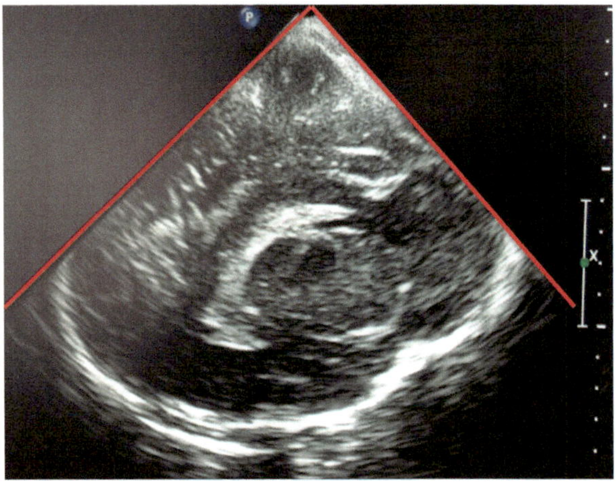

Fig. 4.26 Sagittal-Level Sono

4.6.2.5 CTA Axial Midbrain Level

The duplex-mode (C-Mode), enabling a "sono-Angiography" and a "sono-CTA", is a neuro-application in this sonography, where technique can provide informations in an unique way, cheap, fast and repeatingly whenever needed at bed-side! This tool represents best, that in neurosonography, patho-physiology is presented within the anatomy. This means, a moving imaging in real-time, appearing like a video. The vascularization is visible together with the parenchyma. The imaging of the vessel-tree with visible flow-intensity gives an intuitively presentation, whats going on in this brain at the moment of viewing.

We see here the complete circle of Willis and the over-representation of the insular vessels left side after craniectomy. This represents the over-therapy craniectomy causing craniectomy disease! (Figs. 4.27 and 4.28).

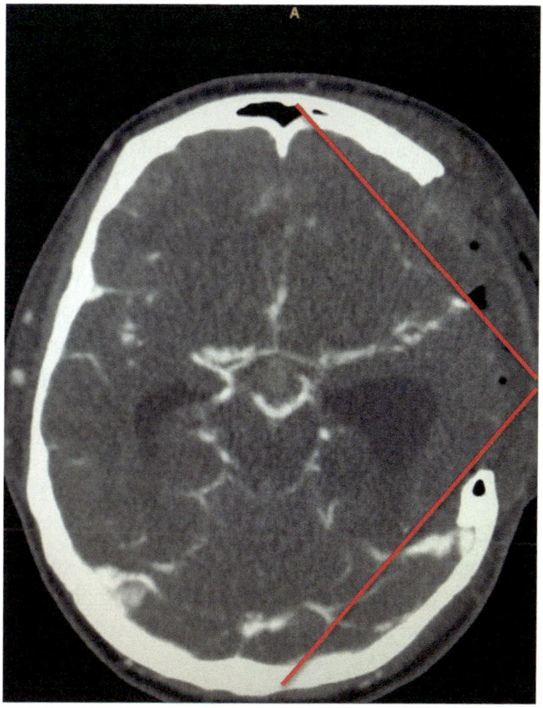

Fig. 4.27 Midbrain-Level CTA

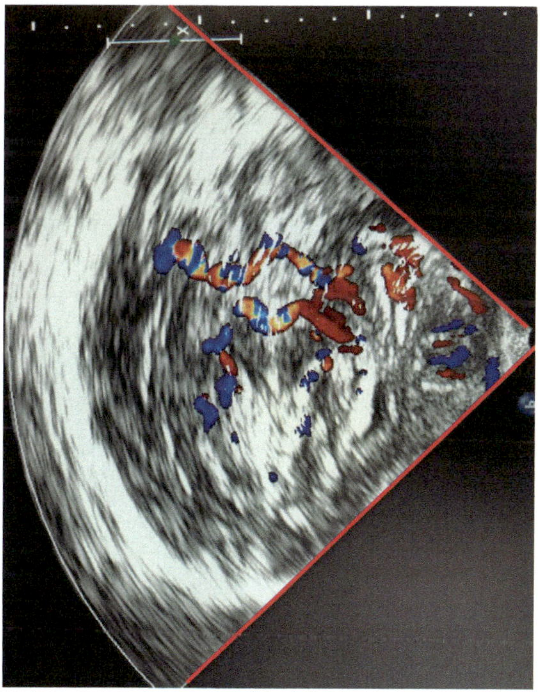

Fig. 4.28 Midbrain-Level Duplex-Sono

4.6.2.6 CTA Axial Hypothalamic Level

The CTA can be correlated with the "Sono-CTA", however, the sono-scan will change according to angling and rotation of the TCD-probe (Graphs 4.15, 4.16 and 4.17).

By that many sub-levels are possible and targeting elected structures to study the morphology and anatomy in real-time. The hypervolämia on the craniectomy side represents the post-craniectomy-syndrome, causing secondary injury to the brain.

The posterior circle of Willis in no more visible, but some parts of the anterior circulation and a temporo-basal vein on the right side.

Pathological vessels, like AV malformation, DVA or cavernomas are excluded here. Subgaleal fluid collection cannot be recognize in this case. The correlation to the CTA is only roughly possible and can be optimized by angling and rotation of the TCD-probe during examination. Edema, shifting and bleeding can be approached, and pulsation is present in the "living" sono imaging. However, hypothalamic level is difficult to evaluate (Figs. 4.29 and 4.30).

4.6.2.7 Insular Vessels

The insular arteries are the landmarks of the sylvian cistern and insular cortex. This means to have a lateral and cortical strong physiological and anatomical marker for the perfusion and shift effects far away from the midline.

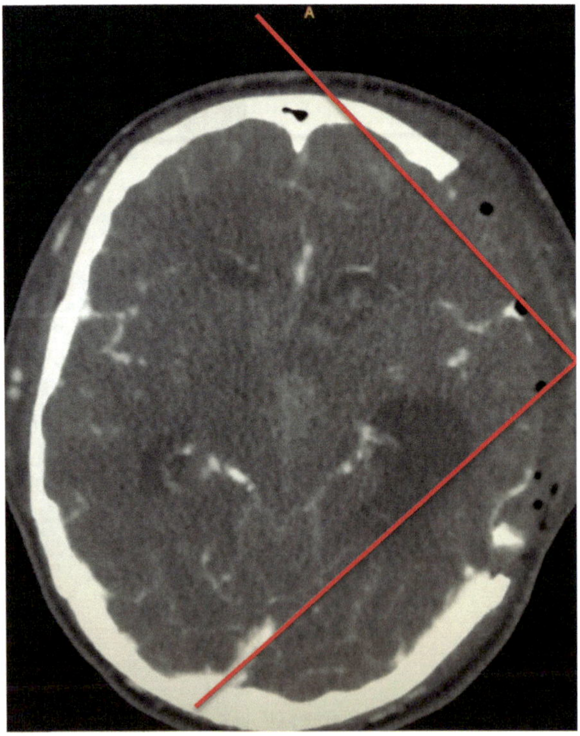

Fig. 4.29 Hypothalamus-Level CTA

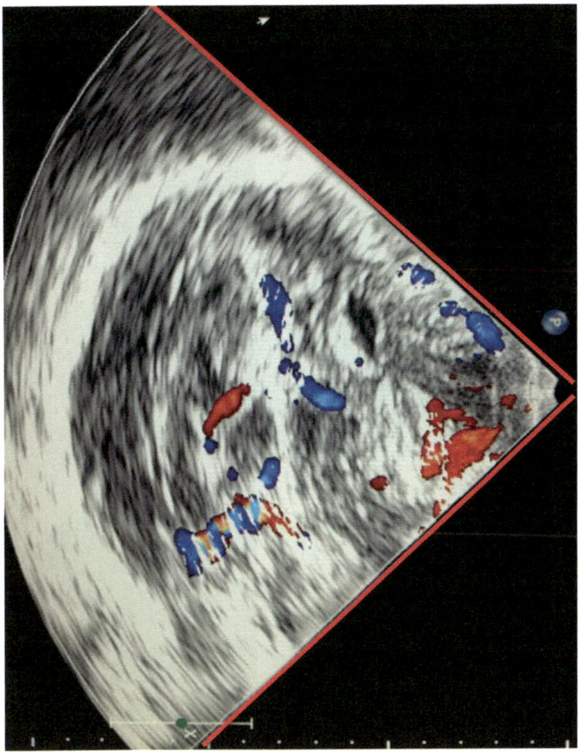

Fig. 4.30 Hypothalamus-Level Dplex-Sono

In this case, after surgery, the patient was considered for organ-donation, due to persistent coma and mydriasis! But during TCD, the opposite insular arteries were present in addition to the CW.

After MIN procedure, due to ultrasound results, the patient recovered completely.

The ispsilateral insular arteries, in this case, are hyper-perfused and present in the TCD. This is a regular finding after craniectomies, representing the over-therapy in the most cases of craniectomy, causing a craniectomy-syndrome and secondary damage to the brain, under the eyes of neurosurgeons.

In addition, the right ICA-bifurcation is visible at the border of midbrain-level and thalamus-level.

The craniectomy-window enables sono-imaging with some features superior to CT and MR! (Figs. 4.31 and 4.32).

4.6.2.8 Briging Veins Level

In this craniectomy case, the briging veins are examined, results representing a safe sign for working of the venous drainage of the cortex and supratentorial brain.

All venous sinuses can be imaged extensively in the craniectomy conditions. The TCD probe may provide better images than intra-operative once, because many artifacts did not occur. However, this is often given after few days (1–3) after sur gery (Figs. 4.33, 4.34 and 4.35).

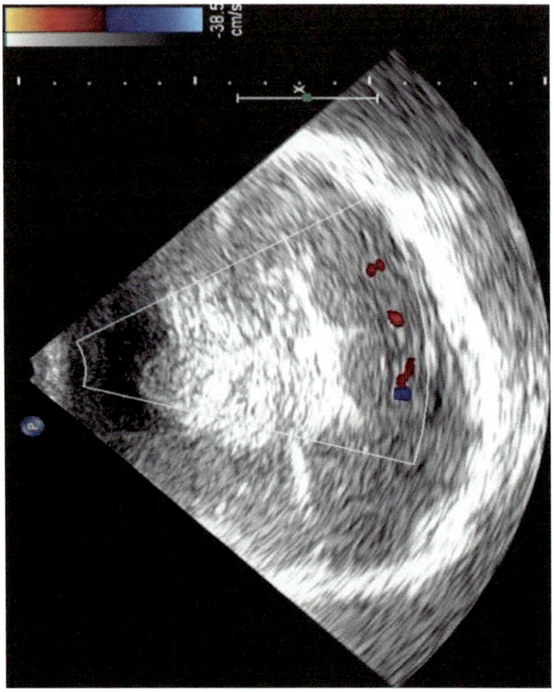

Fig. 4.31 Insular vessels duplex-sono contralateral

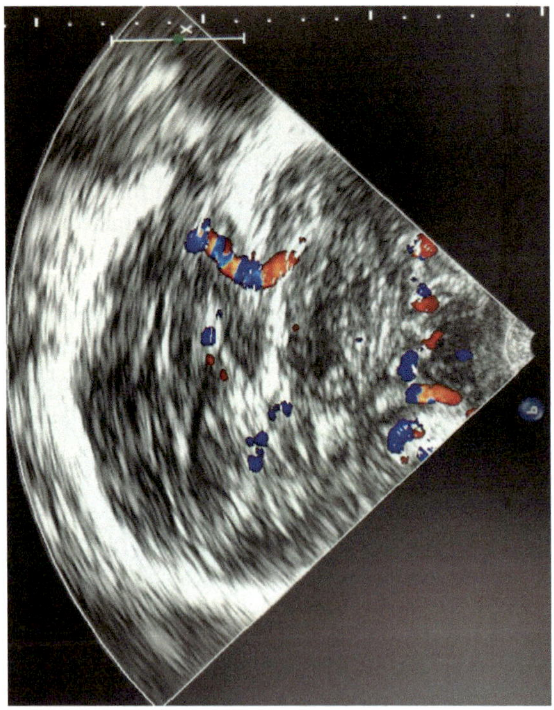

Fig. 4.32 Insular vessels duplex-sono ipsilateral

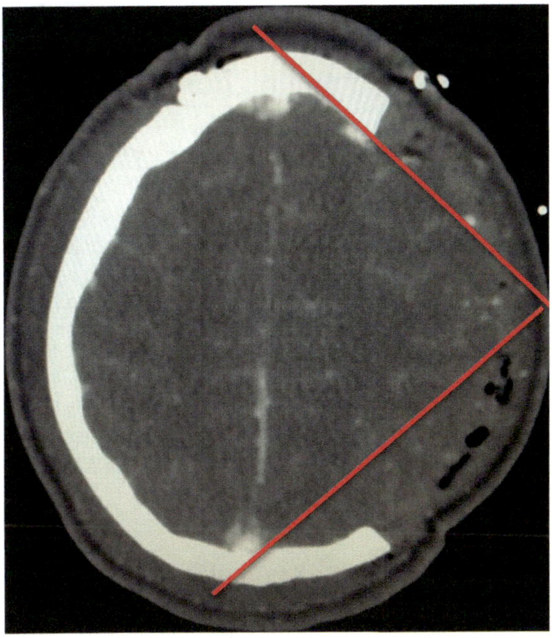

Fig. 4.33 Briging veins-level CTA

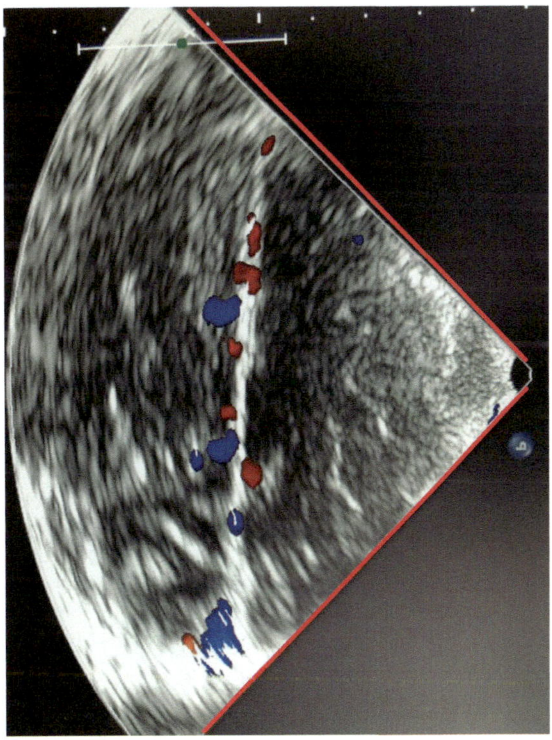

Fig. 4.34 Briging veins axial duplex-sono (+ sag. Sup. Sinus)

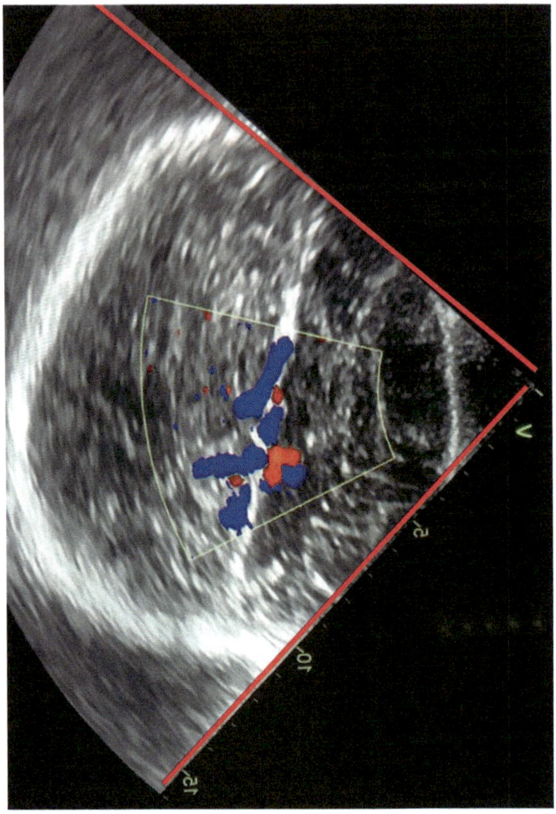

Fig. 4.35 Briging veins axial duplex-sono

4.6.2.9 CTA Coronar Levels

In the craniectomy-window coronar images present the same quality as axial one, in contrast to trans-cranial images. Therefore, the whole circle of Willis can be imaged. The scan must be parallel to the CW-plane and rotation of the probe adapted until the presentation of CW is complete.

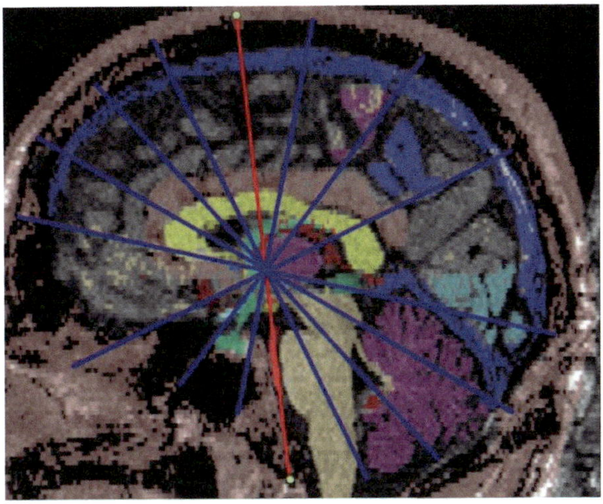

A more fine-tuning of the probe tracking can image the complete basilar head with all 5 main-vessels. The over-representation of insular vessels at the craniectomy side is visible, and shifting can be seen, if present, by frontal falx and frontal horns of the lateral ventricles. Over-representation of sylvian vessels on the craniectomy side is visible also (Figs. 4.36 and 4.37).

The best sono-imaging of the basilar tip is in a coronar plane. Playing with the angling and rotation of the sono-probe can catch the typical figure of the main vessel composition. By the colour-code (c-mode, duplex-mode) the flow-direction in correlation to the probe can be diagnosed.

The over-representation of the insular vessel-goupe impressive on the cranectomy side, representing an overperfusion because necessary ICP cannot be preseved by craniectomy.

At the tentorium-level the trigone area of the lateral ventricles and the cerebellar vessels are visible.

More peripheral over-representation of the sylvian vessels due to the craniectomy is impressive, but rarely mentioned elsewhere.

It is always important to be aware, that an ultrasound image represents morphological and physiological information! (Figs. 4.38 and 4.39).

4.6.2.10 CTA Coronar Basilar Level

In (Graph 4.11) the targeting of the basilar artery course is figured out.

Examinating the brain by ultrasound to track the course of basilar artery and to study the perfusion and presence of flow, was necessary because it was doubtful,

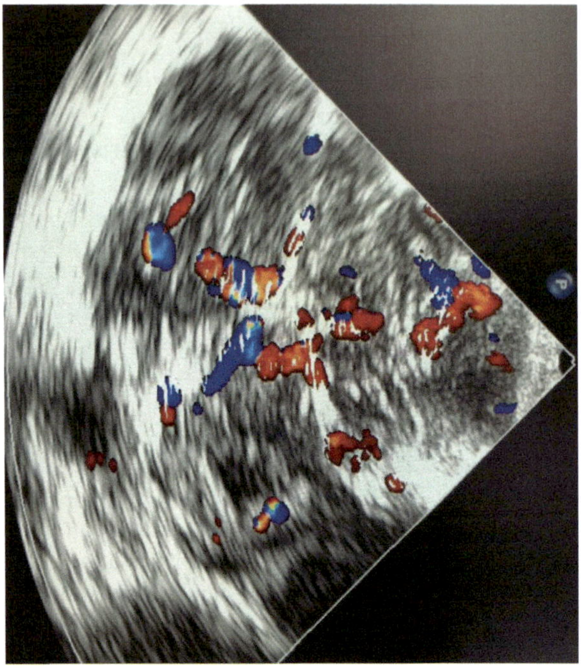

Fig. 4.36 Circle of willis coronal duplex-sono

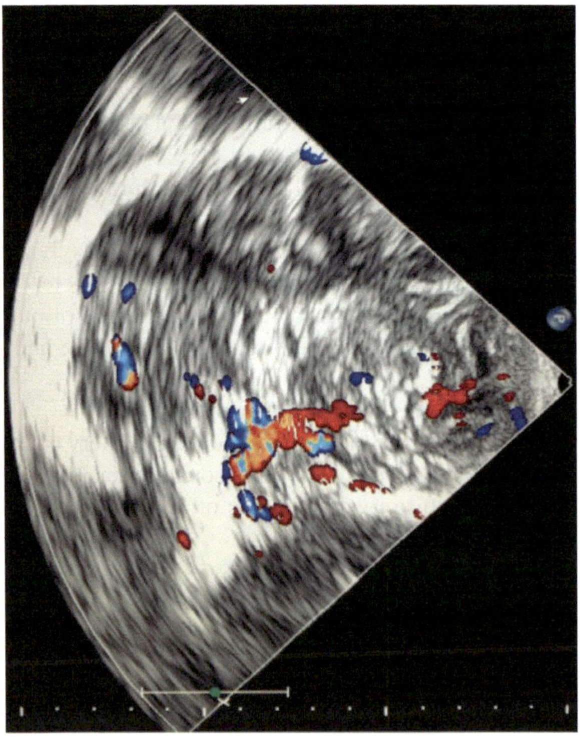

Fig. 4.37 Basilar head coronal duplex-sono

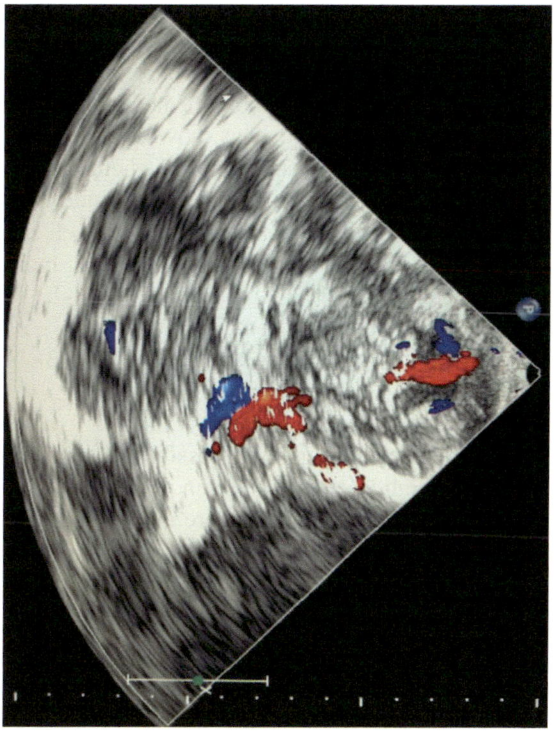

Fig. 4.38 Basilar head coronal (+ sylvian vessels) duplex-sono

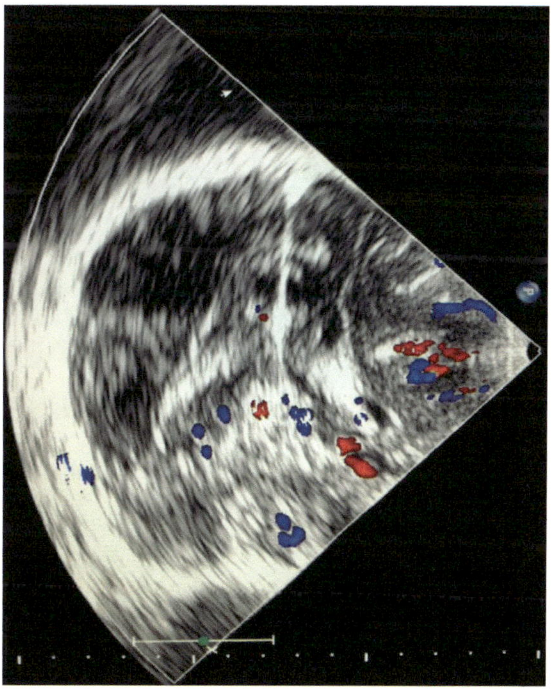

Fig. 4.39 Sylvian and cerebellar vessels coronal duplex-sono

that craniectomy without evacuation of the ventricular hematoma would be sufficient. Transportation was impossible due to the ICP instability. However, at bedside it could be shown, that the left inferior horn of lateral ventricle is obstructed due to blockage of foramen Monroi by the hematoma.

The presentation of flow has shown, that the brain is still in the situation to have a recovery chance by evacuation of the hematoma.

The over-perfusion on the craniectomy side causes additional damage.

Pathophysiologically this damage has a well known progressive and aggressive character. Using ultrasound, this findings can be made regularly (Figs. 4.40, 4.41 and 4.42).

CTA Sagittal Paramed Insular Levels (Fig. 4.43)

CTA Oblique Med.-Sag. Level

The sagittal scanning is only possible in craniectomy-windows or through burr-holes with high level burr-hole sono-probes (ALOKA/Hitachi). The findings in the midline or the sylvian cisterns are unique, but somehow difficult to obtain and to understand (Fig. 4.44).

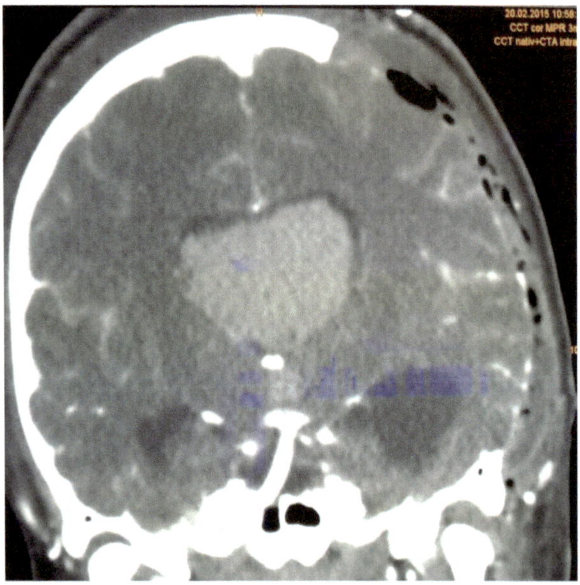

Fig. 4.40 Basilar artery coronal cta (+ ventricular ich)

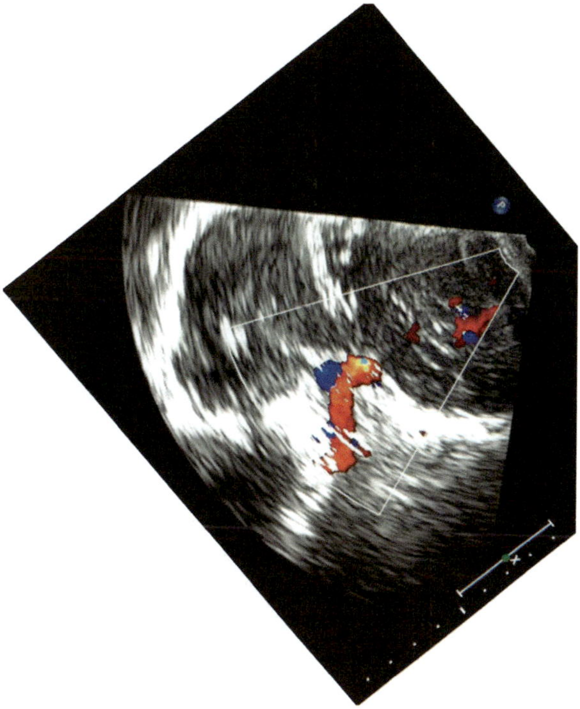

Fig. 4.41 Basilar artery coronal duplex-sono

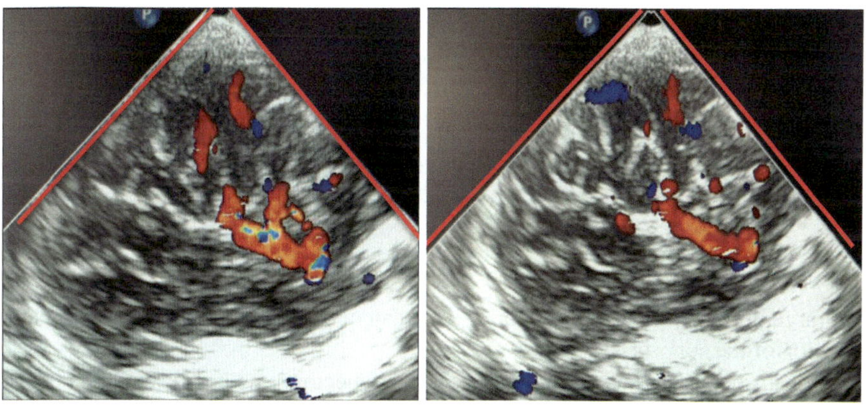

Fig. 4.42 Sylvian vessels sagittal duplex-sono

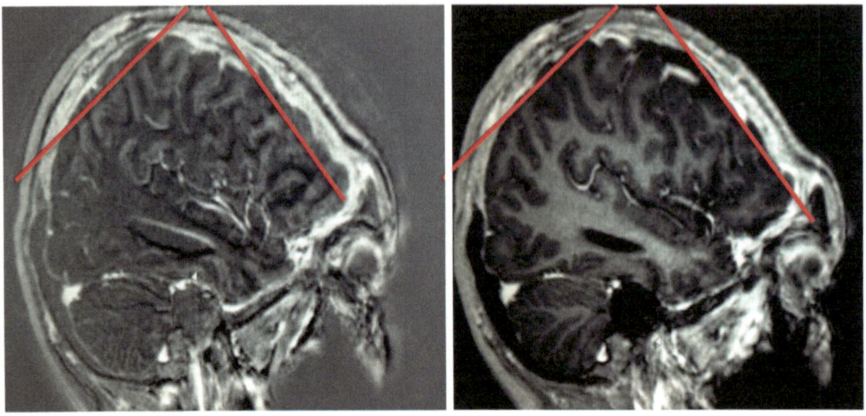

Fig. 4.43 Sylvian-level sagittal MR

Fig. 4.44 Midline vessels saittal (oblique) duplex-sono

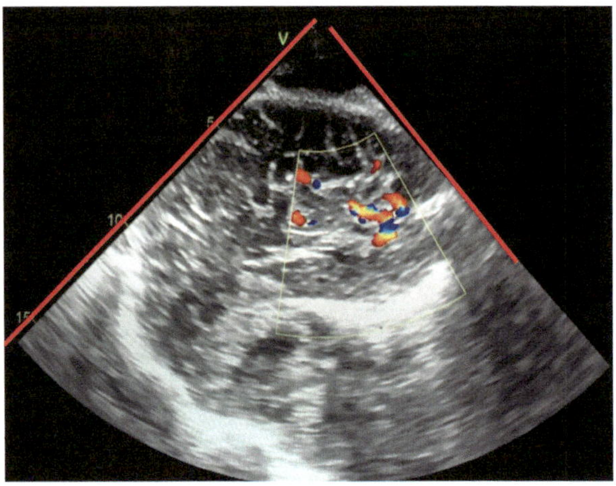

Targeting the basilar artery in a sagittal scan needs experience with angling and rotation of the sono-probe.

At bedside, it was shown, in the case of a craniectomy without evacuation of the hematoma, that perfusion was still present every where supra- and infra-tentorial (Fig. 4.45).

The chance of recovery by evacuation of the hematoma could be visualized preop at bedside.

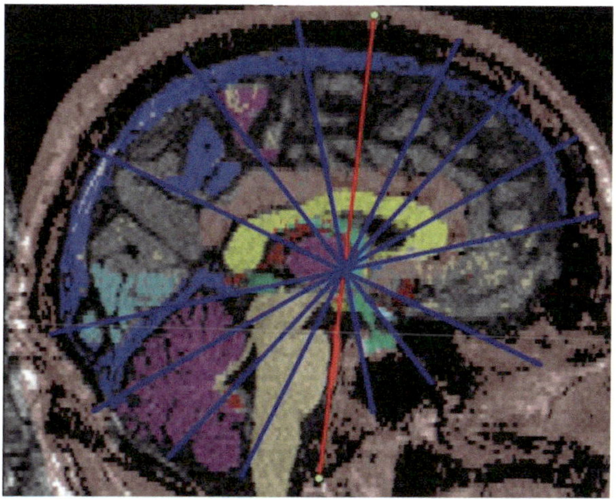

Fig. 4.45 Rotation planes coronal-axial model

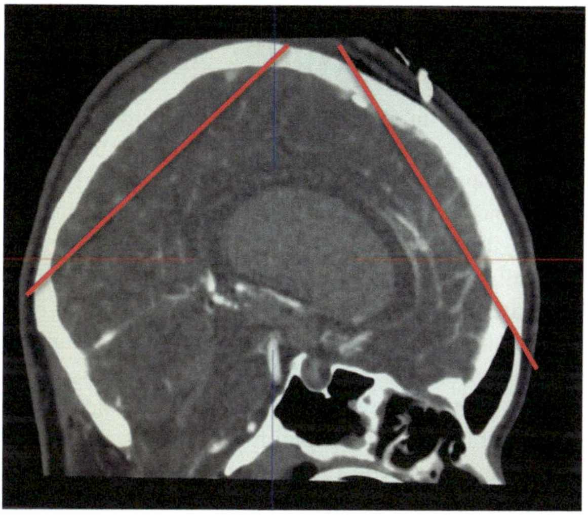

Fig. 4.46 Basilar artery med-sag. Level CTA (ventricle ich)

The understanding of anatomy, physiology in correlation to the ultrasound imaging and geometry, and the ease of use compared to outdoor examination techniques, must be realized to be aware of the extraordinary and underestimated meaning of ultrasound use (Fig. 4.46).

The ultrasound image is not a picture but rather an information carrier and must be analyzed by both brain hemispheres, which must be trained during years in the clinical context.

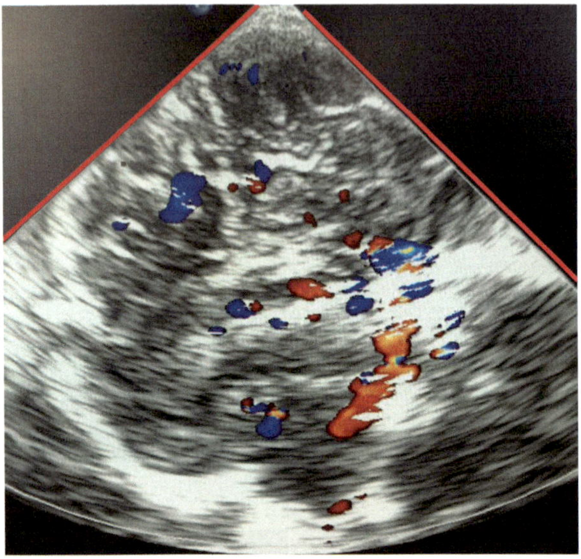

Fig. 4.47 Basilar artery sag. Duplex-sono

The most important features and principles of use is the goal and first learning step to make it a key-technique of MIN (Fig. 4.47).

Intra-operative application in tumor surgery is not limited by imaging the tumor and realizing that it is where it is already known extensively by CT/MR and all the diagnostics long before. But ultrasound transdural can analyze the tumor (and all lesions) precisely to know, what is the character of the pathology and what are the dangers one has to face. Position in relation to the approach is only the beginning of analysis, but vascularisation and main feeder distribution and position within the approach, moreover perifocal edema and shift phenomena as well as next ventricle volume or subarachnoidal cistern to release ICP and relaxing the brain. The information gained by sonography enables to create a strategy that makes surgery faster, and more simple and forewards. To see the pathophysiology within the anatomy through the surgical approach with the perspective of the surgeon is a unique strength of ultrasound. Imaging, navigation and targeting can be started before opening the dura. Never forget to examine by duplex mode and use triplex to identify arteries and veins and the amount of perfusion. Unexpected or even unknown dignity not recognized in the diagnostic imaging may be differentiated in cases of cyst or tumor, tumor or abscess, tumor or angioma and many other uncertain lesions. It is not wise to open the dura before ultrasound exploration (Graph 4.19).

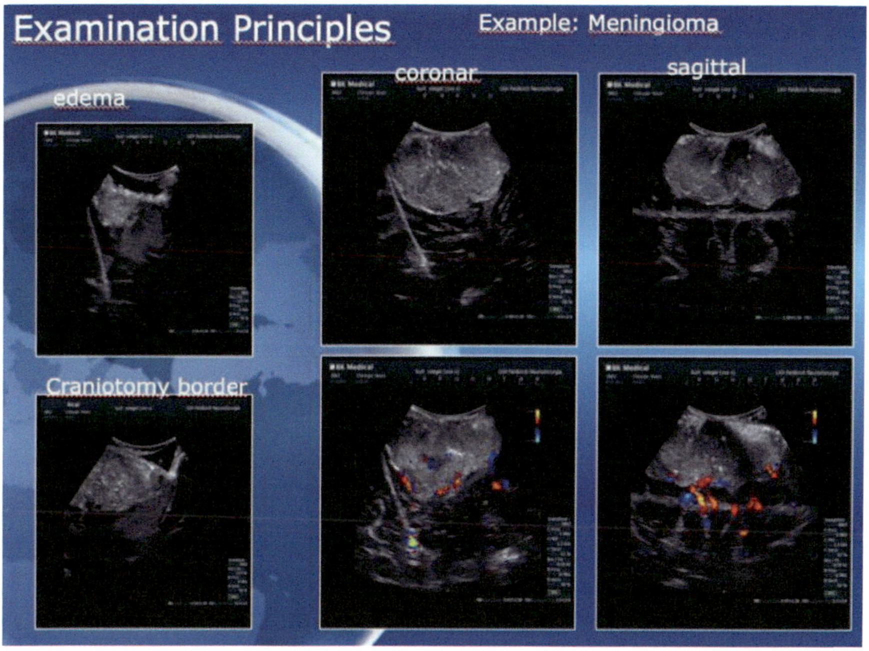

Graph 4.19 Application principles of intr-aop sono

4.6.3 History of Neurosonography (Fig. 4.48)

4.6.4 CTA axial

The complete spectrum of neurosurgical diagnoses presented applicable assistance for therapy in our experience. In 376 intra-operative applications, ultrasound proved to be an excellent neuro-navigation system providing the surgeon with real-time imaging and targeting capabilities. Resection control in 221 tumor cases with targeting in 29 small lesions was very satisfying and in four cases craniotomy correction was possible before opening of dura mater. Compensation of computer-navigation failures was possible in 14 cases preventing possible disasters. The 584 cases of application at the ICU showed a bed-side use, resulting in decrease of risky out-door examination reduce stress for our patients and logistic efforts for the professionals.

Fig. 4.48 Milestones of neurosonography

In 1937, Dussik used high-frequency sound as a means of diagnosis; French followed in 1950 with ultrasonic pulses used to detect cerebral tumors; and Güttner presented the use of ultrasound imaging in the human skull. In 1956 Leksell then detected the displacement of midline structures by 'echo-encephalography,' and Galicich came up with ultrasound B-scanning of the brain in 1965 (Auer and van Velthoven 1990; Grumme 2001; Strowitzki et al. 2000).

The 1960s saw the introduction of intraoperative ultrasound, but it was not widely accepted because of difficulties in image interpretation. In the 1970s sonography started to be used as an intraoperative imaging technique, and special probes for neurosurgery appeared in the 1980s. In the 1990s color Doppler and power mode became available for intraoperative use (Makuuchi et al. 1998).

Neurosonography was first used in pediatric neurosurgery, as it could be used transcranially (Chadduck 1989; Horwitz and Sorensen 1990; Zorzic and Angonesi 1989). In general neurosurgery sonography was used before that, not only as an ancillary technique to assist those performing open microsurgery (Auer et al. 1988; Auer and van Velthoven 1990; Koivukangas et al. 1993; Mayfrank et al. 1994; Reich et al. 1988; Sutcliffe 1991), but also in stereotactic procedures (Kanazawa et al. 1986; Masuzawa et al. 1985; Moringlane and Voges 1995; Slovis et al. 1991; Tsutsumi et al. 1989; Yamakawa et al. 1994). In 1990 Auer and Velthoven presented their *Atlas of Intraoperative Ultrasound in Neurosurgery*.

The clinical benefit of intraoperative ultrasound has become widely known from works published by several authors (Auer and Velthoven 1990; Suhm et al. 1998; Sutcliffe 1991).

Investigations are running in innovative applications, like: brain death diagnosis (54 cases), bedside-sono-CT (92 cases), aneurysm-monitoring (50 cases), bridging-vein monitoring (15 cases), sono-pupillometry (23 cases).

Intra-operatively we examined sono-angiography in tumors and for clipping control of aneurysms. Compensation of computer-navigation pitfalls was frequently necessary and always successful representing the real-time lack of computer-neuronavigation.

With the new trans-endoscopic use, which was introduced by the senior author in 1996 into neurosurgery, it is possible to intuitively navigate endoscopes serving like a "brain-radar". (s. below).

The clinical benefit of intra-operative ultrasound has become widely known from works published by several authors. Since the era of CT and MR ultrasound did not have a chance to be widely accepted in neurosurgery. However, the brain is an ideal organ for ultrasound and neurosurgery can learn today from neighbour disciplines with a large experience. Moreover, we can take over the technical evolution of two decades resulting in a wide field of uncharted territory.

In this chapter the daily use of high-end ultrasound in neurosurgery is presented and a first series of 1084 examinations (2002–2006) evaluated in a broad variety of lesions and indications. The main goal of consequently using ultrasound was to minimize the trauma to the patient and to assist in minimally invasive techniques in neurosurgery.

The transcranial sonography applied via the typical fronto-temporal, temporal or suboccipital windows was only in four cases not possible because of thick bone. The postoperative cases additional presented small bone defects enabling excellent quality of imaging at bed-side. After decompression craniectomy ultrasound scanning was comparable to CT in imaging quality. During 1084 Examinations 1053 diagnoses with a large variety were found and were appropriate for ultrasound technique (Table 4.1). In addition to the diagnosis and the morphology we examined physiological and patho-physiological parameters especially of the blood circulation.

Intra-operatively, in this series, we had in 376 examinations a lot of different applications.

Position of craniotomy was controlled through the first burr-hole to be optimized if needed. Before opening the dura targeting and examination of the lesion in comparison to the pre-op imaging was always done. Blood supply and main feeders of

Table 4.1 (2006)

Diagnose	No
Aneurysm	353
Tumor	221
ICH	148
Trauma	95
SDH	43
Infarction	31
Dysontogen lesion	27
Spinal lesion	28
NPH	24
Cavernoma	16
Aqueduct stenosis	13
Abscess	11
Angioma	12
Epilepsy	2
Others	29
Σ	1053

tumors with the surrounding edema were imaged from the actual surgical perspective. Such information offered actual information to decide or in case change the operative strategy. Details could be localized through the operative approach and targeting was able in real-time. During the procedures, orientation, real-time navigation and resection control of tumor cases and clipping control of aneurysm cases was able in all such examinations.

In a few cases even the pre-op diagnosis had to be corrected with severe consequences for the operative strategy. All emergency cases were assisted by ultrasound for completion of diagnostics concerning the reason and site of bleeding, or because assistance in targeting the lesion was necessary.

Peri-operatively also many applications were seen during 646 examinations: At ICU in 584 examinations monitoring of morphological and patho-physiological parameters were routinely done. Vasospasm and blood circulation follow-ups, obliteration of vessels and developing of hydrocephalus or hematomas were regularly controlled. CT or MRT indications were proven by bed-side sono-CT in 92 examinations decreasing numbers of risky out-doors CTs and all that consecutive logistic disaster (Figs. 4.2 and 4.3).

In 24 investigations a follow-up by high-end ultrasound, during 4 years was used to reduce MR examinations. This was done through small bone defects after craniotomy.

Innovative applications (Table 4.2) were possible in 521 examinations with different indications. Bed-side sono-CT as a substitute for CCT was applied 92 times, Sono-angiography was investigated 65 times to evaluate aneurysms pre- and post-op and to examine the cerebral blood circle in different diagnoses. Aneurysm monitoring was applied 50 times to follow up stress parameters for the aneurysm sac and to estimate the risk of rupture and to get a more precise timing of operation and intensive care measurements. Intra-operatively in 6 cases a clipping control was

Table 4.2 Intra-operative applications

Intraoperatively:
• Imaging (Additional Information about the Lesion)
• Targeting (Find a Smal Lesion, Direction of Approach)
• Neuronavigation (Real-Time Orientation)
• Pathophysiology (Vasospasm, Perfusion, Shift)
• Emergency (Angioma? Aneurysm? Main Feeders?)
• Diagnosis (Vascularisation? Type of Tumor? etc.)
• Reorientation (Complex Lesion, Confusing Changes to preop.)
• Resection Control (Tumor Borders)
• Clipping Control (Patency of Parent Vessels)

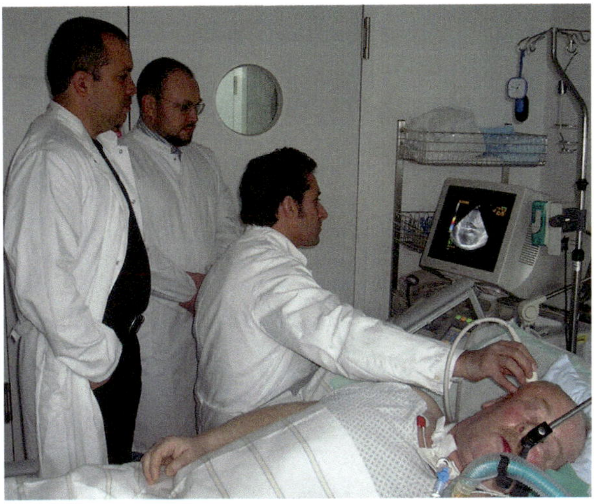

Fig. 4.49 Sono-course group at ICU

done applying sono-angiography pre- and post-clipping to see patience of parent vessels and exclusion of the aneurysm. This intra-operative sono-angiography allows the surgeon to react during operation and correct clip position in the same procedure (Fig. 4.49).

Compensation of computer-neuronavigation pitfalls was able in 14 cases for different reasons of failures of the soft- or hard-ware. Decompression craniectomy effect on the cerebral perfusion was measured in 15 examinations by bridging vein monitoring at bed-side. In normal pressure hydrocephalus shunting procedure effect on perfusion was controlled in 23 cases. Finally, pupillomotor measurements were documented by sono-pupillometry in 23 examinations enabling objective documentation of anisokoria in selected cases of interest.

A new intra-operative application and introduction into neurosurgery was the use of **trans-endoscopic ultrasound** in 75 highly selected cases. In this cases imaging, neuronavigation and targeting in neuro-endoscopy was evaluated clinically after a period of laboratory testing (Graph 4.20). It was shown that navigation of endoscopes is possible in real-time and imaging and targeting of lesions with high resolution was able in each case of application. (s. Case 14 and end of this chapter.)

Ultrasound in neurosurgery is still not widely used in times of CT and MR. A retrospective analysis of a complete spectrum of application in neurosurgery (1084 examinations) using high-end ultrasound technique has not been reported. Few papers presented results of the intra-operative spectrum of applications. Most paper dealt with a special application applied in a very few examinations. In addition of intra-operative application also the peri-operative use is presented, and the first unexpected result of analysis is the broad variety of the applied spectrum. The main goal of application in the present series was to assist in minimally invasive technique of neurosurgery with single high-end equipment. This goal can only be reached by small probes and proper selection regarding size, shape and frequency.

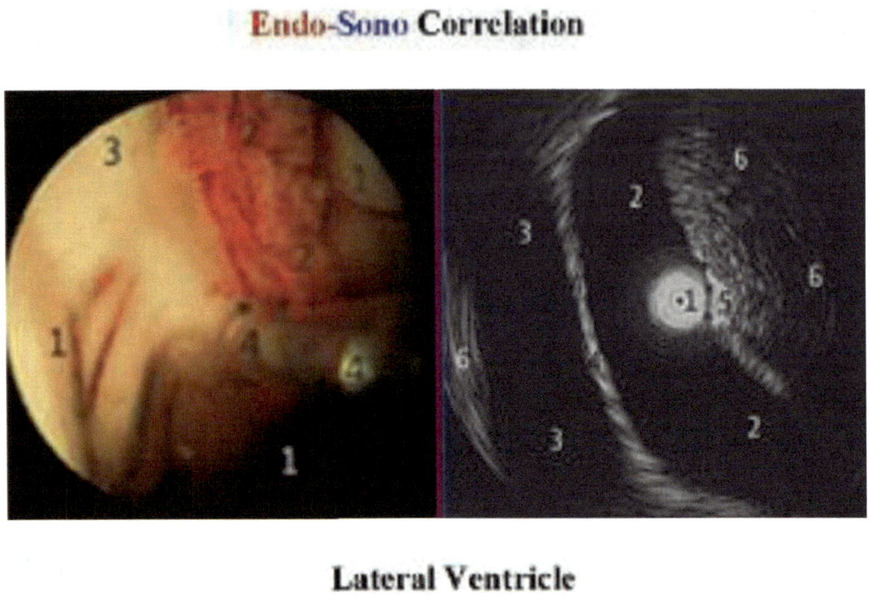

Graph 4.20 The endoscope view (left) shows the right lateral ventricle with the chorioid plexus (2), pellucid septum (3) and the transendoscopic mini -probe (4) in tough with the plexus. The ultrasound scan (right) represents the sonoprobe by a spot (1) in tough with the plexus (5) and shows in addition to the right ventricle (2) the left ventricle (3), the septum (4) and the thalamic parenchyma. This scan is a real-time land map looking like a "mini-CT" and enabling navigation of the endoscope as each moving of the endoscope is visible on the ultrasound scan by the moving of the spot (1) in real-time and in relation to anatomical landmarks not visible in the endoscope. Due to its 360° geometry presentation we call it "brain radar"

Moreover, it does not make sense in our opinion to enlarge the approach and the trauma just for applying a big ultrasound probe.

Recent papers reported mostly on rather complex combination of ultra-sound with other techniques like neuro-navigation. In the present series we focused applications using only technical equipment that compete principles of ergonomics in neurosurgery avoiding "technical overkill". In present series there was no single case in which ultrasound use needed assistance by neuro-navigation system. This is not surprising as the ultrasound is a navigation tool itself and the only technique that compete the definition of navigation, namely a real-time feedback. We had 14 cases in our series of neuro-navigation, where the ultrasound had to compensate pitfalls of neuro-navigation systems (Table 4.2).

In this 1084 examinations, one third were intra-operative and two third were peri-operative use. However, it was seen during daily work that neurosurgical therapy outside the OR can be markedly assisted by ultrasound (Figs. 4.2 and 4.3). These applications differ a lot from neurological applications as there were different problems to be solved and as neurosurgical ultrasound applications can use the bone gaps left in postoperative cases enabling bed-side imaging which even might substitute CT or MR. A convincing learning curve was the integrated investigation of anatomical structures together with physiological parameters which gave a complet different decision making as usual when using only MR or CT.

Intra-operative control of craniotomy position was an easy and helpful use enabling precise dimensioning of craniotomy in a minimally invasive strategy. However consequent epidural examination of lesions created much more safety for consecutive decision making in favour of a minimally invasive strategy. In this series of 376 cases the ultrasound was used like a surgical instrument always at hand and serving like a "personal intra-operative neuro-radiologist". In no single case neuro-navigation was needed for targeting feeders, moreover in contrast to the three orthogonal computer planes the ultrasound did present the vascular supply in relation to the surgeon's perspective through the surgical approach resulting in an intuitively understanding during the targeting. The main disadvantage of neuro-navigation systems is the lack of real-time and actually there are efforts to overcome this by ultrasound integration to neuro-navigation systems. Three different technical realisations are going on: several platforms offer an integration of an ultrasound probe into the navigation system. Another realisation is that ultrasound equipment is supported by integration of navigation components. The third realisation is that navigation system and ultrasound equipment is combined by registration of the ultrasound probe like a pointer in the 3D-data of the navigation system. However, the problem of mathematical description of brain shift is not solved. All these systems can just enable to show the shift on a display in relation to the off-line imaging. In other words, the navigation system serves as an interpretation interface of the ultrasound imaging. In our opinion this solution seems rather expensive just to display a shift which is for pure ultrasound use no problem at all as ultrasound enables the real-time imaging. Moreover, it has been observed by the author, that the imaging management task for the surgeon impairs neuro-surgical ergonomics of the environment as well as mental management. Investigative applications (Table 4.2) have been usual in this series (521 examinations) as neurosurgery can benefit today from neighbour disciplines with large experience and the new developed techniques. However, examination numbers are still low, and more experience is needed to avoid premature reports.

The **transendoscopic ultrasound** however has been used with very strict indication for a long time after introduction into neurosurgery by the senior author. Results in general have been documented and presented already (s. below).

The Limitations of ultrasound imaging are: worsening of the image quality by physical changes of the brain tissue during surgery. In resection control a typical echo density of the resection planes can be regularly observed. Differentiation of the borders of low-grade gliomas is difficult and also resection control, however in this series targeting was always possible and resection control in low grade gliomas is difficult in all imaging techniques. The most used term of limitation in ultrasound is the "strong user dependence". During over 40 years experience with neuro imaging techniques and their neurosurgical use in the daily work it must be recognized that the user dependence is true for all imaging techniques but more known in ultrasound than in the other "anatomistic" imaging methods like MR or CT. Moreover, our experience in this series was also that imaging generation, interpretation and decision making in one hand was a major advantage in the use of ultrasound compared to CT and MR. A special intra-operative condition for using ultrasound was to position the patient with regarding the need of fluid overlay as contact medium. Absolutely clean working field is also essentially, however, these needs are generally prerequisites of good neurosurgery.

In **summary** routine use of high-end ultrasound in neurosurgery enables a broad variety of applications and indications intra- and peri-operatively. Ultrasound contributes to minimally invasive techniques with realisation of good ergonomics in neurosurgery. Most limitations can be avoided by correct usage and sufficient experience and training (Tables 4.3, 4.4, 4.5 and 4.6).

Ultrasound can be used in a wide spectrum of applications and indications. But within **MIN** it is applied as a **key technique** to enable minimally invasive strategies. One of the most obstacles for MIN came from the question: "Can I control the procedure through a small precisely adapted MIN-approach?" To decide the strategy, one has to know which information for planning is needed and which ultrasound application can deliver this information. Once the learning curve gives the experience that the key informations can be acquired regularly, insecurity will decrease and precision of approaching and security of managing difficulties will grow. It is impossible to imagine the MIN approach without this experience. It should become normality by training (Graph 4.21).

In the case above analysis of feeders and vascularitation by duplex-sonography convinced during planning that this strategy will work and can be applied in the handicapped old lady. Standard concepts came to the decision: inoperable due to the trauma and high risk expected for the multimorbid lady.

Table 4.3 Intra-operative targeting

Targeting
• Small lesion
• Tumor remnant
• Slit ventricles
• Distance + angle
• Volume

Table 4.4 Extra-operative monitoring

Monitoring extra-op
• Vasospasm
• Aneurysm remnant
• Vessel-occlusion
• Ventricular size
• Edem and shift
• Rebleeding?

Table 4.5 Intra-operative imaging

Intra-operative imaging
• Size and correction of craniotomy
• Cyst/tumor?
• Tumor borders
• Edem borders
• Tumor dignity
• Resection control
• Sono-angiography
• Clipping control and patency of parent vessels
• Shifting and all intracranial changes
• Ventricular size and changes
• Unexpected findings!
– Cause of bleeding (Aneurysm, Angiom?)
– Actual tumor size (fast growing)
– Perfusion amount of the lesion
• Physiological parameters

Table 4.6 Sono-neuronavigation (Real-Time!)

Neuronavigation
• To the target
• Pitfalls of computer-neuronavigation
– Shift
– Registration failure
– Technical failure (soft, hard)
• Cysts distribution + Parenchymatous topographic anatomy
• Orientation
• Correlations texture
• Early detection of dangers

The spectrum of applications and problem solutions by ultrasound, however, can best be demonstrated by typical cases. Such cases are presented, each standing for a certain capacity of ultrasound assisted solutions:

4.6.5 CTA coronal

4.6.6 CTA sagittal (Graph 4.22)

In this emergency case with a dramatic deterioration of rising ICP to a comatous status during MR, it was not possible to get a sufficient diagnostic about the av-angioma regarding architecture and flow. Within short time it was clear: the reason of bleeding and the life tethering clinical status.

This is one of the most satisfying applications of highend ultrasound giving the surgeon during emergency situation and incomplete imaging the safety to complete the imaging intra-operatively and to target the evacuation precisely. How to

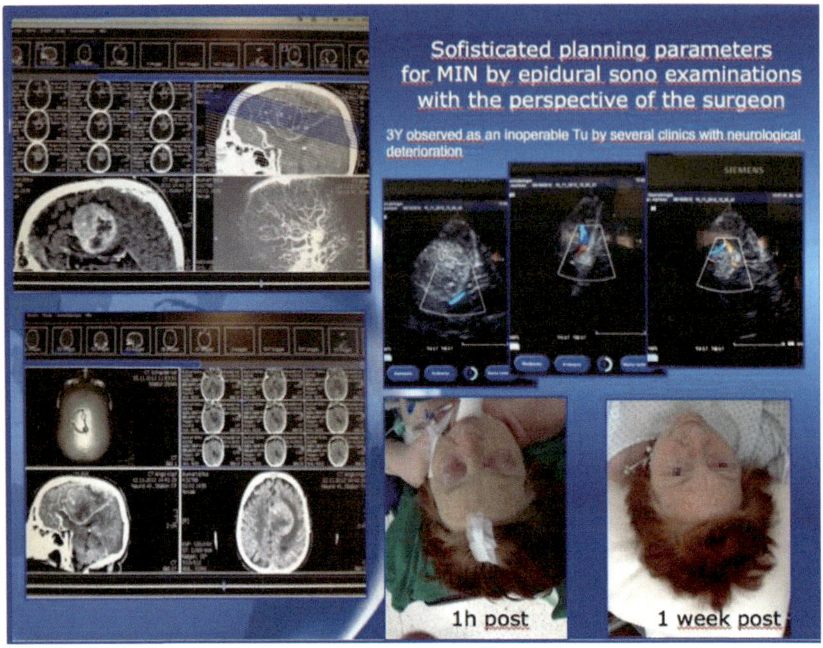

Graph 4.21 Goal: Planning and Controling a MIN-Approach

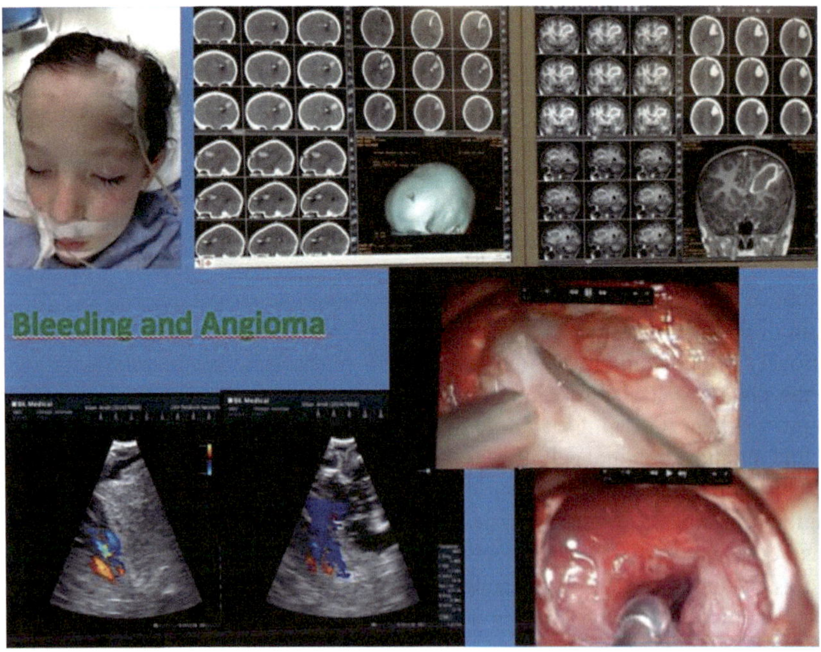

Graph 4.22 Sono-assisted Angioma-ICH Evacuation Emergency

evacuate the hematoma without hurting the angioma and provocating a dangerous rebleeding while releasing the high ICP to normal level and to safe the life? This can only be done fast and safe enough with the assistance of high-end ultrasound in the duplex mode and if needed with triplex mode to identify the danger of the angioma. The boy was fast operated through a single burr-hole and was sent to further investigations by the interventional neuroradiologists in a normal neurological awake status.

4.7 Examination Principles (Graph 4.23)

The CTA in this lateral ganglia-ICH showed close proximity to the ICA bifurcation in coronar and lateral view. Transdural burr-hole duplex-ultrasound showed a plane of cerebral tissue between the vessels and the hematoma in coronar and lateral view, as well as the consistency of the bleeding with a liquid and a thrombosed compartment. The deep hematoma was evacuated through a 1€—approach after the ultrasound findings, that there is no danger reason of the bleeding and that the liquid part of the hematoma will enable a fast and easy release of ICP. This means, that the evacuation of the thrombus will then be possible safe and controlled.

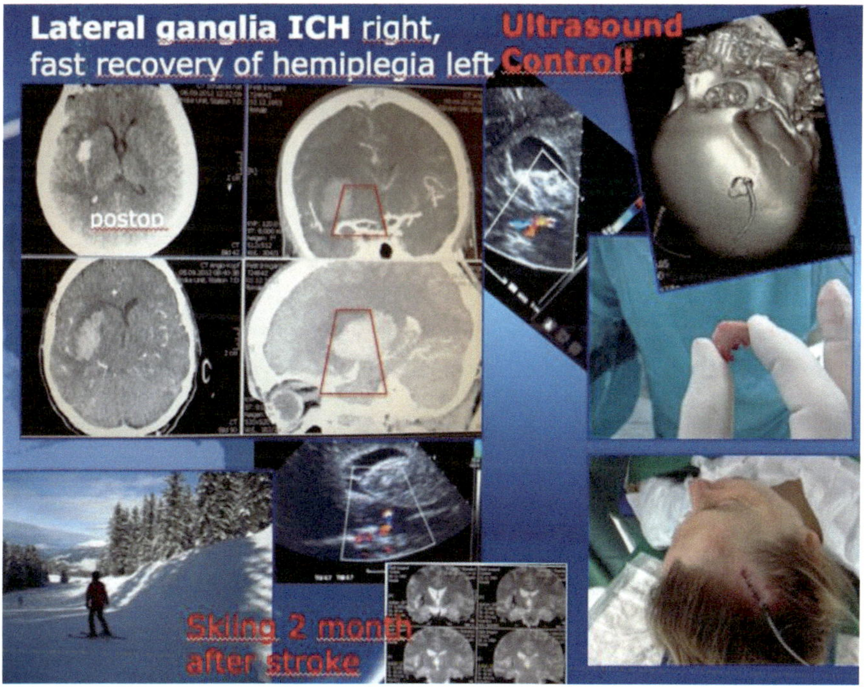

Graph 4.23 Duplex-Sono-assisted ICH Evacuation

The somnolence, headache and sever hemiparesis disappeared within few days and perifocal edema also. After rehabilitation stay, she surprised the author, 2 months since the stroke, with a video clip in which she is skiing already.

The husband, who became desperate during the stroke-unit days, because she got worse every day, urged the neurologists to call the neurosurgeon. This was not usual after the STICH-trial period, a disaster still taking place.

4.8 History of Neurosonography (Graph 4.24)

This emergency case shows that sono-assisted MIN enables operations under bad conditions like high age and severe coagulopathia. While she still was in substitution MIN surgery already was started to prevent brainstem compression. After duplex sonography epidural did not see a danger bleeding pathology intracranial, the procedure could be done safely and fast through a 1€—approach with minimal irritation for the coagulation physiology. Sono imaging moreover a liquid part of the hematoma superficial and a deep thrombosed part compressing the lateral ventricle. After a very fast decompression of the ICP the sono control did not present

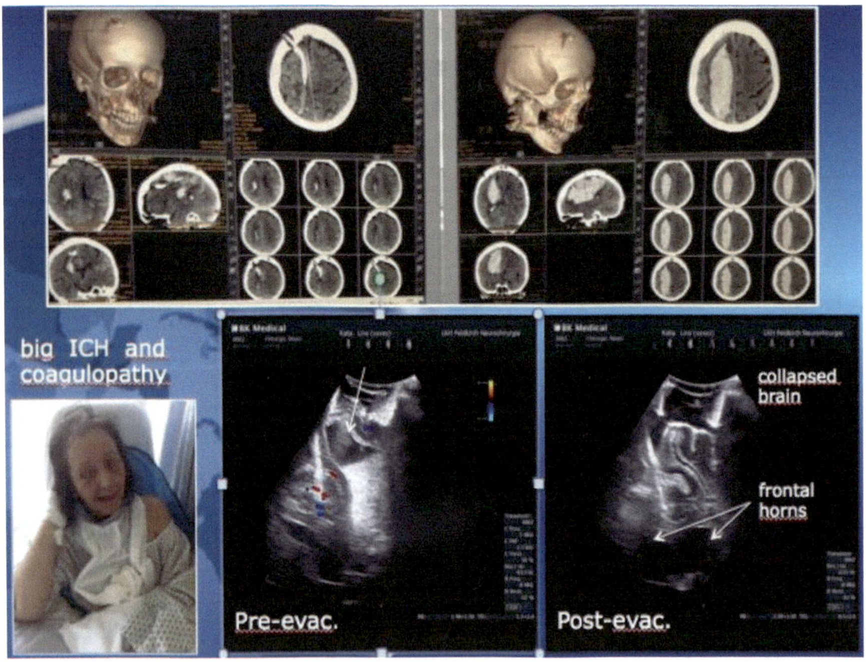

Graph 4.24 ICH evacuation in severe coagulopathy case

pathological vessels in normal ICP conditions. The hematoma was easy and very carefully sucked out but without any explorative actions, the evacuation amount was only checked by ultrasound. This can shorten the procedure and prevent to evoke rebleedings. Opening and closure is maximally fast in such MIN procedure also.

The patient recovered fastly and kept a slide paresis of one arm being full accepted life quality and well cared in her glad family.

4.9 Result of Imaging (Graph 4.25)

This case of a large ICH offered the option to approach it through a lateral supra-orbital window which is a challenging task and need a precise planning. In MIN, volumes should be approached always along the long axis. To be sure not to meet a dangerous bleeding cause, in such a strategy of approaching, ultrasound assistance is absolutely indispensible. You must be aware to manage the complete space into the depth of about 100 mm with safe focus and visualisation. This is only possible with mouth-switch tracking having an online focus with similarly maximum of magnification (s. this chapter). The intra-operative ultrasound imaging promptly

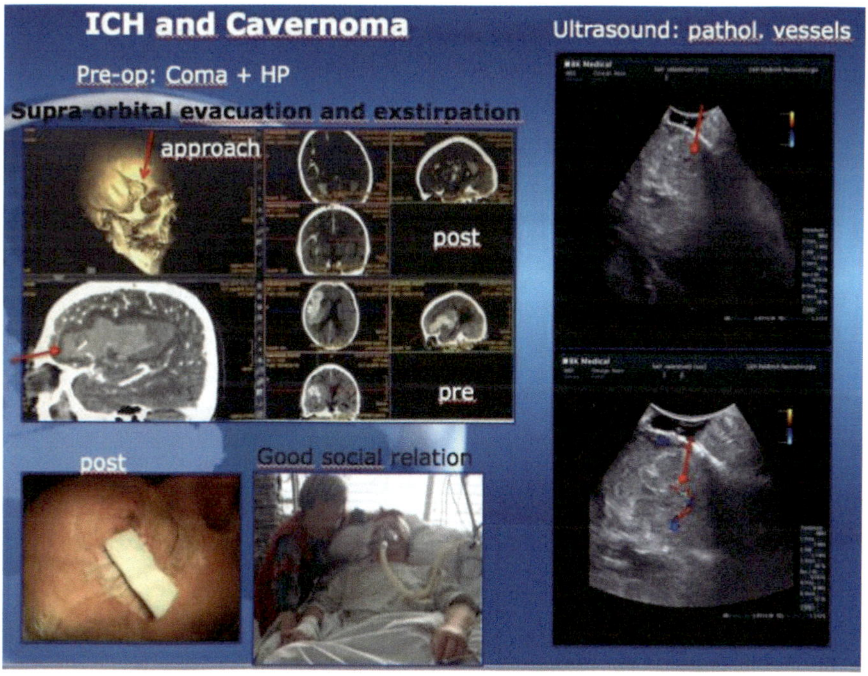

Graph 4.25 Supra-orbital ICH-evacuation and cavernoma exstirpation

showed some clearly pathological vessels which could have been a relative contraindication for a supraorbital approach and which were not visible in the CTA! Ultrasound is much more sensitive for presenting vessels in a raised ICP situation, than CTA or even MRA. A fast second look by duplex-sono will show vessels in a released ICP situation, giving sometimes surprises in negative CTA findings! Here the ultrasound showed a cavernoma before opening the dura, the final information to go on with the supra-orbital approach strategy without choosing an enlarged approach or surgical variation.

4.10 **Cases** (Graph 4.26, 4.27, 4.28, 4.29, 4.30, 4.31, 4.32, 4.33 and 4.34)

In this dramatic case of a near fatal ICH in a 18-year girl, the role of peri-operative ultrasound can be demonstrated. After 3 frustran operations (2 Drainages and craniectomy) coma, mydriasis and high ICP persisted and no therapy was suggested anymore. Next morning through the large craniotomy a precise and extended three-planar duplex examination was undertaken surprisingly presenting a still sufficient real flow

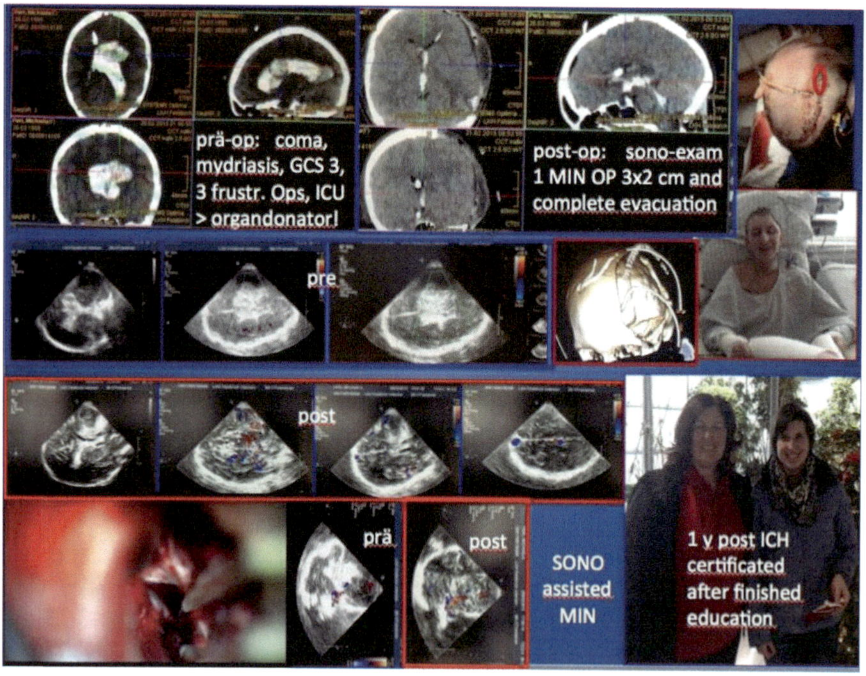

Graph 4.26 Near fatal ICH/IVH with Ventricular Tamponade from left frontal Ganglia

imaging. The contralateral insular vessels and the briging veins could be shown. No bleeding cause was seen at this stage in concordance with MR imaging. This gave the basics for a final MIN strategy: 3 cm of the suture and the dura 2 × 2 cm were re-opened, and with the safety of a real-time imaging assistance by ultrasound (in connection with mouth-switch tracked microscope/s. this chapter) the hematoma was evacuated, and under ultrasound control to detect vascular pathologies and volume reduction an over 90% evacuation was realized within about 1 h.

She recovered completely and went back to her professional education (s. Vol. 4). Post-op sono-control showed near complete evacuation and patent perfusions.

This "swimming"-tumor showed in MRT an appearance which allowed different diagnoses from cystic tumor types or meningioma. The change from angiography as main diagnostic tool to CT and MR caused a decrease of imaging solution and a loss of knowledge about vascular patterns. The capillary finger-print of lesions got lost in only one generation of radiologists and surgeons.

Ultrasound provides, with several modes, to analyze the vascular information and localizations of details about the lesion before opening the dura. In this case the vascular supply presented, the typical pattern of a meningioma. But more important additional, the existence of a pial feeder which has a major impact on operative

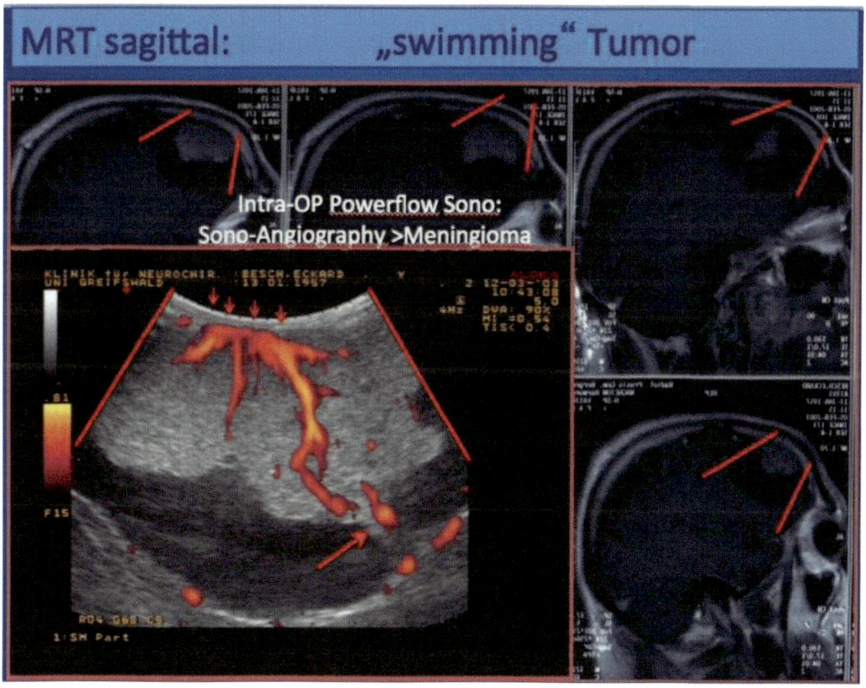

Graph 4.27 Intra-op Sono-Angio of Meningioma

strategy. It was essential for the outcome to approach the pial feeder first and to coagulate and cut it. Otherwise in case of extirpation a bleeding and neurological deficit can occur. Also, the primary coagulation of the arterial supplies from the dura can make surgery fast and easy. Ultrasound can inform about the best tactic and strategy of the surgery before opening the dura. This vascular case shows an a. com. Ant. aneurysm in comparision betwenn radiological CTA and DSA with surgical Duplex-Sono and PW-Sono. The sono imaging provides the imaging through the surgical (pterional, blue arrows) approach and additional with physiological parameters. Ultrasound can intra-operatively give essential information about the blood circle and the aneurysm. Post-clipping these parameters can be compared to pre-clipping, both intra-operatively to see if clip-correction is necessary and if all vessels are patent, but also the reaction of the parenchyma. In contrast to ICG flurescenz-angiography control, the vessels are visible in sono control independently if they are visible through the microscope or not. Ultrasound can compensate

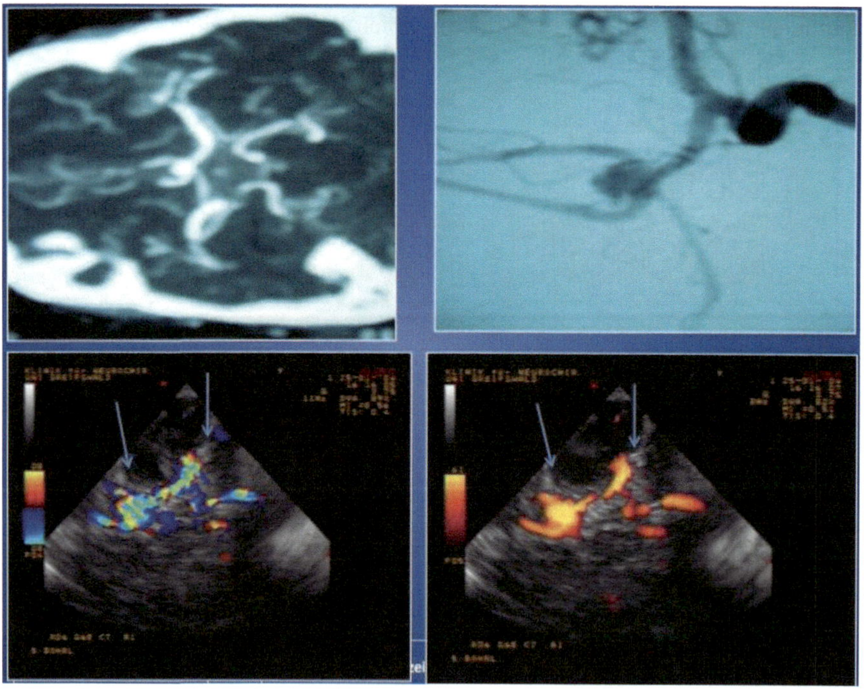

Graph 4.28 Intra-op Sono Imaging Ant-com Aneurysm

ICG control problems. The aneurysm can be seen within a blood-clot and the aneurysm wall can be analyzed as well as intra-aneurysmal thrombus and flow-parameters. All these informations before opening of dura and during the whole operation, making the these challenging procedurs safer, faster and less stressful. In this care of a giant M-bifurcation aneurysm, the sono-examination of aneurysm wall and flow characteristics like tubulances and direction of flow is shown exemplarly which is not so clear visible in medium or small aneurysms. In the duplex the typical yin-yan (blue-red) appearance and screw-turbulance are well demonstrated. The fine calcification of the carrier vessel wall and the thick thrombus inside the aneurysm, are precisely presented. All morphological and physiological parameters are essential to understand the leason and the dangers during therapeutic procedure what ever it should be. This information can be aquired transcranially pre-op and transdural intra-op but also during the procedures. This informations cannot provided by CT or MR. By triplex mode quantitative measurements of flow-parameters can be

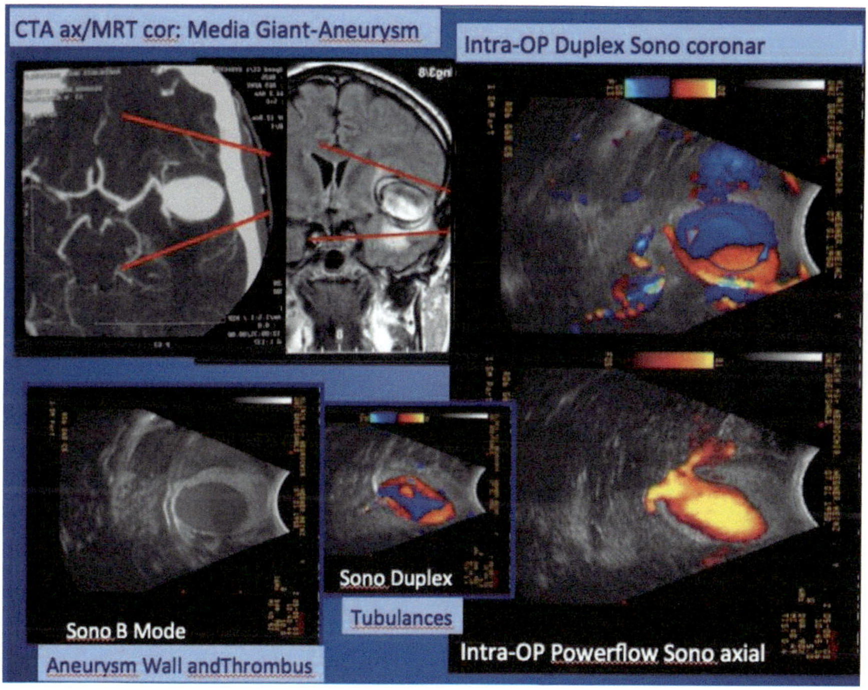

Graph 4.29 Giant Aneurysm: multiple Parameters of Patho-Anatomy and Patho-Physiology

visualized, calculations of stress of aneurysm wall and timing of procedures can be derived. Glioma resection control was one of the first special applications of advanced intra-operative ultrasound in times, when there was no "neuro-navigation" by computer-assistance systems available. Neurosurgeons had to recognize tumor by tissue consistency, colour, morphology, vascularisation, necrosis, perifokal edema and tendency to bleed. Tumor resection control was difficult, as the tissue changes during surgery and shift progress occurs. Imaging in real-time to estimate character of tissue, real-time sono-navigation for re-orientation and targeting tumor remnants, were applications, which came with ultrasound into neurosurgery. The vascularisation pattern as a "finger-print" of different tumors, is only in use anymore by duplex-ultrasound and triplex-sonography. As the only real-time technique without radiation-side-effects ultrasound is independent from shifting of tissue intra-operatively, a promise, that never has been kept by computer "neuro-navigation". Resection control needs a clean surgery and correct positioning of the head, which are preliminaries for MIN. Sono application must not be an excuse to enlarge the approaching for MIN, with high precision. The real-time imaging of ultrasound

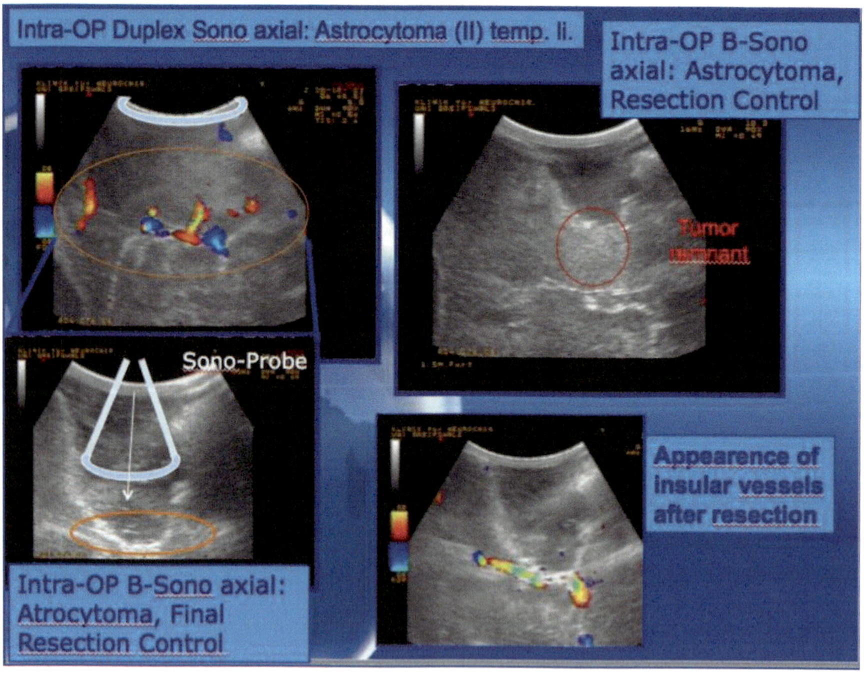

Graph 4.30 Glioma (Astrocytoma II) resection control

makes it the only true navigation-tool, hence it is independent from all changes during surgery, more over it can detect and visualize these changes. Whenever changes cause uncertainty to the surgeon ultrasound can truly navigate during surgery for orientation, correction of direction and distance to the target. In this case of a child, the very experienced surgeon was afraid to enter the brainstem while searching for a pathway to the fourth ventricle. The burr-hole probe (ALOKA) gave safe information how to go on and the small-part probe (ALOKA) showed, comparable to MR-quality, in axial and sagittal, the complete anatomical overview of posterior fossa content. The burr-hole probe with tough-plane 8 × 8 mm, can be used like a "seeing micro-instrument", fastly ad hand throughout the surgery. In cases of less imaging quality, using duplex mode, will show further land-marks by the vessels presented. Moreover, once the liquor pathway is open, the flow will be visible by colour artefacts showing velocity and direction of flow. The small sono probes (s. Graph 4.6) fit elegantly and with good ergonomics into the surgical work-flow. Cavernomas are well visible even when they are very small hence being very safe targets. However due to the small size there is no chance to find them safe by

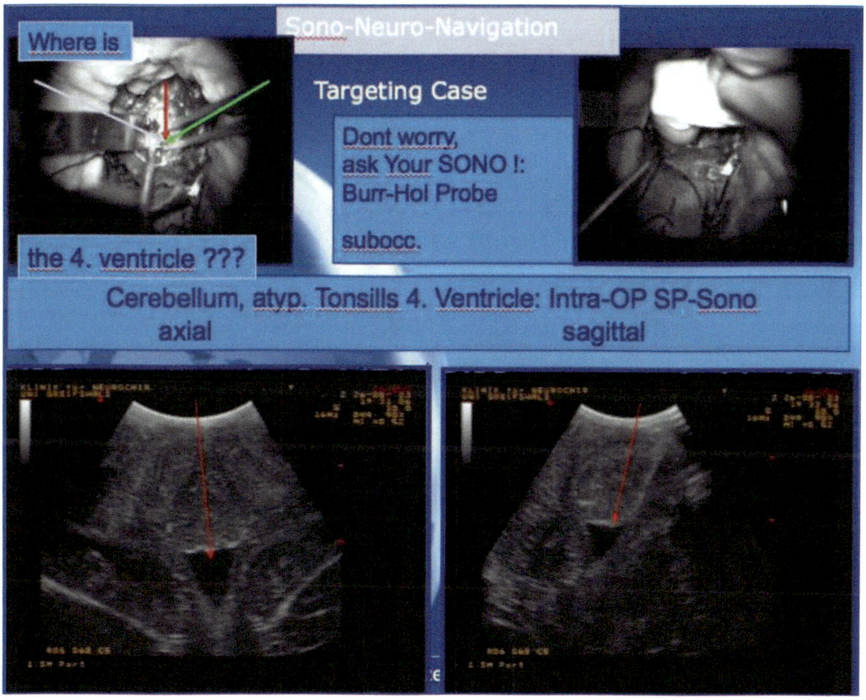

Graph 4.31 Sono-Navigation of the 4. Ventricle

computer—"navigation". Most of these sono-navigation cases are computer-navigation pitfalls, leaving the surgeons in desperation if they have no compensation by ultrasound. The leasons are sometimes below the surgical precision range of the navigation-systems, missleading surgeons into the wrong direction. One cannot rely on the xanthochromy of the perifocal tissue, as the small cavernomas do not always have this sign. Draining veins and DVAs can be identified and preserved as they are easily differentiated from supplying arteries. In this temporo-polare position the cavernous sinus serves as a safe land-mark and final medial border of the surgical field. Also, tentorial notch and mesencephalon as well as the basilar system will be well visible. Standard approaches to the posterior fossa are very invasive because of the deep and strong muscle masses. Only in a small area between superior nuchealis line and transversal sinus the aproach design is minimal invasive regarding approach trauma. Trans-dural duplex sono with burr-hole probe will inform about sufficient

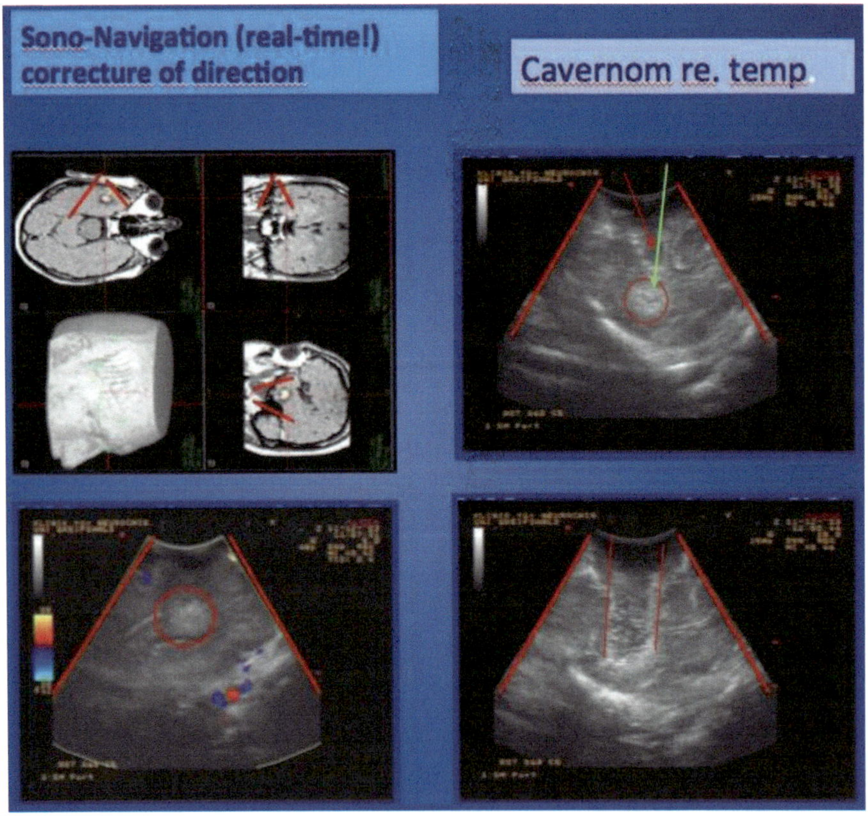

Graph 4.32 Navigation of a small lesion: Targeting

positioning and size of the approach. Morover, the visualization of main-feeders location and the vascularisation will inform whether the leason can be surgical managed safely through a small approach. Sono findings indicate if any adaptations of the planned aproach is finally needed, and sono findings also will show how exactly this adaptation is needed. Surgeons used to standard working overestimate the need of size in general. To the senior authors experience and case analyzes during over 30 years, a MIN approach, detailed by sono assistance, measures one third of a standard approach. Sono assisted correction are in the ranch of some millimeters. Post-op the function of neck muscles will be not impaired with this strategy, which is important for the post-op morbidity and the frequency of complications. Liquor collections in MIN approaches usual need no surgical revision. MIN in preterm is convincing as all condition call for minimally invasive strategies and techniques, ultrasound assistance and guidance is most satisfactory there. All procedures in the

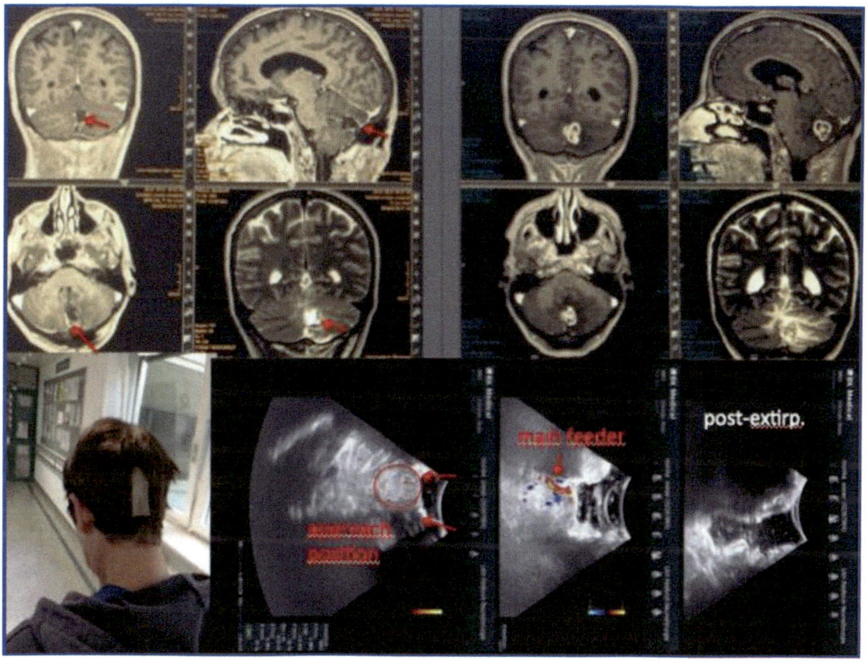

Graph 4.33 Metastasis posterior fossa

youngest of all our patients must be seen with view into future. The prize of standard procedures is much higher in this group compared to grownups. Working space is extremely limited and tissue premature. The problems are often biological banal in this type of hydrocephalus—compared with neoplasma—but the application is most difficult due to the mentioned conditions. Therefore a strong focus must be given to the way of application and to the timing and planning of application. Only ultrasound assistance can realize this challenge, the frequent extraordinary problems that will arise should not be answered by standards rather than by individual decisions and MIN strategies and techniques. Even a simple placement of drainage or reservoir should be sono-assisted and with a minimal opening. Each revision and additional surgery can be the beginning of a disaster in the near future or a life-long burden for the patient. Ultrasound assisted surgeries need an innovative mind and strong education and training (Graph 4.35).

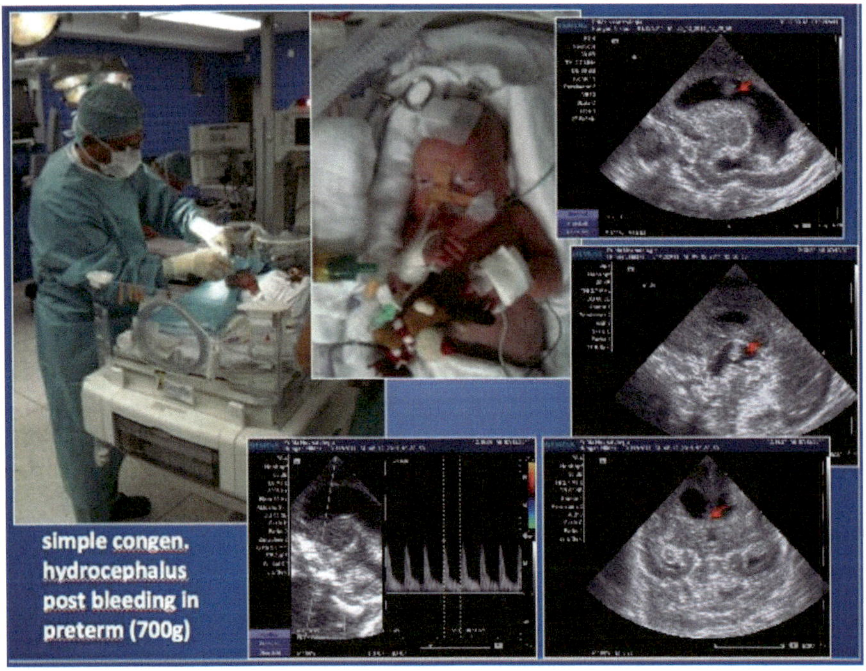

Graph 4.34 Preterm Hydrocephalus post connatal bleeding and sono-assisted procedures

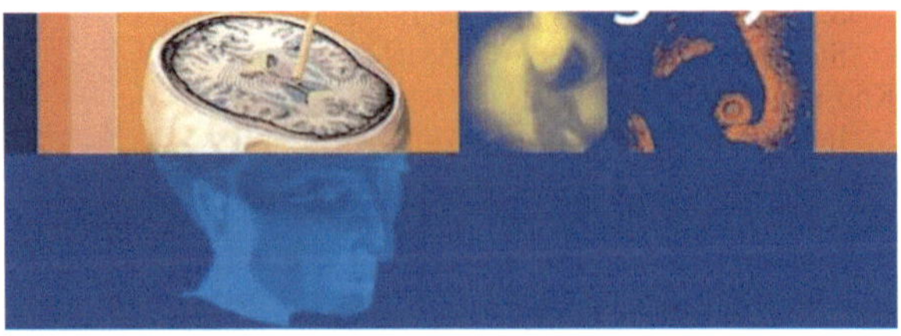

Graph 4.35 Trans-endoscopic Ultrasound for Neurosurgery (2005) (reviewed 2019)

4.11 Trans-endoscopic Ultrasound for Neurosurgery

This very special ultrasound technique is still not well recognized though reported by the author since 1996 and published as a book in 2005, Springer. As it is a valuable tool within neuro-sonography enabling navigation of endoscopes and giving some unique and easy solutions to unsolved problems in neuroendoscopy it is presented here additionally in short. To provide the beginner of neurosonography with the basic for startup this combined technique is in a short booklet integrated to the neurosonography chapter and presented detailed as it differs markedly from usual neurosonography. The book from 2005 is available as e-book. The technique was primarily provided by ALOKA co. but was not certified for commercial reasons. We are working with an alternative company and an evolved technique in the near future. As combined techniques are a central part of MIN—Philosophy it should become a tool for the recent future (Graph 4.36).

The step from microneurosurgery to endoneurosurgery has meant a change to a minimally invasive technique, but at the same time the new technique is less safe (Schroeder et al. 1999), which limits its applicability. Further development have therefore had to be aimed at making neuroendoscopy safer. The current concept used to establish a near-real-time guidance system has involved the use of MR or CT, requiring enormous financial investment and technical effort (Albert et al. 1998; Black and Mehta 2000; Fahlbusch and Nimsky 2000; Grönemeyer et al. 1994; Hayashi et al. 1995; Kobayashi and Okudera 2000; Maciunas 1993; Wirtz et al. 2000). It is still not clear what surgical benefit this approach will have (Kelly 2000; Maciunas 2000; Paleologos et al. 2000; Shekhar 2000) (Graph 4.37).

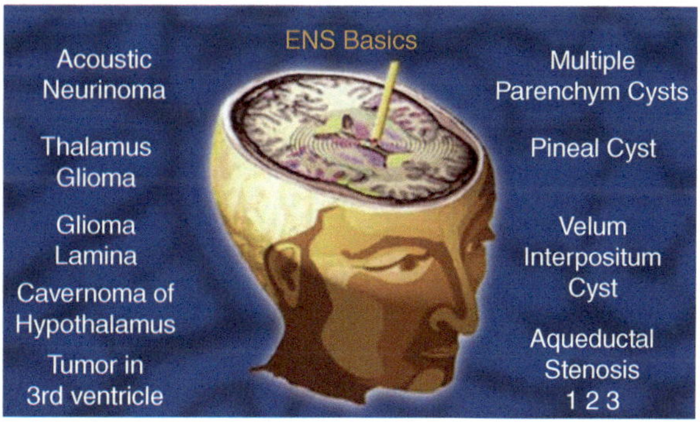

Graph 4.36 Cases of the ENS-Clips (Resch; Springer 2005)

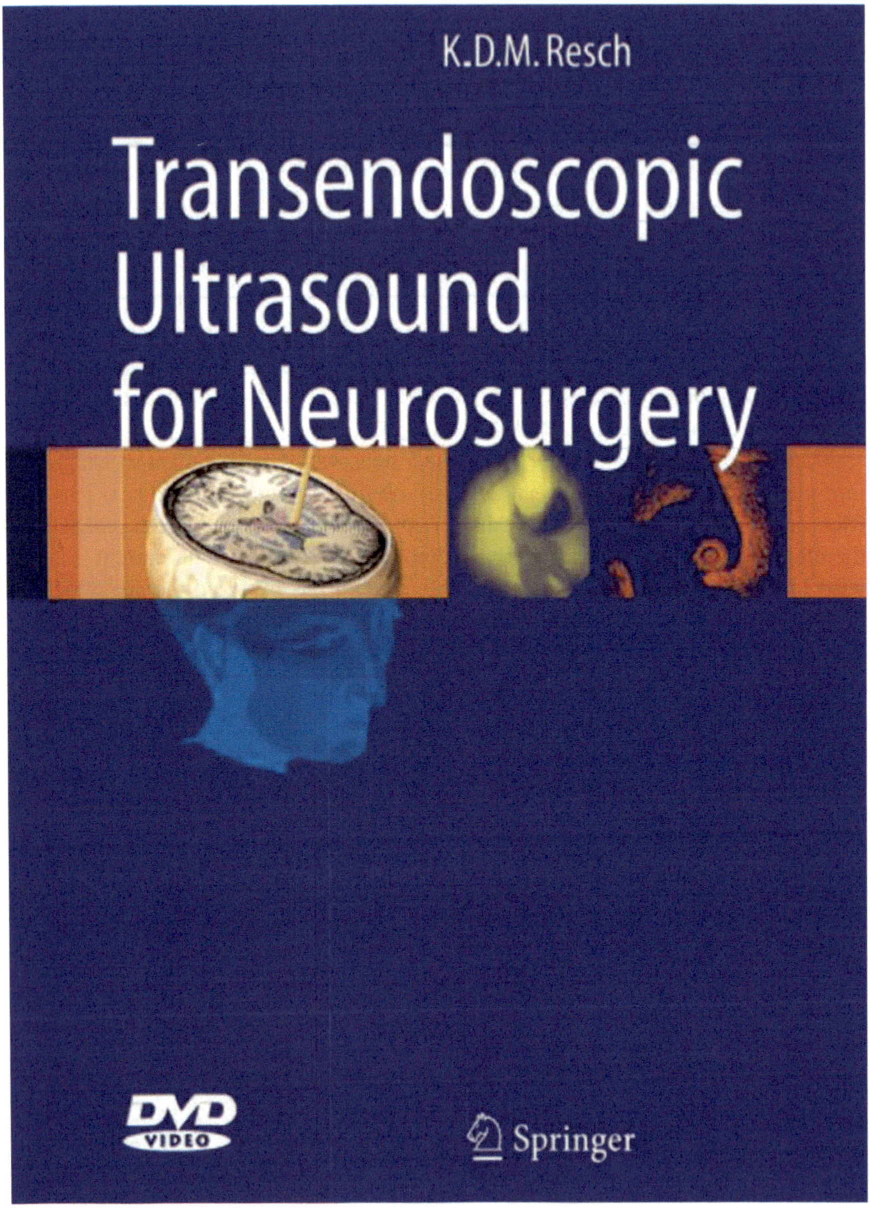

Graph 4.37 First book on ENS

Another idea, therefore, was to equip the endo-scope with a sonographic guidance system. As it stands, endoneurosonography is now a technique that can make a huge contribution to the elaboration of minimally invasive techniques in neurosurgery, as described below (Graph 4.38).

History of Transendoscopic Sonography

History can guide us to the future, but only if we respect such warnings as the one given by Dagi (1997) about the kind of neurosurgical history currently presented: "The obligation was to a neurosurgical ideal rather than to any individual patient; to an incompletely articulated vision of personal virtue rather than to the profession specifically; and to expression through professional activity rather than through philosophical reflection.". A collection of anecdotes about heroes, or a genealogy of the surgeons who did something for the first time, will not be very beneficial. From a neurobiological point of view, history involves documentation and remembering, both of these being essential components of consciousness. Moreover, history can be seen as 'archeology of the brain' and 'phylogenesis of mind' (Resch 1999). The following short summary of past events is not intended to be a data collection, but rather to give a comprehensive update of the current status, and it focuses on the power of conceptual thinking to create the future.

In the 1990s different disciplines started to use the technique of transendoscopic sonography: cardiology (Coy et al. 1991; Müller et al. 1996; Pandian et al. 1993; Roelandt et al. 1993, 1994; Rosenfield et al. 1991; Schwartz et al. 1994), angiology (Aschermann and Fergusson 1992; Aschermann et al. 1992; Cavaye et al. 1991; Delcker and Diener 1994; Isner et al. 1990; Ludwig et al. 1995; Neville et al. 1989), gastroenterology, and urology (Frank et al. 1994; Köstering 1991; Wickham 1993).

One interdisciplinary group, working at the same time as we, were investigating the use of ultrasonography and introducing it in their neurosurgery department, starting to use the probe directly, in combination with stereotactic biopsy, in two patients (Froelich et al. 1996). Another group working in pediatric neurosurgery also started to implement it in their clinical practice; both groups, however, failed to achieve informative imaging. In endoscopy, it was already a given concept that anatomy should be ahead of surgery, so starting with anatomical investigations to learn the typical features of the mini-sonoprobe and their correlation with endoscopic anatomy was quite normal practice (Grunert et al. 1995; Perneczky et al. 1993; Resch and Perneczky 1993; Resch et al. 1994).

In 1996 we examined the anatomical imaging achieved with endoneurosonography (Resch et al. 1996) as seen in the representation of the ventricular system (Resch and Reisch 1997) and the imaging of the basal cisterns (Resch and Perneczky 1997) with 3D imaging (Resch and Perneczky 1997). In a preliminary series, after our first experience of using the technique in clinical practice, anatomical representation of the spinal-cord was included (Resch et al. 1997; Resch and Perneczky 1998).

In March 2002, the first international course on ENS took place in combination with the head-mounted display (HMD) system at university of Heidelberg; Germany (www.ens-surgery.com) (Fig. 4.50, Graph 4.39).

Graph 4.38 Axial sono wave model

Fig. 4.50 Firs course on trans-endoscopic ultrasound for neurosurgery; Univ. Heidelberg 2002

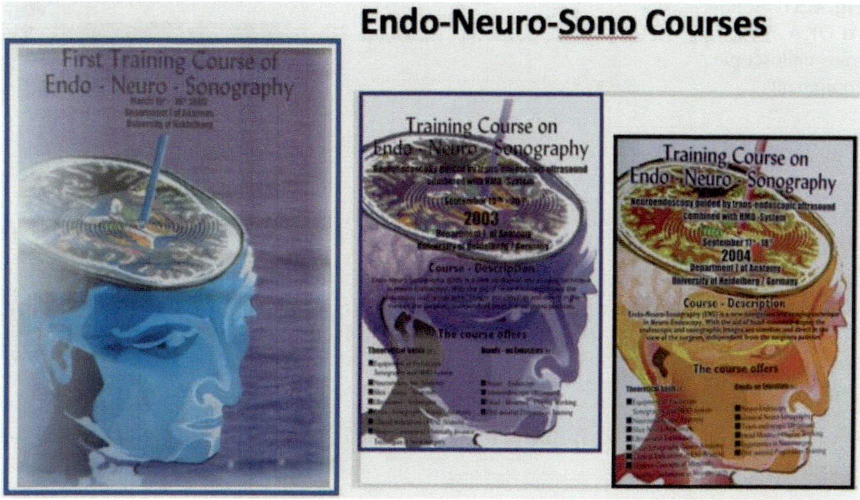

Graph 4.39 Posters of the first 3 ENS courses at Univ. Heidelberg (2002; 2003; 2004)

Sono-system. A small mobile conventional sono-graph machine with an additional arm containing the rotation motor for the sono-catheter. The connector is in contact with the input signal. The keyboard can be used for data input with a variety of imaging parameter configurations. The image is displayed on two monitors. Today highend machines can be equipped with the technique (Aloka 5000) (s. left). Tech. data (Fig. 4.51):

- **MHZ: 10, 20, 12.5**
- **F: / 6; 6.2; 8**
- **L: 192/110 cm**
- **r: 1–6 cm**

Sono-probe. The sono-catheter (*1*) has a diameter of 6 F and can be introduced into the working channel of conventional neuroendoscopes. At the tip of the catheter is a mini-sono-probe (*2*). The catheter is inserted in a sterile plastic sheath that has been filled with aqua (not saline). The thin catheter is the only part that appears in the operative field, so that it does not disturb the ergonomics of the working conditions. Another 8-F catheter is used as a sono-dissector outside the endoscope (Fig. 4.52).

There are two main types of probes, mechanical (a, b) and electronic (c). The ALOKA system is a mechanical type (b), and the probe is rotated by a motor device. The Aloka catheter is an elastic cable with a mini-sono-probe at its tip (Graph 4.40). This catheter is introduced into a sterile plastic sheath. Before the catheter is introduced, the sheath is filled with injectable water by means of a thin plastic trocar with a tiny lumen. While the sheath is being filled with liquid (not saline solution) the trocar is carefully withdrawn from the sheath. The sono-catheter is then slowly

Fig. 4.51 Sono-system ALOKA 5000 with trans-endoscopic equipment

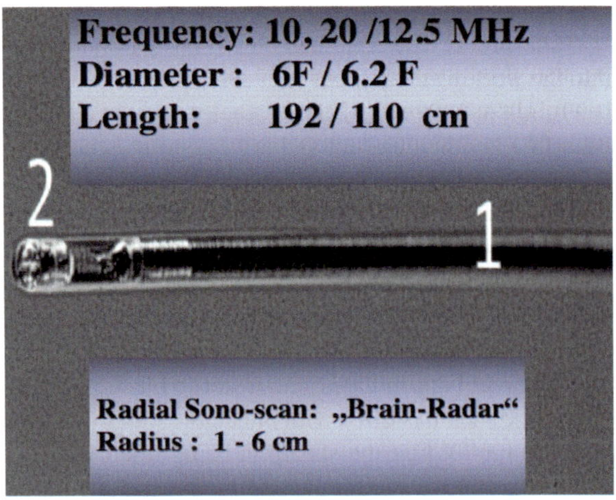

Fig. 4.52 Trans-endoscopic ultrasound catheter

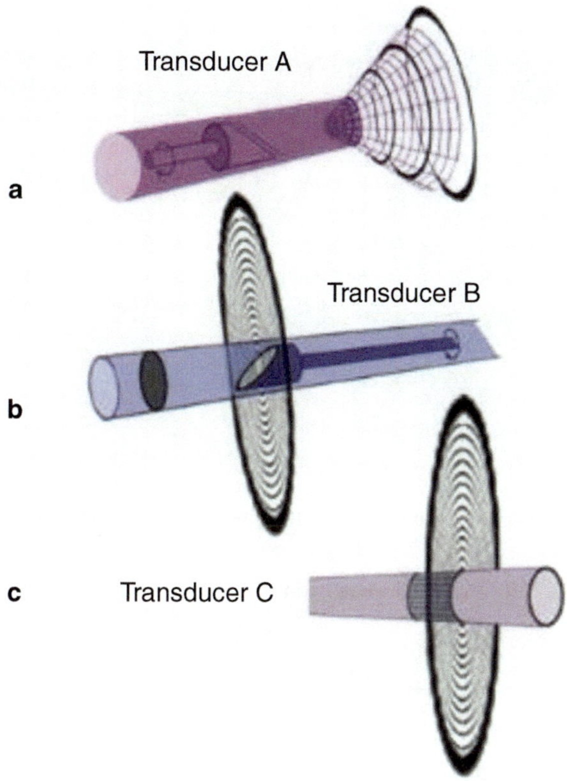

Graph 4.40 Types of transluminal ultrasound catheters

pushed into the sheath, and care must be taken to ensure that there are no air bubbles in front of the sono-probe at the tip. The tip of the probe is smooth, and there is a 1-mm space between it and the end of the sheath. At this point the prepared sono-catheter can be connected with the rotation motor and the signal line. The B&K catheter has a chamber around the mini-sono-probe at its tip, which can be filled with liquid (water) by injection.

The mechanical probes that have been used (ALOKA and B&K) are of type B (Graph 4.40). The image provided by use of the probe is a 360° scan ('brain radar') displayed on a monitor, on which some parameters can be varied to get the best view of different anatomical structures.

The concept of projection geometry is illustrated in (Graph 4.41): the projection of the scan can be orthogonal to the axial, sagittal, or frontal plane of the anatomy, as it commonly is in CT or MR. Often an oblique projection plane oriented to the pathway of the endoscope in the cranium is used. This is one of the main reasons for difficulties in interpretation of the anatomy visualized by the scan.

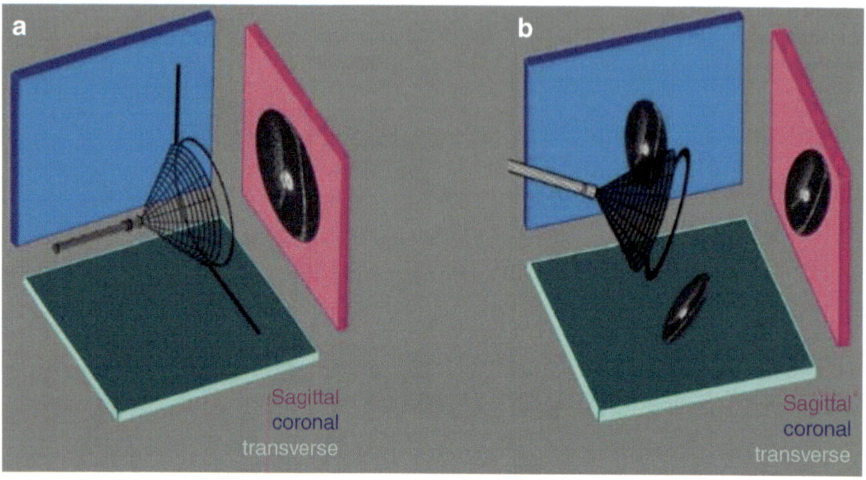

Graph 4.41 Triplanar geomrtry of the trans-endoscopic ultrasound

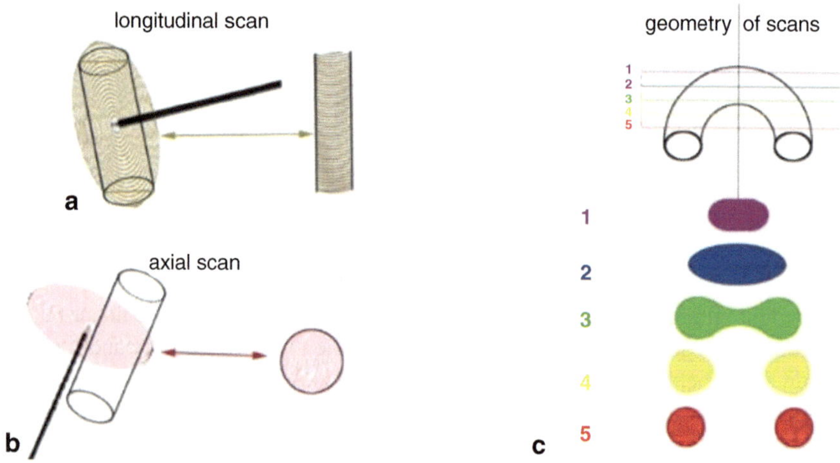

Graph 4.42 Scanning geometry of trans-endoscopic ultrasound

Scanning geometry (Graph 4.42) longitudinal, transverse/axial) is important. The representation of a volume scan must be anticipated mentally according to the geometry of scanning. It must be expected that the result of scanning will be represented in a way that is influenced by the geometry of the volume.

4.11.1 History of Trans-endoscopic Ultrasound

Just as the color, focus, and orientation of an endoscopic image have to be adjusted, a sonographic image also needs fine-tuning if it is to be useful. It is most important

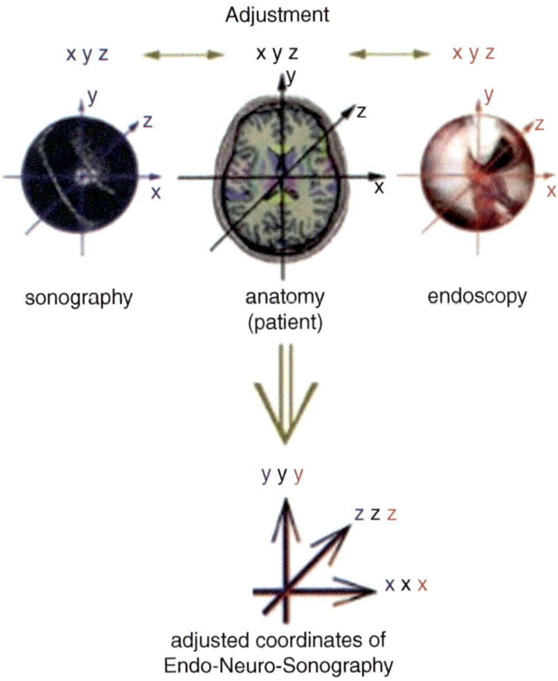

Graph 4.43 Adjustment of trans-endoscopic ultrasound

that adjustment of image orientation be done while the focus is on a well-known typical structure with a strong echo signal. Then the orientation focus must be tested by movements of the scope, which in turn must correspond to changes seen on the display of the sonograph. Actual changes of direction and velocity and those seen on the endoscope monitor and the sonograph display must be exactly the same. If this is not the case, the sono-image cannot be correlated with the endoscopic view or interpreted, and competent navigation cannot be achieved (Graph 4.43).

4.11.2 Imaging Properties

Endoscopic viewing, even with a limited range and without anatomical mapping outside of this view, appears questionable. In combination with mapping given by the sono-scan ('mini-CT') at the tip of the endoscope, and real-time imaging with the sono-probe, however, endoscopic movements can be mapped in the sono-scan. The tip of the endoscope can be seen as a moving spot on the sono-scan when the endoscope is moved. A target outside the endoscopic view can be reached with online control by sono-scan mapping. This real-time feedback by the sono-scan, representing the starting point and target, fulfills the requirements of navigation. It makes endoscopy intuitively safer (Graph 4.44).

Some effects caused by the equipment itself can lead to a wrong interpretation or distort the image on the screen. Such artifacts can occasionally also be useful for

Graph 4.44 Navigation property of trans-endoscopic ultrasound

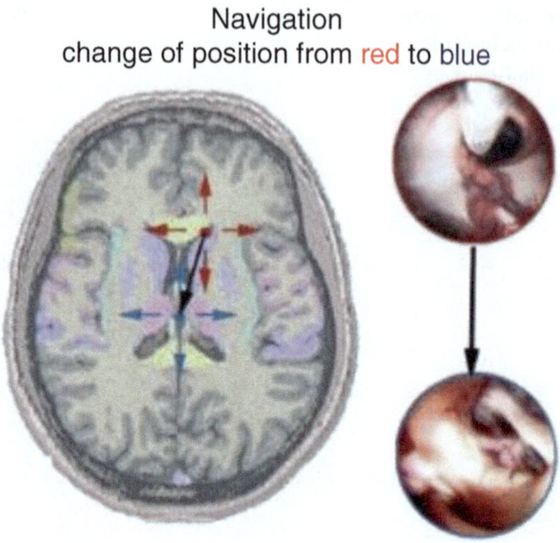

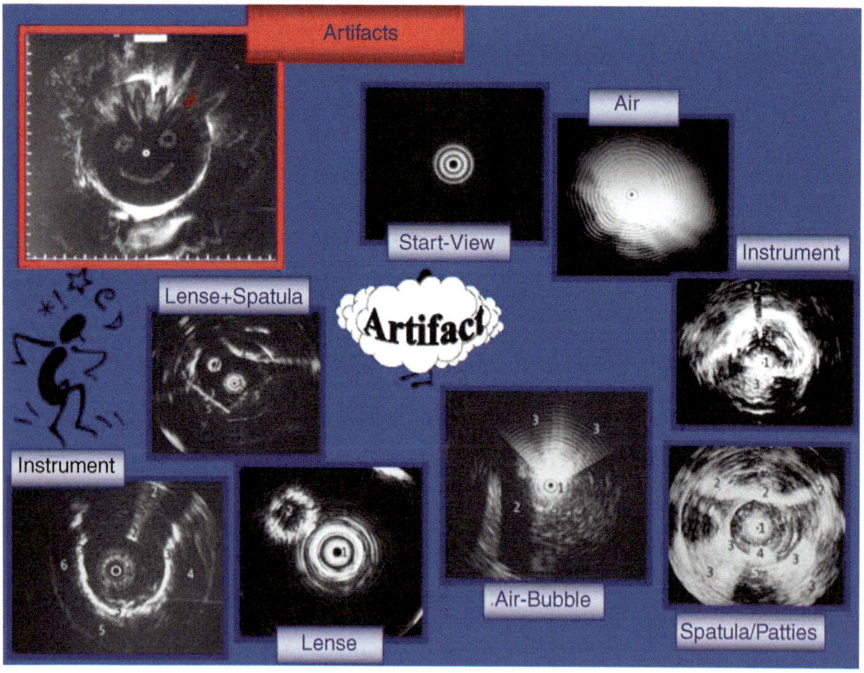

Graph 4.45 Artifacts of trans-endoscopic ultrasound

orientation and adjustment. Some of these artifacts seen in ENS are well known in sonography in general, while others are specific to ENS, as illustrated in (Graph 4.45).

The following sonographic anatomical aspects of ENS were established on 21 specimens. After the first clinical series, the anatomical images were shown from

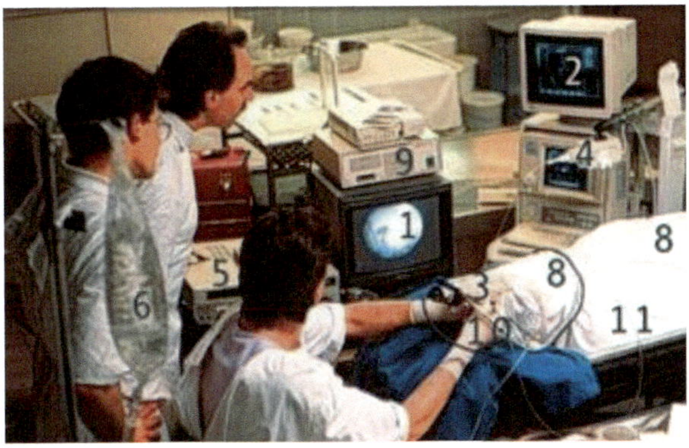

Fig. 4.53 Birthday of trans-endoscopic ultrasound for neurosurgery (laboratory setup)

the viewpoint of clinical relevance and only the most typical and relevant examples are reviewed here, with particular attention to principles of imaging (Fig. 4.53).

The examinations were done on fresh specimens by different neurosurgical approaches and burr holes. The nonfixed specimens offered the best model in terms of surgery and sono-echo characteristics of tissue.

Work on non-fixed specimens is surprisingly similar to actual surgery but cannot be planned. This handicap can be compensated by freezing the specimen, which will lead to conditions close to those in fresh preparations when specimens have been thawed for a planned preparation.

Prints, photographs, and parallel sono- and endoscopy video recordings were used for documentation. Modern equipment can also be used to provide digital documentation ready for multimedia use.

The program for presenting ENS involves:

• Different approaches	• Subarachnoid spaces and cisterns
• Important neural and vascular structures	• Endo–sono-correlation

CT: Cavum of pellucid septum/CT. The CT scan shows a clasp of the midline structures. Frontal horns of lateral ventricles (*1*) are divided by the pellucid septum (*2*), which presents a small cyst (*3*). The septum ends at the fornices, where both foramina of Monro enter the third ventricle. The frontal horns are bordering the caudate nuclei (*4*) laterally and the third ventricle, formed by the thalamus (*5*), on both sides. The hyperdense pineal body (*6*) is clearly visible, marking the dorsal end of the third ventricle (*7*).

ENS: Cavum of septum scan. The sono-scan is concentrated in this close-up of the frontal horns (*1*) with pellucid septum (*2*) between, clearly representing the cyst (*3*) that was visible on CT. Both foramina of Monro (*4*) are visible, plus the choroid plexus (*5*), which appears at higher echo-density than in the CT. The CSF/tissue border is also represented much more strongly on the sono-scan than on the CT scan. The sono-scan shows the typical torsion of geometry according to the angle of the endoscope plus catheter to the CT plane in the sagittal and coronal dimensions. The sono-probe (*6*) is visible in the right frontal horn, while the tip of the endoscope

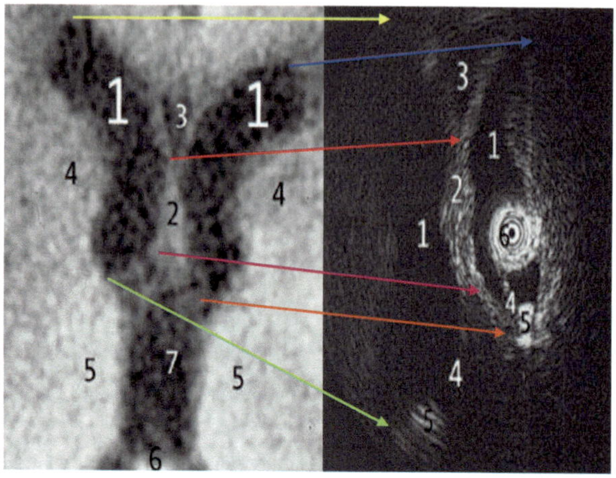

Graph 4.46 CT—ENS Correlation

in the sono-probe is presented in the anatomy precisely and in real time. Movements and direction of movements of the scope are visible online and in real time (Graph 4.46).

Endoneurosonography adds an axial view of the probe tip's position to the forward view of the endoscope, which itself is visible in the scope and on the sonographic view. This axial vision is like an axial 'mini-CT' of the tip plane on which the position of the tip itself can be localized in relation to the axial anatomy. The zoom function makes it possible to adapt the view in relation to the size of structures. It is possible to see an overall picture of some 6 cm or to see only one small cistern of some millimeters in diameter. In addition, it allows anticipation of the aspects that will come into view in the endoscope next, by 'looking through the parenchyma' and viewing a larger overall area than the scope does. The 'online mode' of the sono-view allows changes of structures, such as changing size of ventricles and shifting of the parenchyma, to be observed, as well as pulsation or blood flow in the vessels. When used in combination with the endo-view, it allows safe neuronavigation of the scope online and in real time.

For such problems as intraoperative imaging or navigation of the endoscope, the imaging characteristic for structures 4 cm or less in diameter is comparable to that of CT. A small cyst of the pellucid septum was visible in the sono, for example, as well as retrospectively in the CT. There are, however, important differences between CT imaging and ENS-imaging:

- The geometry of the sono-scan shows **torsion** representing the angle of the scope, as well as the angle of the catheter to the plane of the CT scan in the sagittal and coronary planes.
- The sono-scan shows the **CSF-tissue border** and the choroid plexus more strongly than they appear in the CT scan.

- The sono-scan gives an **online** and **real-time** image, in contrast to intraoperative CT.
- The sono-scan shows movements, such as a change in ventricle shape or blood flow in the vessels, representing **physiological parameters**.

On the keyboard display, **magnifications** of 9–124 mm in 24 increments can be used. This means imaging of a cistern in the display, as well as a hydrocephalic lateral ventricle will be possible at high resolution (Graph 4.47).

Both frontal horn (*1*) and septum (*2*) are zoomed until only structures within 9 mm of the sono-probe (*3*) are visible without loss of optical resolution, but the noise artifact will become more intense.

The ENS catheter detects and displays **volume changes**. The ventricular volume (*1*) is shown becoming progressively smaller from **a** to **c** as a result of CSF evacuation. Eventually the ventricular walls and the choroid plexus (*3*) come close to the sono-probe. This gives possibility to adapt the scan to the anatomy and to the needs of imaging (Graph 4.48).

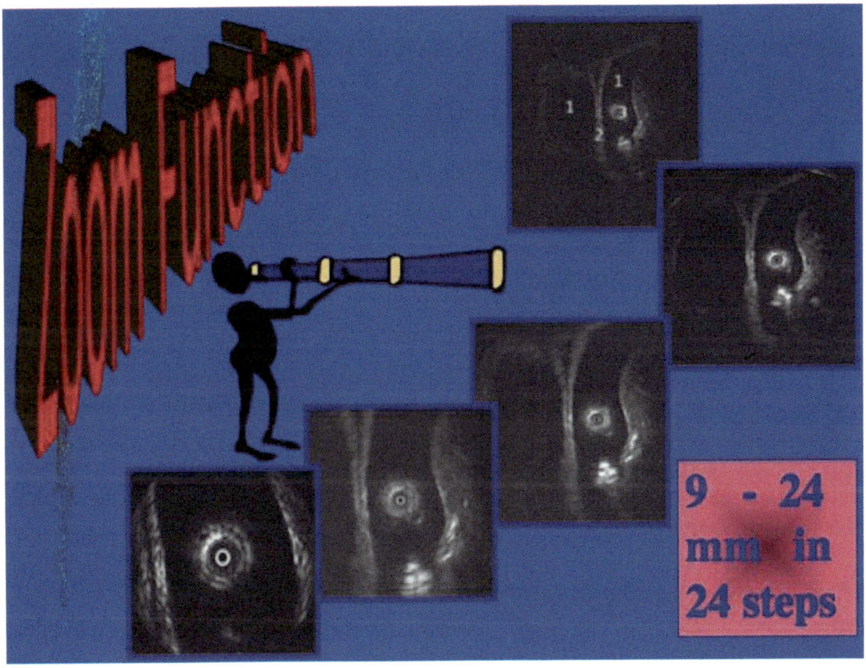

Graph 4.47 Zoom function of trans-endoscopic ultrasound

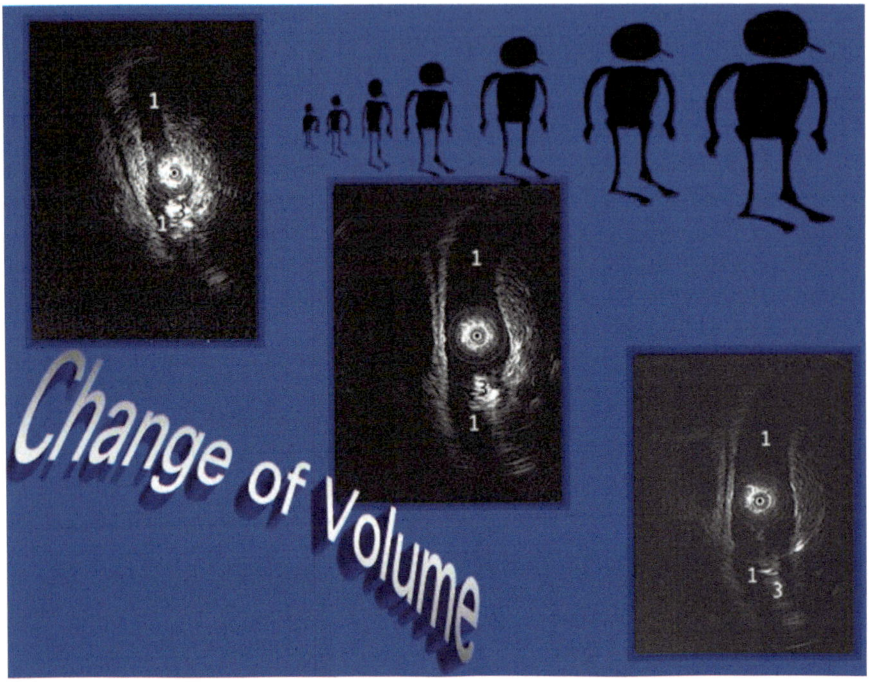

Graph 4.48 Volume monitoring of trans-endoscopic ultrasound

4.11.2.1 ENS-Anatomy

The sono-scan shows the well-known anatomy and endoscopic anatomy in typical slices and shapes with a high significance in recognition and easy correlation to the optical visualization. Once some experience has indicated what can be visualized and what can be expected, intuitive handling and interpretation of the standard images will soon result. Some *typical examples* of imaging are described below (Graph 4.49).

4.11.3 ENS Anatomy

During their progression through the ventricular system, the shape of the scan image varies according to the position of the sono-probe and the tip of the endoscope: the endoscope gives the typical 3D view in front of the lens into a given space showing the typical landmarks of the right lateral ventricle. The sono-scan not only shows the right ventricle, but also visualizes both ventricles in 2D like a CT and also gives a view into the parenchyma. The contact of the sono-probe with the choroid plexus is well controlled by the endoscope and represented exactly in the sono-scan (Graph 4.50).

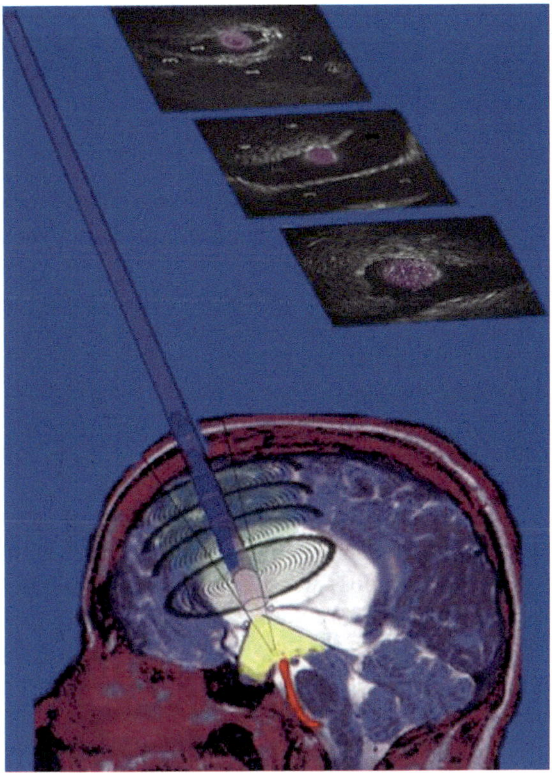

Graph 4.49 Trans-endoscopic ultrasond shaft-system model, Anatomy of sono-levels of the endoscope canal

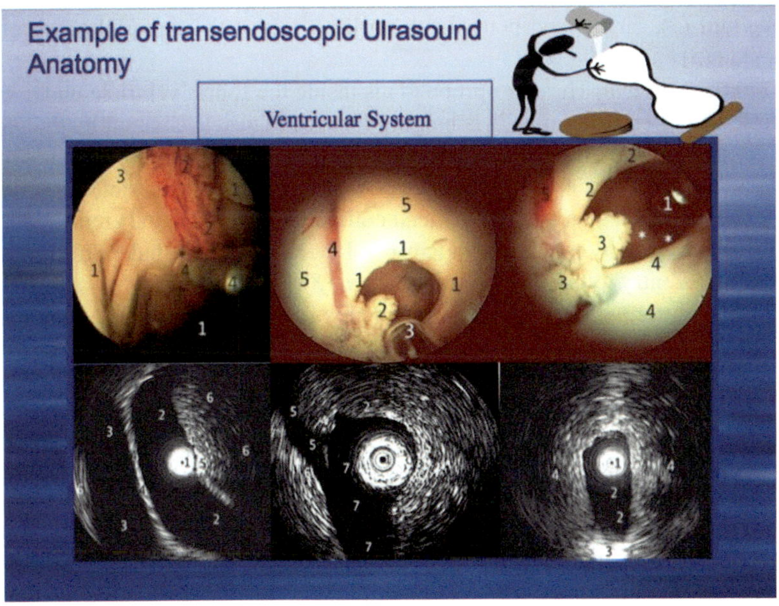

Graph 4.50 Anatomy- and sono-levels of trans-endoscopic ultrasound

Right lateral ventricle: The endoscopic view into the right lateral ventricle (*1*) shows the choroid plexus (*2*) and the pellucid septum (*3*) at the level of the sella media. The sono-probe (*4*) is optically controlled while in contact (***) with the plexus (*2*).

Both, right (*2*) and left (*3*) lateral ventricles, shown by sono-scan. The probe (*1*) itself is visible at the *right*, in contact (***) with the choroid plexus (*5*). Moreover it can 'see' through the pellucid septum (*4*) and into the parenchyma (*6*).

Foramen of Monro. The endoscope is in front of the right foramen of Monro, which is formed by the fornix (*1*) and the choroid plexus (*2*). The sono-probe (*3*) is controlled optically by the endoscope and pushed a little way into the foramen. A septum vein (*4*) crosses over the pellucid septum (*5*).

Foramen of Monro. The sono-probe (*1*) is just in the right foramen of Monro formed by the fornix (*2*) and choroid plexus (*3*). In contrast to the endoscope, the sono-scan represents the left fornix in an axial cut (*4*) and also the left foramen of Monro (*5*). The left choroid plexus too is just in the scan (*6*), and the whole length of the third ventricle (*7*) is displayed.

Third ventricle. Under endoscopic control, the sono-probe (*1*) is pushed into the third ventricle towards the mamillary bodies (*). The fornix (*2*), the choroid plexus (*3*), and the anterior part of the thalamus covered by lamina affixa (*4*) form the foramen of Monro. The septum vein (*5*) crosses the fornix (*2*).

Third ventricle scan. The sono-probe (*1*) is in the third ventricle (*2*), limited by the thalamus (*4*) on both sides and by the pineal body (*3*).

Aqueduct. The sono-probe (*1*) is entering the aqueduct (*) at the posterior third ventricle formed by hypothalamus (*2*) and posterior commissure (*3*) (Graph 4.51).

Aqueduct and posterior fossa. The sono-probe (*1*) is inside the aqueduct of the mesencephalon (*2*). As the zoom is very low, the posterior fossa is scanned with the cerebellum (*3*), clivus, and petroclival border (*4*), plus the tentorial notch on the right side (*5*).

Fourth ventricle. The sono-probe (*1*) is inside the fourth ventricle under visual control. The inferior vermis (*2*) is bulging into the ventricle covered by the variant of a circular plexus (*3*). The superior medullary velum (*4*) limits the inferior view while a view into the depth of the foramen of Magendi (*9*) is possible. The surface of the rhombencephalon shows the facial colliculus (*6*) and medial eminence (*5*) on both sides, divided by the medial sulcus (*). Laterally, the vestibular area (*7*) is visible, as is the entry to the foramen of Luschka (*8*).

Fourth ventricle scan. The sono-probe (*1*) is in the fourth ventricle looking into the parenchyma of facial colliculus (*3*), superior cerebellar peduncle (*4*), and superior medullary velum and superior vermis (*5*).

Inferior 4th ventricle. The endoscope enters the ventricle, controlling the sono-probe (*1*) and touching the plexus (*5*). The left (*2*) and right (*3*) cerebellar tonsils lead to the foramen of Magendi (*4*), where a circumflex tonsillar artery (*9*) is visible. The typical surface of the rhombencephalon contains facial colliculus (*6*), medullary striae (*7*) and hypoglossal triangle (*8*) (Graph 4.52).

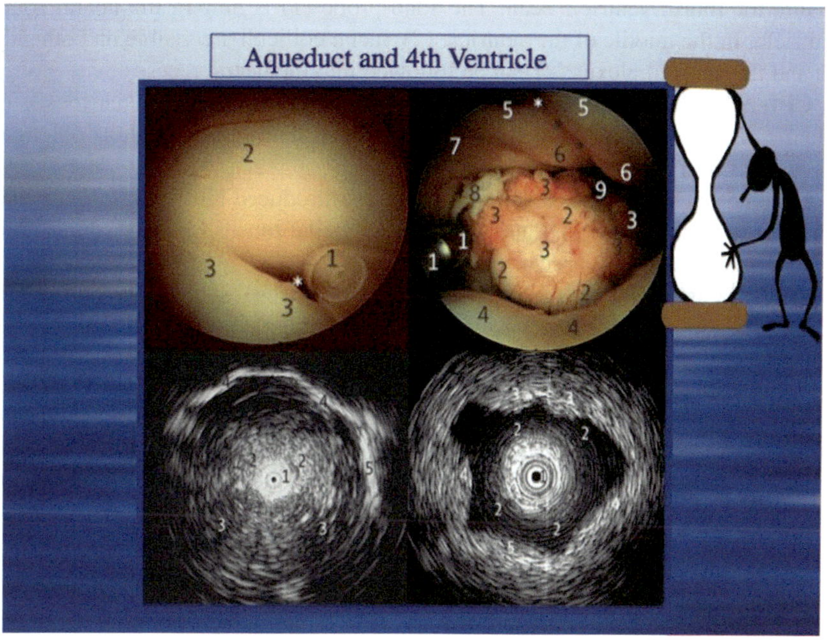

Graph 4.51 Anatomy- and sono-levels of posterior fossa/4th ventricle

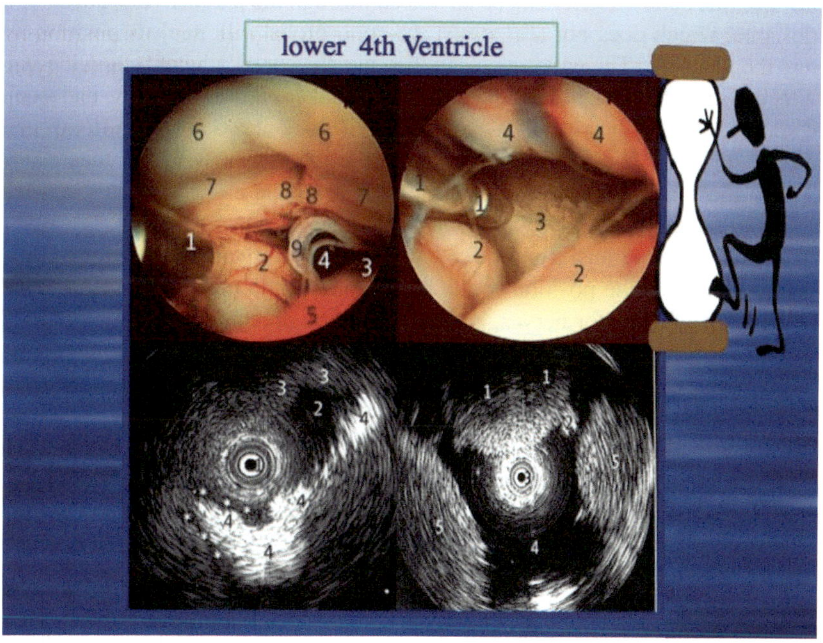

Graph 4.52 Anatomy- and sono-levels of 4th ventricle

Inferior fourth ventricle scan. The sono-probe (*1*) is now in the lateral fourth ventricle. In the middle of the ventricle (2), facial colliculi are visible on both sides (*3*) and the choroid plexus (*4*) can be followed into the lateral recess (*).

Cisterna magna. The sono-probe (*1*) is guided along the cerebellar tonsils (2) into the cisterna magna (*3*) with endoscopic vision. The ventral medulla oblongata (*4*) extends into the foramen magnum.

Cisterna magna scan. The sono-probe (*3*) is introduced into the cisterna magna (*4*) close to the medulla (*1*). Laterally, both cerebellar tonsils (*5*) are visible and the right vertebral artery (*2*) is scanned.

Ventral Cisterns. Like the ventricles, cisterns are ideal subjects for both endoscopy and endo-sonography. The 2D anatomical example presented is the interpeduncular cistern. This cistern is the target area in the procedure of endoscopic third ventriculocisternostomy (ETV).

The endoscope is inserted through a precoronal mediopupillary burr hole and advanced transcerebrally into the frontal horn. It is then passed through the foramen of Monro, and the endoscope is placed above the anterior floor of the third ventricle. The endoscopically visible landmarks for penetration are the mammillary bodies, the tuber cinereum, and the in-fundibular recess. In a translucent membranous cinereum (premamillary membrane) the dorsum of the sella is visible, and at times also the basilar tip. If the floor is bulging the infundibulum is wide open and the stalk is flattened to look like a red spot. The scope should penetrate between the dorsum of the sella and the basilar head.

This route can be navigated by endo-sono instrumentation. Especially in small ventricles, the sono-catheter can be advanced through the parenchyma, ahead of the endoscope, which does not see, whilst the sono-probe will 'see' its position as it enters the ventricle. The endoscope can then use the sono-catheter as a guide-wire.

Once both instruments have entered the frontal horn, both images, endoscopic and sonographic, are visible. The endoscope will look ahead into the infundibulum and onto the floor of the third ventricle, while the sono-probe will 'see' the thalamus and then the hypothalamus, and also the vessels of the circle of Willis and what blood flow there is.

If the floor is not translucent the sono-probe can be advanced as a 'seeing' catheter and will show the position of the basilar artery, the posterior communicating arteries, both oculomotor nerves, and of course the dorsum of the sella before the endoscope is advanced.

With strong zoom function and high-resolution a precise anatomy can be visible being invisible to the endoscope (Graph 4.53).

Sella level. The sono-probe (*1*) is placed at the dorsum of the sella (2). Laterally the petroclival ligament is attached (*3*). Through the bone of the clivus (dorsum) (2) the sella (*4*) is visible in an axial scan.

Prepontine cistern level. The sono-probe (*1*) is in touch with the clivus (2) and the pons (*3*). On both sides, the trigeminal nerves (*4*) with part of SuCA (*6*) and the tentorium (5) are present.

Trigeminal nerve (5) level. The sono-probe (*1*) is in the prepontine cistern (*7*), touching the pons (*4*) and close to the clivus (2). This close-up view shows the

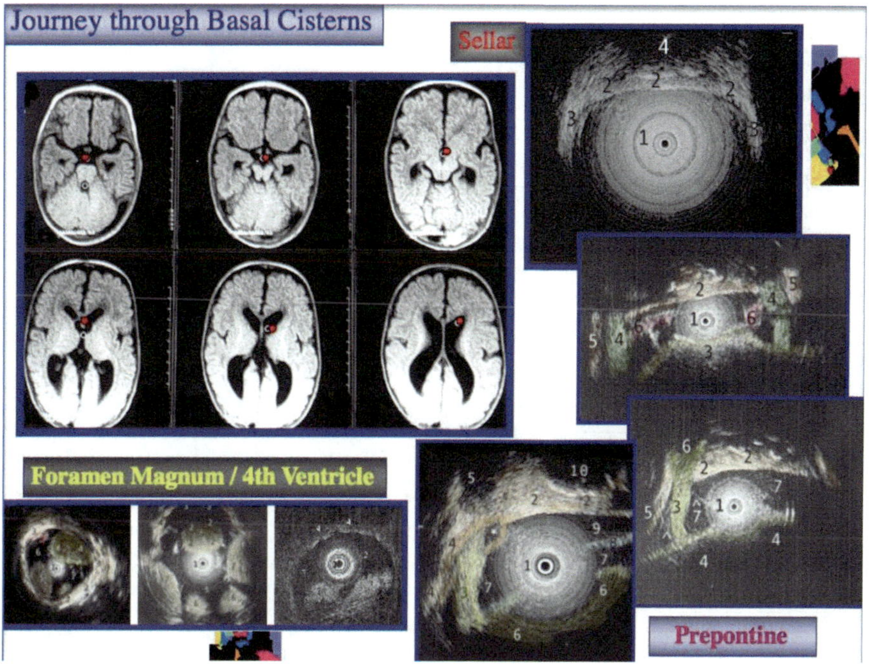

Graph 4.53 CT and ENS-levels of basal cisterns

trigeminal nerve (*3*) crossing the subarachnoid space (*7*) and entering Meckel's cavity (*6*) with its arachnoid sheath (^).

Trigeminal nerve level. The sono-probe (*1*) is located in the prepontine cistern (*7*) covered by the arachnoid membrane (*8*). Outside the membrane (*8*) the epi-arachnoid space (*9*) appears. The clivus (*2*) ends laterally at the cavernous sinus (*5*), where the trigeminal nerve (*3*) enters its dural porus (*) beneath the tentorium (*4*). The sono-probe touches the pons (*6*). In the clivus (*2*) a small sphenoid sinus (*10*) becomes visible.

Fourth ventricle. The sono-probe (*1*) dislocated in the fourth ventricle (*2*). Ventrally the surface of the rhombencephalon with facial colliculus (*4*) is visible, while dorsally the choroid plexus (*3*) is present.

Cisterna magna level. The sono-probe (*1*) is placed at the cisterna magna (*5*) and has made contact with the dorsal surface of the medulla (*2*). Ventral to the medulla (*2*), the premedullary cistern (*3*) and the clivus (*4*) are visible. Dorsally, the tonsils (*6*) and lobus biventer (*7*) are present.

Foramen magnum. The sono-probe (*1*) is placed in the foramen magnum (*2*) and in the subarachnoid space of the cisterna magna (*6*) dorsal to the medulla (*4*) and ventral to the tonsils (*5*). At the bony border of the foramen magnum (*2*), laterally, the occipital condyles (*3*) are visible, as is the vertebral artery on the *left side* (*7*).

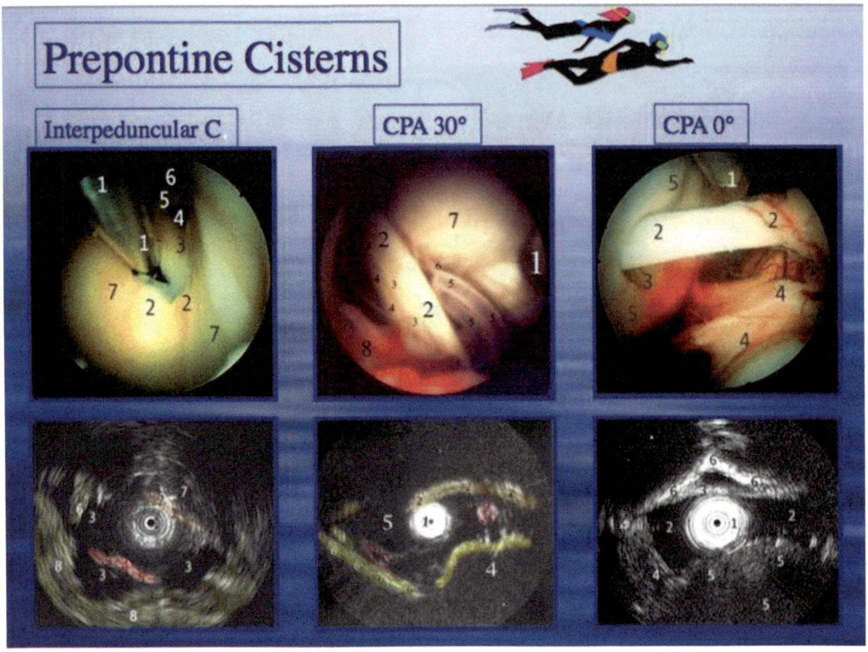

Graph 4.54 Anatomy- and sono-levels of left CPA

ETV. The sono-catheter (*1*) is perforating the tuber cinereum (*3*) anterior to the mamillary bodies (*2*). The endoscopic view does not see now the anatomy of interpeduncular cistern. The further topographical points are the infundibular recess (*4*), the optic chiasm (*5*), the supraoptic recess (*6*), and the walls of the hypothalamus (*7*) (Graph 4.54).

Interpeduncular cistern. In a near axial scan, the sono-probe (*1*) shows the basilar artery (*4*) as a hyperdense spot, both P1 segments (*5*) spreading laterally on both sides, and the dorsum of sella (2) as a strong hyperdense line. These are the main landmarks, which the sono probe may already 'see' even before penetrating the floor of the third ventricle. When placed in the interpeduncular cistern (*3*) it can image the oculomotor nerve (*6*) and the brain stem (*8*) and the view into the parenchyma of the pituitary gland (*7*).

Cerebellopontine angle (CPA). The endoscopic view from superior direction into the left CPA depicts the sono-probe (*1*), inferior to the 7/8 boundle with a facial nerve (*2*), intermedial nerve (*3*), and the acoustic nerve (*4*). An ICA loop (*5*) with a branching labyrinthine artery (*6*) is clearly visualized. The probe (*1*) is in contact with the petrous bone (*7*).

CPA scan. The sono-probe (*1*) is in contact with the lateral clivus (2) positioned in the left CPA (*5*) with the 7/8 boundle (*6*) and the ICA loop (...). Moreover the pons (*4*) and basilar artery (*3*) are visible.

CPA. The endoscopic view into the left CPA represents the sono-probe (*1*) inferior to the 7/8 bundle (*2*), running into the acoustic meatus (*3*) and medial to the petrous bone (*5*). The inferior lateral part of the pons (*4*) is bulging out in this unusual view from above.

CPA scan. The sono-probe (*1*) is in the left CPA cistern (*2*) in contact with the pons (*5*) and an arachnoid membrane (*3*). The 7/8 bundle (*4*) runs towards the petrous bone (*6*).

Jugular foramen. The endoscope looks on the jugular foramen (*3*), while the sono-probe (*1*) is in contact with the jugular tubercle (*2*). Accessory nerve (*4*), vagus fibers (*5*) and glossopharyngeal nerve (*6*) enter their dural pores (*3*). The sono-probe (*1*) is also in contact with 'Bochdalek's' body (*7*) and the medulla (*8*) (Graph 4.55).

Jugular foramen scan. The sono-probe (*1*) is positioned in the lateral cerebellomedullary cistern (inferior CPA cistern) (*7*). In contrast to the endoscopic view (Fig. 2.27), the position is now superior to the vagus fibers (*8*), running to the jugular foramen (*9*) and in contact with Bochdalek's body (*3*), pons (*6*), and medulla (*5*). An arachnoid membrane (*) is hardly visible. The 7/8 boundle (*2*) runs towards the petrous bone (*4*) to the acoustic meatus (*10*).

Jugular tubercle. The endoscope now visualizes the sono-probe (*1*) touching the right jugular tubercle (*2*). The vertebral artery (*6*) and fibers of the vagus nerve (*4*) cross the olive (*5*) and then run together with the accessory nerve (*3*) to the jugular foramen.

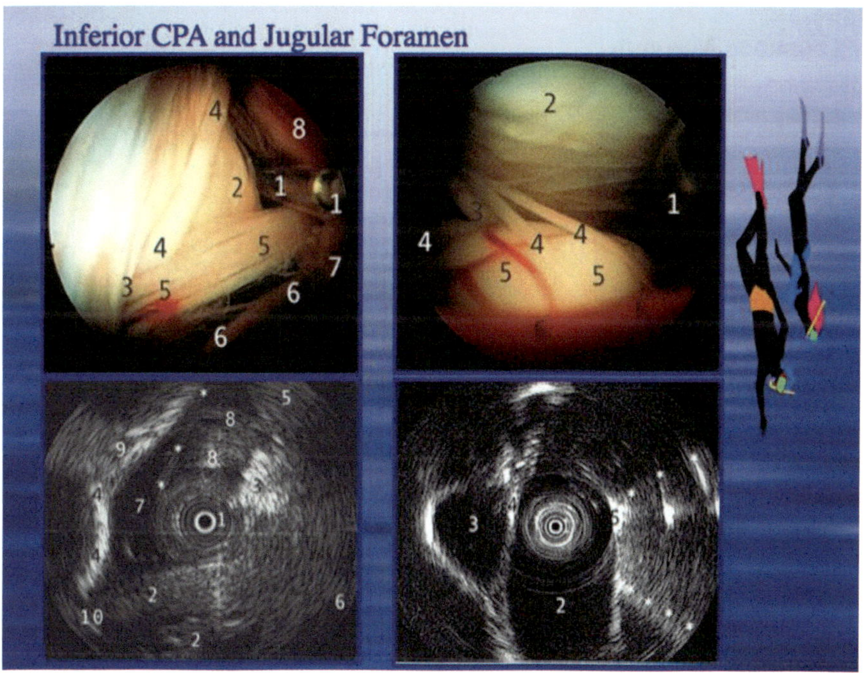

Graph 4.55 Anatomy- and sono-levels of left jugular foramen

Jugular tubercle scan. The sono-probe (*1*) is lateral to the cerebello-medullary cistern (*2*) close to the jugular tubercle (*4*), visualizing the sigmoid sinus at the jugular bulb level (*3*) through the thin bony layer of the jugular tubercle (*4*).

Foramen magnum. The endoscope is looking at the anterior border of the foramen magnum (*6*). The sono-probe (*1*) touches the foramen and the medulla (*2*) at C-1 level. Below this the right C-2 root (*5*), and above it the C-1 root (*3*) are accompanied by a laterally medullary vein (*4*) leaving the medulla laterally (Graph 4.56).

Foramen magnum scan. The sono-probe (*1*) is placed in the lower foramen magnum, ventral to the medulla spinal is at C-2 level (*3*). The dural border of the foramen magnum is present (*2*), as is the ventral bony border (*7*) where the dens axis is hardly visible (*8*). In the spinal subarachnoid space (*6*) the ventral (*4*) and dorsal (*5*) roots are visible on both sides.

Atlas-level scan. The sono-probe (*1*) is placed dorsal to the medulla (*3*) in the spinal subarachnoid space (*2*), touching the dorsal arch of the atlas (*5*) at the posterior tubercle (*4*). The scan is overlaid by the rotation artifact caused by uneven rotation of the ENS cable in the sheath.

Thoracic spinal canal scan. The sono-probe (*1*) is placed in the thoracic subarachnoid space (*4*) dorsal to the medulla (*3*) close to the dorsal subarachnoid septum (<) with some subarachnoid membranes (<). On the *left side* a dental ligament (^^^) within the scan is completely surrounded by the dural and bony border (2) of the spinal canal.

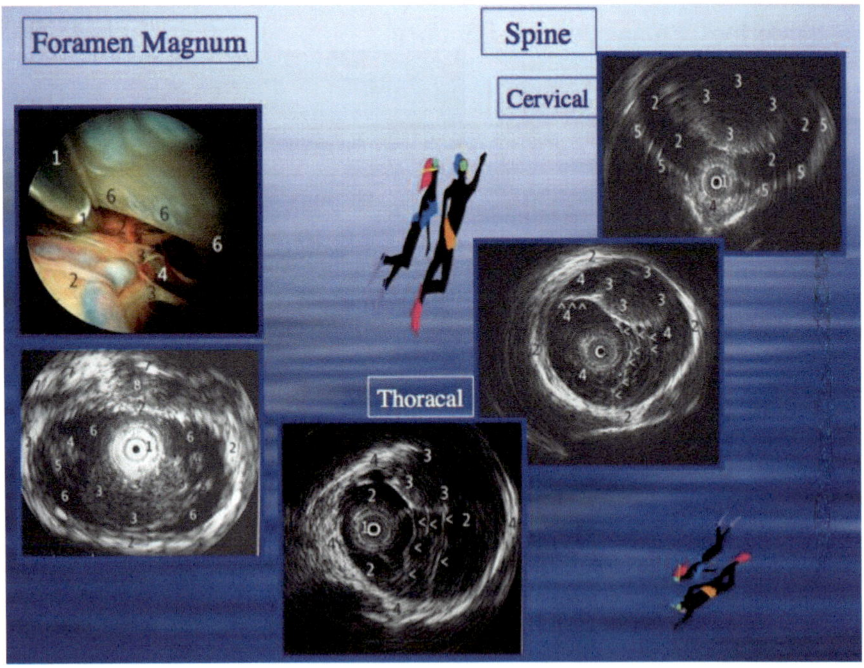

Graph 4.56 Anatomy- and sono-levels of F. Magnunm and spinal canal

Thoracic spinal canal scan. The sono-probe (*1*) is placed in the thoracic spinal subarachnoid space (*2*) dorsal to the medulla (*3*) and close to a network of arachnoid membranes (<), surrounded by the bony border of the canal (*4*).

Suprasellar and Retrosellar Space. When the sub-arachnoid space and the cisterns are to be imaged, endoscopy must be accompanied by irrigation, as the sono-probe needs a liquid medium to show up any findings. In the example presented the scope is inserted through a lateral supraorbital burr hole approaching the chiasmatic cistern with both optic nerves. In this case of a wide-open prechiasmatic window, the pituitary stalk and diaphragm of the sella are visible. Depending on the anatomical circumstances the endoscope can sometimes reach the posterior fossa, but in our case only the sono-probe, which was optically controlled, was advanced so far. This means that the sono-probe scans the suprasellar and sellar space in a semi-sagittal manner.

The sono-scan presents the sella with the entire parenchyma of the pituitary gland. Both, the stalk and the left optic nerve are visible, and in contrast to the endoscopic view the anterior part of the third ventricle is visualized (Graph 4.57).

Suprasellar space. From a right-sided lateral supraorbital burr hole, the endoscope approaches the suprasellar space. The chiasmatic cistern is visible, with left (*6*) and right (*7*) optic nerves, pituitary stalk (*3*) and diaphragm of sella (*2*). Under optical control, the sono-probe (*1*) is pushed along the left posterior clinoid process (*4*).

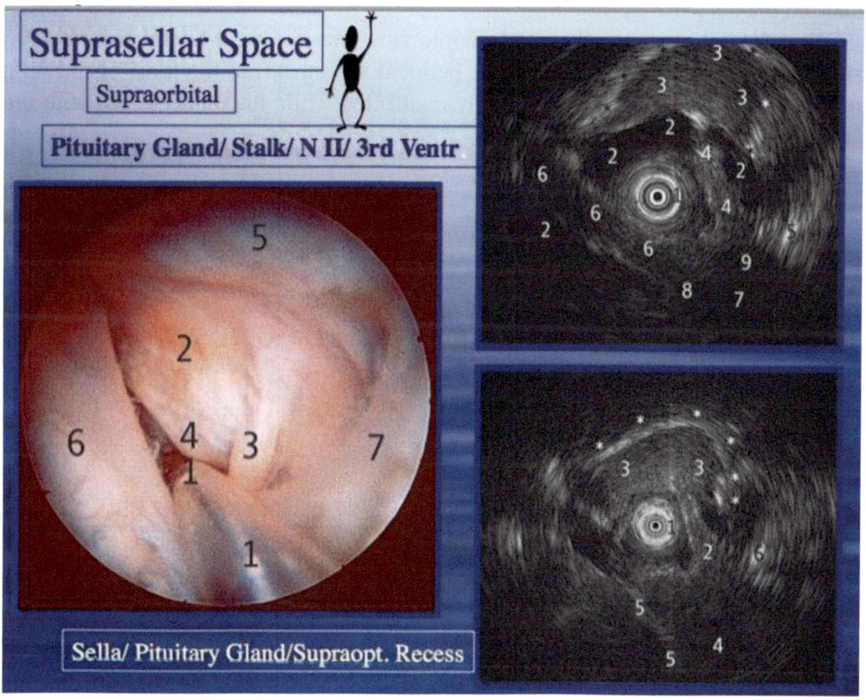

Graph 4.57 Anatomy- and sono-levels of supra/intra-sellar region

Suprasellar space scan. The sono-scan and the sono-probe (*1*) are located in the chiasmatic cistern (*2*), representing a semisagittal imaging plane. The sella (*) with the parenchyma of the pituitary gland (*3*) and the stalk (*4*) with the left optic nerve (*6*) are visible. In contrast to the endoscope, the sono-probe (*1*) can 'see' into the anterior third ventricle (*7*) with infundibular (*9*) and supraoptic recess (*8*). The dorsum of the sella (*5*) is visible.

Suprasellar space scan. The sono-probe (*1*) is in the suprasellar space close to the pituitary stalk (*2*) and the pituitary gland (*3*) in the sella (*). The anterior third ventricle (*4*) with lamina terminalis (*5*) is visible. The basilar artery (*6*) gives a high signal (*6*).

Retrosellar space. The endoscope is looking from the CPA through the trigeminal-pontine window into the retro-sellar interpeduncular cistern into the supra-sellar space, seeing dorsum of sellae with the stalk and the opposite ICA from medial (Graph 4.58).

In the next view, the ENS-catheter (1) is placed superior to the dorsum of sellae (2–3), below the stalk (4) and medial from the opposite ICA (5). It is in tough with the remnant of Liliequist membrane (6****). P1 (9), P com. artery (8) and oculomotor nerve (7) are present.

Retrosellar space scan. The sono-probe (*1*) is placed in the interpeduncular cistern close to the infundibulum and stalk (*2*). The sella (*) and pituitary gland (*3*), mesencephalon (*9*), and pons (*10*) are visible. The anterior third ventricle (*4*) with infundibular recess (*5*), supraoptic recess (*6*), and foramen of Monro (*7*) and the choroid plexus (*8*) are present.

Right lateral ventricle. The endoscopic view into the right lateral ventricle (*1*) shows the choroid plexus (*2*) and the pellucid septum (*3*) at the level of the sella media. The sono-probe (*4*) is optically controlled while in contact (***) with the

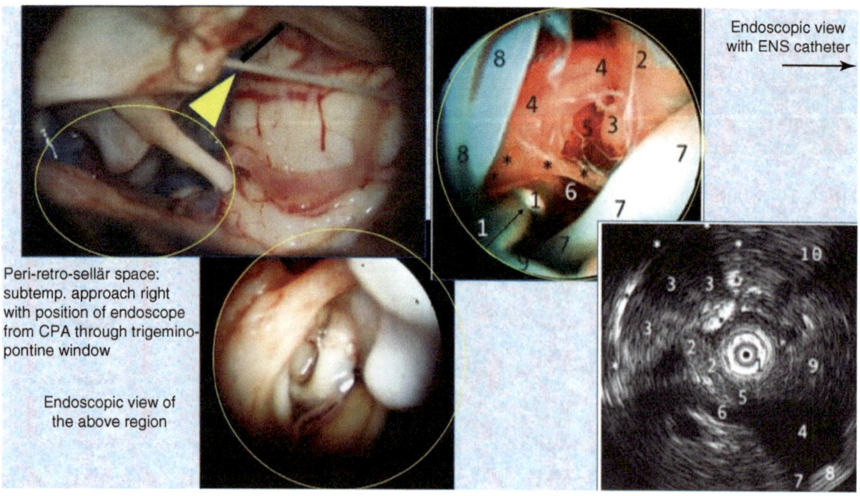

Graph 4.58 Supra-retro-sellar space from CPA through trigemino-pontine window

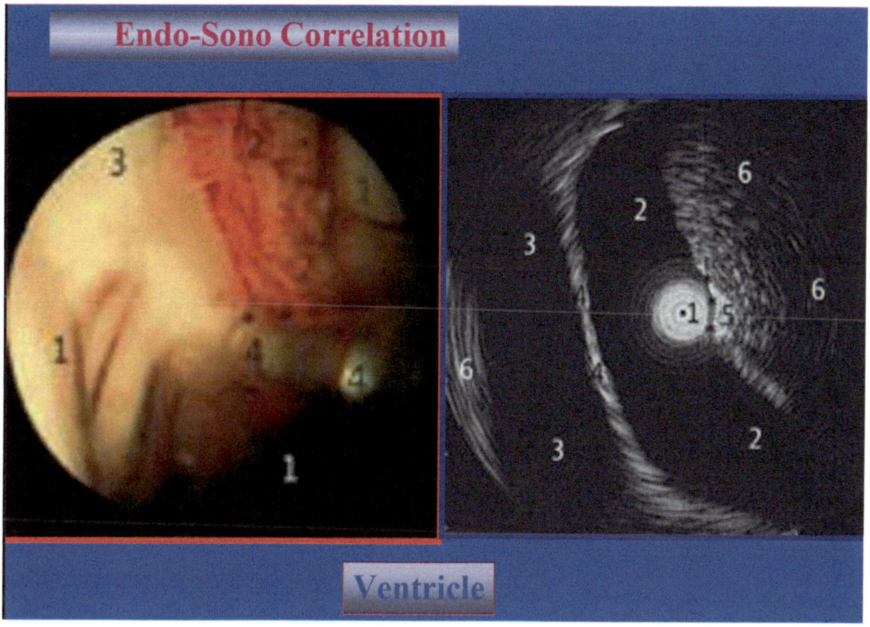

Graph 4.59 Endoscopy-Sono correlation

plexus (*2*) oth right (*2*) and left (*3*) lateral ventricles shown by sono-scan. The probe (*1*) itself is visible at the *right*, in contact (***) with the choroid plexus (*5*). Moreover it can 'see' through the pellucid septum (*4*) and into the parenchyma (6) (Graph 4.59).

Suprasellar space ENS-scan. The sono-scan and the sono-probe are located in the chiasmatic cistern, representing a semisagittal imaging plane. The sella with the parenchyma of the pituitary gland and the stalk with the left optic nerve are visible. In contrast to the endoscope, the sono-probe can 'see' into the anterior third ventricle with infundibular and supraoptic recess. The dorsum of the sella is visible.

The endoscopic view shows the ENS-catheter between the optic nerves attached to the left side and close to the optic chiasm.

The PIP-modus on the monitor or HMD-displays can be varied by voice control. Both images are real-time, independent from shift. The sono-scan provides a navigation road-map, like a "mini-CT" at the tip of the endoscope, the endoscope provides the street-view modus, hence causing a control to one another (Graph 4.60).

4.11.3.1 Per-nasal ENS

The trans-nasal route by ENS enables a coronar scan imaging (Graphs 4.61 and 4.62) of the sella and peri-sellar region. This is only possible in cases of a sphenoid sinus reaching below the sella until the dorsal lamina of the clivus (big sphenoid sinus). The sinus must be filled with destilated water, irrigation or sterile sono-gel. The angel of the catheter defines the angle of the sono-plane ($\angle\alpha = \angle\alpha'$).

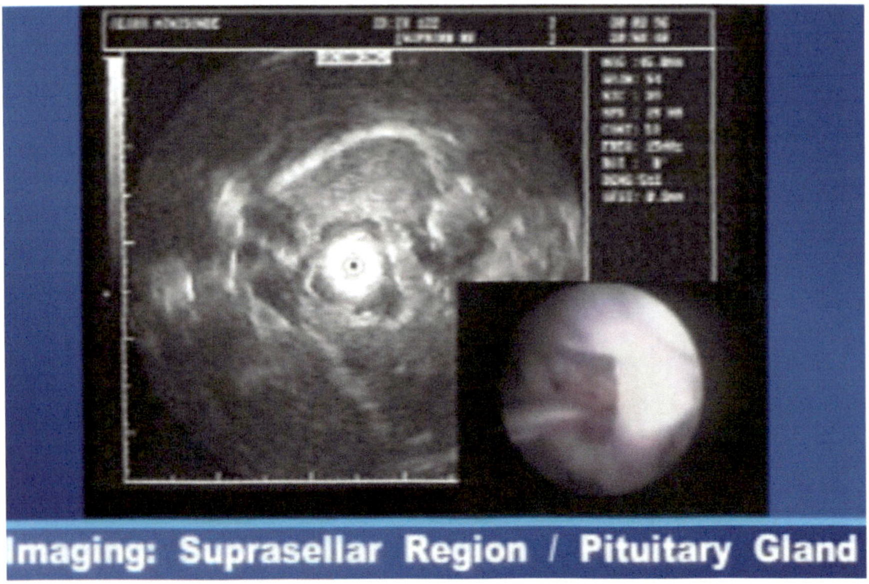

Graph 4.60 Trans-endoscopic sonography suprasellar region (from monitor)

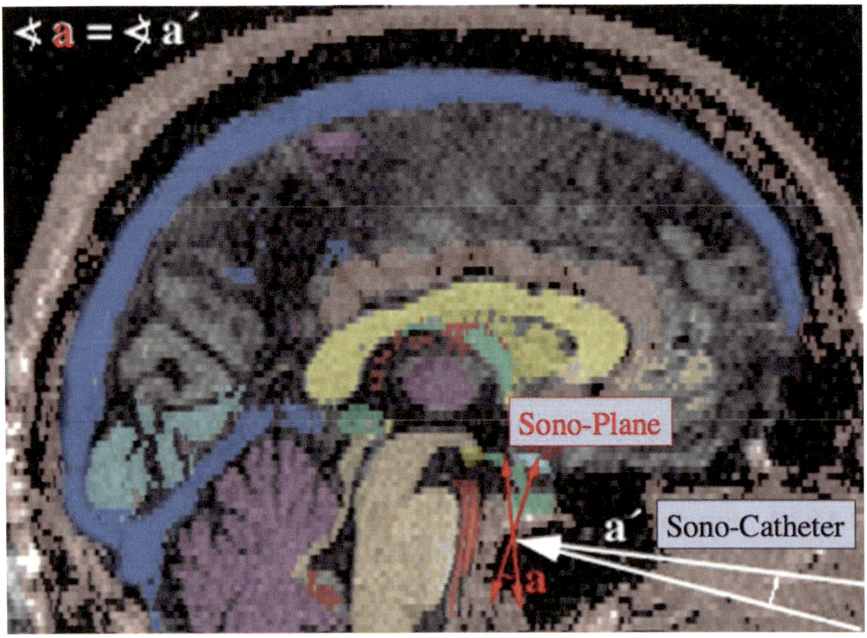

Graph 4.61 Angle of pernasal transendoscopic sono-plane

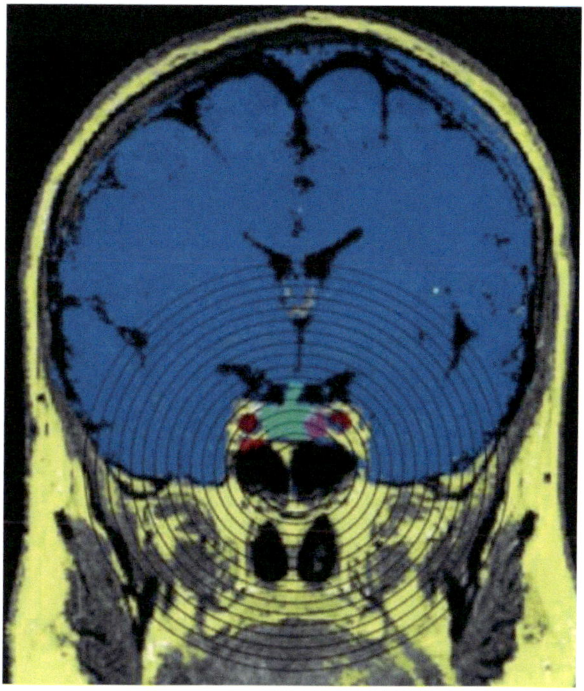

Graph 4.62 CT-Anatomy of pernasal transendoscopic sono-plane sellar coronal (with micro-adenoma)

Frontal MR scans of pituitary region. The cranial MR slices are reconstructed in color and show pituitary gland (*green*), tumor (*violet*), carotid artery (*red*), optic system (*orange*), brain (*blue*), and surrounding tissue (*yellow*) of skull. Penetrating with the sono-catheter transsphenoidally, the sellar region is imaged by the scan, which represents the typical structure. The sono-wave is shown schematically, coming close to the sella on the floor of the third ventricle (Graphs 4.63 and 4.64).

This clinical case of **pernasal ENS** of a pituitary tumor illustrates the application and imaging. The result of imaging is well comparable with the coronar MR-scans but enables a **real-time imaging** intra-operatively by a very small, mobile and trans- or para-endoscopic use, enabling an elegant navigation, imaging and targeting in the sella and para-sella region. Ergonomics and financial efforts are minimal, compared with any other technique. An easy and sufficient interpretation tool is simulated by geometric and colour-coded workout of the MR- and sono-images.

However, this application is only used in few cases and need to be elaborated to a combined new endo-sono equipment. For detailed information and clinical cases see the e-book: "Trans-endoscopic Ultrasound for Neurosurgery", Springer 2005 (by the Senior-Author) (Graph 4.35). Many neighbour disciplines are using combined techniques of ultrasound and endoscopy since decades and need to be adapted to neurosurgery.

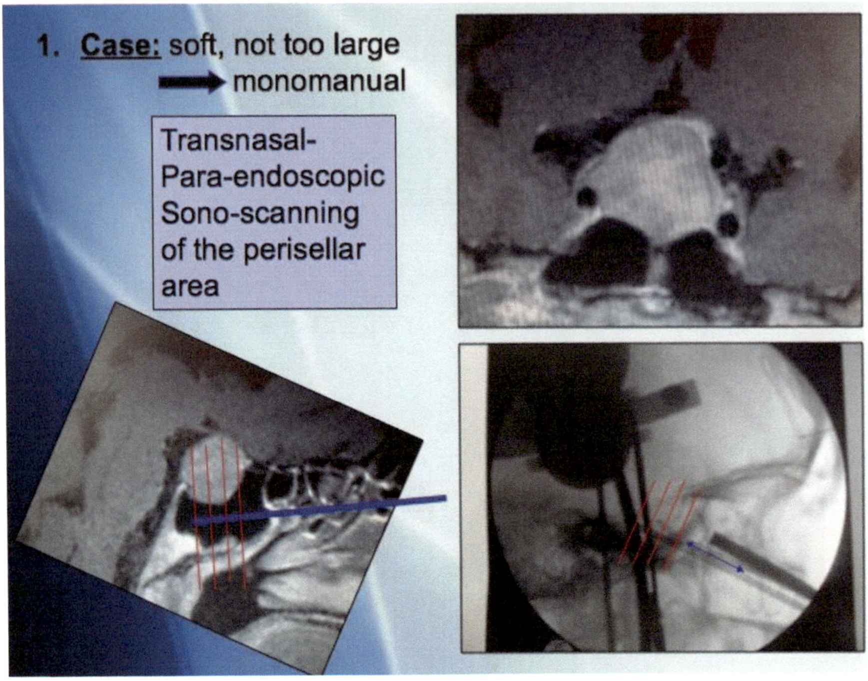

Graph 4.63 First clinical case of transendoscopic ultrasound-assisted pituitary surgery

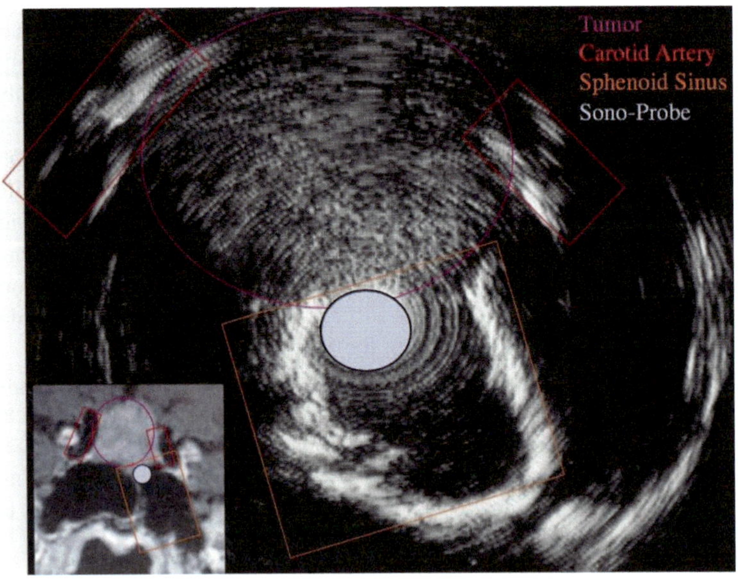

Graph 4.64 Sono-view from Graph. 4.63

During pernasal endoscopic pituitary surgery the trans-endoscopic ultrasound can be used to image the peri-sellar region in coronar scans. The sphenoid sinus needs to be large enough and the sella floor thin, which however is usually the case when peri-sellar navigation became important. During the real-time imaging the para-sellar carotid arteries in the cavernous sinuses are visible as well as the blood flow though it is only a B-mode image. Being a high-resolution image with sono frequencies 10–30 MHz the intra-sellar/-tumoral fine structures become visible. The radius range of the system can reach far intra-cranial and even foramina Monroi level can be visible (s. Chap. 7 and Transendoscopic Ultrasound for Neurosurgery p. 39–44; Springer 2005).

Changes during surgery will be visible in contrast to neuro-navigation systems, resection control, shift of cavernous sinus wall and decendence of diaphragma are visible in real-time. Of course, the sphenoid sinus needs to be filled with water.

4.11.3.2 View of the Surgeon in ENS

Using monitor or HMD, the position of the two real-time images can be varied as convenient. The images substitute one another causing a synergistic effect (Graphs 4.65 and 4.66).

The **endoscope has followed the sono-probe** (*1*), finding it lateral to the basilar artery (*2*) on the left, as expected from the sono-scan image (Graph 30). The basilar bifurcation (*3*) into the two P1 segments (*4*) is visible, as is some arachnoid tissue. The light gets lost in the depth of the prepontine cistern (*5*) and clivus (*6*).

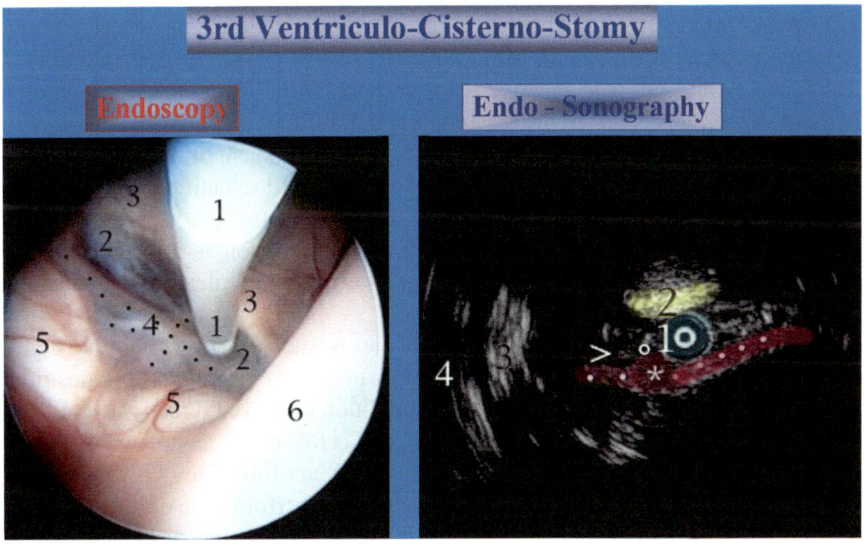

Graph 4.65 First ENS controlled ETV 1996

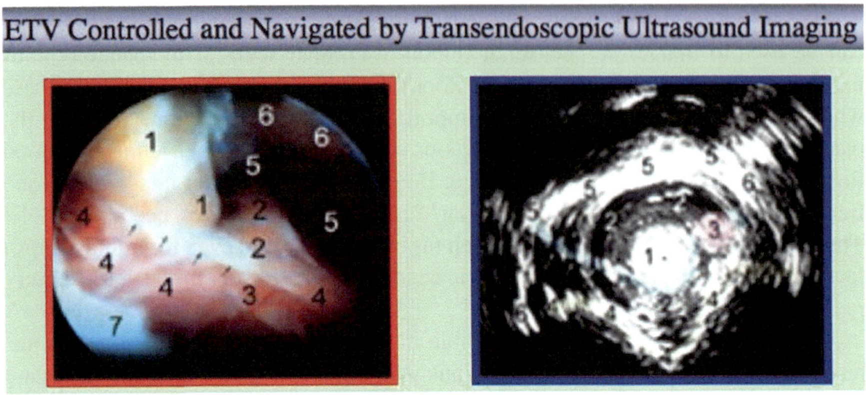

Graph 4.66 ETV controlled and navigated by transendoscopic ultrasound imaging

After perforation of the third ventricle the **sono-probe (*1*) 'sees' what the endoscope does not see**: the clearest landmark signal of the sellar dorsum (*5*), the typical peduncular shape of the ventral mesencephalon (*4*), the interpeduncular fossa, (*2*), and basilar artery (*3*). On the moitor all pulsations, flows, and movements of arachnoid membranes (2), and also the sono-probe itself (*1*) are visible in real time. Laterally on both sides the tentorial notch (*6*) is seen.

The best and most ergonomics visualization technique is the head-mounted display system, presenting picture in picture (PIP) technique with voice control (Graph 4.65). This setting enables extreme difficult endoscopies using two real-time imaging which support one another compensating weaknesses. The new exoscope equipment actually coming up (s. this chapter) will not be able to overtake the spectrum of ENS. Especially the complex hydrocephalus cases in children, the most sever and frequent problem worldwide in pediatrics, will take advantage of ENS. Combined techniques with minimal invasive capacity are from the core-philosophy of MIN. However, this is one of the reasons, why RCTs studying mono technique will not be of any benefit or scientific support, but the development of this route of solving surgical problems of hydrocephalus, cannot wait until scientific handicaps have been solved (Graph 4.67).

Our youngest patients deserve at least the same attention as military and economic advances. This is an ethical imperative for medical profession, and also for neurosurgery (s. Chap. 7).

The HMD-system. The head-mounted display-system (*1*) allows multiple imaging information to be displayed on one screen in an ergonomic working position. The surgeon (and the assistant) sees all information displayed directly at eye level (*2* and *3*), regardless of his/her head position, and is free to concentrate on the endoscope (*4*) and the patient's head (*5*). The assistant can follow the procedure even without a monitor and in a position that is helpful for performance of the procedure regardless of the monitor's position. The kind of information displayed can be changed and selected by voice-control. With this system the surgeon can perform time-consuming procedures while maintaining a relaxed position and a high level of concentration (Fig. 4.54).

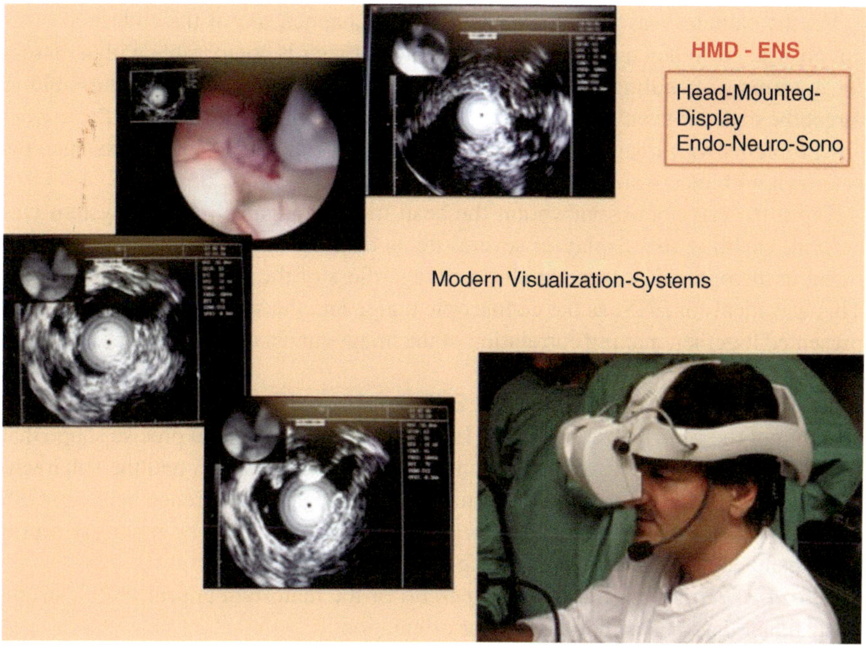

Graph 4.67 Head mounted display ENS with PIP-monitoring

Fig. 4.54 HMD-ENS operation (1999), ENS-navigated ETV in a 2-month baby

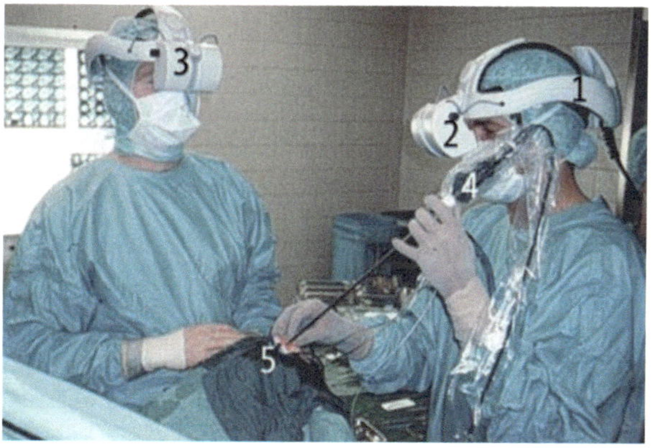

The sono equipment does not take up a lot of space. The machine itself was placed outside the sterile area, and a nurse handed the long thin catheter to the surgeon to be fixed near the patient's head. This meant that the equipment did not in any way hamper the operating procedure or the surgeon's movements. Preparation of the equipment took only minutes, with some differences between the two systems used. The sono system was easier to prepare than an endoscope.

As the catheter was inserted in the working channel, use of the endoscope was not obstructed. When used freehand, the sono-catheter is comparable with a dissector in terms of handling and of space requirements. Therefore, it was possible to direct the catheter even in small deep gaps to obtain optimal imaging. If there were any problems with imaging quality decreasing as a function of known distance (as is known with macro sono-probes), these were not apparent.

From the ergonomics viewpoint the head-mounted display (HMD) system was helpful, enabling the display of several items of imaging information at once and being in the optimal position at all times regardless of the surgeon's head position. This was ideal for ENS, as the endoscopic image and the sono-picture were always presented together, making correlation of the images and navigation easier and safer.

4.11.3.3 Summary Remarks on ENS Anatomy (Graph 4.68)

Imaging of the ***ventricular system*** by ENS is very clear and has a precise shape that is easily recognized, giving good orientation. The *basal cisterns* and the subarachnoid space are also clearly visible. This is due to the echo-difference between CSF and neural tissue. Therefore, all parts surrounded by CSF can be precisely represented and changes in size and shifting are reliably detected.

Dural structures such as the falx cerebri or the tentorium appear with a strong echo signal that makes them useful for orientation.

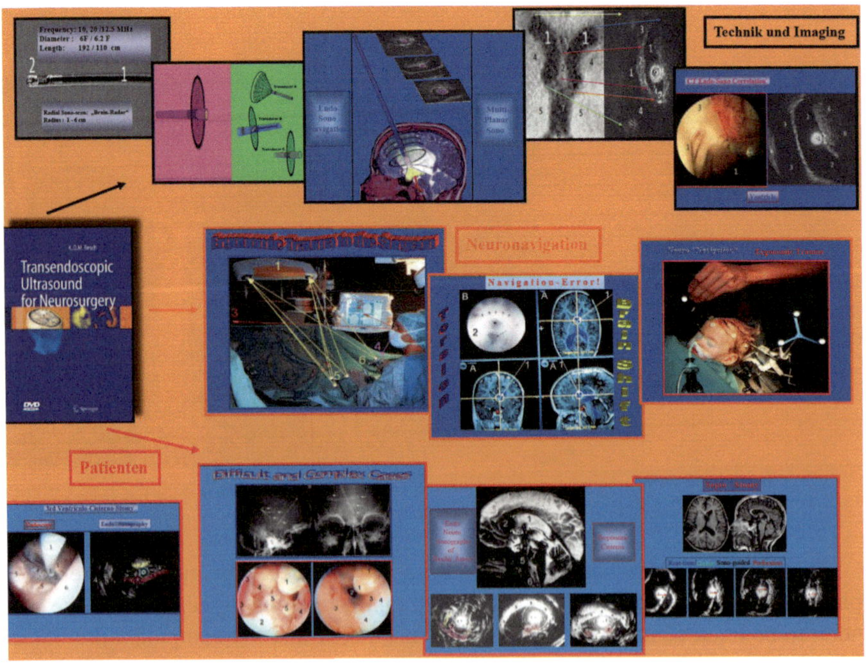

Graph 4.68 Features of ENS: technique/imaging; navigation; cases

Of course, all ***bony structures*** give a high-echo signal and are detected even when the parenchyma is already outside the detecting depth. The skull borders will be shown on the display even when the parenchyma is only present in a 2-cm radius around the probe.

Vessels give a higher signal than the neural tissue but will be recognized in clinical use even better from the clear appearance of the blood flow and pulsation.

The ***choroid plexus*** also has a strong signal, showing up even the finest villi spreading in the same way as fingers, which can be seen during clinical application of the technique as movements due to the irrigation of the endoscope and pulsation of the vessels. The different fibers and nuclei of *neural tissue* do not show up distinctly.

Since ultrasound imaging is in real-time and online, it can detect all such changes in the anatomy as size of ventricles, shifting of tissue, pulsation of liquids, and movements of the displayed sono-probe itself. All this means that the movements of the endoscope within the anatomy are visible in the sono-scan leading to the specific ***neuronavigation*** characteristic. Neuronavigation has become a major topic in actual neurosurgery regarding numerous publications. Some authors even believe that it could change Yasargil's concept of the subarachnoid approach.

However, despite all efforts, intraoperative CT and MR have not yet reached online and real-time characteristics. Only X-ray and sonography fulfill the navigation requirements of real-time feedback. Sonography can moreover be used as an online technique without negative side effects.

Training in sonography is necessary before some of the difficulties in interpretation of the images can be mastered. If endosonography is not to lead to frustration it is most important to learn how to get a competent image. The first step will be to know the imaging characteristics of the anatomical structures so as to be aware of what to expect and what the typical features of an anatomical structure look like in the endosonographic scan. Therefore, this work was started with anatomical examinations. The next step was to learn how to adjust the image of the sono-scan in terms of orientation, frequency, and zoom.

The benefits of sonography have already been documented (Auer et al. 1988; Auer and van Velthoven 1990; Sutcliffe 1991; Van Velthoven and Auer 1990), The benefits of endosonography will be confirmed as time goes on but depend primarily on the user's patience in following the above steps. Experience is limited as yet but has shown that these are essential.

Precise imaging quality and correct adjustment are indispensable for neuronavigation by ENS. Few clinical examples will show what can be done and which problems can be solved or at least simplified. (s. e-book "Trans-endoscopic Ultrasound for Neurosurgery". Springer 2005).

- Precise imaging of anatomical structures
- Ability to 'see' into the parenchyma
- An axial overview: 'mini-CT' (in addition to endoscopic viewing)
- Compensation of 'blind angles' of the endoscope

- Compensation of problems that can hamper vision, such as bleeding, no space, dirty lens, technical pitfalls
- Real-time-imaging for neuronavigation 'brain radar'
- Online observation of all movements and physical changes (size of ventricles, shift of tissue, blood flow, pulsation).

Few **clinical examples** will show which problems can be solved by ENS during endoscopy illuminating the abilities which are unique for this combined technique. Especially the navigation capacity for endoscopes, real-time imaging and targeting to small structures, impossible to find with endoscopy alone.

A broad variety of lesions and locations can be approached.

4.11.3.4 General Remarks on Variety of Lesions

In clinical use the sono-catheter has superior imaging and navigation abilities to those seen in anatomical laboratory work. Real-time and online characteristics represent changes such as shifting, pulsation, CSF flow, and changes in size and form of structures. When confronted with clinical problems, this technique still has some limitations. The sono-catheter has nevertheless proved to be a valuable tool for sorting out unresolved problems and to be much easier to use than other instruments. Moreover, there appears to be an interesting potential for further developments.

Thus, the sono-catheter has been tested intra-operatively in imaging of a variety of lesions (Tables 4.7, 4.8 and 4.9).

Table 4.7 Application spectrum of ENS

Treatment	Problem	Advantage of ENS
Third ventriculocisternostomy	Localization of basilar artery/perforation	Visualization through floor of third ventricle
Septostomy	Perforation point	Navigation, targeting
Catheter placement	Small ventricle	Navigation
Catheter deletion	Bleeding	Compensation of poor vision
Ventricular hematoma evacuation	Poor sight	Compensation of poor vision, orientation, monitoring of evacuation
ICB evacuation	No landmarks	Compensation of poor vision
	Dark view	Evacuation control
	Difficult orientation	Navigation in the dark
Resection control	Remnant of tumor	Tumor echo and location
Biopsy	Detection of lesion	Navigation
Cystostomy	Orientation	Navigation
Aneurysm clipping	Blood clot, aneurysmal wall	Flow, targeting

Table 4.8 Limitations of ENS and solution

Limitation	Consequence/solution
Little penetration depth (radius)	3 cm parenchyma, 6 cm bone
High background signal	Interpretation only >0.5 mm
Geometrical torsion by radial scan	Experience necessary
Geometrical torsion by extraplanar scan	Experience necessary
Tip protrusion 1 mm ahead to scan plane	Planning and visual control
No image in air	Irrigation, clear working field
No scan in anterior direction	Planning, future research

Table 4.9 Indication spectrum of ENS

Indication	Solution
Ventricular puncture	Navigation + subependymal target
Ventriculocisternostomy	Transventricular imaging of intepeduncular cistern
Fenestration of cysts	Navigation + imaging of size
Tumor resection control	High-resolution imaging intraoperatively
Tumor biopsy	Targeting + bleeding control
ICH evacuation	Evacuation control, navigation
Aneurysm examination	Neck, rupture side, pulsation, etc.
Endoneuronavigation	Sono-navigated neuroendoscopy in real time, simple, not expensive

Ventricles. The easiest application was *navigation of the endoscope into the ventricles*. The catheter was inserted into the working channel of the endoscope, and after the endoscope had penetrated into the brain the catheter was pushed ahead, scanning the area in front of the endoscope. The falx cerebri was used for orientation; the sono-scan first had to be adjusted to the anatomical orientation based on the falx cerebri and also adjusted to the endoscope's view. Once this had been accomplished, the catheter functioned as a scanning tool and literally spoke, acting in the same way as a 'seeing' Cushing cannula. Once the typical shapes of the ventricles were on the display the catheter could be used as a guide along which the scope was advanced into the ventricle without loss of any cerebrospinal fluid (CSF). This strategy was helpful in small ventricles, where punctures were made in unusual positions, and also in ventricles displaced by edema, hematomas, tumors, or congenital lesions.

Another use in the ventricular system was for assistance in **third ventriculocisternostomy**. Various changes to the ventricles, and especially to the floor of the third ventricle, can make it difficult, if not impossible, to target the perforation point (s. Chap. 6). In such cases the sono-catheter can be used for blunt perforation. Before the endoscope sees the features of the interpeduncular cistern, the scan of the sono-probe will show the dorsum sellae, the prime landmark, on the display. This can be used to re-adjust the sono-scan orientation if necessary. The next most important landmark will be the basilar artery, but the tentorial notch, the oculomotor nerve, and the ventral surface and parenchyma of the midbrain are also discernable. All these structures can be used for orientation and navigation, and they can be

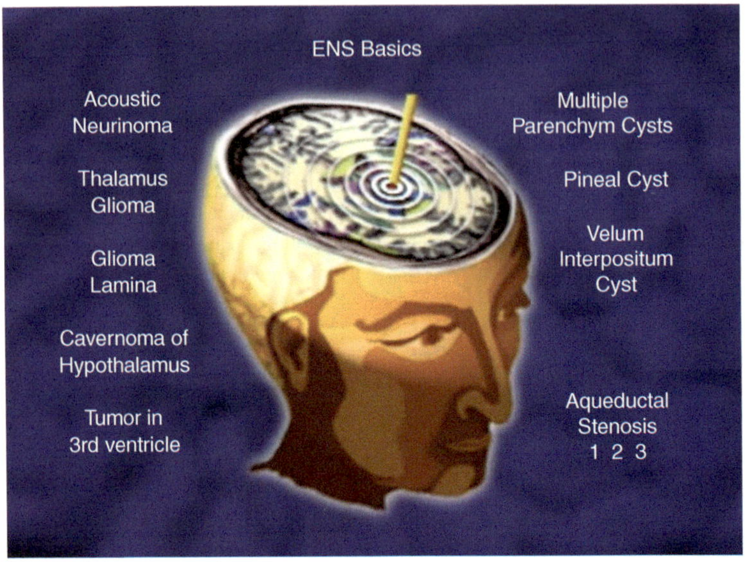

Graph 4.69 Case selection (clips of the book 2005/springer)

perceived while still invisible to the endoscope. In many cases the infundibulum was deep enough to allow visualization of parts or all of the circle of Willis, and of the blood flow inside its vessels.

When there are problems with visualization, the possibility of damaging these structures is markedly increased, and measures must be taken to ensure that the risk is managed in an acceptable manner (Table 4.7, Graph 4.69).

4.11.4 Cases (Graph 4.70, 4.71 and 4.72)

Application of combined techniques is the core of MIN-"philosophy". Trans-endoscopic ultrasound was introduced by the senior author into neurosurgery and first published 1996 and presented 2005 in Springer. It works, so to speak, like a "brain-radar", regarding imaging geometry. It allows to navigate neuro-endoscopes intuitively in real-time and can moreover realize high solution imaging. As the sono probe (sono-catheter) is introduced into the working canal of the endoscope it is a single system in application.

This technique enables extremely complex cases of hydrocephalus and similar cystic lesions to operate when vision endoscopically is not sufficient and navigation the endoscope is not safe.

In the above emergency case orientation by endoscopy alone was impossible, already recognizable in the MR-images. The identification of correct perforation point for ETV was not possible without ultrasound navigation. More complex navigation problems during impaired vision can be solved with this technique.

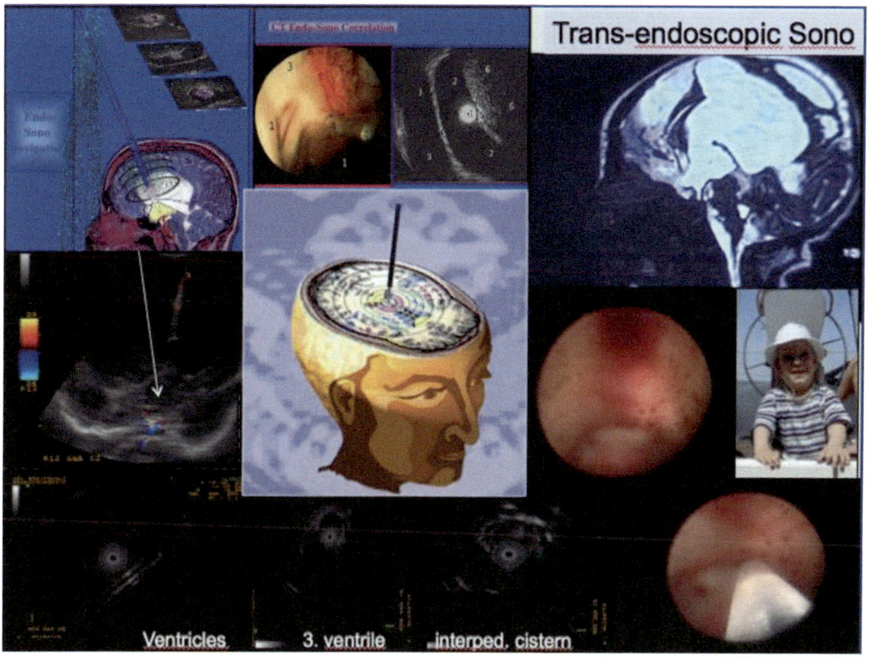

Graph 4.70 Trans-endoscopic sono-assisted ETV

Another phase of postoperative **sagittal MR** shows air (*2*) in the frontal horn of lateral ventricle (*1*). There are strong flow signals (→) through the foramen of Monro (3), ventriculocisternostomy (*4*), along the brain stem (*5*) and basilar artery (*7*), and into the foramen magnum (*6*).

The sono-probe (*1*) in the epiarachnoidalspace is pushing the arachnoid against the basilar head with the basilar artery (*4*) both P1 (*5*) and right SuCA (*6*). The oculomotor nerves on both sides (*7*) and the interpeduncular fossa (*3*) with its cerebral peduncle (*2*) are visible.

The sono-probe (*1*) is placed at the vertebrobasilar junction, where both vertebral arteries are just branching (*2*). The median pontine vein (*3*) is hardly visible, running in the median pontine sulcus of the pons (*4*). The clivus (*5*) is visible at the level of its inner mold.

The sono-probe (*1*) is positioned epi-arachnoidally at the superior margin of the atlas (*4*). Occipital condyles (*5*) are scanned on both sides. Both vertebral arteries (*2*) are divided to the lateral sides of the medulla (*3*).

This child was found on the axial MR to have an asymmetrical ventricular system (*1*, *2*) with deviation of the pellucid septum. Part of a shunt system was also seen (*4*). The planned route for ETV and septostomy was transfrontal-transcallosal (*3*). The lateral aspect of MR shows a normal third ventricle and some irregularities of the foramen of Monro. The aqueduct is not clearly visible. The lateral ventricle (*1*) is slightly hydrocephalic.

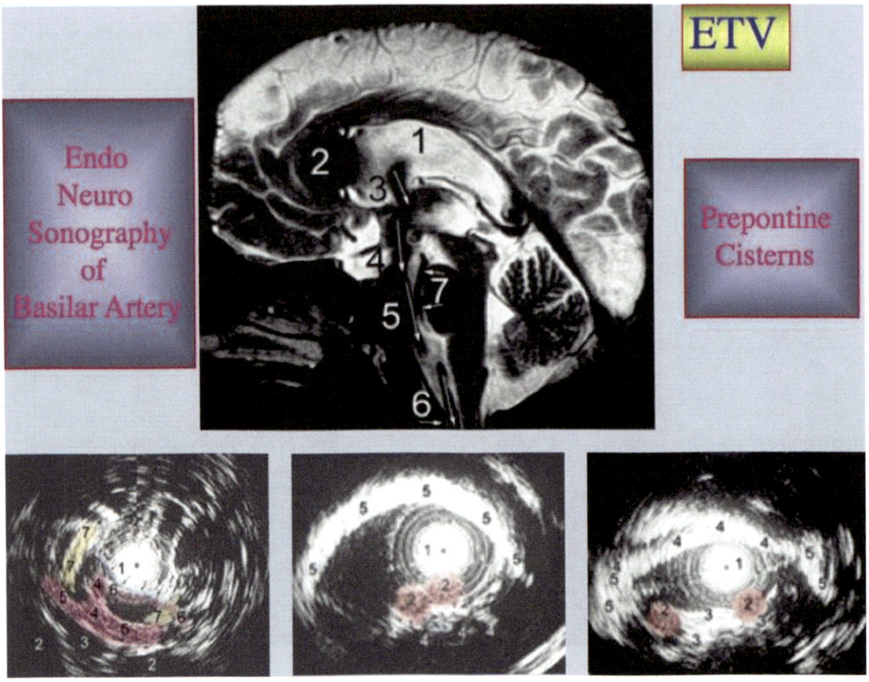

Graph 4.71 ENS-navigated ETV (no anterior view in the endoscope due to subarachnoid scarfs)

The sono-probe (*1*) is placed in the big left frontal horn (*2*) in contact with pellucid septum (*4*) searching for the right perforation target to make access into the small left frontal horn (*3*) possible and create a communication.

The sono-probe (*1*) starts the perforation from the left frontal horn (*2*) through the pellucid septum (*4*) into the right frontal horn (*3*). The septum is bent, almost touching the lateral wall (*5*) of the right ventricle (*3*). The lateral wall (*6*) of the left ventricle (*2*) and choroid plexus (*7*) on both sides are visible.

The sono-probe (*1*) is just passing through the septum (*6*) to create a communication between the left (*2*) and right (*3*) ventricles. The right (*4*) lateral ventricle wall and left choroid plexus (*5*) are shown.

The sono-probe (*1*) has passed from the left ventricle (*2*) through the septum (*4*) and is now in the contralateral ventricle (*3*). The small right ventricle is now enlarged by irrigation, and the right lateral wall and left plexus (*5*) are also visible.

4.11.5 Summary and Final Reflections

To sum up the clinical results of transendoscopic ultrasound imaging in neurosurgery, we can conclude that while in other fields a great deal of experience has

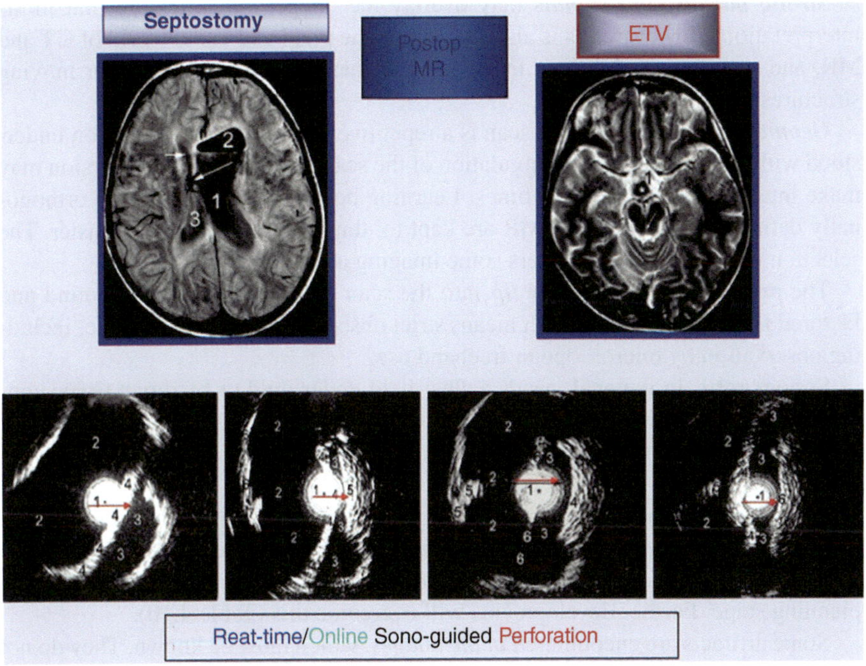

Graph 4.72 Transendoscopic sononavigated septostomy and ETV in assymmetric ventricles

already been collected, in neurosurgery we are still just on the threshold. We have one additional benefit of intra-operative imaging over all other disciplines using transendoscopic ultra-sound, which is the **neuronavigation capacity**.

The **routine use of ENS** has shown the capabilities of this technique in different lesions and problems and highlighted its benefits and limitations. The three cases above provide illustrations of few lesions. Tables 4.7, 4.8 and 4.9 give a summary overview of the applications of ENS. Practical work has made it clear how surprisingly many different applications there are. (for more cases see e-book "Transendoscopic Ultrasound for Neurosurgery". 2005 Springer).

The use of **additional sono-scanning** in endoscopy and microsurgery has made most of the procedures safer or easier. It has been rewarding to see how, once the learning period is over (best done in the laboratory; see Vol. 2), endoscopy becomes safer. Moreover, if the sono-catheter is not available endoscopy is more time consuming and the lack of sonographic information in difficult cases leaves the user in an uncomfortable position and the patient in danger.

The *shallow* ***penetration depth (3–6 cm)*** of ENS is a problem, giving sight of only a limited region. In large regions of interest, CT and MR will give a complete view. The catheter has to be moved through the whole field to monitor resections and evacuations. Especially ICH-evacuations are very satisfying controlled by ENS, because inside the hematoma vision becomes very dark, but the ENS-scan will be impressive as the hematoma gives a strong signal.

Strong background signals may overlay the image with a texture that limits interpretation. However, this is also the case in the pixel and voxel world of CT and MR, and imaging resolution is in any case higher in ENS, especially for moving structures.

Geometric torsion by radial scan is a repetitive phenomenon and is soon understood with training. When the angulation of the scanning planes is used torsion may make interpretation difficult at times. Learning becomes easier when the orthogonally defined planes of CT or MR are kept to, thus making correlation easier. The selection of each plane also offers some imaging advantages.

The ***protrusion of the probe tip*** into the scan plane must be kept in mind and factored into calculations, which means strict observation of the endoscope, including observation by microscope in freehand use.

Sonography, in general, needs a clear field under fluid or by direct tissue contact. Gas insufflation by CSF release results in an *air artifact* or total lack of any image at all. Positioning and clear working must be planned and coincidence with MIN strategies.

Conventional sono-scanners ‘sound’ forward, while the ENS catheter ‘sounds’ radially and laterally, as if it were a **‘brain radar**.’ Therefore, it is *not possible* to *target a structure in a straight working direction*. This must be calculated during the planning stage. Further developments will overcome this (Table 4.10).

Some artifacts are encountered *in the images*, which must be known. They do not always disturb or change the image, but may at times be helpful in orientation and adjustment:

An ***air artifact*** is an area of no imaging and can be only a segment or comprise a whole radius. It occurs when there are air bubbles in front of the probe inside the sheath or intracranially, or when the probe comes into air. Air artifacts such as air bubbles, however, reflect the surrounding area and can mimic a tumor mass in the endoscope. This could appear in the endoscope image as an irregular mass with bubbles in the CSF acting in the same way as a mirror. The sono-scan will discover it by an air-artifact signal.

Table 4.10 Future Program of Developments

More clinical evaluation and experience
Enhanced adaptation to neurosurgery
3-D ENS in real time
Better integration of physiological parameters
Compatibility with other techniques
List of approved indications
Probes with forward sound direction
Integration into a general design of miniaturization
ENS courses for training in the new technique

A ***metal artifact*** produces a sound shadow originating from the instruments used, such as spatulas or clips. Such artifacts do not only disturb the imaging but can also be helpful in targeting and orientation.

A ***rotation artifact*** occurs when the rotation of the catheter in the sheath is decelerated by force. Radial beams disturb the image, and the whole image may even rotate, making new adjustment of the orientation necessary.

A special effect is the ***lens artifact***, which represents the tip of the endoscope as a strong echo circle, mimicking an axial scan of a vessel.

4.11.5.1 Indication

A variety of indications have to be discussed. **Ventricular puncture** is usually not a problem but becomes one if the ventricles are small or there is some shifting, or as a result of a change in normal endoscopic anatomy. In these cases, use of a simple imaging technique to solve the problem, rather than stereotaxy or neuronavigation, seems justified. The usual ultrasound technique is adequate in areas that are not too deep. For deep-seated lesions ENS is far superior to the more common sectoral sono-probes in terms of optical resolution of the depth.

Ventriculocisternostomy (ETV) is one of the main indications in ENS that still present some problems, such as patency of the ETV stoma, visual control, and criteria for recognizing when it is indicated. However, there are reports of fatal or near-fatal complications in such cases (Handler et al. 1994; Lowry et al. 1996; McLaughlin et al. 1997; Schroeder et al. 1999). The sono-catheter makes orientation and navigation with the endoscope safer by additional real-time imaging and high optical resolution.

Fenestration of cysts may become a problem when they are small, located intraparenchymatously or inside small ventricles, or multiple. Such cases present a navigation problem, which has not yet been solved by current neuronavigation systems (Caemaert et al. 1992; Hopf et al. 1998; Schroeder and Gaab 1998). With the aid of the sono-catheter these lesions can be treated much more easily and safely with a minimum of effort. Especially in close proximity to vessels or important neurological structures, the additional imaging makes the risk calculable, owing to online observation.

Tumor biopsy is difficult in small or deep-seated lesions. In the case of intraventricular cases the release of CSF can cause shifting, and neuronavigation is then unreliable. A combination with stereotaxy is possible, especially if the imaging is insufficient because of a low-echo signal or limited experience. If the tumor gives a strong echo signal, as in the case of metastases, the sono-catheter can target the lesion in real-time. Radiated metastases may present weak in ultrasound in general.

It is well known that **aneurysms** are challenging and high-risk lesions. There are many items of information that the sono-catheter can display about them, which again can make the approach easier. Blood flow before and after clipping, an aneurysm inside a blood clot, a thrombosis inside an aneurysm, the constitution of the aneurysmal wall, and the intraluminal result of vessel reconstruction by clipping can all be seen online. This information cannot be obtained by endoscopy alone

(Perneczky and Fries 1998; Taniguchi et al. 1999) and is dependent on regular post-operative control by means of DSA, MRA, or CTA. The intraoperative information provides an opportunity for direct correction if needed.

Neuronavigation has come into quite wide use, but it has yet to be determined whether its benefits justify the investment it entails in terms of effort and cost.

Moreover, computer-'navigation' systems do not fulfill the true definition of navigation, as they lack real-time feedback, which the sono-catheter facilitates by providing real-time and online intra-operative imaging with true navigation capabilities. As indicated above, the sono-catheter can solve most navigation problems in neuroendoscopy with a minimum of effort and at low cost. Moreover, the sono-catheter can assist in microneurosurgery as an intra-operative imaging tool.

There are two indications that have not yet been examined but are possible application candidates according to the laboratory results:

- **Morphometric ICP monitoring** at the bedside by imaging of ventricular sizes and changes.
- *Bedside imaging of space-occupying lesions*, such as *hematomas*, as a *substitute for CT* in instabil ICU-patients.

With regard to the enormous logistic and financial effort involved in diagnostic imaging in ICU-patients, the last two suggested indications, which have still not been investigated, could have a significant impact on the healthcare system: the technique comes to the patient and not vice versa.

Future developments of ENS will depend on more documentation and analysis of clinical results. This experience should provide a user-profile for better adaptation of the system to the needs of neurosurgery. Display of physiological parameters and real-time 3D reconstruction, which are now possible with general sonography, should also be developed for ENS. Combination with other techniques, such as laser navigation or the neuronavigation interface, may make new applications and interpretations easier.

ENS is not just a tool; it may rather become the cornerstone of a new concept behind a new technical generation of neurosurgery. This concept is presented in Chap. 3 and this chapter. The ultimate goal of ENS is to assist in minimizing the trauma to our patients by ergonomic real-time imaging.

A list of investigations for the future is displayed in Table 4.10.

In terms of lowering the level of risk in brain surgery further, ENS appears to be a desirable and useful tool in the hands of experienced neurosurgeons, being relatively low cost and easy to apply.

The importance of the stage of laboratory-training must not be underestimated; this, however, will pay off handsomely in the form of added confidence and better results in the operating room. It will release surgeons from a multitude of tasks that currently fall to them, and this combination of simplicity and added safety warrants further exploitation of the merits of the technique and the possibilities it offers.

In summary, ENS is a tool for:

- Application in a wide variety of indications
- Intraoperative real-time and online high-resolution imaging
- Neuronavigation of endoscopes with a working channel at least 2 mm in diameter
- Making neuroendoscopy easier and safer

ENS is:

- Small, mobile and easy applicable in endo- and microneurosurgery
- Limited by the need for laboratory work and clinical experience
- Open to further development
- A cornerstone of a new concept of MIN (s. Chap. 3 and this chapter)

4.11.6 **Final Reflections on ENS (2005)** (s. Chap. 3)

ENS is understood within an environment of ergonomic evolution of the whole process in the operating room. It is part of the minimally invasive concept of the future: para-endoscopic intuitive computer-assisted operation-system. That means that the ergonomic system, the microhabitat, is not made up just of the technical equipment, but also of the patient and, especially, the surgeon. The major blind spot of neurosurgeons is their own brains (Paradigm 1: "Neuro-psychological-Interaction"). Consequently, technical equipment and today's mental training are at odds with what is actually known of neuropsychology. An athlete being trained to compete for an Olympic medal, or a pilot being trained to fulfill a military purpose is mentally better prepared for his or her task than neurosurgeons usually are when faced with a human brain in trouble.

The neurosurgeon's brain is a high-potential VR (Vitual Reality/narrative) system. The most intense connections involving the brain are not those with the environment or the body it is housed in, but those that take place within itself. Each VR technique should be seen as in competition with the human brain in training efforts. It does not seem prudent to neglect this capacity of the human brain and to invest indolently in technical solutions only.

The understanding of ENS imaging and visualization, as of imaging in general, depends on the size of the anatomical image library in the surgeon's brain and all the virtual surgical clips on which he can draw as he anticipates the procedure mentally. A technical tool that interrupts or disturbs the relation between the patient's brain and the surgeon's brain mentally or physically should have no future, as it handicaps the surgical art (Paradigm 2: "Ergonomic Zones"). ENS fits in smoothly with endoscopy and microsurgery, but from the microsystem point of view it is still not mature and needs to be better integrated in a single system.

It seems to give us a good, and perhaps necessary, feeling about ourselves if we can think that the surgical process is rational and completely controllable. However, this is not the case as far as the function of the surgeon's brain is concerned. It might be more helpful to view the surgical procedure as a 'chaotic system' that can be described by fuzzy-logic. With this concept we will focus on the peripheral parameters that are the only links that have any influence on the outcome within such a chaotic system (Paradigm 3: "Chaotic System"). The author's qualitative experience in 25 years of laboratory work and his study of neurosurgeons over a long period (one series of very eminent neurosurgeons and one of 'standard' neurosurgeons) has led to the hypothesis that the outcome can best be influenced by controlling the peripheral parameters in advance. Too much concentration and energy are lost in trying to control badly prepared peripheral parameters, which prevents any surgeon from performing at an optimum level. Such peripheral parameters are described by the three paradigms of ergonomics in neurosurgery. The future focus in the evolution of minimally invasive neurosurgery must be the synergistic mechanism and interdependence of all three paradigms; if one is disregarded the others cannot be fulfilled.

We have progressed through three major stages of visualization in neurosurgery as far as technical tools are concerned: macro-, micro-, and endovisualization. We can symbolize this by illustrations of a well-known building visualized in each of these different ways. Each visualization has its own specific implications for ergonomics, which affect the entire procedure and all the manual and mental abilities necessary for success.

The last step in technical change in viewing is a completely different dimension from the three mentioned and illustrated above. We change from visualization to imaging. The ergonomic implications of this 'virtual' imaging are still not clear, as the mental and neuropsychological effects are not sufficiently well understood to allow us to estimate all the effects on both the surgeon and the procedure.

This paradigmatic journey is presented as a series of technical steps, but the same evolution can be viewed with an historical perspective.

Since the time of Leonardo da Vinci, the word 'imago' has come to mean something quite different from our 'image,' and even more different from 'imaging.' Da Vinci did not believe in citations but tended to translate earlier sentences into sayings that fitted better in his own time. Today he would probably no longer say, "*To see is to understand,*" perhaps preferring, "It's not possible to understand everything we can see." Da Vinci's pictures have never been nothing but images; rather, he always understood them as symbols and concepts. His main vision in his art and his scientific work was directed at symbolizing the invisible.

Imaging is completely different, as it has no symbolic characteristic; there is no 'message' in an imaging. Moreover, imaging has lost the intuitive context for the spectator, while da Vinci produced his pictures with his own hands and was conscious of each detail of technique and methodology and of how his pictures related to reality. Da Vinci had the ability to write with both brain hemispheres, and his pictures were writings from a neuro-psychological point of view.

Imaging is made up mainly of information, and only a small part of the information present in an imaging is seen and used by neurosurgeons. The amount of imaging output does not increase the benefit, because it is limited by the ability of the spectator to manage all the imaging masses mentally. In addition, many images can make the brain blind to their meaning, even if there were any chance of understanding the real meaning of such complex artifacts as fMRT or PET, for example. Da Vinci created a concept-guided science and not an image-guided one, and we can assume that he would prefer to speak to us if he realized our blindness to the meaning of what we call an image and what we believe that imaging can mean in neurosurgery. He would realize that conceptuality and awareness of context are lacking in our technical images and would change from painting to writing to convey his ideas to us and make the invisibility of a message visible.

Graph 4.73 is entitled "*The Vision of the Cat*", but it really has nothing to do with cats, rather symbolizing the relation between endoscopy and ultrasound in ENS. It is not an imaging but an 'imago', and to see it, is to understand it.

ENS imaging can only be understood if the images are seen in conceptual relation to anatomy, geometry, ultrasound physics, and operative imaging problems such as targeting and orientation.

In future, the emphasis will have to move away from image-guided and back to concept-guided neurosurgery if the masses of images are really to culminate in any benefit to our patients (Fig. 4.55).

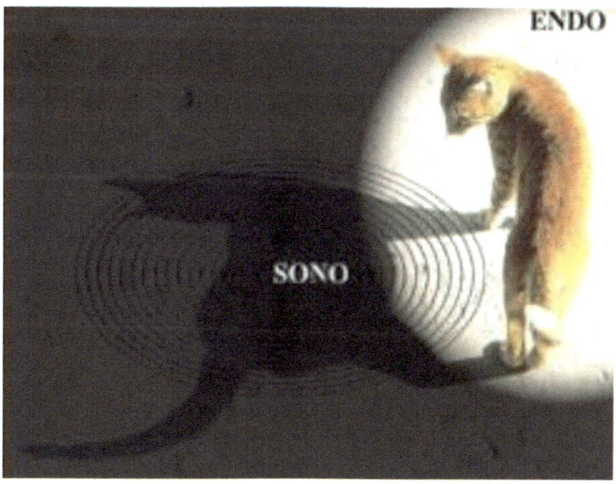

Graph 4.73 "The Vision of the Cat"

4.12 Final Summary on Ultrasound as Key-Technique in MIN (Graph 4.74)

Ultrasound belongs to the **key-techniques for MIN**, because without, one will not reach high level of MIN. Looking at any results, what so ever in medicine, it became usual not to ask about the prize in terms of costs, efforts and side-effects. What is the prize for never changing a common routine technique, strategy or therapy? What are the costs of "heavy metal medicine" and technical overcill medicine? The costs of neurosurgery without neuro-sonography are much higher and not only financially. Moreover, the costs are a decrease of ergonomics intra-operatively (s. Chap. 3), high logistical efforts in personal and space, high time consumption, decrease of patho-physiological thinking, decrease of strategical thinking, decrease of experience and understanding workflow and mental flow.

Yes, ultrasound is user-dependent, which can be made to one of its major advantages by training and experience. You do the imaging, you interpretate the images and you act along your imaging, which you understand because you know, how the images were produced. Radiologists do not see what they image, a surgeon using ultrasound does and will soon see what the image can tell about morphological and physiological reality, causing a fast learning process.

Fig. 4.55 Endoneurosonography paper (overview)

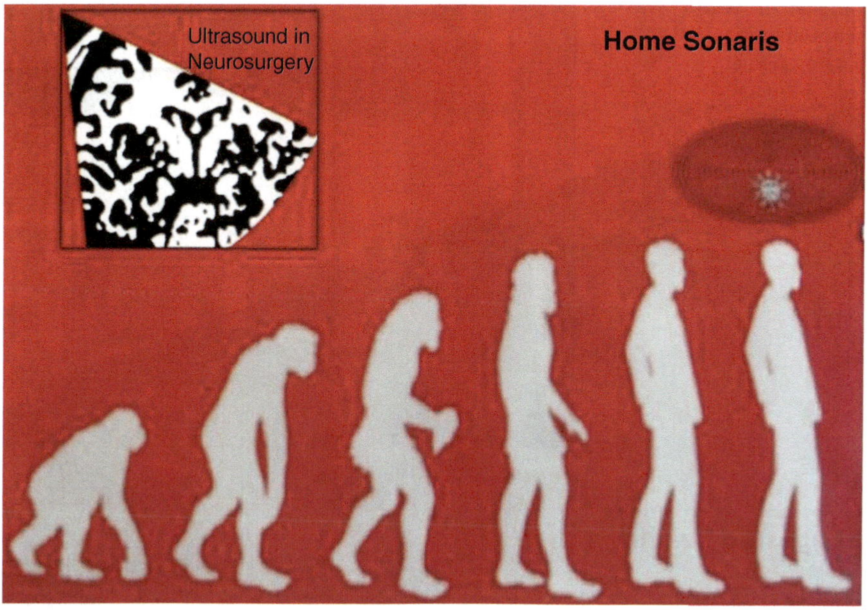

Graph 4.74 Evolution of ultrasound

The costs are paid first by the patients, but also by all professionals due to work conditions and lack of education, and finally, by the societies health care systems.

Ultrasound is one of the **future pathsways for MIN** enabling new effector systems. Focused ultrasound in tumor therapy, sono-thrombolysis or agency delivery are recent developments (Graph 4.75).

As already usual in many neighbour disciplines, combined systems, like ENS, do have the best capacity to be valuable effector-systems for MIN. Endoscopy has a stagnation after an innovation height 30 years ago. Without ultrasound in combination endoscopy will fall into the role of an exotic speciality. The combination must be realized in single rack systems, intuitively to use, small and cheap.

Combinations with neuro-navigation systems does not make sense, because ultrasound is a real-time navigation by itself and would becoming expensive or not usable intuitively by this combination. It will make more sense to add interpretation aid systems to the interface of the sono-systems.

Combinations of ultrasound with the other key-techniques like tracking, endoscopy and laser in single rack systems and open for additional developments (modularity) might be the basis for MIN of the near future.

Thereafter the aera of non-mechanic systems might appear as the next revolution.

Graph 4.75 Wide perspective of ultrasound future

Suggested Reading

Auer LM, van Velthoven V. Intraoperative ultrasound imaging in neurosurgery. Heidelberg: Springer; 1990. p. 11–21.

Bauer BL, Hellwig D. Minimally invasive neurosurgery I. Acta neurochirurgica supplement, vol. 54. Wien, New York: Springer; 1991.

Bauer BL, Hellwig D. Minimally invasive neurosurgery II. Acta neurochirurgica supplement, vol. 61. Wien, New York: Springer; 1994.

Chadduck WM. Perioperative sonography. J Child Neurol. 1989;4(Suppl):91–100.

Frank EH, Horgan M. An endoscopic aneurysma clip applicator: preliminary development. Minim Invas Neurosurg. 1999;42:89–91.

Kanazawa I, Shiraishi K, Kamitani H, Sato J, Masuzawa H. Intraoperative ultrasonography through a burr hole: clinical trial of ultrasound-guided stereotactic surgery. No Shinkei Geka. 1986;14(3 Suppl):295–300.

Linke DB. Hirnverpflanzung. Die erste Unsterblich-keit auf Erden. Reinbek near Hamburg: Rowohlt; 1993. p. 11–309.

Linke DB. Cognitive neuroscience foundations for a theory of neuronavigation. Comput Aided Surg. 1997;97:03–016.

Linke DB. Das Gehirn. Munich: CH Beck Wissen; 1999. p. 7–36.

Lowry DW, Lowry DLB, Berga SL, Adelsen PD, Roberts MM. Secondary amenorrhea due to hydrocephalus treated with endoscopic ventriculocisternostomy. J Neurosurg. 1996;85:1148–52.

Ludwig M, Wetzig H, Sauer A, Vetter H. Experiences with use of an intravascular 6 French endosonography catheter in vivo. Klin Wochenschr. 1995;68(11):570–5.

Maciunas RJ, editor. Interactive image-guided neurosurgery. Park Ridge, IL: American Association of Neurological Surgeons Publication Committee; 1993.

Maciunas RJ. Comment. In: Paleologos TS, Wadley JP, Kitchen ND, Thomas GGT (2000) Clinical utility and cost-effectiveness of intracranial image-guided craniotomy: clinical comparison between conventional and image-guided meningioma surgery. Neurosurgery. 2000;47:40–8.

Makuuchi M, Torzilli G, Machi J. History of intraoperative ultrasound. Ultrasound Med Biol. 1998;24(9):1229–42.

Manwaring KH, Krone KR. Neuroendoscopy, vol. 1. New York: Mary Ann Liebert; 1992.

Masuzawa H, Kanazawa I, Kamitani H, Sato J. Intra-operative ultrasonography through a burr-hole. Acta Neurochir. 1985;77(1–2):41–5.

Mayfrank L, Bertalanffy H, Spetzger U, Klein HM, Gilsbach JM. Ultrasound-guided craniotomy for minimally invasive exposure of cerebral convexity lesions. Acta-Neurochir Wien. 1994;131(3–4):270–3.

McDicken WN. Invasive ultrasound transducers. In: Lees WR, Lyons EA, editors. Invasive ultrasound. London: Martin Dunitz; 1996. p. 1–16.

McDonald JM. Mental readiness and its link to performance excellence in surgery. Ottawa: Kinetek; 1993.

McLaughlin MR, Wahlig JB, Kaufmann AM, Albright AL. Traumatic basilar aneurysm after endoscopic third ventriculostomy. Case report. Neurosurgery. 1997;41:1400–4.

Mintz GS, Pichard AD, Satler LF, Popma JJ, Kent KM, Leon MB. Three-dimensional intravascular ultrasonography: reconstruction of endovascular stents in vitro and in vivo. J Clin Ultrasound. 1993;21(9):609–15.

Moringlane JR, Voges M. Real-time ultrasound imaging of cerebral lesions during "target point" stereotactic procedures through a burr hole (technical note). Acta Neurochir. 1995;132(1–3):134–7.

Mueller G. Microsensors. In: Wickham J, editor. Minimally invasive therapy, vol. 3. Berlin: MEDTECH 94; 1994.

Müller A, Hatfield ARW. Miniature endoscopic ultrasound. In: Lees WR, Lyons EA, editors. Invasive ultrasound. London: Martin Dunitz; 1996. p. 235–42.

Müller S, Bartel T, Baumann G, Ebel R. Preliminary report: evaluation of three-dimensional echocardiographic volumetry by simultaneous thermal dilution in coronary heart disease. Cardiology. 1996;87:552–9.

Neville RF, Bartorelli AL, Sidawy AN, et al. An in vivo feasibility study of intravascular ultrasound imaging. Am J Surg. 1989;158(2):142–5.

Nunez G, Kaufman H. Ergonomic considerations in the design of neurosurgery instruments. J Neurosurg. 1988;69:436–41.

Oi S, Samii A, Samii M. Frameless free-hand maneuvering of a small-diameter rigid-rod neuroendoscope with a working channel used during high-resolution imaging. J Neurosurg. 2005;102:113–8.

Paleologos TS, Wadley JP, Kitchen ND, Thomas GGT. Clinical utility and cost-effectiveness of intracranial image-guided craniotomy: clinical comparison between conventional and image-guided meningeoma surgery. Neurosurgery. 2000;47:40–8.

Pandian NG, Sugeng L, Vogel M, Marx G. Three-dimensional echocardiography: the future in cardiac imaging (highlights 6). Boston, MA: Learning Center; 1993.

Patkin M. Ergonomics applied to the practice of microsurgery. Aust NZ J Surg. 1977;47:320–8.

Patkin M. Ergonomics in microsurgery. Aust NZ J Obstet Gynaecol. 1981;21:134–6.

Perneczky A. Planning strategies for the suprasellar region. Philosophy of approaches. Neurosurgery. 1992;11:343–8.

Perneczky A, Fries G. Endoscope-assisted brain surgery: evolution, basic concepts, and current technique. Neurosurgery. 1998;42:219–25.

Perneczky A, Tschabitscher M, Resch KDM. Endoscopic anatomy for neurosurgery (atlas/video). Stuttgart: Thieme; 1993.

Perneczky A, Müller-Forell W, van Lindert E, et al. Keyhole concept in neurosurgery. Stuttgart, NewYork: Thieme; 1999.

Poggio T. Wie Computer und Menschensehen. In: Ritter M, editor. Wahrnehmung und visuelles system. Heidelberg: Spektrum der Wissenschaft; 1987. p. 78–89.

Prada F, Solbiati L, Martegani A, Di Meco F, editors. Intraoperative ultrasound (IOUS) in neurosurgery. Heidelberg: Springer; 2016.

Reddy K, Fewer HD, West M, Hill NC. Slit ventricle syndrome with aqueduct stenosis: third ventriculostomy as a definitive treatment. Neurosurgery. 1989;23:756–9.

Regan D, Beverly K, Cynader M. Die Wahrnehmung von Bewegunen im Raum. In: Ritter M, editor. Wahrnehmung und visuelles System. Heidelberg: Spektrum der Wissenschaft; 1987. p. 90–103.

Reich J, Onik GM, Maroon J. Intracerebral biopsy hemorrhage: monitoring and intervention guided by intraoperative sonography. AJNR Am J Neuroradiol. 1988;9(6):1240–1.
Resch KDM. MIN: transoral transpharyngeal approach to the brain. Neurosurg Rev. 1999;22:2–25.
Resch KDM. Postmortal inspection (PMI) for neurosurgery: a training model for endoscopic dissection technique. Neurosurg Rev. 2002;25:79–88.
Resch KDM. Endo-neuro-sonography (ENS). First clinical serie (52 cases). Childs Nerv Syst. 2003;19:137–44.
Resch KDM. Transendoscopic ultrasound for neurosurgery. Heidelberg: Springer; 2005.
Resch KDM, Perneczky A. Endoscopic approaches to the suprasellar region: anatomy and current clinical applications. (Advances in neurosurgery), vol. 22. Berlin, Heidelberg, New York: Springer; 1993.
Resch KDM, Perneczky A. Endoneurosurgery: anatomical basics. In: Samii M, editor. Skull base surgery. Basel: Karger; 1994.
Resch KDM, Perneczky A. Endo-neuro-sonography: anatomical aspects of the basal cisterns. Minim Invasiv Ther Allied Technol. 1997;6:332–9.
Resch KDM, Perneczky A. Endo-neuro-sonography: basics and current use. In: Hellwig D, Bauer BL, editors. Minimally invasive techniques for neurosurgery. Berlin, Heidelberg, New York: Springer; 1998. p. 21–31.
Resch KDM, Reisch R. Endo-neuro-sonography: anatomical aspects of the ventricles. Minim Invasive Neurosurg. 1997;1:2–7.
Resch KDM, Perneczky A, Tschabitscher M, Kindel S. Endoscopic anatomy of the ventricles. Acta Neurochir Suppl. 1994;61:57–61.
Resch KDM, Reisch R, Hertel F, Perneczky A. Endo-neuro-sonographie: eine neue Bildgebeung in der Neurochirurgie. Endoskopie Heute. 1996;2(3) [whole issue].
Resch KDM, Hopf N, Kessel G, Perneczky A. Endo-neuro-sonography: new imaging technique in neuro-surgery. In: Proceedings of the 11th International Congress of neurological surgery, Amsterdam, 6–11 June 1997; 1997a, p. 1969–74.
Resch KDM, Atzor KR, Perneczky A. Anatomical phantom CT study of surgical approaches for 3-D and VR in neurosurgery. Minim Invasive Ther Allied Technol. 1997b;6:228–34.
Resch KDM, Perneczky A, Schwarz M, Voth D. Endo-neuro-sonography: principles and 3D technique. Childs Nerv Syst. 1997c;13:616–21.
Rieger A, Rainov NG, Sanchin L, Schopp G, Burkert W. Ultrasound-guided endoscopic fenestration of the third ventricular floor for non-communicating hydrocephalus. Minim Invasive Neurosurg. 1996;39:17–20.
Roelandt JRTC, ten Cater FJ, Brunning N, et al. Transoesophageal rotoplane echo-CT. A novel approach to dynamic three-dimensional echocardiography. Thoraxcentre J. 1993;6(1) [whole issue].
Roelandt JRTC, ten Cater FJ, Vletter WB, et al. Ultrasonic dynamic three-dimensional visualization of the heart with a multiplane transoesophageal imaging transducer. J Am Soc Echocardiogr. 1994;7(3):217–29.
Rosenfield K, Losordo DW, Ramaswamy K, et al. Three-dimensional reconstruction of human coronary and peripheral arteries from images recorded during two-dimensional intravascular ultrasound examination. Circulation. 1991;84(5):1938–56.
Roth G. Das Gehirn und seine Wirklichkeit. Frankfurt/Main: Surkamp; 1999. p. 258–311.
Sacks O. Der Tag an dem mein Bein fortging. Reinbek bei Hamburg: Rowohlt; 1989a. p. 205–22.
Sacks O. Stumme Stimmen. Reinbek near Hamburg: Rowohlt; 1989b. p. 117–82.
Schlöndorf G. Idee und Entwicklung der "Computer Assisted Surgery" an der RWTH in Aachen. In: Aachener Workshop on Navigierte Hirnchirurgie, Aachen, 4–5 September 1998, abstract book. p. 1–3.
Scholz M, Tombrock S, Konen W, et al. Application of a newly developed visual navigation system in humans. First results. Minim Invas Neurosurg. 2005;48:67–72.
Schroeder HWS. Historische Entwicklung der endoskopischen Neurochirurgie. In: Grumme T, editor. Neurochirurgie in Deutschland. Berlin, Vienna: Blackwell Wissenschaft; 2001.

Schroeder HWS, Gaab MR. Endoscopic management of intracranial arachnoid cysts. In: Hellwig D, Bauer LB, editors. Minimally invasive techniques for neurosurgery. Berlin, Heidelberg, New York: Springer; 1998.
Schroeder HWS, Warzok RW, Assaf JA, Gaab MR. Fatal subarachnoidal hemorrhage after endoscopic third ventriculostomy (case report). J Neurosurg. 1999;90:153–5.
Schwartz S, Cao Q, Azevedo J, Pandian NG. Simulation of intraoperative visualization of cardiac structures and study of dynamic surgical anatomy with real-time three-dimensional echocardiography. Am J Cardiol. 1994;73:501–7.
Shekhar LN. Comment: in: Paleologos TS, Wadley JP, Kitchen ND, Thomas GGT (2000) clinical utility and cost-effectiveness of intracranial image-guided craniotomy: clinical comparison between conventional and image-guided meningioma surgery. Neurosurgery. 2000;47:40–8.
Slovis TL, Canady A, Touchette A, Goldstein A. Transcranial sonography through the burr hole for detection of ventriculomegaly. A preliminary report. J Ultrasound Med. 1991;10(4):195–200.
Steinmeier R, Fahlbusch R, Ganslandt O, et al. Intra-operartive magnetic resonance imaging with the Magnetome open scanner: concepts, neurosurgical indications, and procedures: a preliminary report. Neurosurgery. 1998;43:739–48.
Strowitzki M, Ultrasound in Neurosurgery Study Group. Ultraschall in der Neurochirurgie (abstract). Zentralbl Neurochir Suppl. 2000;1:6.
Suhm N, Dams J, van Leyen K, Lorenz A, Bendl R. Limitations for three-dimensional ultrasound imaging through burr hole trepanation. Ultrasound Med Biol. 1998;24(5):663–71.
Sutcliffe JC. The value of intraoperative ultrasound in neurosurgery. Br J Neurosurg. 1991;5(2):169–78.
Taniguchi M, Takimoto H, Yoshimine T, et al. Application of rigid endoscope to the microsurgical management of 54 cerebral aneurysms: results in 48 patients. J Neurosurg. 1999;91:231–7.
Tisell M, Almström O, Stephensen H, Tullberg M, Wikkelsö C. How effective is endoscopic third ventriculostomy in treating adult hydrocephalus caused by primary aqueductal stenosis? Neurosurgery. 2000;46:104–9.
Tronnier V, Staubert A, Wirtz R, et al. MRI-guided brain biopsies using 0.2 tesla open magnet. Minim Invasive Neurosurg. 1999;42:118–22.
Tsutsumi Y, Andoh Y, Sakaguchi J. A new ultrasound-guided brain biopsy technique through a burr hole (technical note). Acta Neurochir. 1989;96(1–2):72–5.
Van Roost D, Clusmann H, Urbach H, Helmstaedt C, Schramm J. Transcortical versus transsylvian approach for selective amygdalo-hippocampectomy. In: 52nd annual meeting of the German Society of Neurosurgery, 27–30 May 2001, Bielefeld, Abstract Book. p. 50.
Van Velthoven V, Auer LM. Practical application of intraoperative ultrasound imaging. Acta Neurochir. 1990;105(1–2):5–13.
von Weizäcker V. Der Gestaltkreis. Stuttgart: Thieme; 1950.
Weinberg R. Future direction in neurosurgery visualization. In: Apuzzo MLJ, editor. Neurosurgery for the third millennium. Park Ridge, IL: American Association of Neurological Surgeons; 1992. p. 47–64.
Wickham JE. Treatment of urinary tract stones. BMJ. 1993;307:1414–7.
Wirtz CR, Kunze S. Neuronavigation: computeras-sistierte Neurochirurgie. Dtsch Arztebl. 1998;95:A-2384–90.
Wirtz CR, Tronnier VM, Bonsanto MM, et al. Image-guided neurosurgery with intraoperative MPI: updated frameless stereotaxy and radically control. Stereotact Funct Neurosurg. 1997;68:39–43.
Wirtz CR, Knauth M, Staubert A, et al. Clinical evaluation and follow-up results for intraoperative magnetic resonance imaging in neurosurgery. Neurosurgery. 2000;46:1112–20.
Wurm G, Wies W, Schnizer M, Trenkler J, Holl K. Advanced surgical approach for selective amygdalo-hippocampectomy through neuronavigation. Neurosurgery. 2000;46:1377–82.
Yamakawa K, Kondo K, Yoshioka M, Takakura K. Ultrasound guided endoscopic neurosurgery-new surgical instrument and technique. Acta Neurochir (Wien) Suppl. 1994;61:46–8.
Yasargil MG. Microsurgery applied to neurosurgery. Stuttgart: Thieme; 1969.
Yasargil MG. Microneurosurgery, vol. 1. Stuttgart: Thieme; 1984a.

Yasargil MG. Microneurosurgery, vol. 2. Stuttgart, New York: Thieme; 1984b.
Yasargil MG. Microneurosurgery, vol IVA: neuroradiology, conclusion. Stuttgart: Thieme; 1994a. p. 209.
Yasargil MG. Microneurosurgery, vol IVA: CNS tumors: surgical anatomy, neuropathology, neuroradiology, neurophysiology, clinical considerations, operability, treatment options. Stuttgart, New York: Thieme; 1994b.
Yasargil MG. Microneurosurgery, vol IVB: CNS tumors. Stuttgart, New York: Thieme; 1994c.
Zorzic C, Angonese I. Subependymal pseudocysts in the neonate. Eur J Pediatr. 1989;148(5):462–4.

Key Techniques of MIN: Neuroendoscopy 5

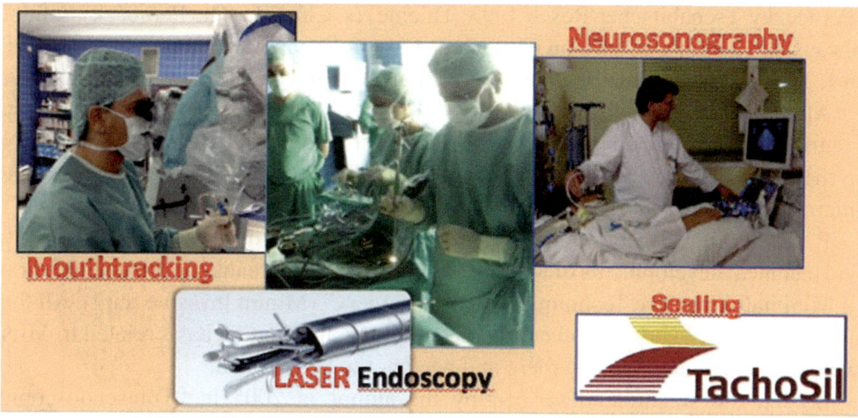

5.1 Introduction: The Story

Endoscopy once was the "flag-ship" of MIN, but anyway became an unloved child of neurosurgery like the other key techniques of MIN. In contrast to ultrasound, which never had a hype in neurosurgery, endoscopy was the leading surgical technique within MIN in the 1990-ies.

Most neurosurgeons being interested, and going into this "new" field, triggered in growing by technical revolutions of the video—chain, saw endoscopy just as a tool. (Marburg-/Greifswald-school) But only Perneczky had the vision to see it as one corner—stones of the new concept of MIN, which he called at that time, unluckily, "key—hole"—concept, provoking many misunderstandings by that (s. Chaps. 1 and 2).

K. D. M. Resch, *Key Concepts in MIN - Intracerebral Hemorrhage Evacuation*, Key-Concepts in MIN 1, https://doi.org/10.1007/978-3-030-46513-1_5

The brake-through however came with the following concept of "endoscopy-assisted micro-neurosurgery". This concept was born in the anatomical lab 1990 (s. Chap. 3 Vol. 2) and was taken over fast to the OR after a convincing series of some 200 dissections in our simulation model. Endoscopy fitted perfectly into the MIN concept and due to its excellent visualization-, illumination- and magnification qualities became the major technique surgically within MIN. It is of vast importance to understand, that it came into an overall concept and was not only "just a tool", a frequent phrase in congress presentations and papers at that time. Today for MIN we need to say, that without the excellent concept with the tool alone you may become like a fool.

However, the chronology was formed to history by the important paper of P. Grunert together with other leaders of endoscopy of that time like D. Hellwig and M. Gaab: "German neuro-endoscopy above the skull base" 2009, Neurosurgical Focus. One mile-stone was not mentioned in the paper, namely the first anatomy atlas book for neuro-endoscopy: "Endoscopic Anatomy for Neurosurgery" (Perneczky/Tschabitscher/Resch 1993; Thieme) (s. Chap. 1). The Perneczky-school always had a strong culture on clinical anatomy following the historical imperative of J. Tandler and E. Ranzi from Vienna school: anatomy should be ahead to surgery!

After the sudden death of Perneczky the paper of P. Grunert et al was a necessary contribution to keep MIN alive, and not to bury it with the body of Perneczky. MIN in the sense of Perneczky did not reach the gold standard and is not accepted like micro-neurosurgery in the sense of Yasargil.

P. Grunert in 2013 presented another major paper on the exotheric philosophy and technical developments of MIN: "From the Idea to its Realization: The Evolution of Minimally Invasive Techniques in Neurosurgery" (Minim Invasive Surg). All the classical fields of surgery and neurosurgery and their scientific basis related to MIN were described meticulously there and need not to be repeated.

So far, the situation, described by the author in 2005 in "Transendoscopic Ultrasound for Neurosurgery, has not really changed:

"The Step from micro-neurosurgery to endo-neuro-surgery has meant a change to a minimally invasive technique, but at the same time the new technique is less safe (Schroeder et al. 1999), which limits its applicability. Further developments have therefore had to be aimed at making neuro-endoscopy safer. The current concept used to establish a near-real-time guidance system has involved the use of MR or CT, requiring enormous financial investment and technical effort (Albert et al. 1998; Black and Mehta 2000; Fahlbusch and Nimsky 2000; Grönemeyer et al. 1994; Hayashi et al. 1995; Kobayashi and Okudera 2000; Maciunas 1993; Wirtz et al. 2000). It is still not clear what surgical benefit this approach will have (Kelly 2000; Maciunas 2000; Paleologos et al. 2000; Shekhar 2000)."

In addition to P. Grunerts' papers among some others, and with the vision of a testimony, it must be questioned, why endoscopy is far behind of its technical and clinical possibilities, even in Europe. Since 1990 there are still only few centers in Europe that use endoscopy routinely and regularly with an amount of more than 1 case per month and with all of the broad spectrum of indications. Moreover, very few centers do education and hands-on courses in neuro-endoscopy. The mainstream is not going towards MIN, after more than 30 years-experience in over 10 high level clinics, the author has the concern that MIN has an existence in a hostile territory. Compared with number of endoscopy procedures in developing countries Europe has lost its leading position. The so called “EBM” (s. Chap. 6) has worsening the situation as it is becoming difficult to get big series of procedures for RCTs in an adequate time, and multi-level science does actually not gain a beneficial acceptance in the professional community (s. Chap. 6).

It became an irony of history, that patients with an indication for a neuro-endoscopic procedure have a better chance to get it done in some low-income countries, like Mexico for example, than in central Europe, and that unusual clinical situations are answered with not recommended routine procedures. Many complications are accepted with shunting procedures before an endoscopy is considered and many prolonged EVDs became usual, instead shortening such therapy time by an endoscopic procedure is considered (see below).

Example 1: This child had to wait 3 years and went through an Odyssey of complications and shunt-revisions, before an endoscopic procedure stopped the disaster, and she went home (Graph 5.1).

Example 2: This child with a typical temporal sub-arachnoidal cyst right was considered for an endoscopic procedure by an experienced neurosurgeon, finally a shunt was applied, as usual, by a trainee. Even if one may see the result as an accident, it is regular use, to prefer a shunt as standard procedure for each liquor problem, no matter, what risks are encountered. Shunt procedures are mostly seen as a training opportunity while endoscope procedures are seen as dangerous adventures (Graph 5.2).

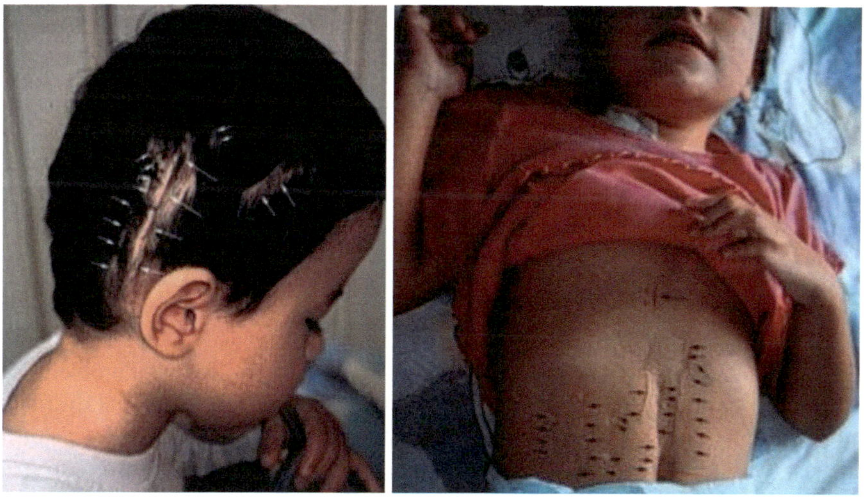

Graph 5.1 Battlefield of scarfs by the standard concept of shunting every type of hydrocephalus

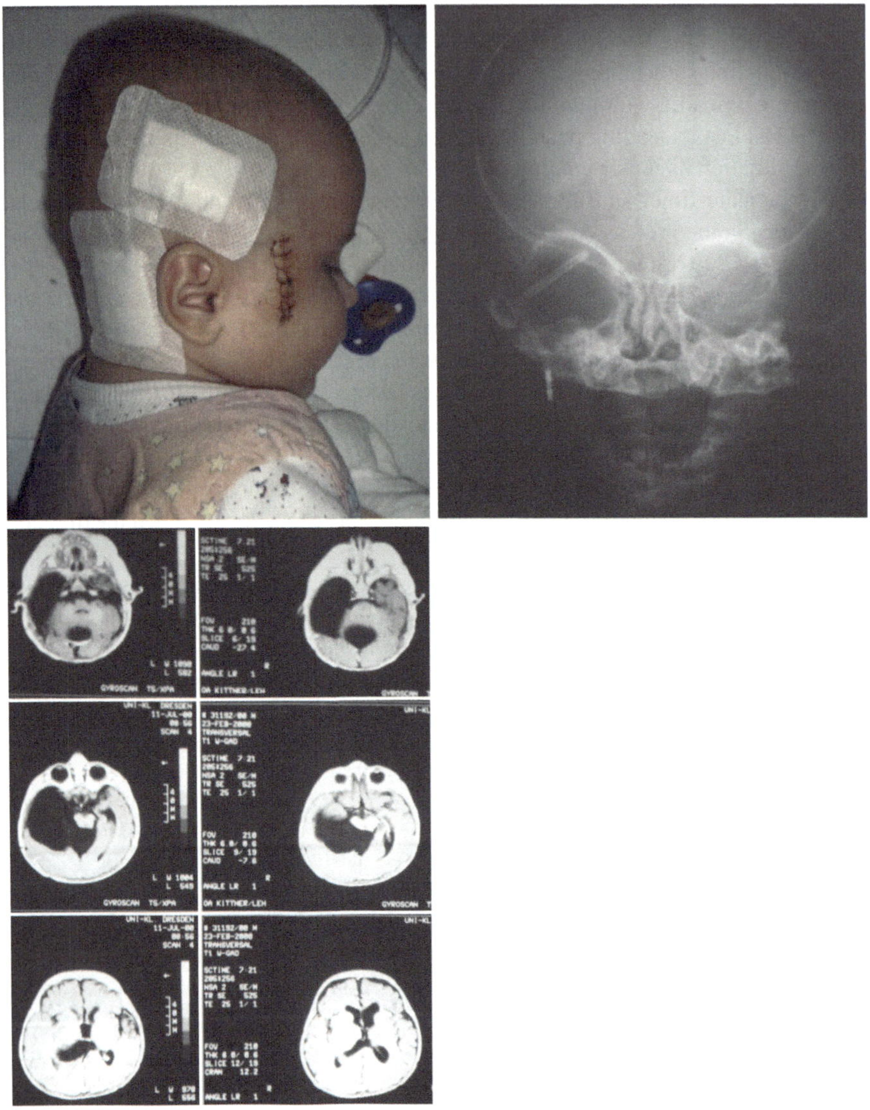

Graph 5.2 Shuted child with an endoscopy indication

Example 3: This This case of a 78-year lady presents two frequent problems of endoscopy indication and understanding of endoscopy results. The shunt dysfunction was overseen for many years (shunt since 18 years). The primary aqueduct stenosis was neglected. After ETV the patient recovered fast. But during recovery of vigilance and motility, side effects of risperdal-medication therapy became prominent. These symptoms were misinterpreted as failure of the ETV though MR showed patency of ETV. During the course two contra-indicated ventricular drainages did

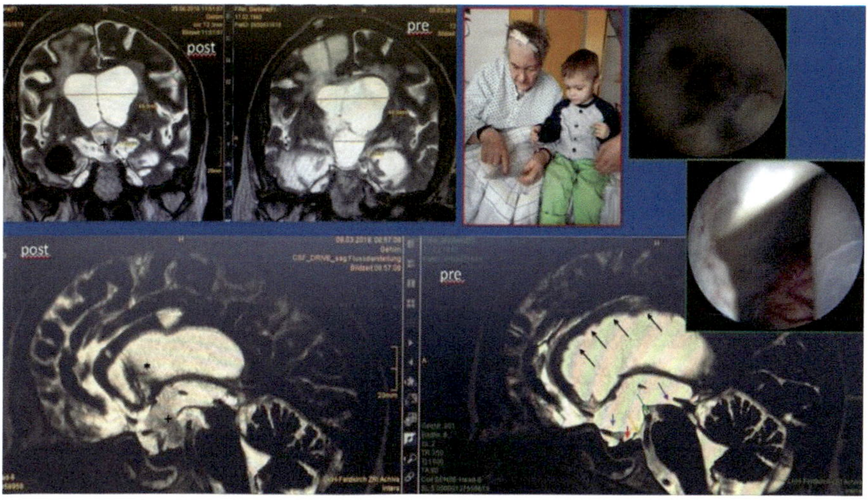

Graph 5.3 Shunt desaster over 18 years in a ETV indication case

not improve the symptoms. Even the sever malignant neuroleptic syndrome (MNS), discussed by the neurologists, was not accepted and a drainage was anyway placed without recovery.

Only stop of risperdal and amandatin-medication ended this dangerous MNS and she started to wake up, began to speak and rigor, trismus and meningism as well as hyperthermia disappeared within days (Graph 5.3).

This story tells us: to judge the postop result of ETV you must be familiar with the varieties of postop courses after ETV and, psycho-pharmacology must be taken into account, especially psychiatric drugs are misused in elderly patients to calm them down. Over 50% of senior-home residents worldwide show pharmacologically induced parkinsonism! In this case even the feared NMS (Malignant Neuroleptic Syndrome) was mixed-up with post ETV failure! Such happens easily and usually in a hostile environment towards neuro-endoscopy.

On the **5th Congress of International Federation of Neuro Endoscopy in Athens 2009** (Graph 5.4), due to the "data evidence" the ETV was raised to "gold standard" in obstructive hydrocephalus. However, the educational situation and the ignorance about endoscopy in neurosurgery in central Europe, did not allow to realize this gold standard.

The benefit of scientific results is depending on more than evidence by RCTs. Endoscopy can teach us: **once you focus your vision you need to widen your mind!** (Graph 5.5).

But progress and profound understanding must not follow this lasting pitfall of scientific working by RCTs, just to collect myriads of data worked out by statistical refinements and statistical power.

The real problem in endoscopy indications is, that liquor dynamics is still one of the top secrets of neurosurgery. It is typical for present time to interpret everything

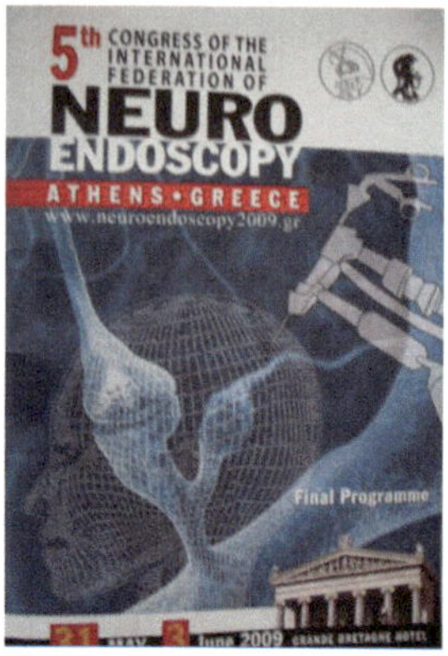

Graph 5.4 IFNE 2009

Graph 5.5 Mental navigation

according to imaging, neglecting pathophysiology and viewing liquor-dynamic through the glasses of static parameters like volume of the ventricles on images or ICP values by inserted sensors. Understanding of liquor-dynamics needs to have cognitive pleasure in "symptom criminology". One needs to become a "Liquor Sherlock Holmes" to get each single case. The few categories, different hydrocephalus types can be brought in an order, are far away to understand the individual case.

However, there are few but important physical and system-analytical rules one needs to consider, for actual routines in clinical reality:

1. Physical function of the liquor is, that it guarantees, so to speak: the blood-flow-"tsunami" from the heart shall not kill the brain inside its hard cover by the **water-hammer effect**, occuring each second. This force destroys very slowly the endothelia of the periventricular capillaries causing sub-ependymal atrophies (Graph 5.6).
2. The spinal canal is (especially in children) the most effective liquor peak **compensation path-way**. Direction of liquor flow longitudinal into this space compensates the peaks of liquor pulsation (Graph 5.7).

Graph 5.6 Liquor dynamics protects the brain against the Water-Hammer of the heart

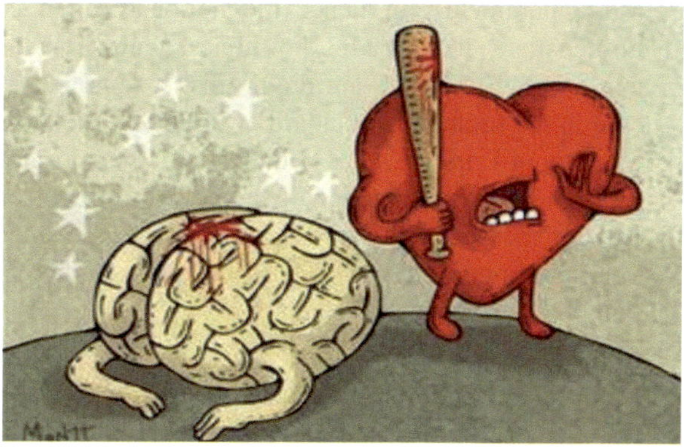

Graph 5.7 Spinal canal: compensation path of liquor P-peaks

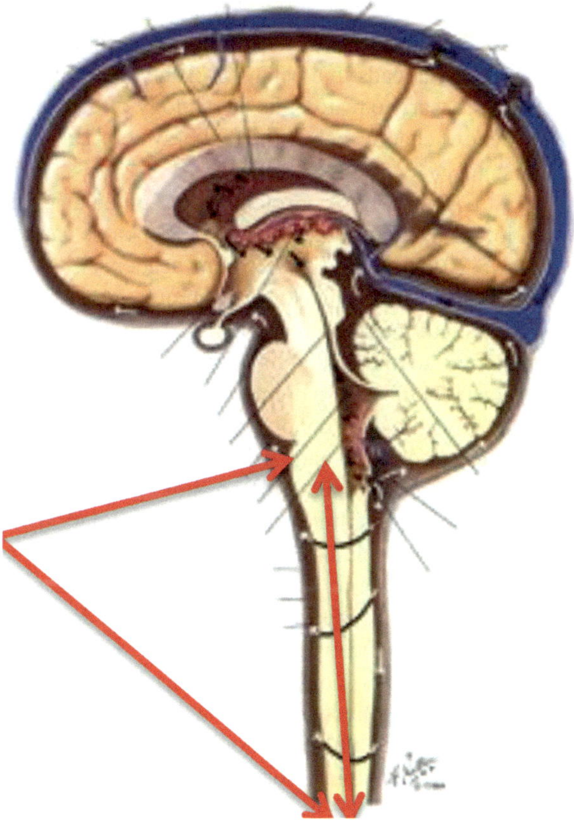

3. Many parameters influence the liquor-dynamic continuously: All factors of **homeostasis**, like proteins, hormones, neuro-transmitters etc. as distribution system parallel to blood, are doing this. It must be analyzed individually in each case in the future as a standard of liquor diagnostics.
4. The **law of Laplace** helps to understand the correlation of volume and force that works against the ependyma. We know the effect from blowing up a balloon. The smaller the ventricles are the more force is needed to enlarge them. In small ventricle condition the ICP can be much higher in liquor-dynamic pathology than in large ventricles situation.

$$\boxed{\Delta p = \frac{2 \cdot \gamma}{r}}$$

5. The ventricle system can change its behavior paradox in some conditions like megalo-ventricles or post-bleeding conditions. Such paradox changes of systems that are influenced by different parameters are described by the **theory of "hysteresis"**, which cannot be explained in short. The loss of turgor of the brain parenchyma leads to a behavior of the ventricular system that makes a negative ICP necessary to restore the physiology of the cerebral fluid compartments. This somehow chaotic behavior of the system can be described by the theory on Hysteresis.

   At EANS 2017, A. Grotenhuis presented an excellent lecture about the application of hysteresis theory in hydrocephalus cases. (s. A. Grotenhuis, lecture at EANS 2017 Venice; EANS online academy).
6. The concept that only lowering liquor pressure and/or decrease of ventricular volume will make hydrocephalus patients symptom-free is wrong and is in contradiction to the pathophysiological parameters mentioned. The real-loss of liquor from the CNS volume via shunt systems changes the complex regulation from physiological condition into a shunt-regulated condition causing a syndrome, which was named **"shunt-disease"**. Even the most expensive shunt-systems cannot simulate the physiological regulation. Handicapped brains with less compliance suffer from the shunt-regulation, not to speak about shunt complications. Complains of the patients are not taken serious and children cannot communicate them. Interpretation of present symptoms in children without regard on the conditions for their future developing is not sufficient. Shunt-disease can handicap the normal course, which is often ignored, instead of thinking of an endoscopy indication. In about 60% cases of symptomatic communicating hydrocephalus will do well, if an ETV, supporting mentioned experience and pathophysiological knowledge, optimizes the velocity of CSF.
7. Radiological imaging does have difficulties to judge different conditions of hydrocephalus, like megalo-ventricles or slit-ventricles, and it even does not accept the diagnosis of **"functional" aqueduct-stenosis** in cases of symptomatic three-ventricular hydrocephalus. Deterioration of clinical course and symptoms are ignored if morphology stays more or less similar in large- and even mega-ventricles. The primate of clinical signs and symptom is still not accepted.

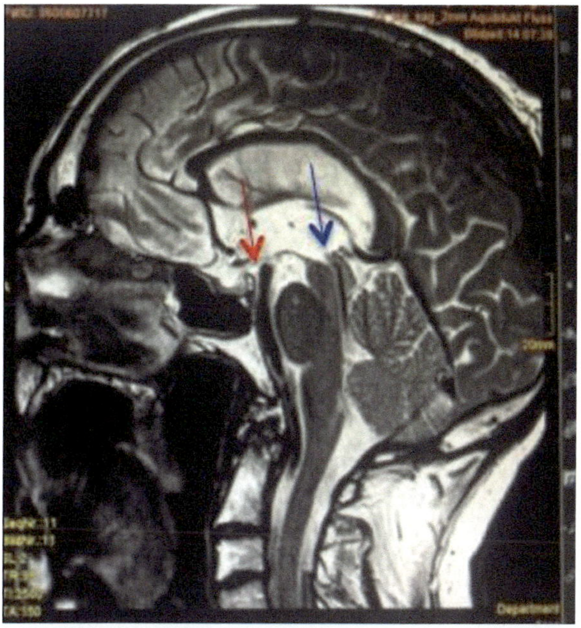

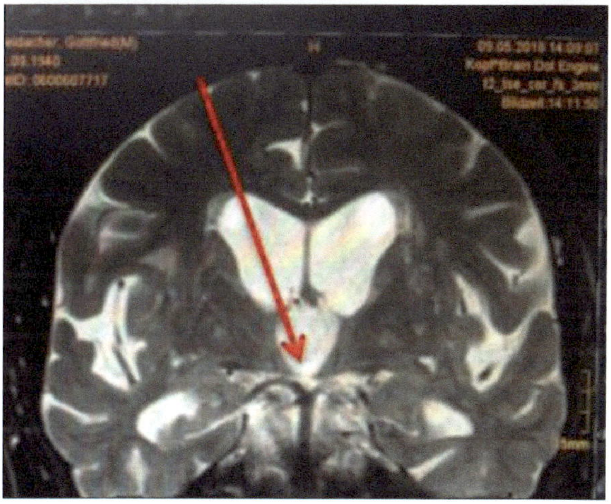

Example 4: This is a typical case of functional aqueduct-stenosis with a flow-signal but in connection with symptomatic tree-ventricular hydrocephalus. It is obvious (in German: evident) due to morphology, that the functional liquor flow has a partial obstructive problem in the aqueduct.

The ETV resolved all symptoms caused by liquor-dynamic problems. Frequent neuro-endoscopic experience gives knowledge on the quality of the flow: A "jet-flow" or sudden stop of the flow signal indicates a functional stenosis.

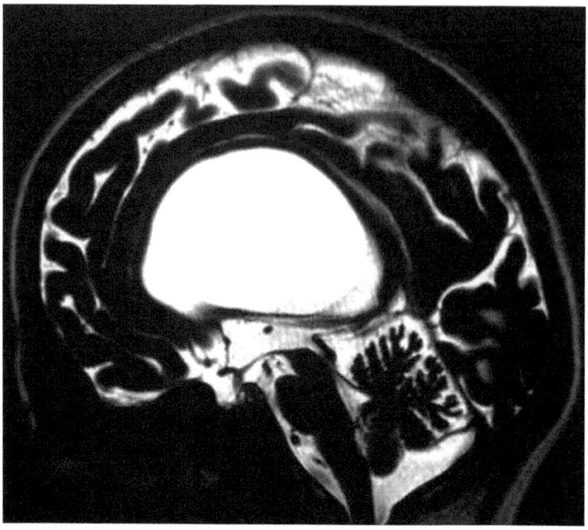

In this morphological decompensated case of three-ventricular hydrocephalus with sever symptoms was diagnosed stable course of ventricular size followed by "no indication for ETV". A constant struggle during 1 year for ETV by the patient was followed postop by disappearance of the most disturbing symptoms. The scarfs from prior shunt revisions now made the rest of symptoms.

This **critical introduction** is indispensible to mark the **reality of endoscopy**, here only in few types of hydrocephalus cases, and typical inacceptable indication strategies ignoring of knowledge, experience and even so-called evidence. Beside few promising spots in Europe one can discover, what are the real problems of clinical endoscopy.

This chapter shall describe the main techniques, strategies and applications of endoscopy as key-technique within MIN.

History is to form a vector from the past into the present by understanding chronology making it comprehensive for changing present processes into future (Graph 5.8).

Primarily there was the insight to minimize the trauma to have better outcomes. The problem was first to better see with more light and magnification and to converge the vectors of light, sight and instruments, which was started by microscopy.

- Endoscopes were first used as tools to solve specific problems in hydrocephalus (Graph 5.9).
- After 50 years of neuro-endoscopy of the ventricular system and cystic lesions, Guiot used endoscopy also by the trans-nasal approach and Prott, via the CPA, but it was still seen as just a tool.
- After 80 years of neuro-endoscopy, Perneczky's team developed a new concept, which was followed by publication of the first anatomical atlas as revealed by neuro-endoscopy. In endoscopy-assisted micro-neurosurgery the endoscope is

1910	Lespinasse	Hydrocephalus	Urethroscope
1918	Dandy	Hydrocephalus	Cystoscope
1919	Payr	Hydrocephalus	Encephaloscopy
1922	Dandy	Obstetric hydrocephalus	Ventriculoscopy, ETV
1923	Volkmann	Hydrocephalus	Encephaloscope
1923	Mixter	Hydrocephalus	ETV with monitor
1923	Fay/Grant	Hydrocephalus	Endoscopic photography
1934	Putnam	Hydrocephalus	Rod lens, diathermy
1936	Scarf	Hydrocephalus	70° lens
1939	Pool	Spinal pathology	Myeloscope
1952	Nulsern/Spitz	Hydrocephalus	Shunt system
1959	Hopkins		Rod-lens endoscope
1963	Guiot	Hydrocephalus colloid cyst	Endoscopic light
		Transnasal inspection	Biportal approach
1973	Fukushima	Hydrocephalus, tumores	Fiberscope, ciné-endoscopy
1974	Prott	CPA	Cisternoscopy
1975	Griffith	Hydrocephalus	Hopkins endosope
1977	Apuzzo	3rd Ventricular lesions	Stereotaxy and CT guidance
1978	Vries	Hydrocephalus	Large series of ETV
1979	Oppel	Cerebellopontine angle	CPA endoscopy
1983	Auer	Intracerebral hemorrhage	5-Channel endoscope
1987	Gaab	Multiple lesions, tumors	Endoscopy set
1987	Mayer/Brock	Spinal disc lesions	Spinal endoscopy set
1989	Manwaring	Complex cystic lesions	Large series of endoscopic shunting
1989	Bauer/Hellwig	Tumors	Stereotactic fiberscopic biopsies
1990	Caemert	Cysts, ventricles	Stereotactically guided lens-scope
1990	Resch	Endoscopic pathological anatomy	Endoscopic postmortem inspection
1990	Perneczky	All CNS lesions	Endoscopically assisted micro-NC
1992	Karakhan	Ventricular + subarachnoid	Fiberscope
1993	Perneczky et al.	First atlas	Endoscopic anatomy for neurosurgery
1993	Cohen	Ventricular lesions	Large series
1996	Resch	ENS	Transendoscopic sono-guided endoscopy
1996	Grotenhuis	Clinical book	Endoscope-assisted craniotomy
1997	Jho	Skull-base lesions	Extended endoscopic pernasal transsphenoidal neurosurgery

Graph 5.8 Chronology of neuro-endoscopy

by no means just a tool, but has become the corner-stone of a new concept (Perneczky et al. 1993; Resch 1999). This new vision has opened up a wide variety of indications for endoscopy.

- Finally, Jho took over the ENT concept of endoscopy for per-nasal skull-base surgery, in which the microscope is substituted by the endoscope (Cappabianca 1998, 1999; Jho and Carrau 1997; Jho et al. 1996; Resch 1999, 2002). This concept not only enhanced the technical development itself, but also expanded the indications. Interestingly, the history of microneurosurgery shows the same characteristic: not the tool but the new concept led to the brake through.

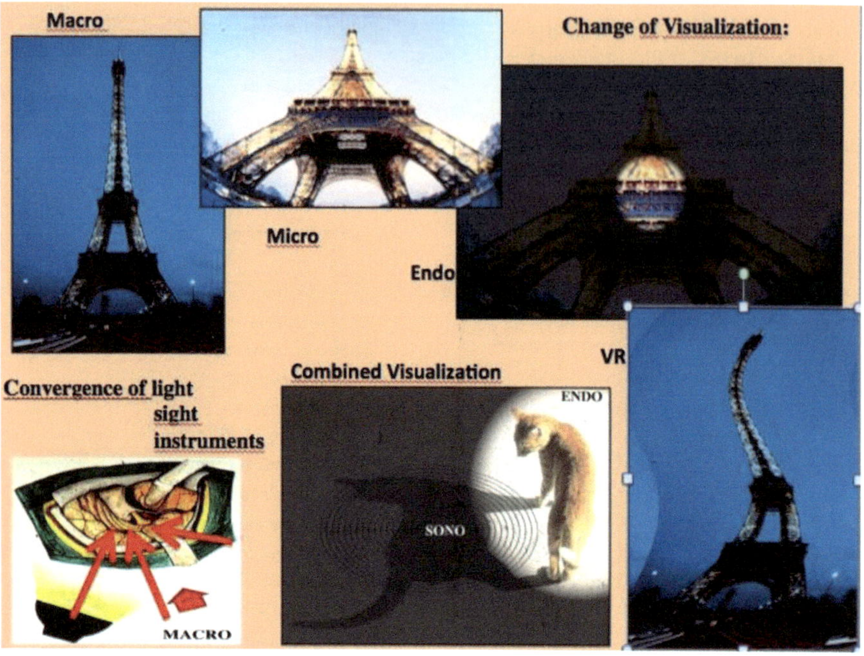

Graph 5.9 Evolution of visualization

With increasing experience more advantages during application were understood and communicated. The optical superiority of endoscopes regarding light, magnification, deep focus range, angled-panorama view and 3D effect by paralaxis. But also disadvantages and needs of further development appeared.

The most important difference to the microscope was, that the key-hole had to be seen as inside the cranium causing sever safety problems. A safe and effective holding device did not exist at that time and is still a major problem. The best solution by Olympics Co was not a commercial success and not available. However a prototype was used by Perneczky and later on became the label "Exoscope", but had its brake through 20 years later on the marked, today offered by many companies (s. Chap. 3).

However, it became also obvious, that endoscopy can assist microsurgery to overcome visual problems in MIN approaches. The concept by Yasargil of contralateral approach could be made to a principle as it was not a problem anymore how deep the vision needs to reach. It became also clear, that the major superiority of the endoscope was the brilliant visualization. All other surgical applications of microsurgery became handicapped like coagulation and resection of volumes. In parallel results of endoscopy simulations in the lab provided the next big step into the field (s. Vol. 2; Chap. 3).

The concept of para-endoscopic dissection was tested in the simulation from the lab in 60 aneurysm cases post-mortem, when it was impossible to dissect aneurysms

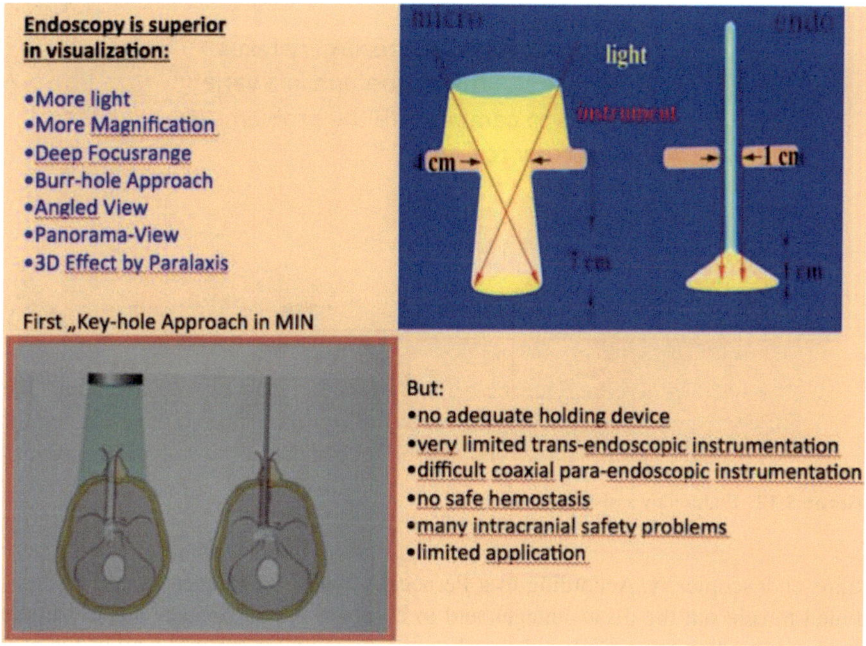

Graph 5.10 Light and sight

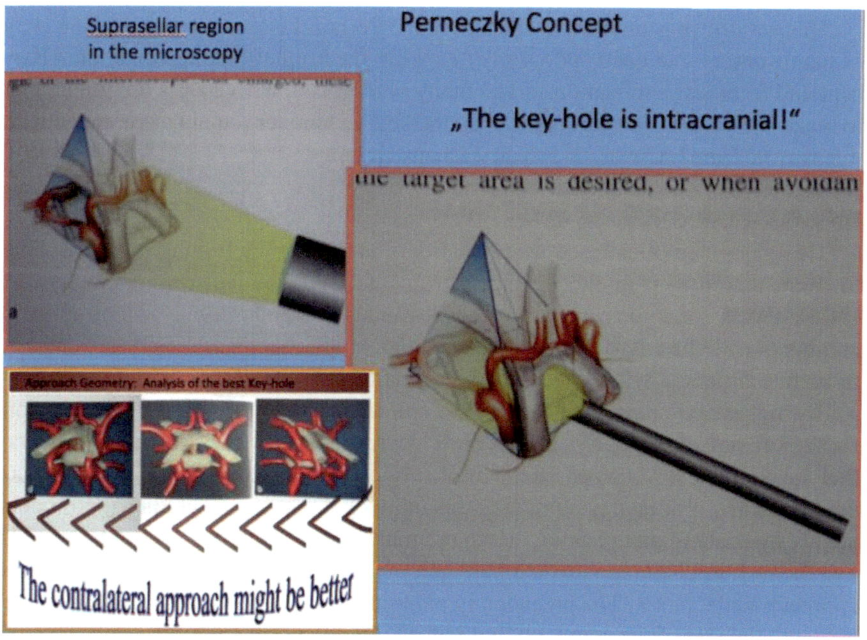

Graph 5.11 Key-Hole concept (Perneczky)

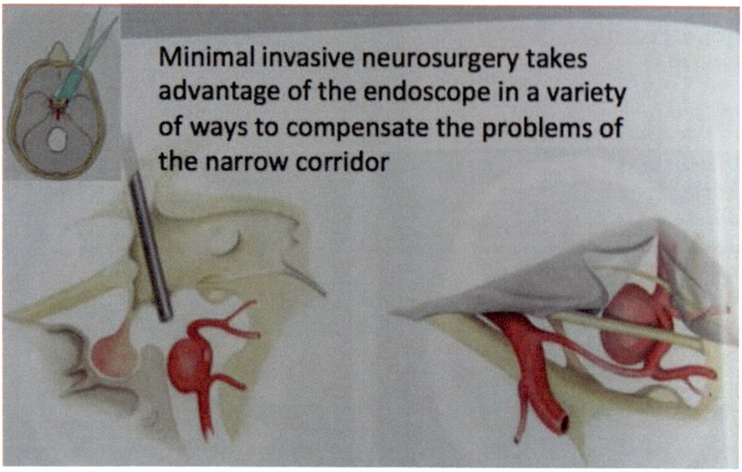

Graph 5.12 Endoscopy assistance concept

trans-endoscopically. According to a Perneczky "rule" techniques should be combined to rule out the disadvantages and to combine the advantages. So it was the next step to dissect parallel to the endoscope (para-endoscopically) microsurgical and to visualize by the endoscope. The superiority of this strategy was very convincing and was clinically fast developed to "endoscopy-assisted micro-neurosurgery" (Graph 5.12).

This was the most important concept of the Perneczky school causing a new way of neurosurgery and open endoscopy to a wide spectrum of applications. This concept and technical application solved many problems of MIN and is the fundament to make MIN a gold standard. However, the key-techniques remain a prerequisite on this way (s. Vol. 1; Chaps. 2–5; Vol. 2 Chaps. 1, 2).

Actually it remains unclear if the new series of "exoscopes" can face this challenge (see Chap. 3).

The major step of endoscopy application was the change from trans-endoscopic instrumentation to para-endoscopic instrumentation. This led to the new field of endoscopy-assisted microsurgery. The advantage of endoscopic visualization is combined with the microsurgical instrumentation, thus compensating disadvantages of both techniques. MIN is such combining of techniques (Graph 5.13).

The next step remains one of the most critical point though it was the inspiration for the key-hole concept. Systematic pre-closure analysis of the surgical field shows, that the really necessary and used part of surgical field is in average 1/3 of the standard approach. With todays' imaging this individual part can be predicted pre-operatively by analysis and planning. Each approach can be seen as a sum of burr-holes and the task in planning is which burr-hole or burr-hole group is the optimal MIN approach to the target. The precision in planning and design of the procedure is the goal, not the size of the approach. Individual approach-designs, however, will usual be 1/3 of the standard approaches.

Graph 5.13 Endoscopy assisted microneurosurgery

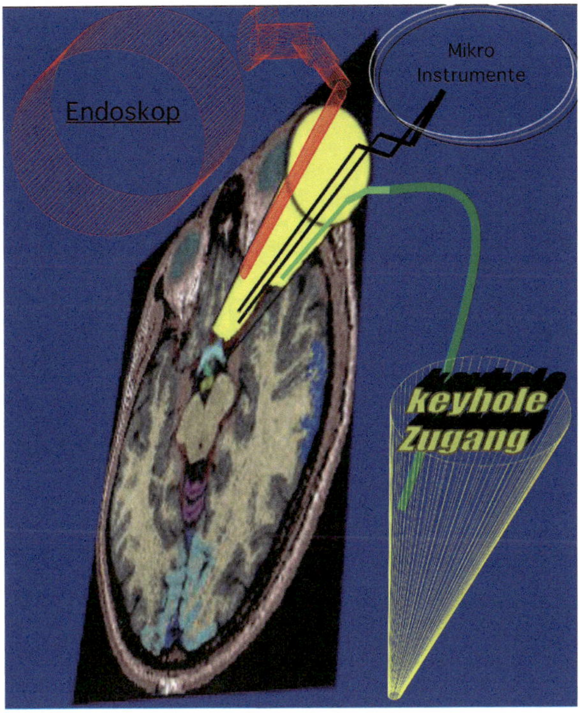

To open and then to see what one can do is out. The results of the vast evolution of imaging must be transformed to a benefit for the patient if this imaging should not be just a "fun-imaging". The possibilities endoscopy can provide, can be used in a wide spectrum in daily micro-neurosurgery. The "exoscopes" cannot see around corners and behind structures, leaving a future field of application for endoscopy, if perhaps exoscopy should be able to replace endoscopy in the major field of MIN.

These principles of MIN are derived from advanced micro-neurosurgery, but they are adaptations to the technical abilities and may adapted further on with technical evolution (Graph 5.14).

5.2 Surgical Endoscopic Anatomy for Neurosurgery

The main anatomical features shall be remembered, like those shown in the first atlas book 1993 (Perneczky/Tschabitscher/Resch, Thieme), but mainly the once done during surgery. The majority of images are surgical anatomy, which provide us with life-anatomy from the different regions most needed for endo-surgical procedures. Moreover, the endo-approaches and endo-aspects of the posterior fossa are supplied and classified.

It is an absolute "must" to take part in an endoscopy course before entering a patient endoscopically. Endoscopy needs a strong mental navigation: you know a

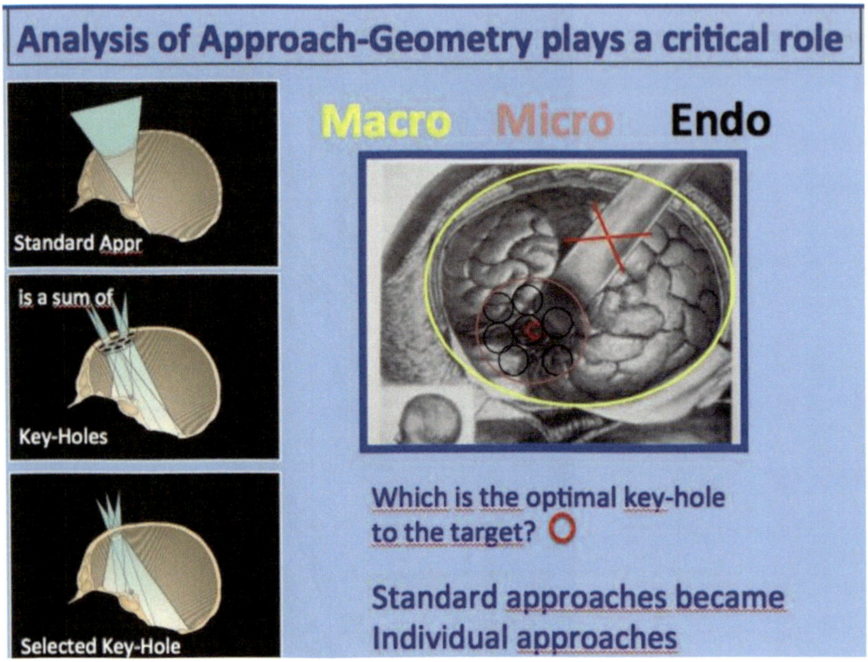

Graph 5.14 Approach analysis in MIN

broad perspective of anatomy but you see only a focused target area. It is a video-anatomy and needs a perfect adjustment of the video chain (Chap. 3).

The indication on endoscopic procedures should be decided on the basis of patho-physiology, but which procedures and how, will be decided according to anatomy. The main features should be kept in mind to be able for orientation precisely and to judge the changes in pathological cases. Orientation in endoscopy is possible by very detailed anatomy that can be called an "anatomical fingerprint". The usual landmark structures might not be visible for orientation, but the fingerprint of the focus you see must be known and recognized. Vascularization, color and texture may serve as landmarks then.

27 years after appearance of the atlas-book those images, acquired within the surgical simulation concept (KDMR), are still well comparable with surgical anatomy (s. also Vol. 2, Chap. 3).

Anatomical orientation is an absolute question on safety as there is not much space and the tissue is very fragile. A wrong decision regarding target and space might result in a disaster. Endoscopic target regions are critical regions with sever functional outcomes in case of a complication. The endoscopic neurosurgeon must be able to control navigation failures and video chain pitfalls. Wrong side of ventricle or wrong level of depth must be recognized by anatomy, in case even in contradiction to navigation data. Estimation and judgment of distances, angles and diameters must be learned, a library of endoscopic experienced anatomical

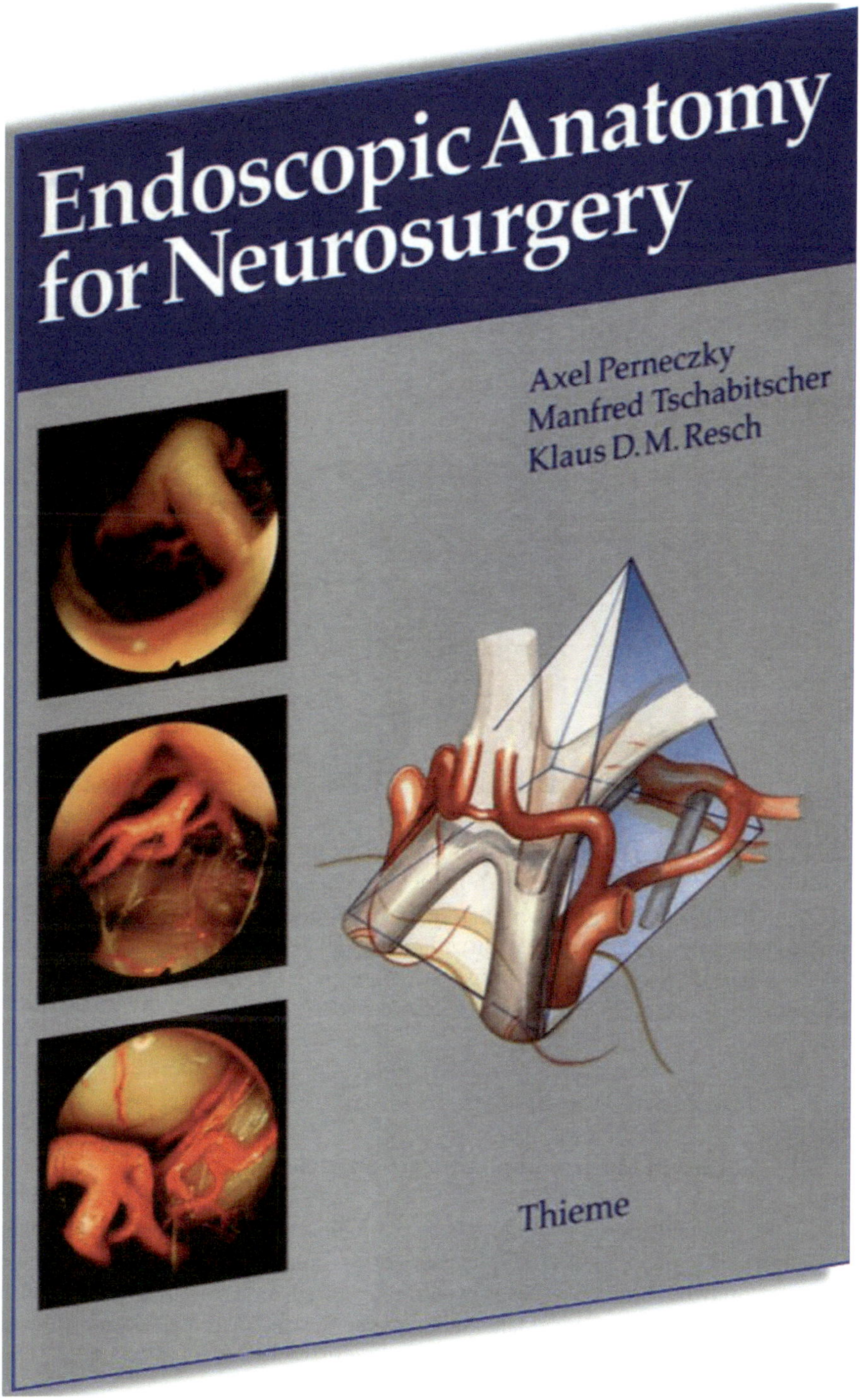

Fig. 5.1 First atlas on endoscopic anatomy for neurosurgery 1993

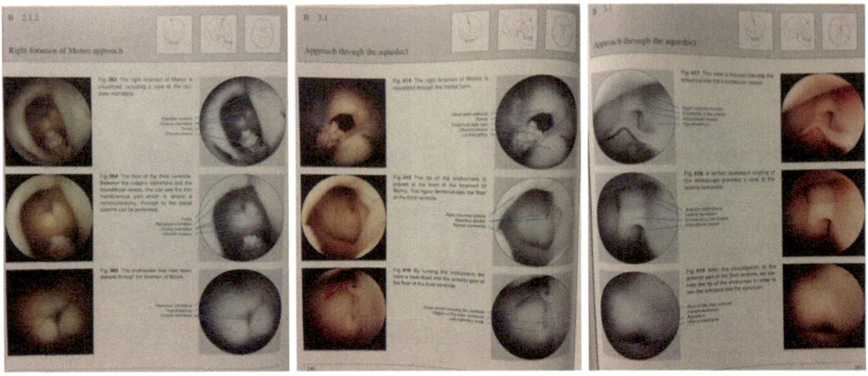

Graph 5.15 1993 (thieme; stuttgart)

Graph 5.16 Angeling and size F. Monroi

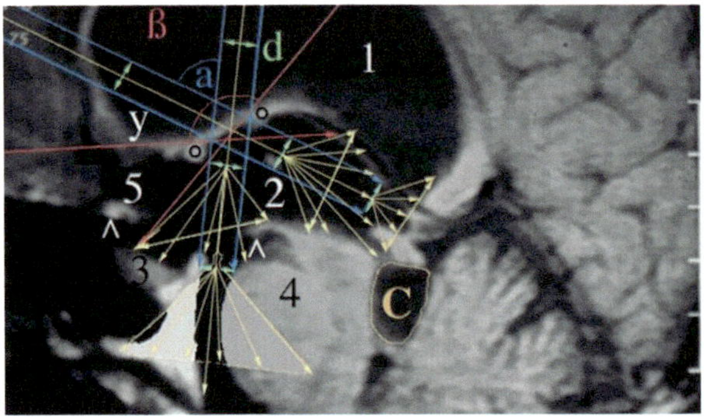

landscapes needs to be acquired. You must control the assisting- and technical systems by anatomy and should never rely on the technique alone (Graph 5.13).

Anatomical appearance differs markedly from microscopic view (Graph 5.17). There is the fish-eye effect, size of structures and all distances are difficult to understand. Angle of view and perspective are completely changed to the well-known experience (Graph 5.16).

5.3 The Ventricular System

5.3.1 Foramen Monroi and Lateral Ventricles

Usual the first endoscopic procedure is a ventriculoscopy confronting the beginner with a spectrum of variations of typical structures. Presenting the aspect of foramen Monroi it may be difficult (in cases even for advanced surgeons) to understand if it is left or right side. The typical landmarks are memorized in a steep learning curve, but the surgical meaning is not mentioned enough.

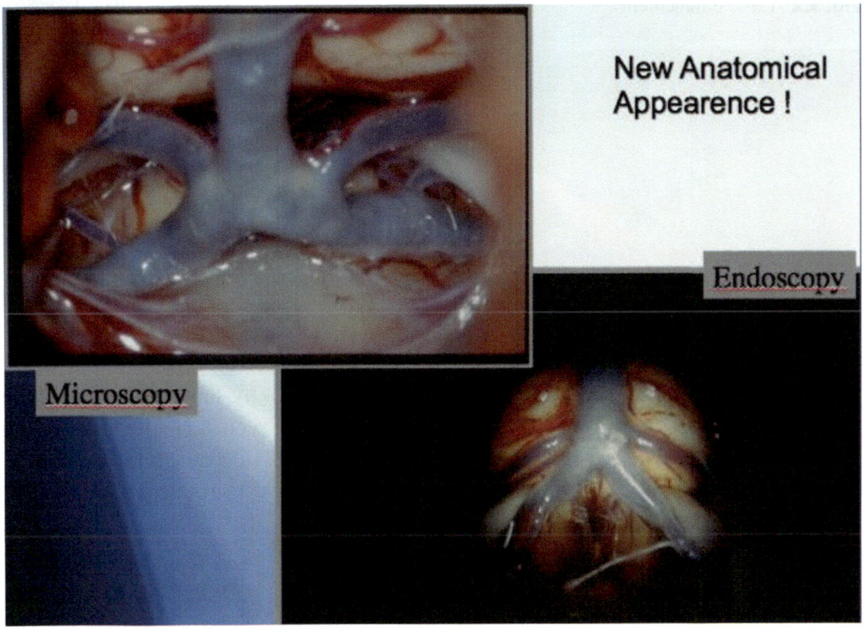

Graph 5.17 Micro view and endo view

It must be kept in mind, that, according to recent neuroscience results, this region contains the structures causing the abilities of human, which are known in history, religion, philosophy and neuro-psychology as the soul of humans. Therefor it is impossible for neurosurgery to judge the functional results alone, some groups of patients are not able to go through the batteries of neuro-psychological tests. However, it is so far clear, that, as easy the procedures seem to be, as big is the disaster one can produce with failures or complications. The label easy or difficult should be given according to complication management possibilities rather than to the simple ease of mechanical aspects. The first software package for these procedures is anatomy.

5.3.2 Foramen Monroi (FM)

This is a little enlarged foramen, presenting "standard-anatomy" and with a view trajectory to the well visible anterior third ventricle. The cranial entry pint in such a view is para-median-medio-pupillary and coronary. The venous angle is hidden by plexus and running dorso-laterally. A tiny vein at the lamina affixa side is visible, easily overseen and a mayor point of bleedings during endoscopy. The fornix is well-developed and the usual strong septal vein runs in the typical route (Fig. 5.2).

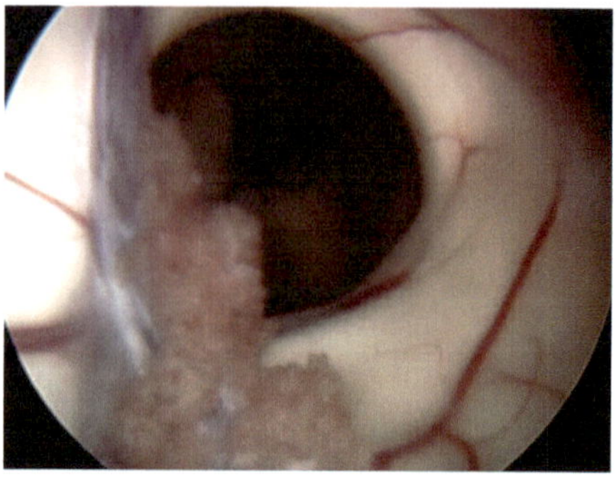

Fig. 5.2 Little enlagement

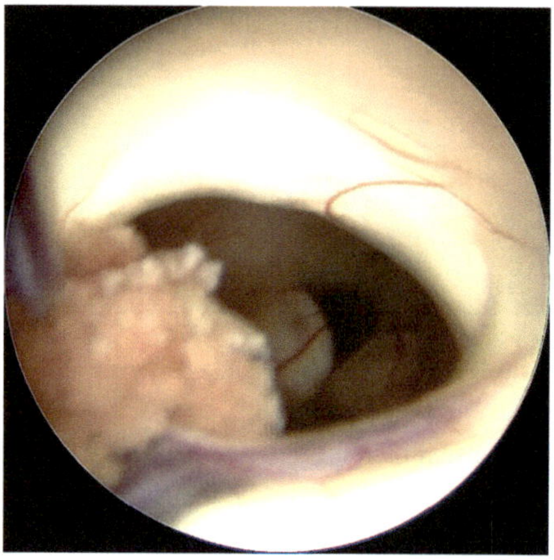

Fig. 5.3 Oval transvers FM

This foramen is oval transversal oriented with a well-developed plexus. The typical trajectory of view gives view to the normal anterior third ventricle (Fig. 5.3).

The venous angle is here formed by a slim septal vein and the good visible foraminal base vein close to the lamina affixa and the thalamo-striate vein. The ependyma shows a bit a xanthochromia. Such a good visible foraminal base vein will be preserved by visual control, but mostly it is tiny and covered.

In the 30° view anteriorly a strongly enlarged foramen with an atypical triangle shape belongs to a mega-ventricle system. The veins are small and the ependyma appears opaque. View into the third ventricle is obscured by remnants of bleeding and infection.

A ventricular catheter is present, the plexus colored by the sediment of the mentioned pathological process (Fig. 5.4).

The 30° view into a mega-ventricle presents some variations. The foramen is very big giving view on the anterior third ventricle and the roof of the mid-brain (Fig. 5.5).

The plain of the foramen is nearly horizontal and fornix is stretched. The veins are present with a big caudate nucleus vein and two accessory veins parallel to

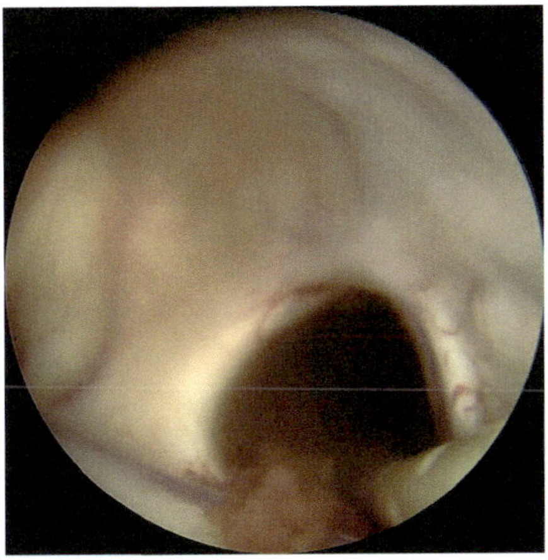

Fig. 5.4 Xanthochromia; basal foramen vein

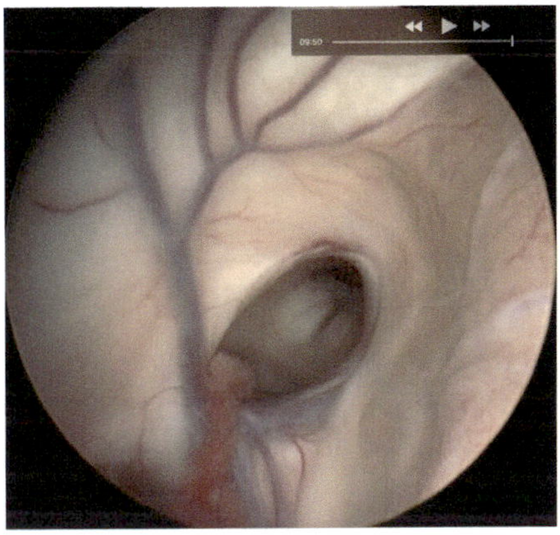

Fig. 5.5 30°, large FM

thalamo-striate vein. The strong system of the septal vein is impressive and the periventricular and subependymal forms and colors are rather nice preserved.

In the 0° view and close-up the variations of the veins are well present. The accessory veins running on the lamina affixa and a typical colored plexus can be seen (Fig. 5.6).

The big thalamo-striate vein is covered laterally by the striae of the lateral ventricle and is becoming nearly invisible.

Another close-up and change of angle, gives view to the until now hidden foraminal base vein and in the depth of third ventricle on the inter-thalamic adhesion (Fig. 5.7).

The typical colored and shaped plexus becomes stronger at the roof of third ventricle.

The fornix is well differentiated and floor of third ventricle, starts to run into the depth. From this trajectory the posterior third ventricle is not accessible. Only a 30°

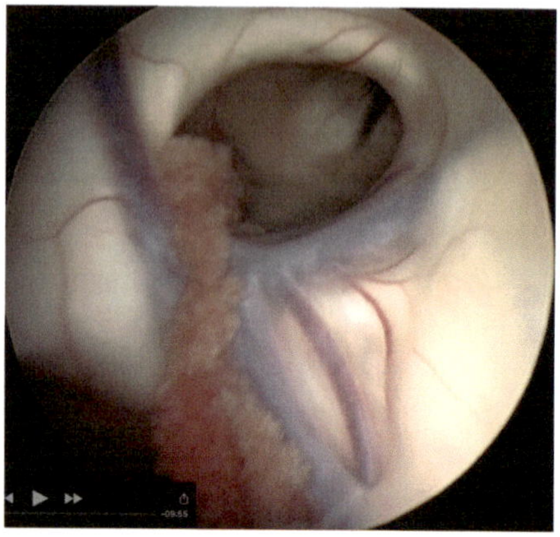

Fig. 5.6 Strong accesory veins

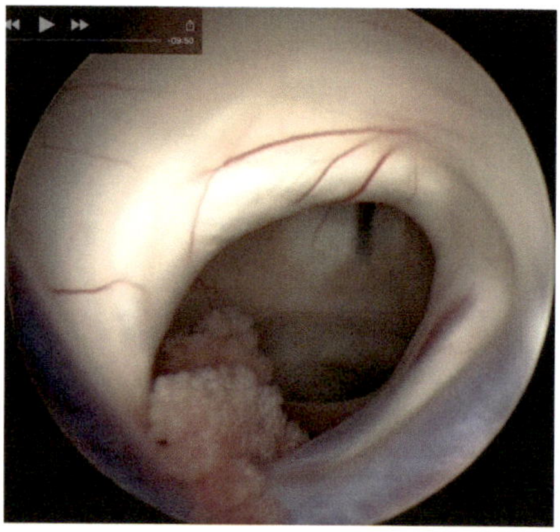

Fig. 5.7 FM base vein and Thalamo-Striate vein

endoscope looking dorsally by 180° rotation could look there but anyway the angle between plexus and inter-thalamic adhesion is closed obstructing the access there.

This case contains a rare variation often overseen in the MR: the mammillary bodies and the floor of third ventricle is lifted by a high basilar artery head to the level of the base of foramen Monroi (Fig. 5.8).

Between the mammillary bodies the basilar tip nearly perforates the floor of third ventricle. Additionally, the plexus is large obstructing more than $^1/_3$ of the foramen.

This is a kind of "missing foramen", when the atrophies after some decades of no therapy in sever chronic hydrocephalus. A fornix cannot be differentiated, and plexus is completely atrophic and not recognizable. Nearly the whole floor of third ventricle can be seen (Fig. 5.9).

A good preserved thalamo-striate vein and an opaque septum vein is present.

The ependyma is changed and vascularization atrophic.

Due to the large space intra-ventricular, orientation is not difficult, and each part can be approached within a directly overview.

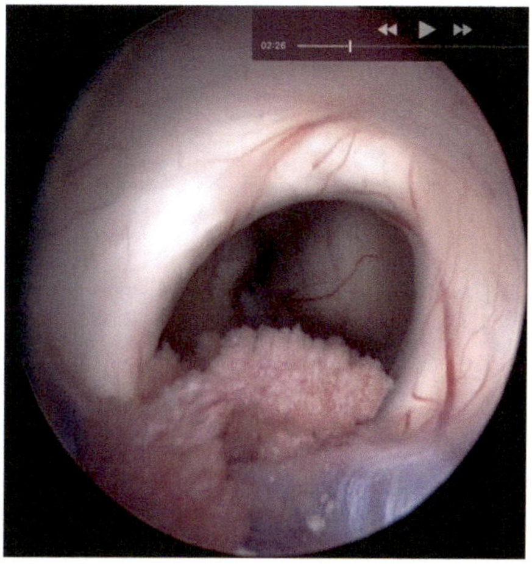

Fig. 5.8 Prodruding basilar head (intermammillar)

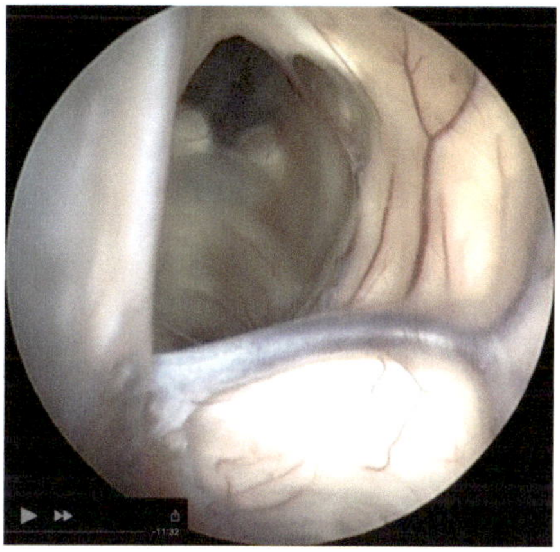

Fig. 5.9 Mega FM by atrophy

This mega-foramen in a horizontal position still presents a visible fornix framing the overview of the third ventricle from the pre-mammillary membrane anteriorly to the aqueduct in the depth and dorsally (Fig. 5.10).

These are the cases where one trajectory through one burr-hole is sufficient to reach anterior and posterior part of the third ventricle floor.

But the trajectory and elasticity of the parenchymatous canal of the brain must be taken into account.

In this 30° view we see transversal oriented foramen in an enlarged lateral ventricle (Fig. 5.11).

A strong septal vein and thalamo-striate vein are forming a perfect venous angle. The septal vein seems to dive subependymal where the lateral striae of the ventricle cross over. The plexus has a typical color and mammillary bodies are visible, framing the pre-mammillary membrane, the cinereum. The visual trajectory is perfect for ETV.

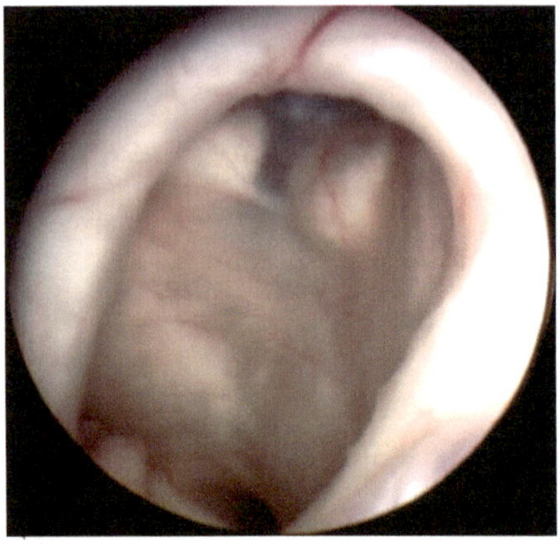

Fig. 5.10 Mega FM

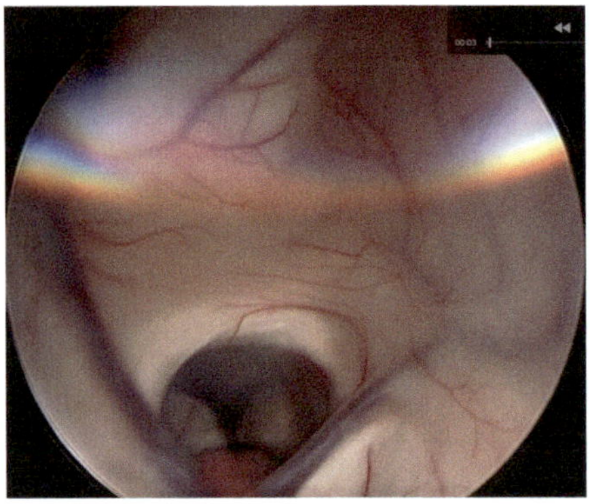

Fig. 5.11 Large FM + Air bubble effects

Superior we see two optical artifacts: mirror-effect by an air bubble and a rainbow-effect.

This is in a 30° view through an enlarged foramen to the anterior floor of third ventricle after an ETV in the past. The well present plexus follows unusually the thalamo-striate vein. The septal vein enters not the venous angle but more dorsally. The ependyma is changed by typical post IVH process with spread micro-spots over whole visible region (Fig. 5.12).

The lateral ventricle is severly enlarged and the foramen rotated horizontal.

In 0° view with slightly opaque post-bleeding liquor, we see no septal vein and a strong thalamo-striate vein. The foramen is a little enlarged and a well-preserved plexus is seen. The View into third ventricle is obscured by remnants of a bleeding in this ventricle (Fig. 5.13).

Striae of the lateral ventricle are well seen where they crossover the caudate nucleus vein.

This foramen is obstructed by a cyst. The Size is only little enlarged, and the form and orientation is normal. At the base, a typical small foraminal base vein can be seen, but septal vein is rather slim. The venous angle is variant and not visible

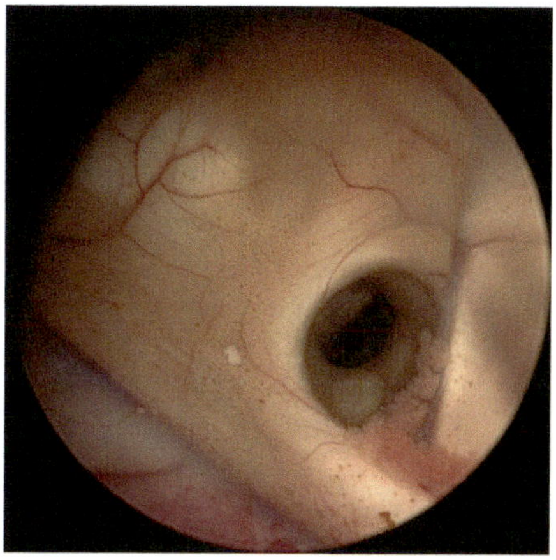

Fig. 5.12 FM post IVH and ETV

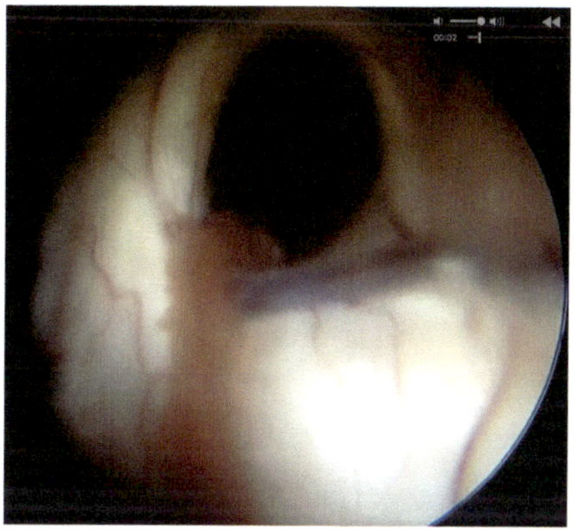

Fig. 5.13 Enlarged fm post IVH

due to a strong plexus. The visible caudate nucleus vein is small, and the form of peri-ventricular structures are expressive (Fig. 5.14).

This foramen shows a narrowing by the septal vein, but even the plexus is absent in the foramen, it is anyway narrow. The foraminal basal vein is obscured, making conditions of endoscopy more difficult. There is a small plexus vein entering the parenchyma (Fig. 5.15).

In such foramina a mini-endoscope or a ventriculoscopy only with the optic might be necessary.

This is a post-infection foramen, extremely narrow and borders changed with opaque remnants. The septal vein and thalamo-striate vein are covered by the ependymal changes of the infection. There is also no plexus visible and the view into the third ventricle obscured (Fig. 5.16).

Tissue in such cases tend easy to bleedings and the ependyma may be edematous or even tuff like a scarf.

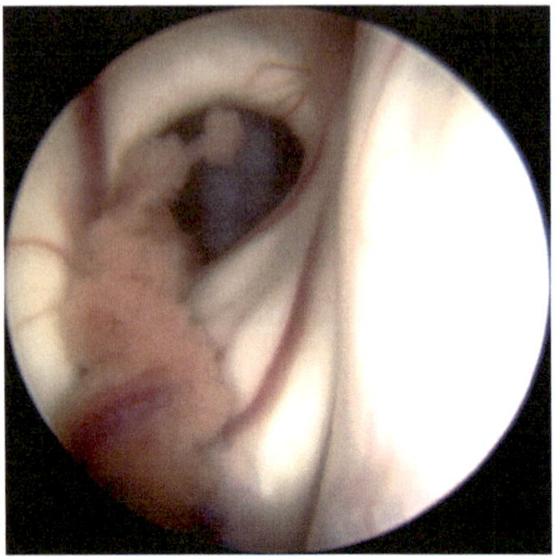

Fig. 5.14 FM and cyst of the 3rd ventricle

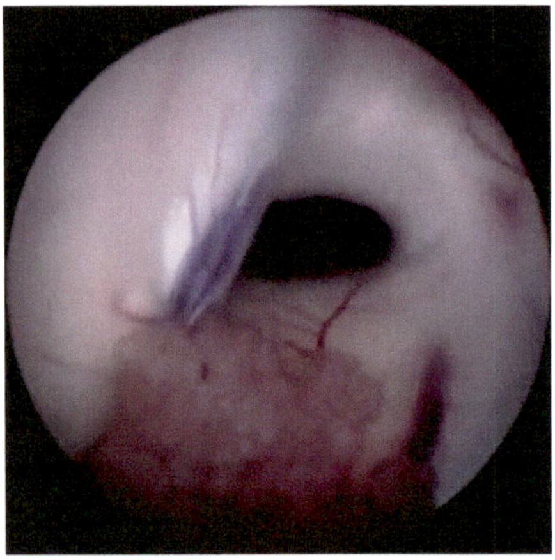

Fig. 5.15 Large septal vein narrowing FM

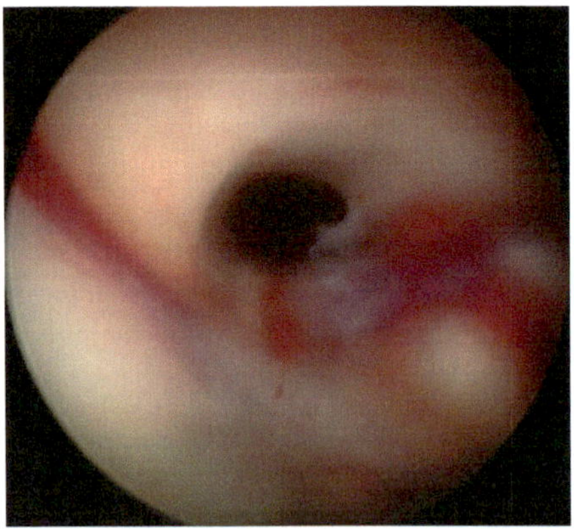

Fig. 5.16 Narrow FM post infection

In this enlarged foramen the strong thalamo-striate vein starts under the lamina affixa and lateral strial fibers are hardly visible. The septal vein is not in the view, being more dorsally. Also, the thalamo-striate vein is un-visible, very small or covered by the well-developed stria terminalis (Fig. 5.17).

In the depth of third ventricular floor the diaphragm of sella can be seen by a reddish color, and the mammillary bodies are very close to a strong dorsum of sellae, causing a narrow space for ETV. The pre-mammillary membrane is transparent and might be tuff and is typical for chronic hydrocephalic conditions.

This is one of the few left-sided foramina, enlarged, with a strong fornix. The thalamo-striate vein divides early and septal vein is slim. The plexus stops at the foramen, giving view to the inter-thalamic adhesion. The window to the posterior third ventricle is closed, and this condition needs a steerable fiber-scope to reach posteriorly (Fig. 5.18).

But the approach angle for ETV is un-comfortable, making a very slow procedure necessary.

This foramen has two foraminal basal veins and a slim septal vein. The thalam-striate vein is strong and there is no connection to the septal vein visible.

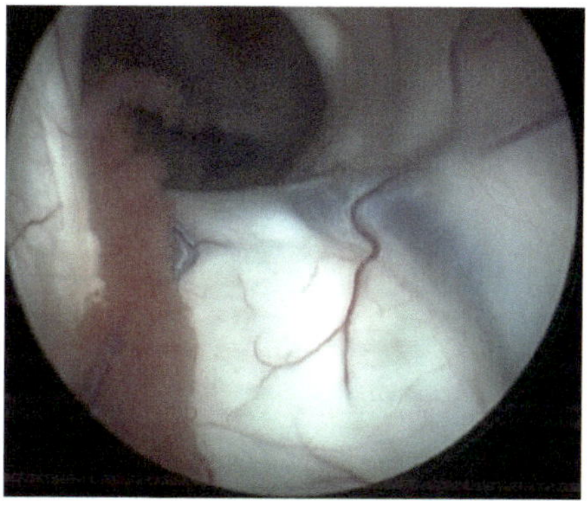

Fig. 5.17 FM with strong stria terminalis of the lateral ventricle (limbic system)

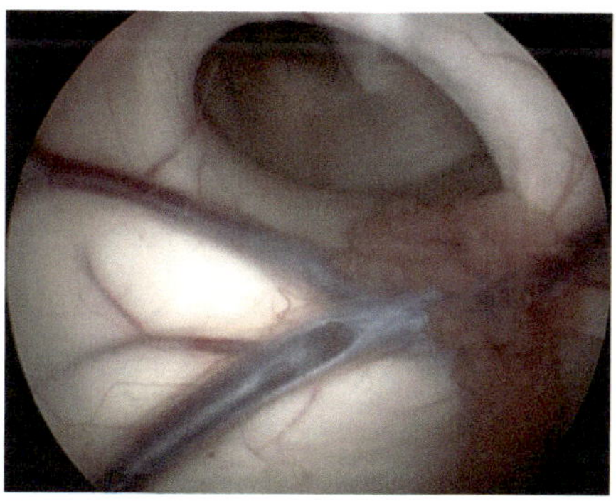

Fig. 5.18 Left fm with early devited Thalamo-Striate vein

Under the plexus, making curiously figures, a venous angle may anyway be present and being an origin of bleedings, because this angle can easily tear (Fig. 5.19).

In this case the view into third ventricle is not given.

This foramen is completely closed after infection (Fig. 5.20).

There is a strong septal vein involved in the adhesion and a small caudate nucleus vein.

The plexus is compressed inside the foramen.

ETV in such cases are master pieces and should be done by masters. It is no reason to place a shunt, rather to send the case to a center with enough experience.

In this mega-ventricle the enlarged foramen horizontal and gives view into third ventricle (Fig. 5.21).

Septal vein and thalamo-striate vein form a perfect venous angle. The plexus follows the septal vein and the caudate nucleus vein also, before running its' typical route.

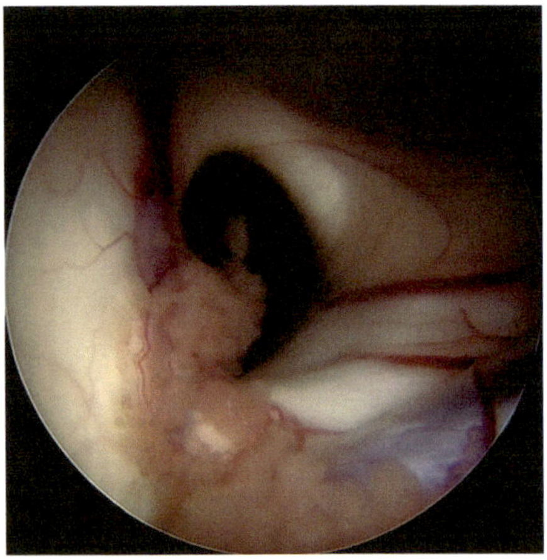

Fig. 5.19 FM with 2 foraminal basal veins

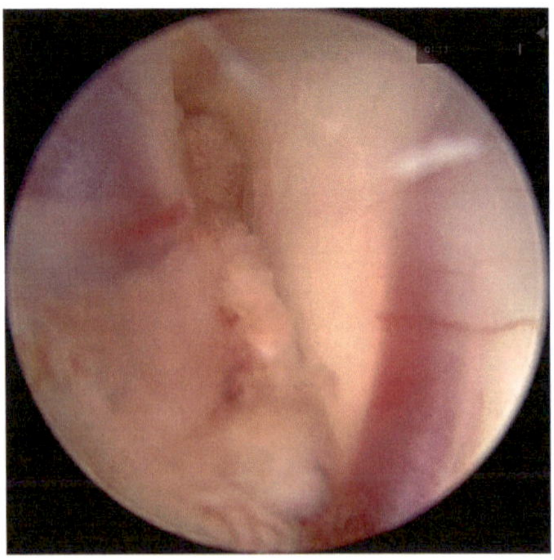

Fig. 5.20 Complete obstructed FM post infection

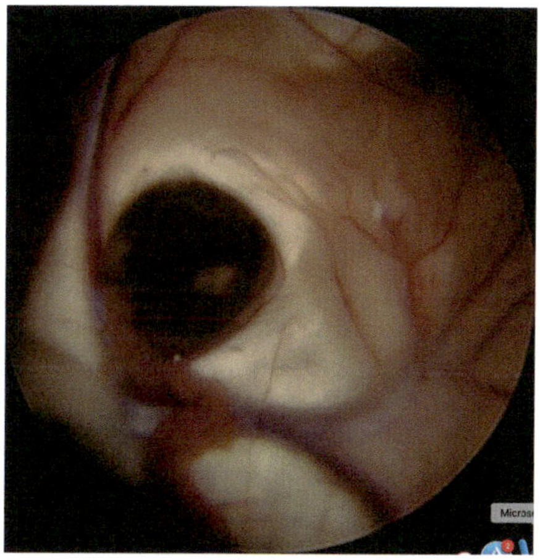

Fig. 5.21 Horizontal FM in a mega ventricle

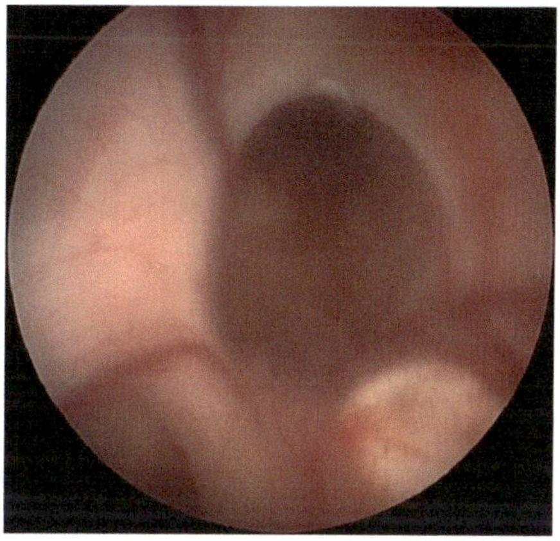

Fig. 5.22 Post bleeding FM

Inter-thalamic adhesion can be hardly seen and the enormous distance of mammillary bodies indicate a large third ventricle as well, as a chronic hydrocephalus. Foraminal basal vein is atrophic and just visible.

This enlarged foramen after a bleeding, causing an opaque liquor hardly gives view into third ventricle, but mammillary bodie are visible. Septal vein is absolute slim and at its' origin hidden by the fornix, which runs its' typical route dorsally, well visible as a cable, because septum is absent. The plexus is rather small, but thalamo-striate vein appears strong. The venous angle is hidden and therefor in danger during ETV (Fig. 5.22).

This is a narrow foramen by strong plexus and large veins, but also and primary by parenchymal construction (Fig. 5.23).

The trajectory of view is untypically from rather posterior causing a view on the anterior commissure.

This unfortunate angle of approach and the hidden venous angle makes ETV difficult. The procedure needs to be done very slowly and, if possible, should be done

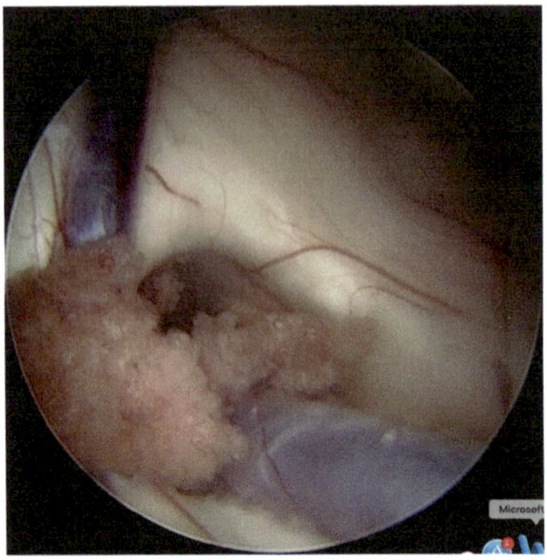

Fig. 5.23 Narrow FM with strong veins

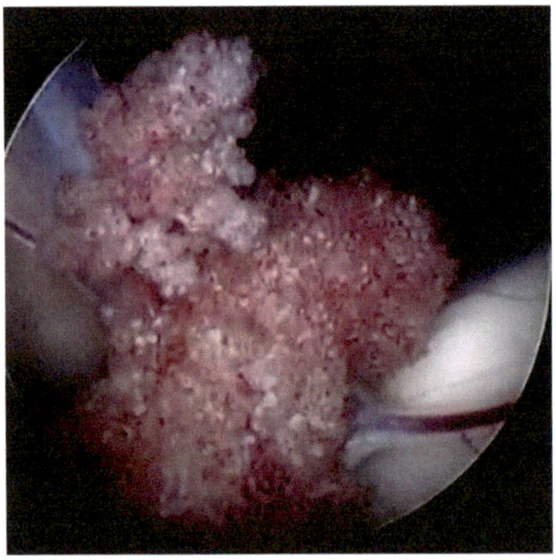

Fig. 5.24 FM with strong plexus

with a pediatric scope or an optic without shaft only. Irrigation in such cases can be done through a thin catheter.

In this foramen we see a nice developed plexus and a accessory foraminal basal vein (Fig. 5.24).

Anyway, in contradiction to literature, the plexus is rarely the origin of bleeding during endoscopy and if, it is easy to be safely coagulated.

Also hypertrophic plexus can be coagulated always, but with guaranty of sparing the veins, which can be done by very low energy and short time-spots.

In this immature tissue, there is no real foramen. Only the landmark caudate nucleus vein and the plexus give orientation where we are looking at (Fig. 5.25).

The surface of the walls shows some irregularities and the ependyma is whitish and a rarefication of visible vessels is given. Only the plexus shows a mature morphology.

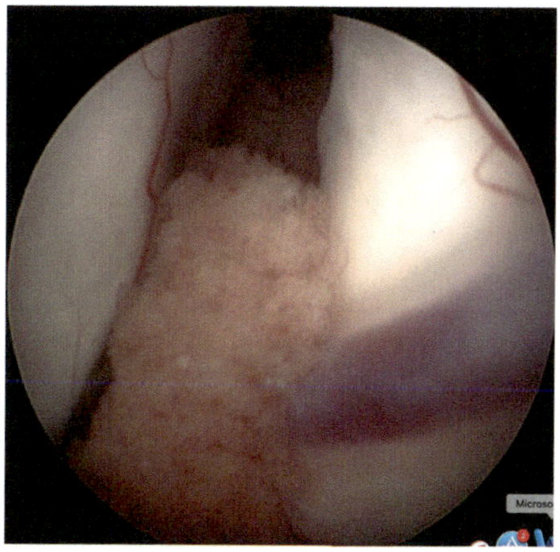

Fig. 5.25 Immature FM

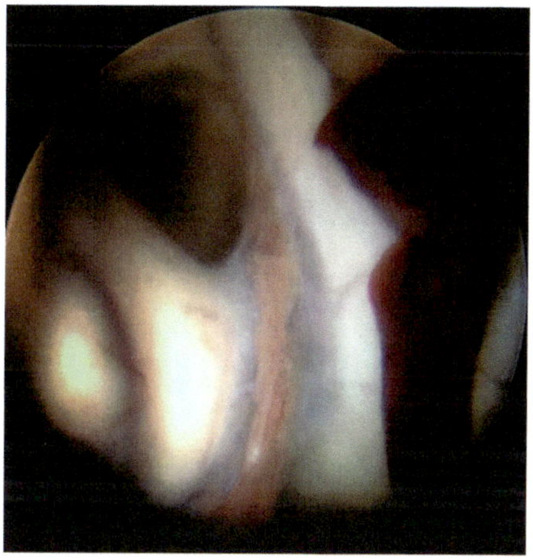

Fig. 5.26 Left Side of monoventricle (+ blood clot right side)

The form of the ventricles can be quite confusing making orientation difficult. Embryology and ultrasound aid is necessary to understand, and to do any safe procedure.

We are looking in a dysplastic mono-ventricle giving view on the left foramen and base-line of the missing septum, placed on the unified fornix of both sides. The view into the right side is obscured by a blood-clot, caused from a shunt catheter (Fig. 5.26).

The plexus running parallel with thalamo-striate vein is the best landmark in this case. The view into the third ventricle falls into the depth of posterior ventricle floor.

Additional to the immature tissue and dysplastic morphology in this case signs of chronic hydrocephalus were given.

This is a megalo-foramen and the fornix is pushed aside by the lateral ventricle enlargement (Fig. 5.27).

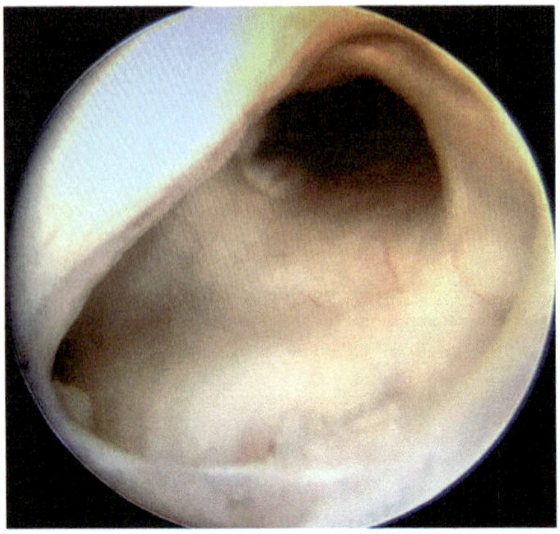

Fig. 5.27 Mega FM

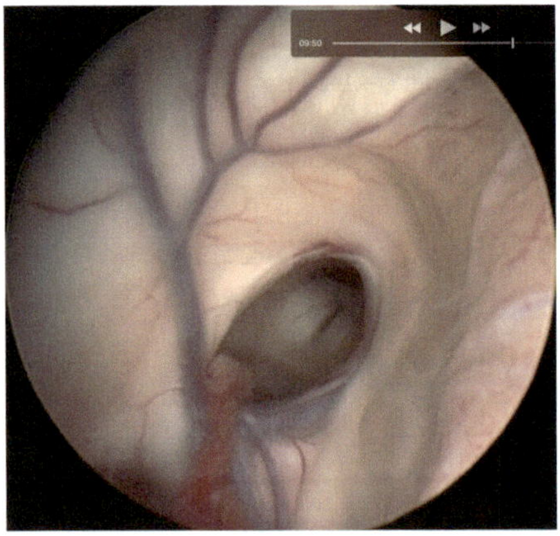

Fig. 5.28 Mega LV

There are no sufficient landmarks of the foramen itself, only the mammillary bodies in the depth of third ventricle gives some orientation. The ependyma is severely atrophic and the width of the third ventricle absolute to big, caused by atrophy of peri-ventricular tissue due to chronic hydrocephalus.

5.3.3 Lateral Ventricle (LV)

In the 30° view of the frontal horn the typical forms and colors of the ventricular walls are present (Fig. 5.28).

The blueish color of the septum and the whiteish color of the rostrum of corpus callosum can be differentiated, and caudate nucleus is also visible in its' typical shape. Only the superficial veins of caudate nucleus are covered by the opaque ependyma, indicating, that there is a chronic process going on. Venous variations have been described already above.

The view from frontal horn to the cella media region presents a typica aspect of the right lateral ventricle (Fig. 5.29).

A well-preserved plexus medial to the lamina affixa and covering typically thalamo-striate vein is the main landmark here. The atrophic and perforated septum indicates the border of frontal horn and cella media, and the hydrocephalic conditions. The corpus of caudate nucleus can by safely differentiated by its' shape. The ependymal veins in the cella media region are in contact with corpus callosum, the can rupture by catheters and endoscopes.

This view, trans-septal into the left lateral ventricle, shows the typical venous system of caudate nucleus (Fig. 5.30).

The course of thalamo-striate vein over lamina affixa and the along thalamo-striate sulcus.

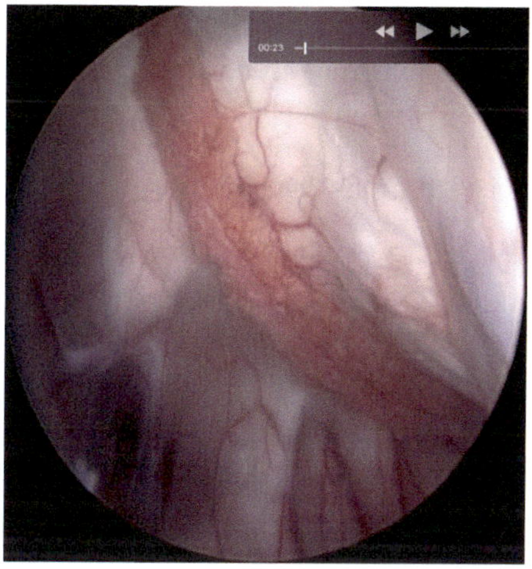

Fig. 5.29 Cella media LV

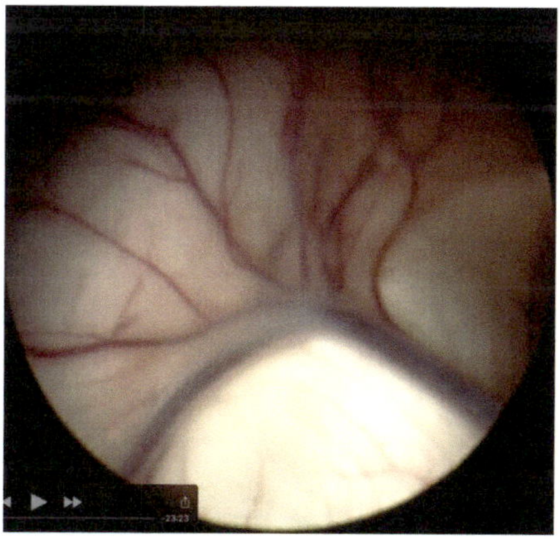

Fig. 5.30 Trans-septal left LV

The caudate nucleus has a more reddish color, the lamina a whitish one.

The overview of left side trans-septal into the cella media region and trigone region of this large lateral ventricle shows the ependymal veins in an unusual fashion (Fig. 5.31).

The so-called fish-eye effect can be experienced, a kind of broad angle panorama view. This is possible in such big ventricle as it needs some distance to appear.

Also, we see in such enlarged ventricle the absorption of light in the depth by the liquor. In large ventricles we need very strong light machines.

However, distances are difficult to estimate in endoscopy.

The cella media of right lateral ventricle can be easily seen in large ventricles. The view shows the typical parallel subependymal veins, which can bleed when perforating into. This bleeding of floating veins can be difficult to coagulate. The plexus is well developed here (Fig. 5.32).

The trigone region, here right side, contains the glomus of the plexus and can obscure the view into the posterior horn of the lateral ventricle (Fig. 5.33).

It is important to remember, that radiatio optica and capsula interna are very close, especially in hydrocephalic conditions.

This view posterior to the foramen into the region of cella media we see an accessory and strong thalamo-striate vein (Fig. 5.34).

Fig. 5.31 Trans-septal left trigone LV

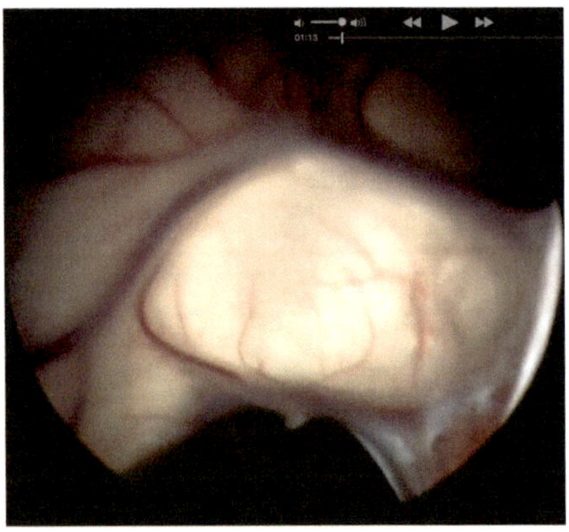

Fig. 5.32 Cella media right LV

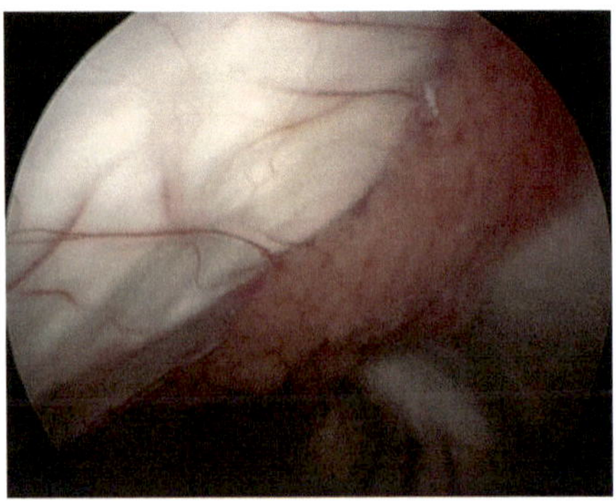

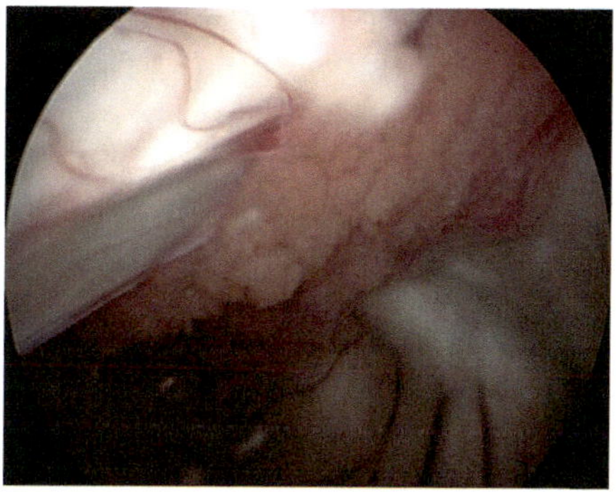

Fig. 5.33 Trigone right LV

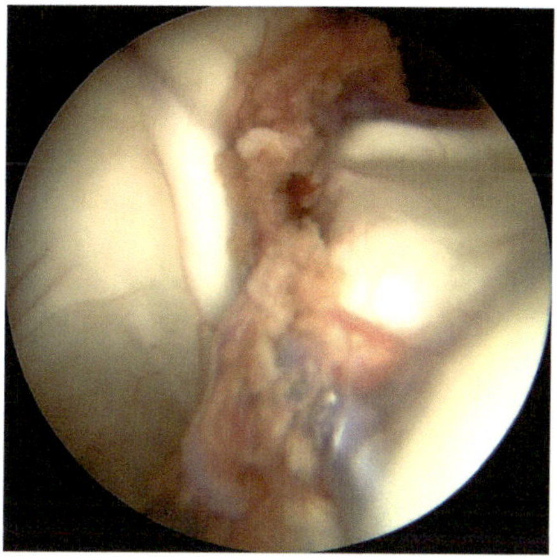

Fig. 5.34 Accessori Thalamo-Striate vein right LV

The slightly atrophic plexus is running from the foramen through the cella media to the trigone region (here not visible), on its' way to the inferior horn in the temporal lobe.

The ependyma is xanthochrome, even the usual white lamina affixa.

One small artery of choroidal artery is visible, one possible origin of sever bleeding, nearly impossible to stop endoscopically. They may supply tumors also.

This is a plexus with a strong perfusion giving the typical color (Fig. 5.35).

The septal vein and a small accessory vein of the thalamo-striate system over the white lamina affixa.

Anyway, the plexus rarely is origin of bleeding during endoscopy. If it bleeds you see a pathology, a tumor or a coagulopathy.

This is a close-up of the glomus of the plexus. We see the typical strong plexus-vein with the calcification spots. These veins are big and look some-how like ilium loops in the abdomen (Fig. 5.36).

If necessary, they can be coagulated, but floating vessels in the liquor an in the flow of irrigation, may be difficult to be coagulated, and one must act very fast to control the situation.

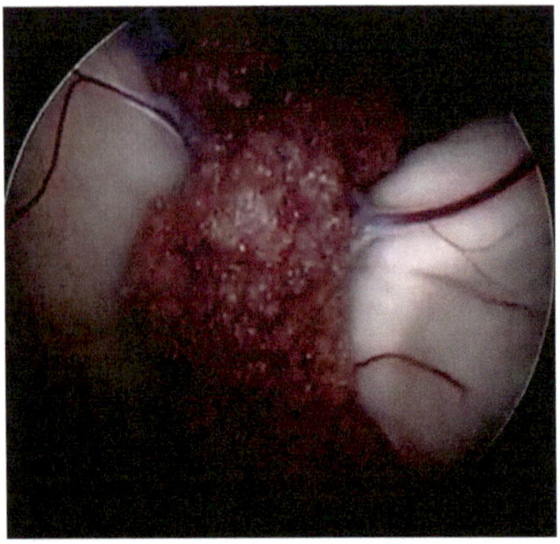

Fig. 5.35 Strong perfused plexus LV

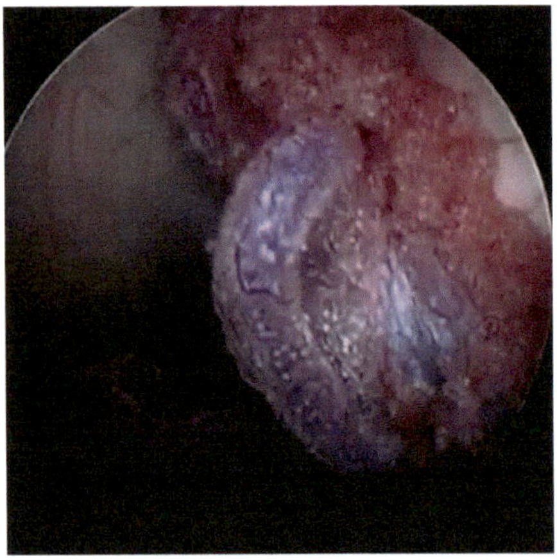

Fig. 5.36 Glomus of plexus LV

A typical colored plexus, coming from the foramen and crossing over thalamostriate vein, is visible, running to the trigone region (Fig. 5.37).

There the plexus veins forms loops. The depth into the inferior horn is dark, as liquor is absorbing light strongly, making strong light source necessary. This region is accessible by the endoscope in large ventricles.

This is one of the classical morphological findings in chronic hydrocephalus, a thin atrophic septum, transparent or even missing, with some remnants of veins (Fig. 5.38).

The functional outcomes of such cases are usually not assessed, but careful examinations and clinical histories use to give allot of signs, and postop changes after ETV presents often unexpected but not surprisingly outcomes in the neuropsychological and neuroendocrinological field.

This septal atrophy, with remnant of venous trees in the liquor give view transseptal into the contralateral ventricle and anatomy. For geometrical reasons the lateral structure, like caudate nucleus, can be seen better from contralateral side. This may influence the decision of approaching (Fig. 5.39).

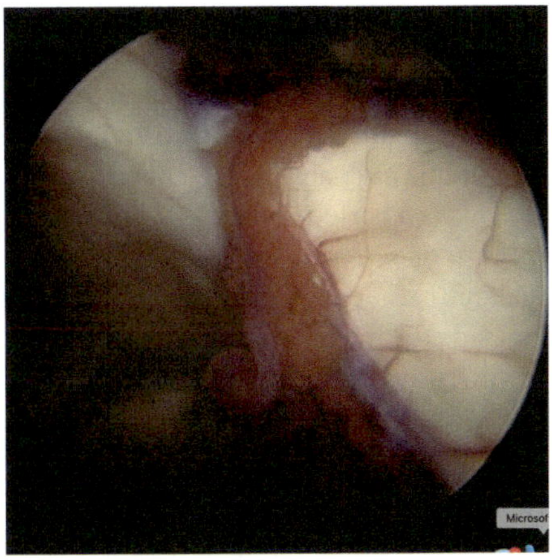

Fig. 5.37 Beginning glomus of plexus LV

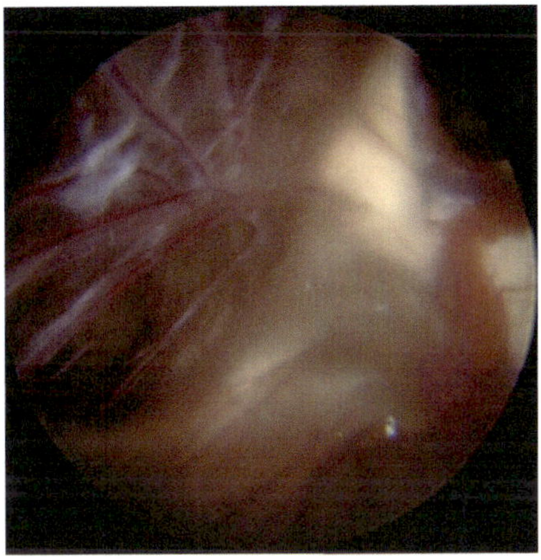

Fig. 5.38 Atrophic septum LV

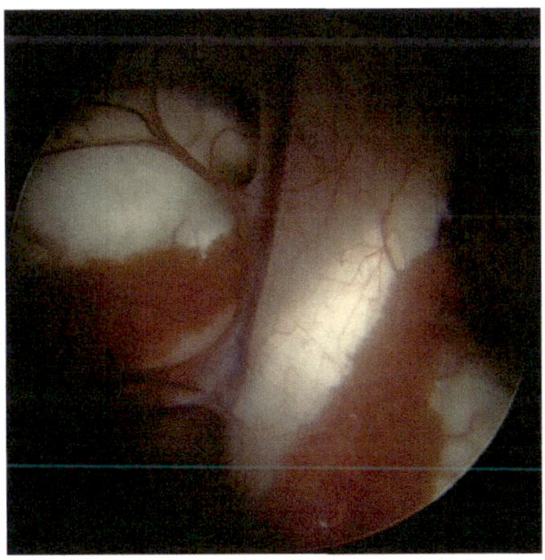

Fig. 5.39 Remnant of septal veins

The close-up gives a very good view trans-septal into the contralateral ventricle: We see the contra-lateral plexus and lamina affixa (Fig. 5.40).

This approach is excellent for approaching thalamus lesions.

The septal veins can be coagulated if needed, however, the windows in the septum in this case are far big enough to use endoscopes within the ventricle.

This is a perfect example to see, that coming from the right side and entering this mega-mono-ventricle, in the contra-lateral side left the lateral structures like plexus, thalamo-striate vein and lamina affixa with thalamus are seen better. On the ipsilateral side the medial structures like foramen Monroi, septal vein and fornix are better seen. In such a mega-mono-ventricle one can get nearly every-where in the ventricular space (Fig. 5.41).

In this dysplastic case there is no orientation possible at all. Even the only landmark, which is a remnant of plexus did not lead into the foramen Monroi. In such cases a ultrasound navigation is mandatory (see Chap. 5) (Fig. 5.42)

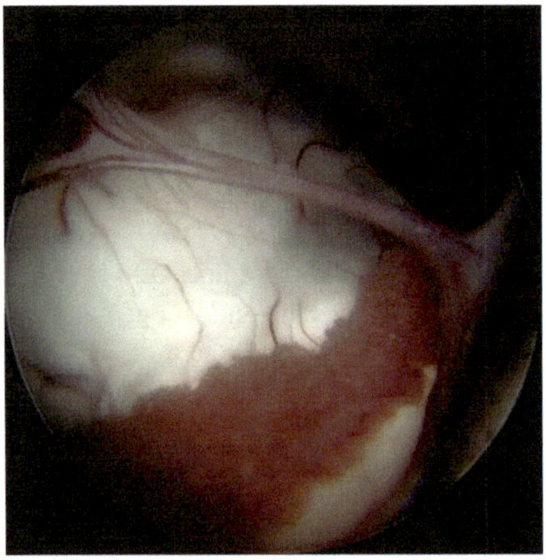

Fig. 5.40 Contralateral view LV right to left

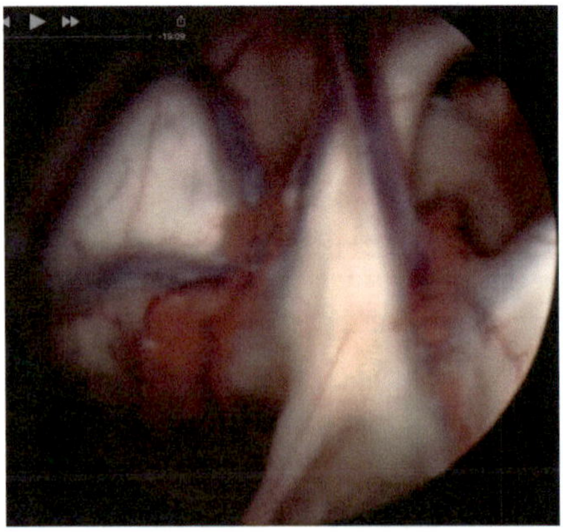

Fig. 5.41 Mega-mono ventricle

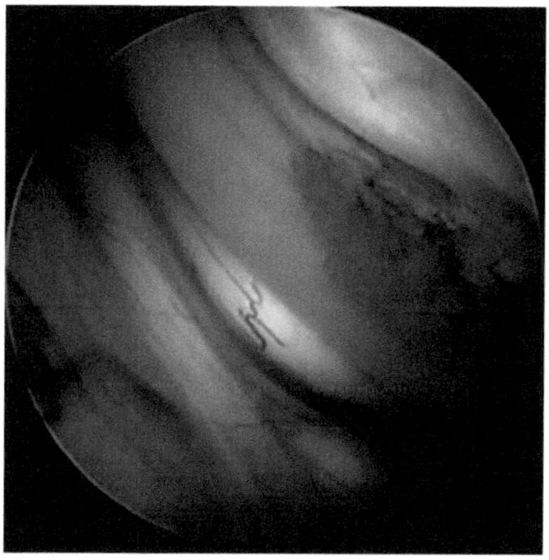

Fig. 5.42 Immature dysplastic monoventricle

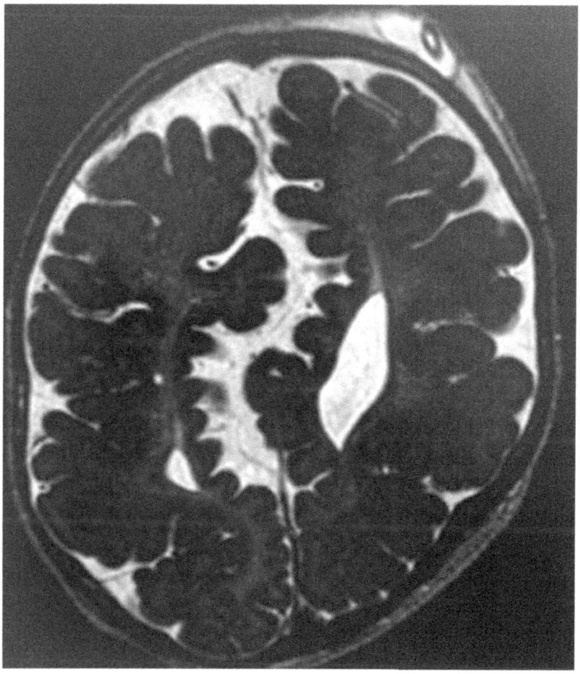

In the depth there is no chance for any identification of structures. Finally, with irrigation it was possible to open one of the gaps and to enter very slowly until mammillary bodies became visible. Here we see a mono-fornix and striate ganglia both-side (Fig. 5.43).

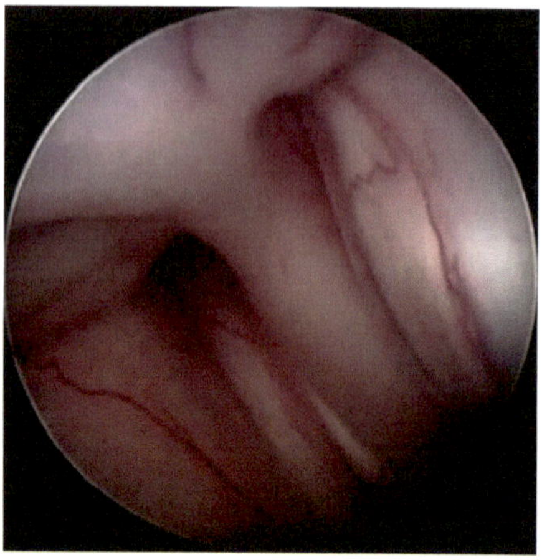

Fig. 5.43 Immature dysplastic monoventricle

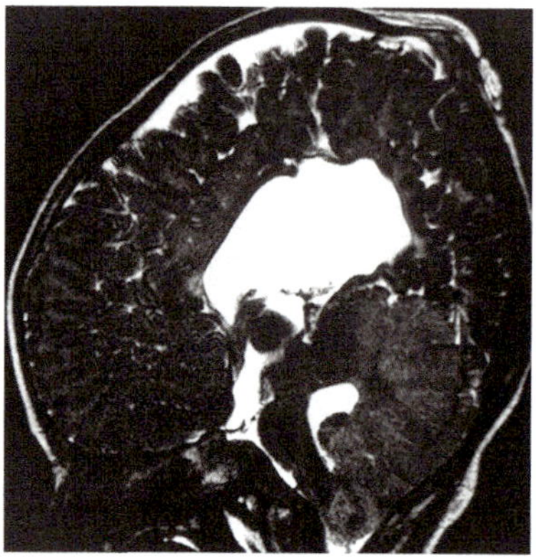

5.3.4 Third Ventricle (TV)

See Graphs 5.18, 5.19, 5.20a, and 5.20b.

This hydrocephalic case shows well preserved mamillary bodies and a parenchymatous premamillary area (cinereum), however, the diameter of third ventricle is too large, and the distance of the bodies starts to grow. Choroid plexus is strong developed and inter-thalamic adhesion is big, both causing closure of the window to the dorsal region of third ventricle, dorsally of the inter-thalamic adhesion (Fig. 5.44).

Contralateral the hypothalamus is well visible, the color is typically slightly reddish. The paraventricular nucleus is there subependymal, invisible: the “switch” of awakeness!

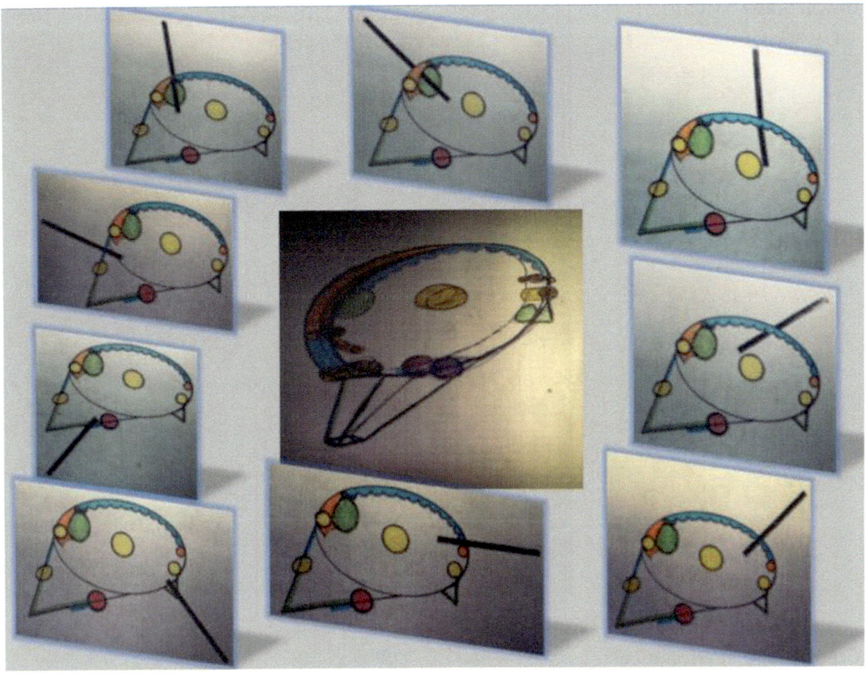

Graph 5.18 Endoscopic approaches of 3rd ventricle

Graph 5.19 Periventricular structures of 3rd ventricle

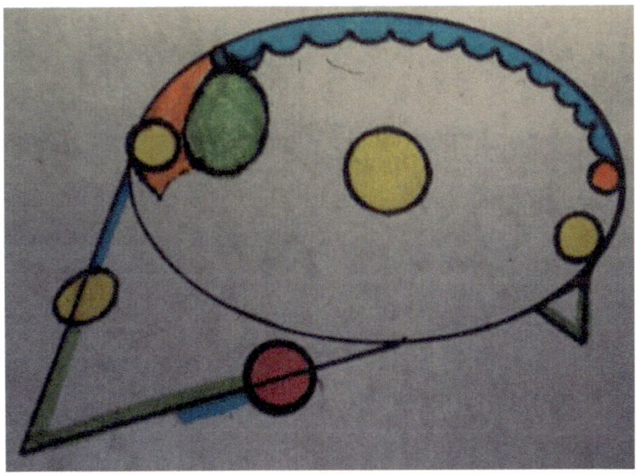

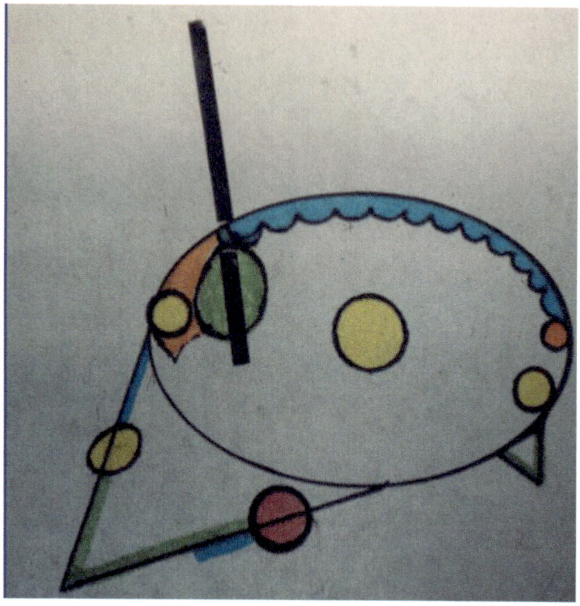

Graph 5.20a Trans-FM approach into TV

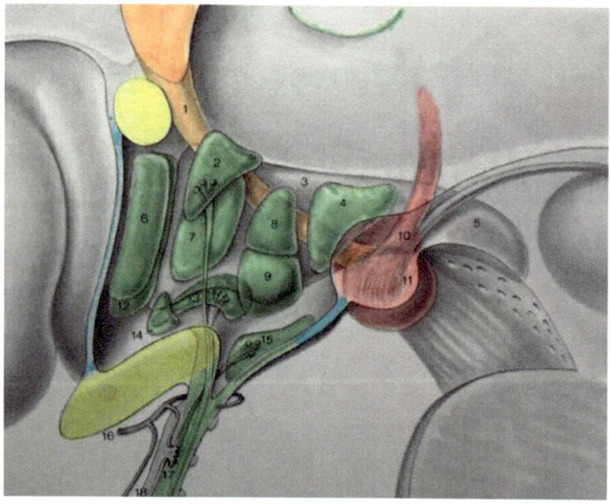

Graph 5.20b Periventricular structures hypothalamus

The close-up shows the complete cinereum and the optic chiasm. The infundibular recess at the rostral end of cinereum is leading into the stalk. The slightly and typical reddish color compared to the white optic chiasm and the mamillary bodies (Fig. 5.45).

The paraventricular nucleus is subependymal and therefor not visible. But, its' injury can cause disturbance of vigilance!

This is also the neuro-endocrine region, regulating also behavior and neuropsychological functions.

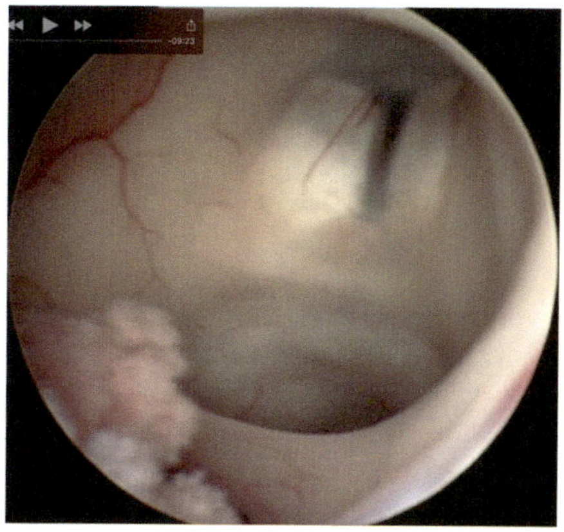

Fig. 5.44 Typical floor of 3rd ventricle

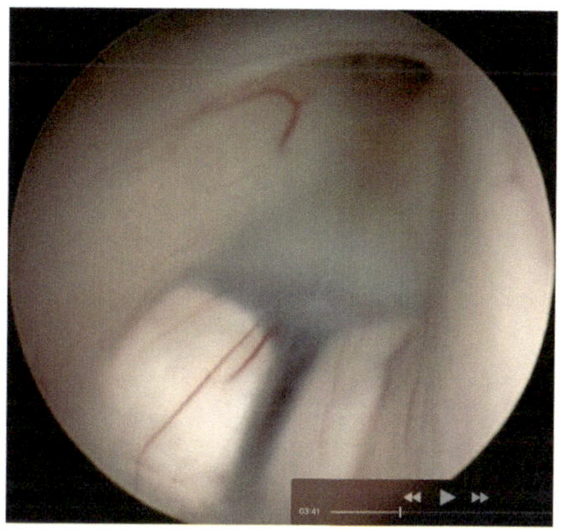

Fig. 5.45 Hypothalamus, zinereum and mamillary bodies

This is a 30° view into the third ventricle, therefor the floor is present from mamillary bodies until the terminalis lamina. We see a strong optic chiasm and a reddish colored infundibular recess. The cinereum shows a parenchymatous floor with a bluish color in contrast to the reddish hypothalamus and the white optic chiasm and mamillary bodies (Fig. 5.46).

The 30°-view into the third ventricle after resection of a colloid cyst, looking to the plexus of the contralateral foramen Monroi. Some calcifications and coagulation remnants are visible. Additional a collection of air bubbles are adherent to the anterior commissure (Fig. 5.47).

There is a nice overview of the third ventricle floor from inter-thalamic adhesion, the roof of the midbrain, mamillary bodies, cinereum and infundibular recess with the optic chiasm rostrally. The ependyma shows a slightly reddish color.

In this case of a chronic hydrocephalus with sever high levels of albumin and proteins in the liquor samples due to an acoustic neurinoma, we see a very pale, slightly xanthochrome ependyma. The edema of the optic nerve could be seen only

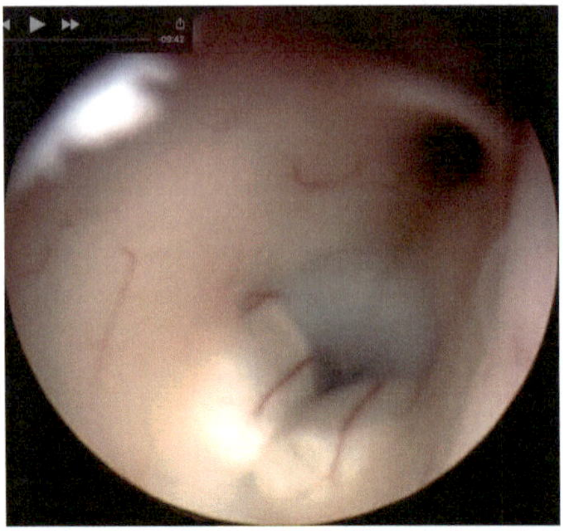

Fig. 5.46 30° view floor 3rd ventricle

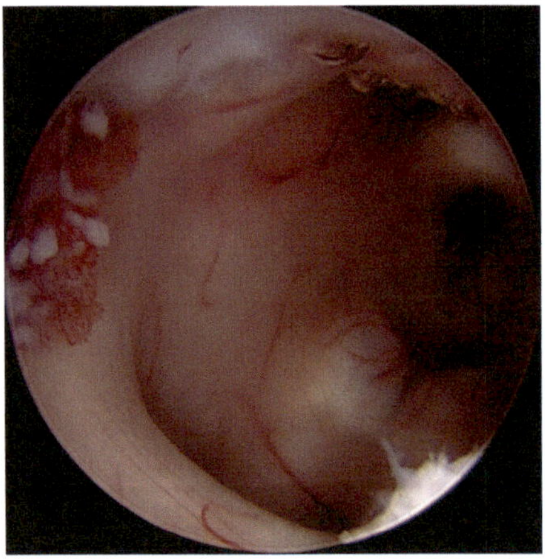

Fig. 5.47 30° view to contralateral side (left) 3rd ventricle

in the ophthalmoscope, but decompensation occurred suddenly. In addition, the old patient was multi-morbid (Fig. 5.48).

The close-up more dorsally shows a normal inter-thalamic adhesion. Even in NPH situation, the normal pressure may preserve the parenchymatous consistency of the cinereum. The best perforation point must be planned by imaging (Fig. 5.49).

3D OCT (Optic Coherence Tomography) shows massive papilledema due to high protein-level of Liquor.

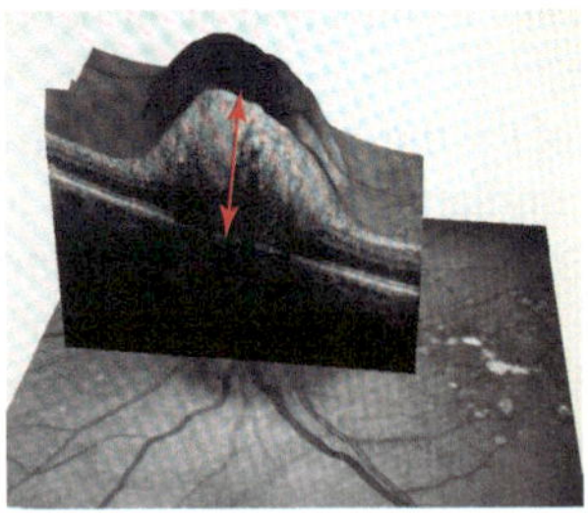

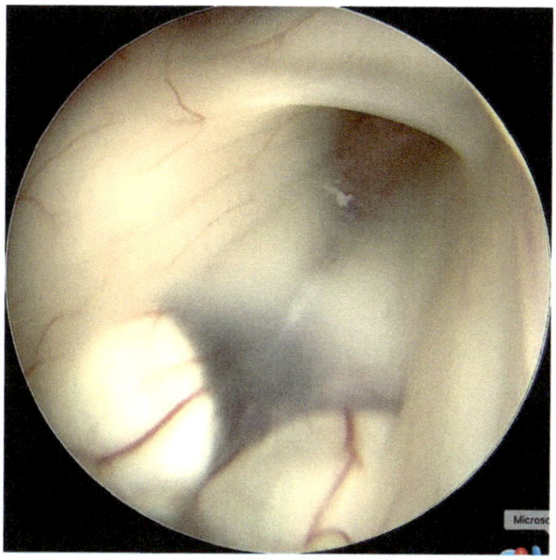

Fig. 5.48 Xanthochromie TV in high protein level of liquor

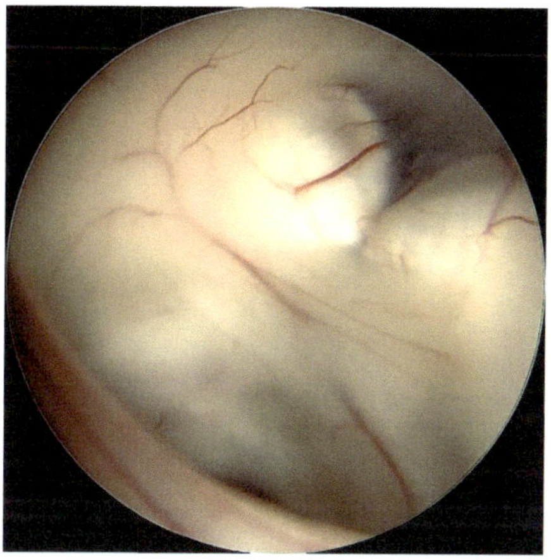

Fig. 5.49 Dorsal view TV

This is a trans-callosal midline approach and view into a narrow third ventricle. The mamillary bodies are slim and also is cinereum and small pre-mamillary area (Fig. 5.50).

This way is "the kings road" because it does not transgress the frontal lobe, but only corpus callosum.

However, it is difficult and only for experienced neuro-endoscopists (100 cases).

It is the gold standard in slit ventricles syndrome, and should be preferred also in handy capped brain with decreased functional compliance.

The 30° view of a NPH-case shows the change from a parenchymatous to a membranous state. The mamillary bodies start to drift away from one another (Fig. 5.51).

The optic chiasm is well visible in this perspective indicating the correlation between CSF pressure and visual function, which is of course always the case.

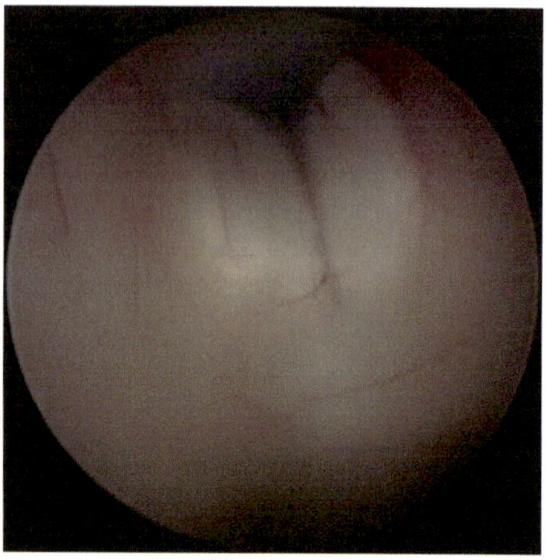

Fig. 5.50 Trans-callosal approach in a small TV

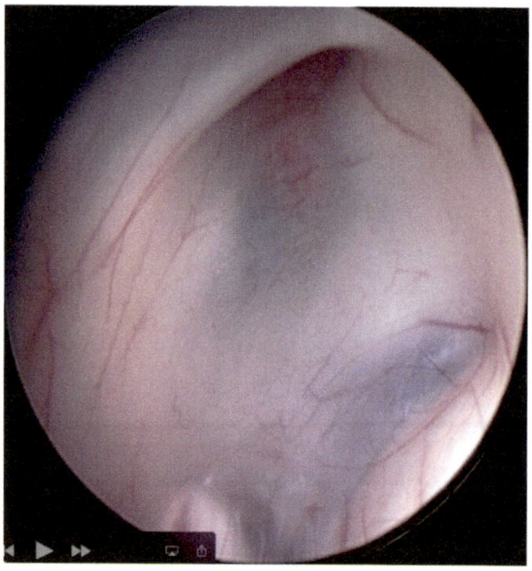

Fig. 5.51 NPH TV 30°

Drawing the endoscope back into the foramen Monroe the complete mamillary bodies come into sight. We see the distance the development of hydrocephalus has already reached. But, most impressive and rare in such expression, we see the basilar head protruding through and lifting up the third ventricle floor. In such conditions, visible on the sagittal MR, the floor will not bulging basaly, as usual expected in aqueduct-stenosis cases (Fig. 5.52).

In this case the distance between Infundibular recess and mamillary bodies is short and a very small area for perforation was given (Fig. 5.53).

In such cases the planning must be very precisely and anyway, the perforation should be rather over the diaphragm of sella than over the basilar bifurcation or above the perforator area. But the, under visual control the further perforation through Lilliquist membrane is necessary to get a free low of CSF.

In this case the pre-mamillary area is already a membrane. There is a thin layer of arterial blood on the ependyma causing the suspicious of a bleeding pathology

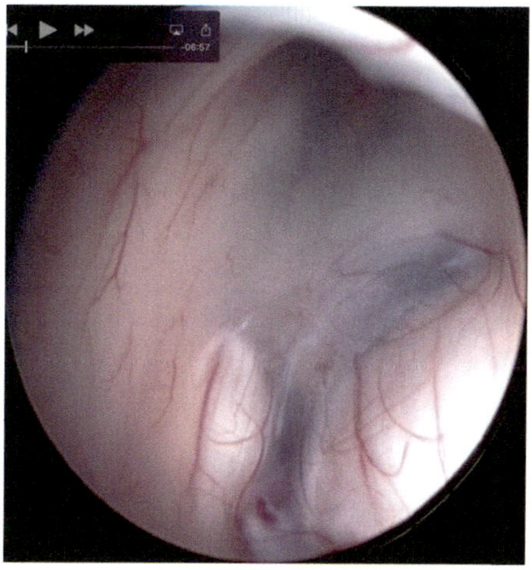

Fig. 5.52 Prodruding basilar head lifting floor of TV

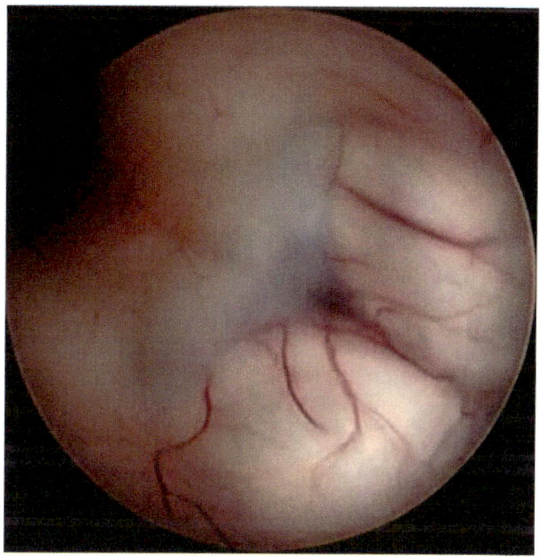

Fig. 5.53 Short cinereum with small perforation area of TV floor

there. But here we see a bleeding effect coming from the approach canal. This must be considered before getting confused. Sucking this away endoscopically can cause injury to the parenchyma (Fig. 5.54).

This is a sever chronic hydrocephalus with a subependymal atrophy causing a complex NPH symptomatic far beyond the Hakim-trias. The optic chiasm is hardly visible, the hypothalamus pale and decreased, distance between mamillary bodies large and premamillary membrane milky due reactions to mechanic stress. Endocrinological, neurological, ophthalmological, neuro-psychological functions may be impaired. Most of them are ignored in neurosurgery. Shunt-systems address only small part of this complex pathophysiology (Fig. 5.55).

This case shows isolated fibers where the tissue becomes transparent between the mamillary bodies. There also small perforators from the P-com artery right side are recognized. Such loss of parenchyma has functional consequences, which appear

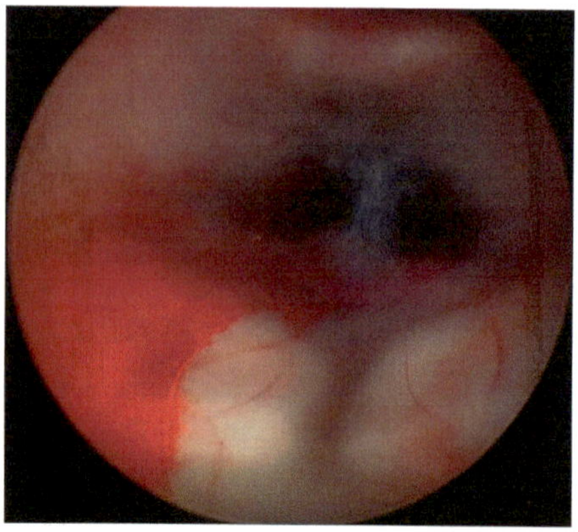

Fig. 5.54 Membranous cinereum of TV

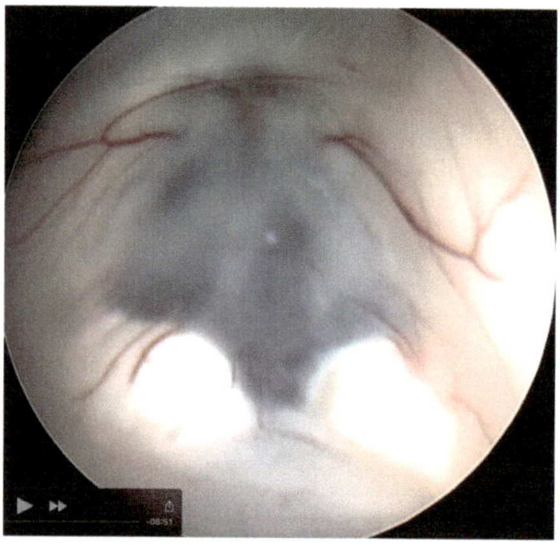

Fig. 5.55 Sever NPH TV

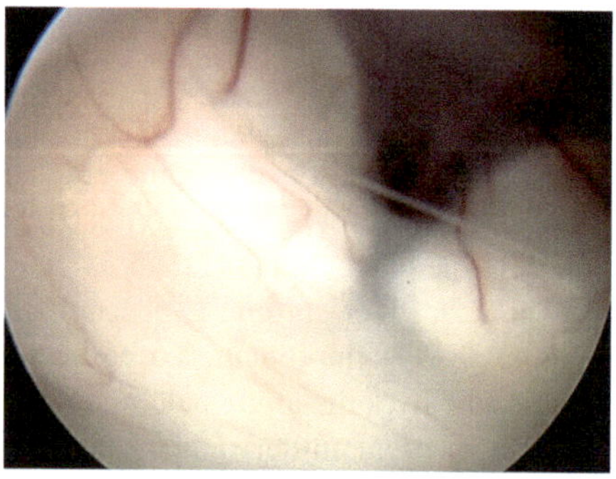

Fig. 5.56 Remnant fibers of floor TV

clinically step-wise rather than continuously. This atrophy of tissue can be seen also in imaging. But clinical history is ahead to imaging! (Fig. 5.56).

This case of a subependymal atrophy of third ventricle presents a transparent pre-mamillary region affecting mamillary bodies. The precapillary network becomes visible and dorsum of sella with the diaphragm and its' opening for the stalk. In this stage of atrophy the stalk has become a "red spot". Such cases occur under compensation conditions of the NPH. Here it was a small hole in the membranous stenosis of the aqueduct (Fig. 5.57).

This view into third ventricle from the left side shows a milky cinereum and a small yellow spot in the midline. It can be a remnant of an infection or caused by a bleeding (Fig. 5.58).

However, it belongs to the hydrocephalus course. The milky color of the premammillary area is not a normal parenchymatous color, but has a bluish nuance.

Such condition need a careful analysis of the sagittal MR to determine the perforation point.

This view shows a sever change of the third ventricle floor indicating a long-time course of the liquor dynamics disturbance. Through the transparent floor the reddish diaphragm and dorsum of sellae is visible and the stalk has become a red spot. Even the oculomotor nerve can be seen left, and the right side of the posterior circulation with P-com, P_2 and a strong thalamo-perforator trunk. Mamillary bodies and hypothalamus are also atrophic. Such courses happen in disabled patient when changes are identified late within the main illness (Fig. 5.59).

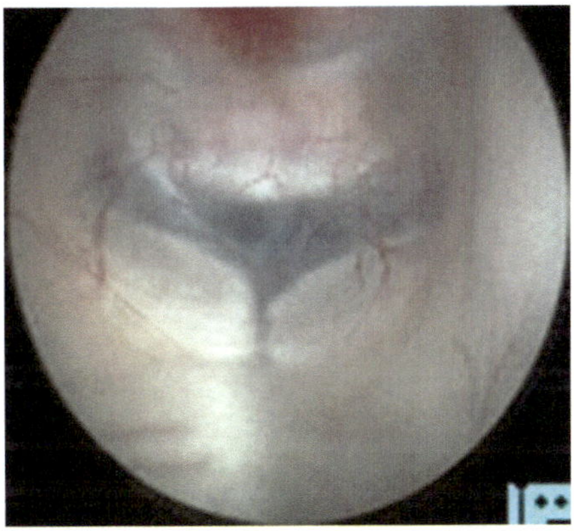

Fig. 5.57 Transparent cinereum and visible dorsum of sellae and diaphragma

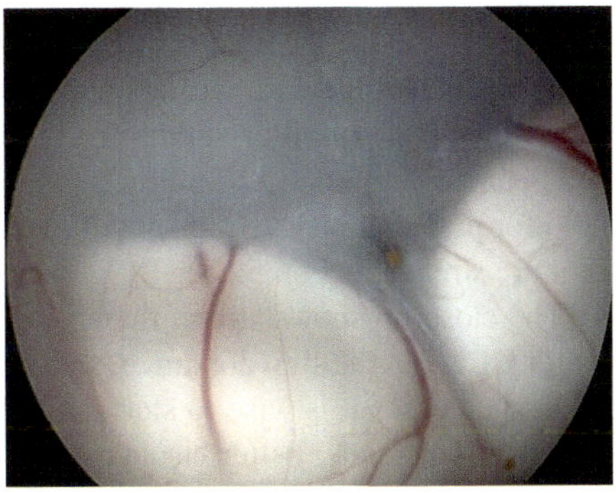

Fig. 5.58 Opaque cinereum of TV

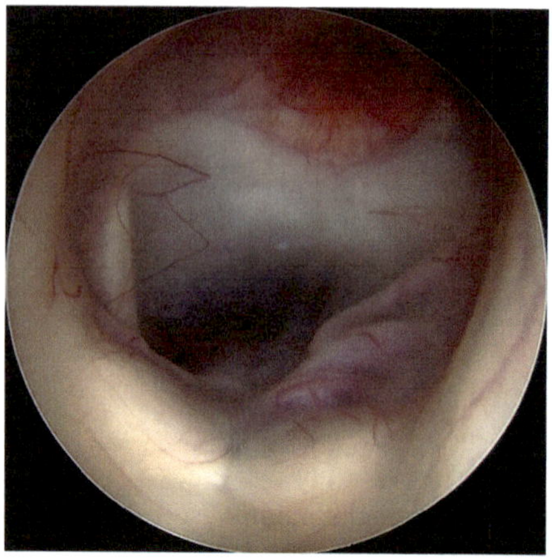

Fig. 5.59 Membranous deepseated floor of TV

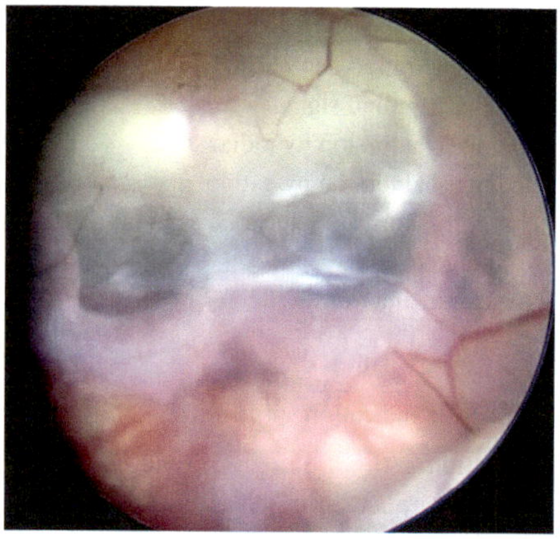

Fig. 5.60 Transparent floor of TV

View on the floor of the third ventricle which is rather well transparent. The posterior circle of Willis can be seen an two strong thalamo-perforator-trunks (Fig. 5.60).

The yellowish dorsum of sellae and the red spot of former stalk present.

This is from a 21 years woman with a minimal symptomatic. A small hole in the membranous stenosis of aqueduct, allowing the course to be very slow.

This is a chronic case with extremely changes of the whole third ventricle floor. The distance of mamillary bodies is, even for such cases, extraordinary. The basilar head can be suggested and a thin dorsum of sellae is present marking a very small field for perforation. The red spot of former stalk is large and fibers of the midbrain roof are stretched (Fig. 5.61).

The close-up shows strong thalamo-perforator system, indicating the area of no go for ETV. But the slim dorsum of sellae mark a small field only for ETV (Fig. 5.62).

In such cases it is wise to start the perforation on the diaphragm sellae and in a following step under better visual control progress through the Lilliquist membrane, the roof of interpeduncular cistern. Such condition is a good LASER indication.

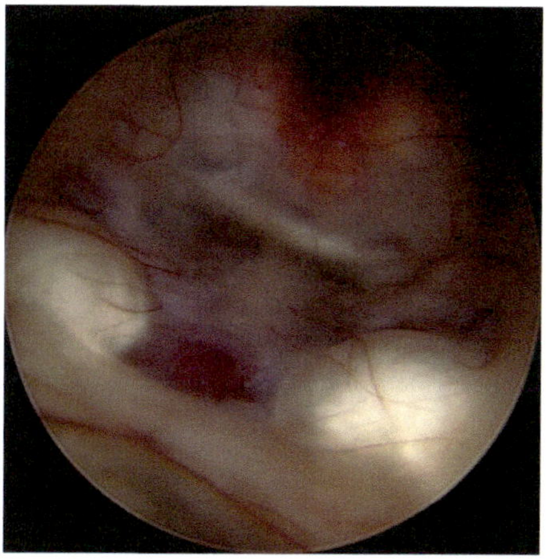

Fig. 5.61 NPH TV with high distance of mamillary bodies

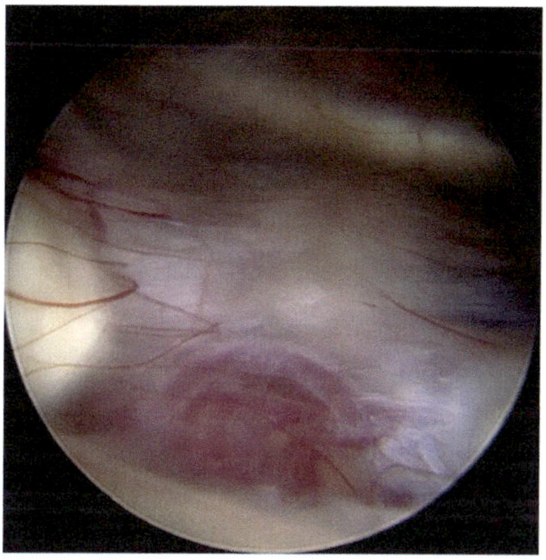

Fig. 5.62 Thalamo perforators of basilar head under transparent floor of TV

The close-up supra-sellar shows the red spot of flattened stalk and the sella region. The atrophic optic chiasm is visible indicating the close relationship of optic chiasm and liquor-dynamics. At this point a compression effect results in the most sever way at a small volume, in very slowly occurring blindness (Fig. 5.63).

After perforation on the diaphragm sellae, ETV must be continue through the Lilliquist membrane under direct endoscopic control. It can exact be seen now that the space between dorsum and "no-go area" is only 2 mm (Fig. 5.64).

The different functional impairments will recover in a very different time-course. This imbalance may even result in new neuro-psychological complains of the patient. This should not be misunderstood as failure of the ETV!

After release of liquor pressure by ventriculoscopy with no or mild irrigation, the floor of third ventricle might be bulging into the ventricle, changing the whole aspect. In such condition it is recommended not to perforate blunt rather than to open by bipolar coagulation or LASER. This gives view through the parenchymal

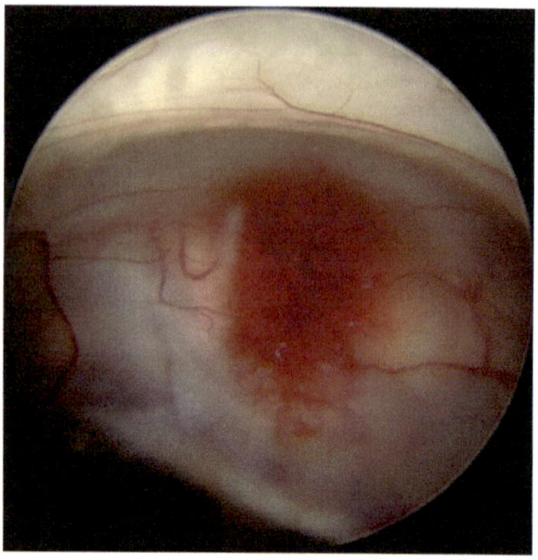

Fig. 5.63 The red spot, flattened pituitary stalk at floor of TV

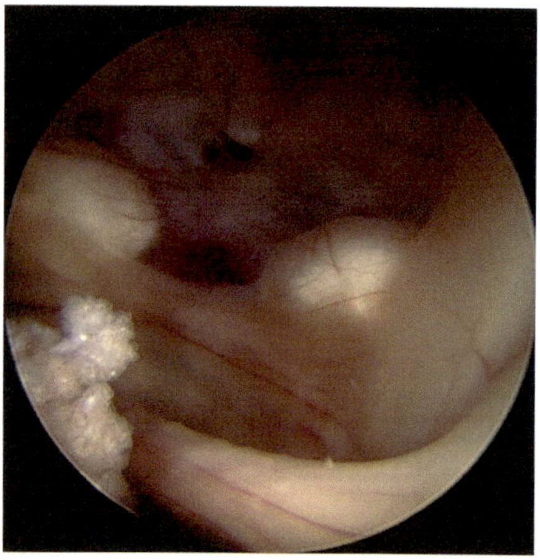

Fig. 5.64 Very small perforation area at floor of TV

tissue on Lilliquist' membrane and control of vessels. Coagulation of a floating vessel of the tissue after perforation is quite difficult endoscopically (Fig. 5.65).

The changes, of third ventricular floor, are well visible. The stretching of fibers and the enlarged distance of mamillary bodies are typical consequences of dilatation of the third ventricle. Here the aspects of cinereum and pre-mamillary area are nearly disturbed and hardly recognizable. ETV in such cases must be planned precisely and opening of the floor need to be done stepwise (Fig. 5.66).

Overview into the third ventricle through remnants of a foramen Monroi in a patient who had to wait decades for diagnosis and therapy. Under the anterior border of the foramen, in the depth mamillary bodies are visible. The extremely stretched inter-thalamic adhesion and posterior all the way down towards aqueduct entrance. There is no real plexus visible and subependymal tissue is extremely atrophic. Such cases are usually neglected completely (Fig. 5.67).

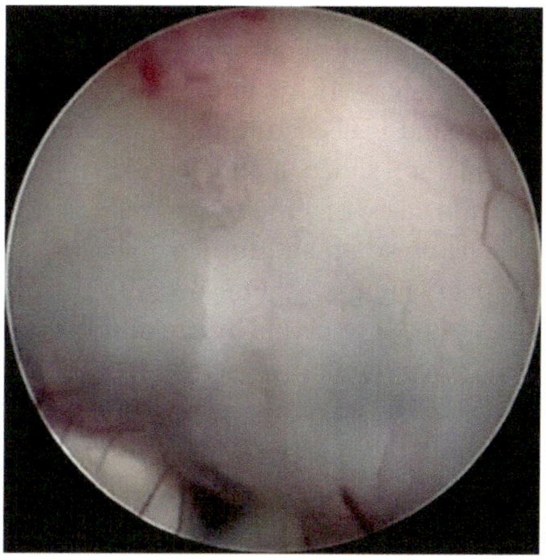

Fig. 5.65 Cranial bulging of floor of TV after entering the ventricle

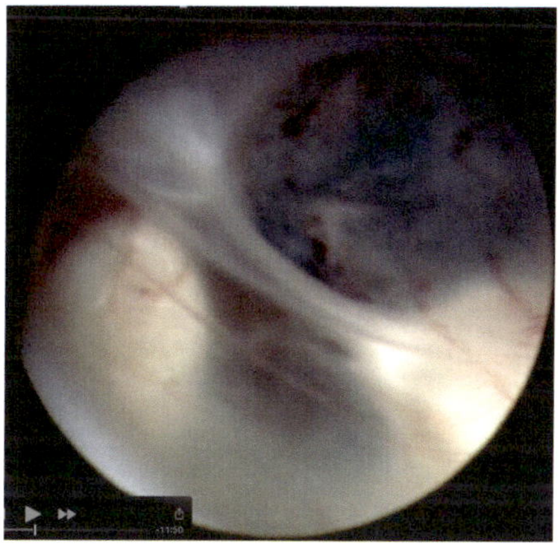

Fig. 5.66 Remnant fibers and changes of TV floor

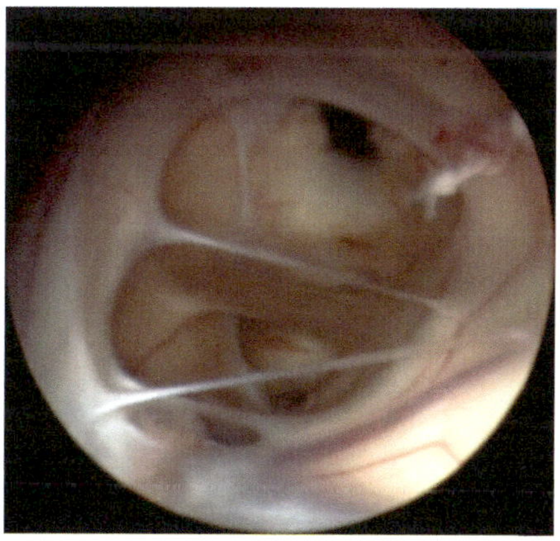

Fig. 5.67 Extremely NPH, Mega-FM and view into the whole TV

This is a dysplastic brain and we are looking into a distorted third ventricle. Such cases are challenging and force even experienced neuro-endoscopists to give up. Three-planar MR planning is mostly not enough, but a real-time navigation, like ultrasound is needed to reach the target. The posterior floor of third ventricle with pre-mature subependymal tissue and vessels, extremely vulnerable, are visible. Even in cases of a mono-ventricle targeting may be very difficult (Fig. 5.68).

The close up and further journey to the mamillary bodies shows not only anatomical landmarks but especially the premature tissue. It is characterized by wavy lines at the surface and by irregularly vessels. These cases may need mini-endoscopes and indication is always an individual decision. However, mostly it's the last chance in desperate cases, especially after multiple shunt-failures. Endo-neuro-sonography is helpful (see Chap. 5) (Fig. 5.69).

This view into a third ventricle shows a nearly unrecognizable floor without any safe landmarks. Such aspects may be seen after bleedings and infections. Such

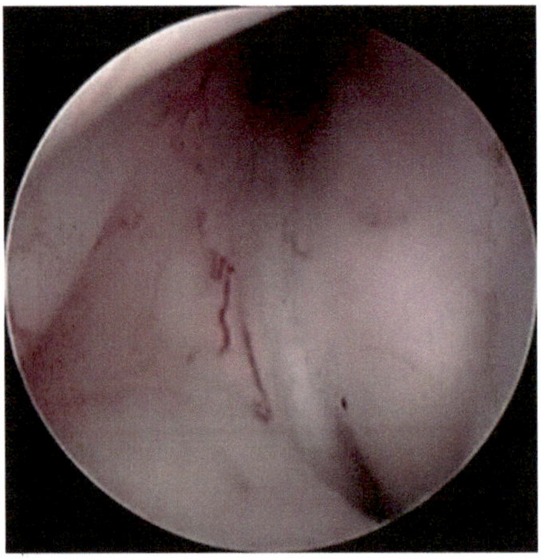

Fig. 5.68 Dysplastic premature floor of TV

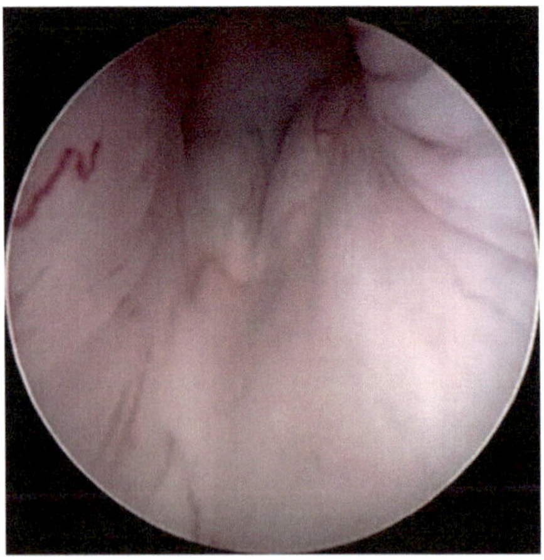

Fig. 5.69 Dysplastic premature floor of TV with mamillary bodies

conditions need an ultrasound navigation (1) (ENS; see Chap. 5) and perforation must be stepwise. The atrophic plexus (2) and fornix (3) is hardly visible as well as hypothalamus (4) and mamillary bodies (5). The tissue tends to bleed, however, shunt may be the reason of such results and do not stop the secondary pathology like post-infection changes (Fig. 5.70).

Here we see a large third ventricle after a former ETV. The inter-thalamic adhesion is stretched and slim. The ETV window and entrance of aqueduct are visible from one view-point. The liquor is slightly opaque and subependymal vascularization is impressive making a reddish appearance even for the mamillary bodies (Fig. 5.71).

This case shows a third ventricle floor after a former ETV. The subependymal tissue is pale and atrophic, it has signs of intraventricular bleeding. The optic chiasm is nearly un-recognizable due to atrophy. The infundibular recess does not exist, but also no red spot has developed. The view through the old ETV gives sight on diaphragm and dorsum sellae (Fig. 5.72).

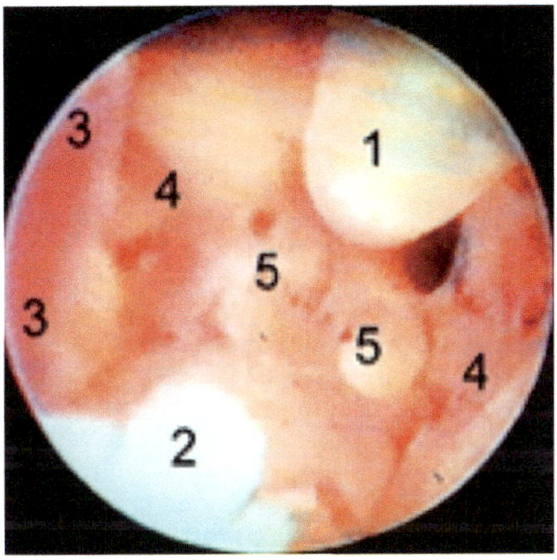

Fig. 5.70 Chronic post bleeding/infection floor of TV

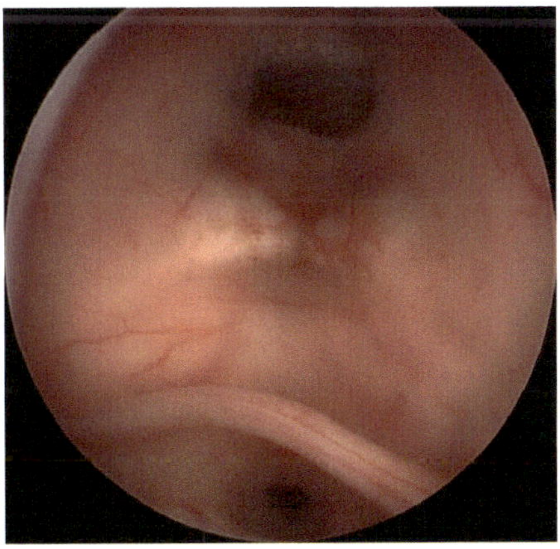

Fig. 5.71 Ependymal inflamatory vasculary changes; stretching of interthalamic adhesion TV

This view into third ventricle after a former ETV shows an atrophy at the borders of ETV and through it a strong dorsum of sellae. Mamillary bodies have an enlarged distance and parenchyma of cinereum dorsally. Infundibular recess is fattened to a red spot. Subependymal tissue is atrophic presenting vessels prominent (Fig. 5.73).

Ventriculoscopy and Re-ETV was necessary, perforating arachnoid membranes and scarfs.

This is a third ventricle floor after actual ETV, showing the vulnerability of the tissue. Cinereum is torn widely in this narrow premamillary region and the midline indicates the torsion of this immature third ventricle. The subependymal tissue is pale and infundibular recess has not the typical reddish color (Fig. 5.74).

This typical view of a premature third ventricle with the wavy lines of the ependyma shows spots from former intraventricular bleeding (Fig. 5.75).

The cinereum is changed markedly infundibular recess does not exist.

Premamillary membrane looks scary and subependymal tissue is pale with irregularly vessels.

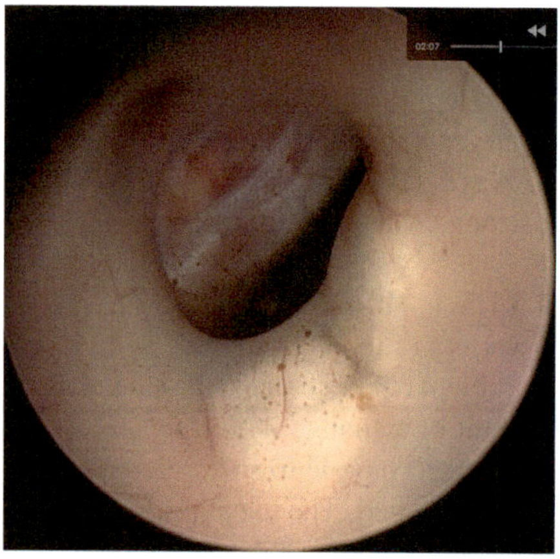

Fig. 5.72 Secondary changes after ETV

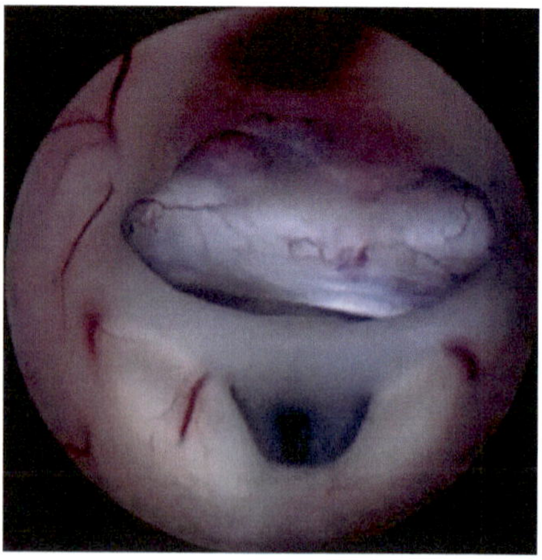

Fig. 5.73 Strong dorsum of sella through ETV

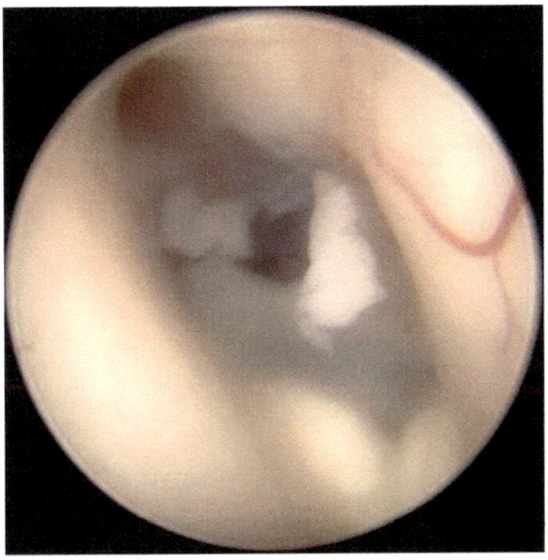

Fig. 5.74 Premature tissue with torsion and fresh ETV

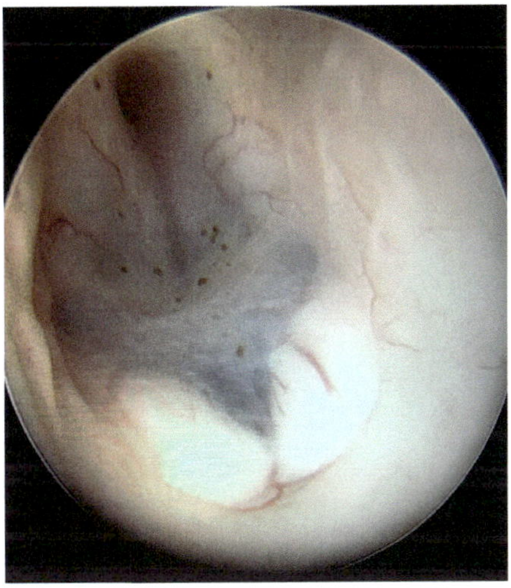

Fig. 5.75 Bleeding and NPH remnant of ependyma TV

Overview of the whole third ventricle gives an impression on the dysplastic brain. The thalamus is asymmetric and with irregular form on the left side (Fig. 5.76).

The entrance to the aqueduct looks not normal, posterior commissure is immature. All the subependymal tissue is pale and only few vessels are present.

This case had several shunt complications, additionally disturbing the maturation and functions.

This posterior third ventricle surprises with a nearly obliterated entrance of the aqueduct (Fig. 5.77).

Posterior commissure is atrophic. This constellation bears difficulties for imaging and diagnosis by flow void signals.

This floor of the third ventricle does not exist anymore but is a thin and tuff and completely transparent membrane (Fig. 5.78).

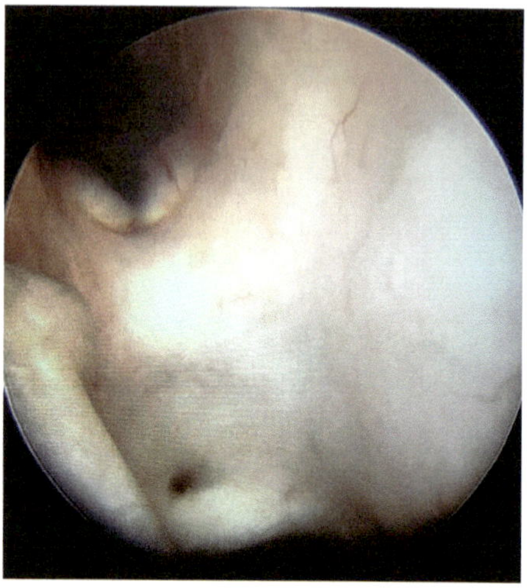

Fig. 5.76 Dysplastic TV

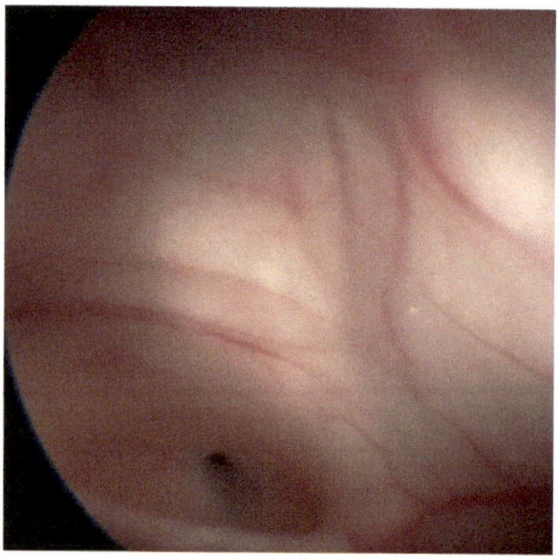

Fig. 5.77 Incomplete obstruction of aqueduct (fuctional aqueduct stenosis)

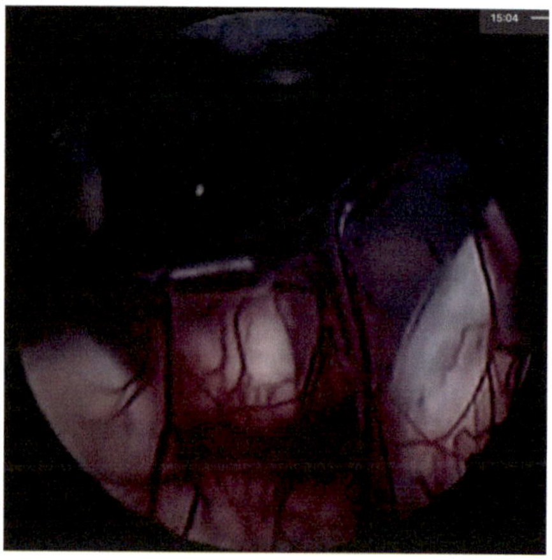

Fig. 5.78 Complete transparency of tv cyst, interpeduncular fossa and basilar head

The Basilar bifurcation proximal and all thalamo-perforators are visible. The pre-basilar part of the membrane is very deep, and the light is absorbed there strongly.

Oculomotor nerve on the left side is hardly visible. The here recognizable part of the third ventricle floor and interpeduncular cistern is just the "no-go area" for ETV.

5.4 Posterior Third Ventricle, Aqueduct and Fourth Ventricle

See Graphs 5.20c and 5.20d.

The posterior part of the third ventricle is difficult to be reached. Often, there is no patent trajectory for lens-endoscopes, and this area is therefore one of the few

Graph 5.20c Trans-FM approach for posterior TV (model)

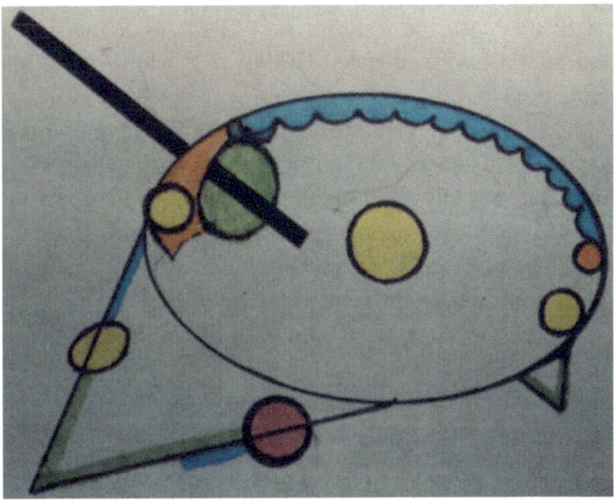

Graph 5.20d Peri-TV and peri-aqueductal structures

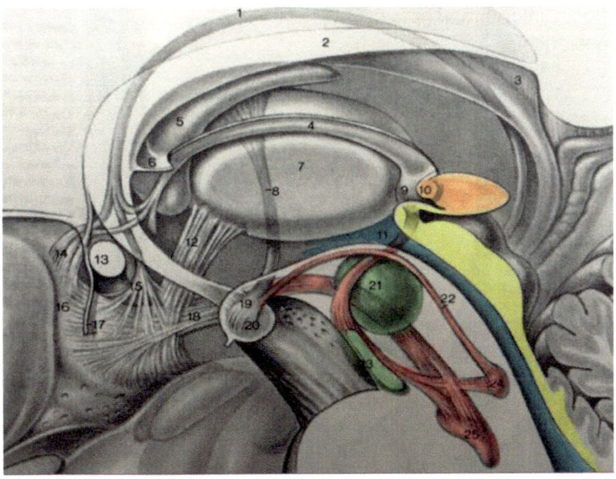

indications for fiberscopes. The correlation between size of inter-thalamic adhesion, size of the foramen, position of the venous angle and size of the plexus determine if there is a trajectory given. The position of the burr-hole must be adapted for this target. If one needs two targets, one anterior and one posterior, it must be planned precisely, if this is possible with one route only. But it is better to use fiberscopes and not two burr-holes and approach-canals. Also, it is necessary in the target of posterior third ventricle to use 30° endoscopes, but higher degrees are not recommended, because it is very difficult to guide them by visual control, and damages might occur easily (Fig. 5.79).

Approaching the posterior third ventricle towards the entrance of aqueduct is a critical task, and it must be remembered that it is possible to produce a coma vigil as a complication. Normally the shape of aqueduct starts oval, becomes triangle and then pentagonal, the course is not straight but curved in the median sagittal plane. Only in very widened cases the view can reach the fourth ventricle, which must be planned in the sagittal MR. We see here a slightly enlarged case at the triangle level viewing medullary velum of cerebellum (Fig. 5.80).

In this case the superior part of the aqueduct is wide, and we can look on the entrance of 4th ventricle. The level of stenosis defines the form of the aqueduct (Fig. 5.81).

In this case the subependymal is atrophic and vessels presented are not normal.

As the tissue of the ventricular system is very bright, the light source regulation automatisms allows to see the structures of the fourth ventricle late, when the endoscope enters the volume of the ventricle space.

The cause of the stenosis defines also the changes in the subependymal tissue, most common are infection and bleedings.

Here we see typical anatomy of the posterior third ventricle close to the entrance of the aqueduct (Fig. 5.82).

The strong posterior commissure forms the posterior border of aqueduct entrance. At this level the transversal shape is oval. More posteriorly the pineal recess is

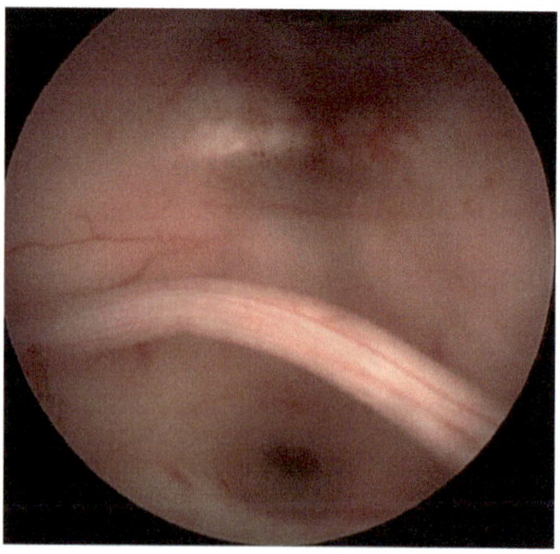

Fig. 5.79 Overview of TV floor (anterior and posterior)

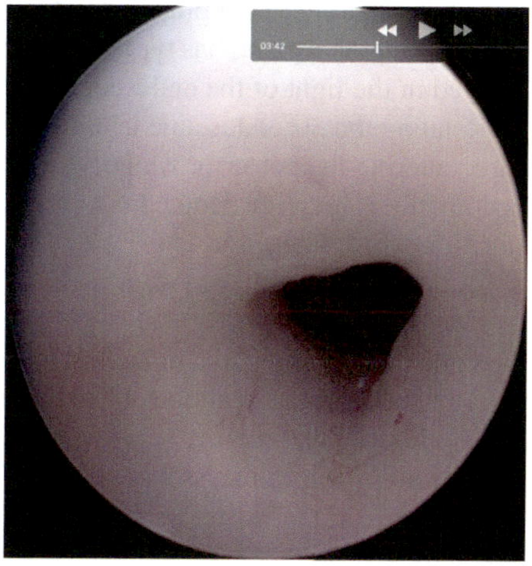

Fig. 5.80 Aqueduct at triangular level

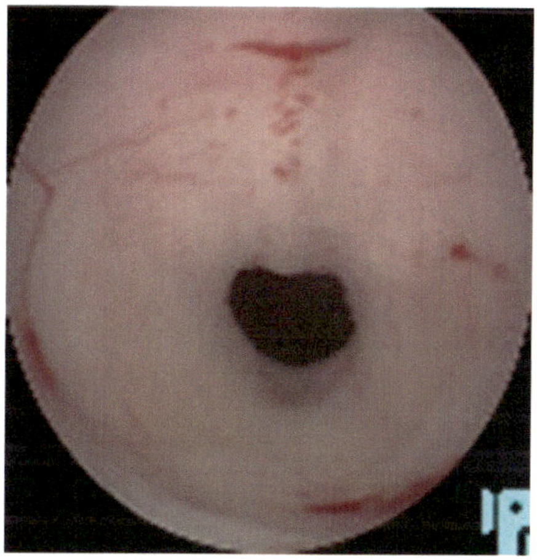

Fig. 5.81 Aqueduct at oval-level (enlarged)

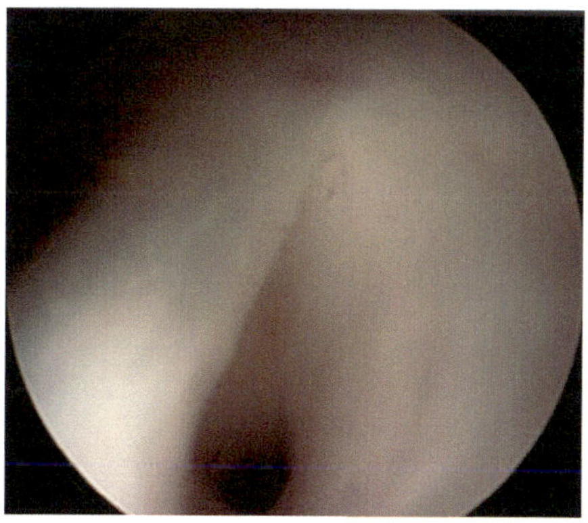

Fig. 5.82 Posterior TV from anterior-lateral (entrance of aqueduct)

beginning leading to the pineal gland. This is only visible, when using a 30° endoscope. The pineal gland has typically sand-like calcifications which may shine gold-colored in the light of the endoscope.

Rotating the 30°endoscope laterally gives view on the lateral part of the commissure, entering the posterior wall of the third ventricle. This area contains many connection fibers between deep parts of the limbic system with behavior- and vegetative functions (Fig. 5.83).

But also, visual reflex system is close in the midline, and ventral to the aqueduct.

Complications here are always a catastrophe for the patient. Therefor the aqueducto-plasty was abandoned by most groups.

A further rotation with the 30° endoscope allows to see the roof of the third ventricle with the two internal veins and the plexus. Beneath the left habenula and the lateral end of pineal recess can be seen. The strong posterior commissure entering lateral wall of the posterior third ventricle with typical vessels became visible (Fig. 5.84).

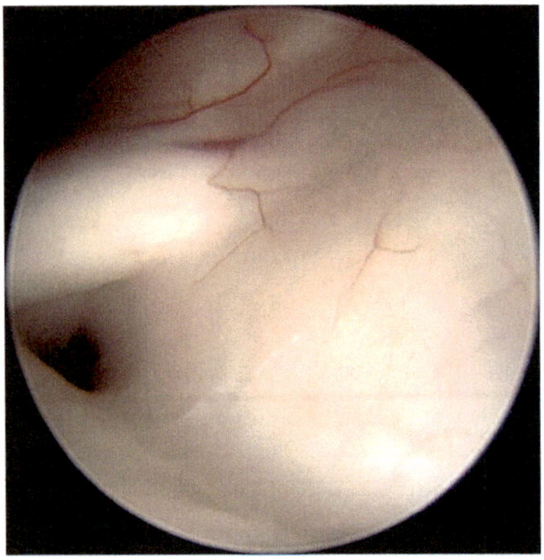

Fig. 5.83 View on the lateral opposit left side of posterior TV

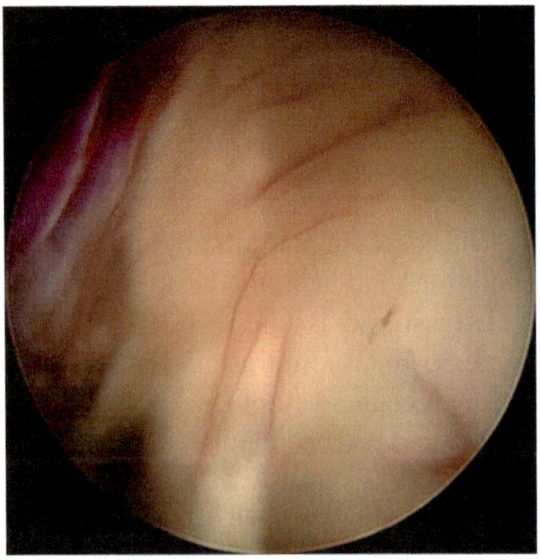

Fig. 5.84 The internal cerebral veins and plexus at the roof of TV with superior-posterior lateral view

This is a critical region because in case of complications the results are terrible for the patients.

In this case the entrance of aqueduct is narrowed by a strong posterior commissure. In such cases even a flexible fiber-scope may not be able to give sight into the aqueduct (Fig. 5.85).

Injury of this commissure may cause failures of consensual pupil reflexes.

This is an extreme case of chronic hydrocephalus. We look through a big opening, that does not remember to a foramen Monroi, giving an overview over the posterior third ventricle. We see the stretched posterior commissure, and dorsally a pineal recess with remnant of pineal gland. The entrance of aqueduct shows a perforated scary membrane (Fig. 5.86).

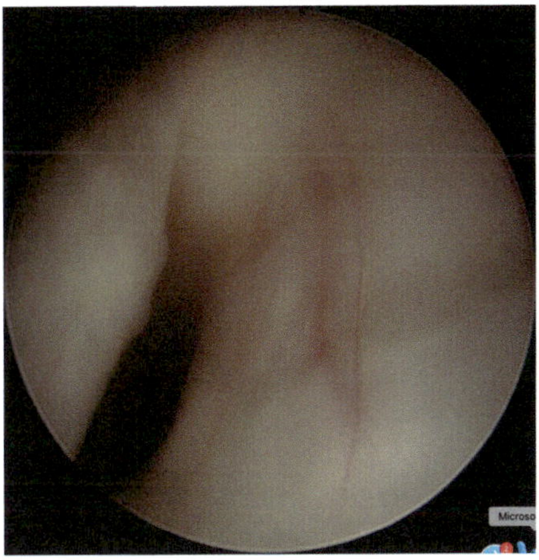

Fig. 5.85 Narrow entrance of aqueduct

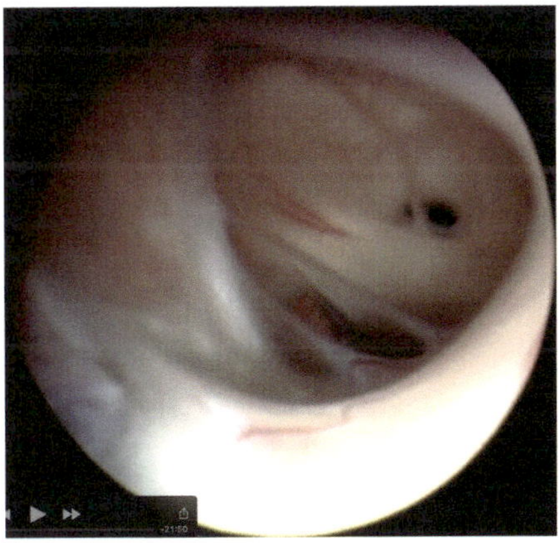

Fig. 5.86 Posterior TV in extrem nph and view into obstructed aqueduct by scarfs

This is an absolute rare condition, giving sight directly into this part of the wide open third ventricle.

The close-up makes the conditions of the aqueduct entrance more visible. The inter-thalamic adhesion is just a stretched band, and all the subependymal tissue is massively atrophic (Fig. 5.87).

This patient had to await diagnosis and therapy for more than 5 decades (s. Chap. LASER).

After ETV she improved and came out of bed after 3 years, however, sever impairments could not be changed, but live-quality and handling became better. These type of patients have no compliance for the shunt maladaptation and develop a shunt-disease, which does not occur after ETV.

A stronger close-up of the aqueduct make the situation more clear, and we see the membrane, that is opened at two small areas devided by a scary band. Traction of the scar-activity narrows the right border of the aqueduct (Fig. 5.88).

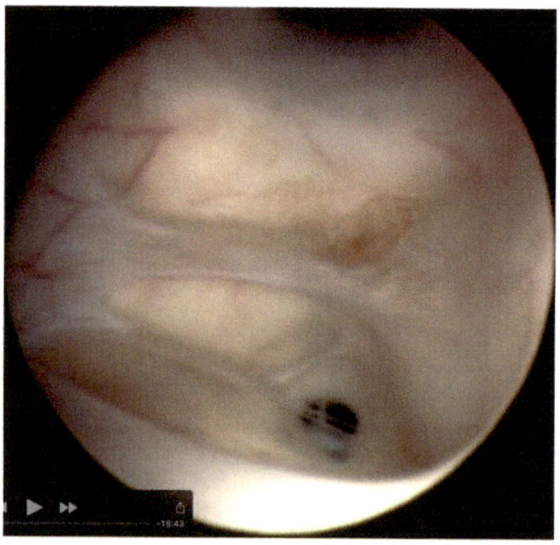

Fig. 5.87 View into posterior TV and aqueduct with obliteration by scarfs

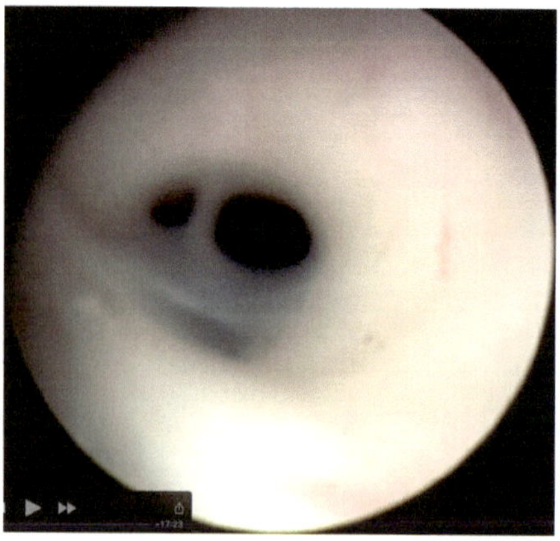

Fig. 5.88 Narrowed aqueduct by scarfs

Such morphological conditions are challenging radiologist and clinical causes a slow chronic destruction of subependymal tissue. Though there is still the compensation by partially open aqueduct with flow-signals, the resistance is too much and the long-time effect too strong.

This is the aspect of an obstructed aqueduct from lateral right by an astrocytoma (Fig. 5.89).

The lumen of the superior part is nearly blocked, and only a small rest of lumen is visible. The results in developing a three-ventricular hydrocephalus is the same, like in other reasons, but progression may be faster. Posterior to the entrance we see the posterior commissure. If a biopsy can be done without functional risk must be planned meticulously. Closure of the ETV occur more frequent compared with other reasons of obstructions.

Another case of tumor, obliterating the aqueduct. Here the tumor is coming from dorsal, which means from quadrigeminal lamina. Signs and symptoms bring the precision into the imaging in such cases. This is important for planning a biopsy (Fig. 5.90).

As mentioned above, election of trajectories is challenging.

In any case, ETV is necessary, and stenting bears more functional outcome risks.

Such premature cases present a high-risk situation, as the anatomy is dysplastic and the tissue very vulnerable. Even soft shunt catheters cause injuries of the ventricular walls (Fig. 5.91).

Aqueduct entrance is absolute changed and hardly to be identified. Mostly, as here, there is an obstruction or aplasia of the aqueduct.

In such cases ultrasound navigation is necessary and best done by trans-endoscopic technique, if ad hand. With normal ultrasound probes it is much more difficult. Due to shift-effects computer-neuro-navigation is not recommended, making the procedure time-consuming and complicated.

This is the posterior third ventricle in a case of supra-sellar cyst. Additionally, there is a membranous obliteration of the aqueduct, that was nearly invisible due to

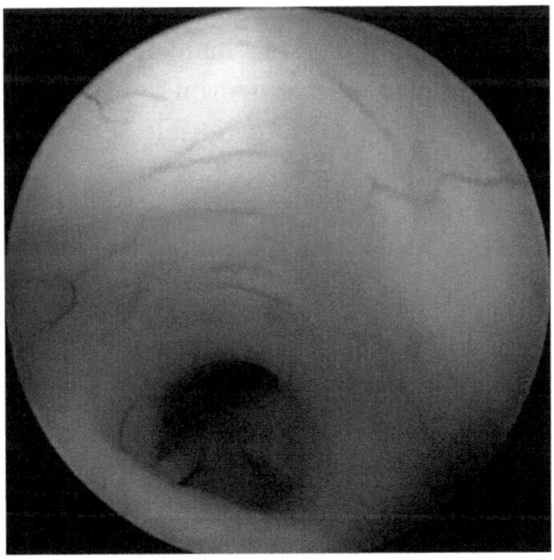

Fig. 5.89 Obstructed aqueduct by tumor

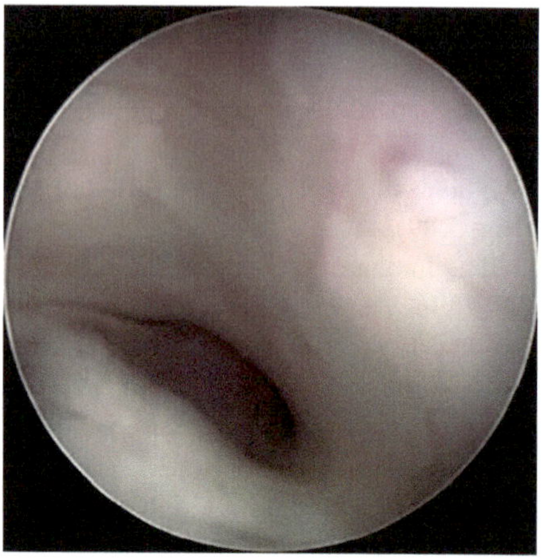

Fig. 5.90 Nearly obliterated aqueduct by tumor

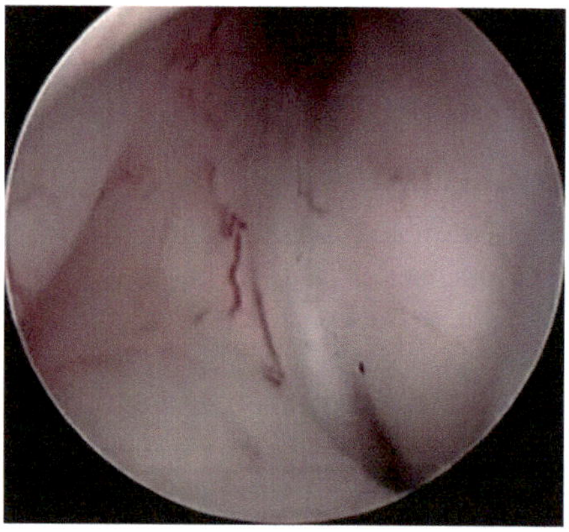

Fig. 5.91 Obliterated aqueduct by torsion in premature tissue

the huge cyst. The membrane had a small opening through which the obstruction could be mostly compensated, and this was super-formed by the behavior of the cyst. This complex mechanism caused a complex symptomatic. We see the entrance of the aqueduct and a strong posterior commissure. A remnant of the cyst-wall is bulging into the third ventricle (Fig. 5.92).

Again, a very small rest of the aqueduct opening by a parenchymatous membrane (Fig. 5.93).

Here it's the same problem with radiology, that they cannot accept the diagnosis functional stenosis of aqueduct.

A flow, they believe mostly, means no stenosis. This causes an "odyssey" for the patient and also mostly a not indicated shunt.

In this dysplastic brain we see a stenosis of the aqueduct. This is often combined with other dysplastic changes (Fig. 5.94).

The therapeutic tactic in such cases should encounter an ETV, even the concept of combined strategies to spare operations may be discussed. In each case of shunt revisions ETV should be considered.

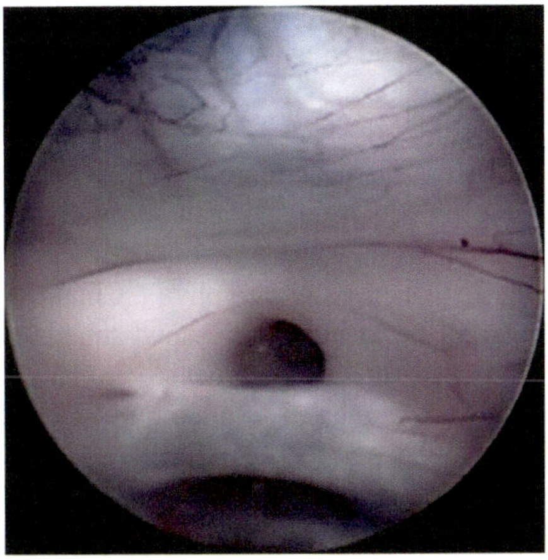

Fig. 5.92 Incomplete membranous obliterated aqueduct (functional aqueduct-stenosis)

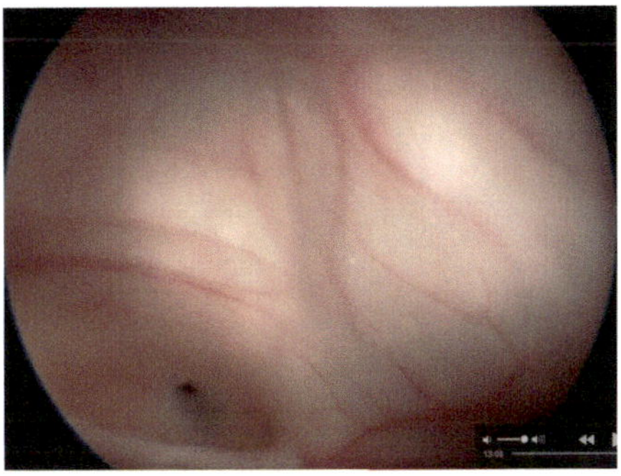

Fig. 5.93 Incomplete obliteration of qaueduct (functional aqueduct stenosis)

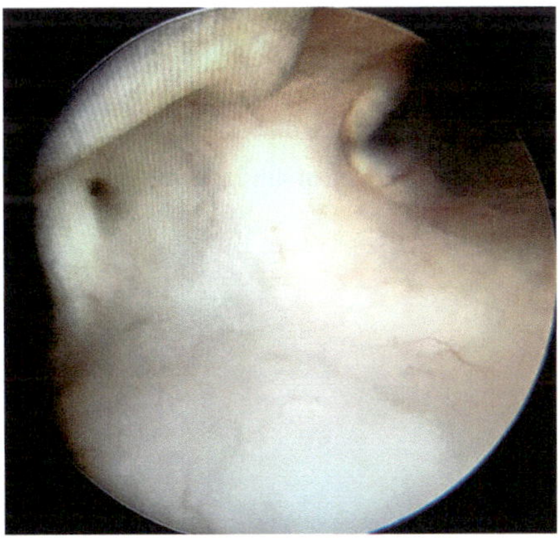

Fig. 5.94 Congenital dysplastic aqueduct-stenosis

The view goes into a mega-aqueduct, the aqueduct is wide open until the entrance to the 4th ventricle. There the pentagonal shape is already present (Fig. 5.95).

The subependymal tissue is premature and shows the typical wavy surfaces in such conditions.

This tissue is extremely vulnerable and the absolute space is anyway very small. Trajectory-planning is mandatory. This can be really done by neuro-navigation-systems, as they are in real planning systems. For navigation a real-time technique, like endo-neuro-sonography (s. Chap. 7) is recommended.

After the pentagonal shaped aqueduct-level, the fourth ventricle comes in sight. Of course, this is a “no mans’ land” also, like the aqueduct, and approached in rare cases, when anatomy shown by MR or planning-systems, a proven and safe trajectory (Fig. 5.96).

The pentagon is given by the brainstem ventrally, the cerebellar peduncles laterally and by the cerebellum dorsally. The tissue in this case shows premature irregularities, like the exophytic tissue of the dorsal wall.

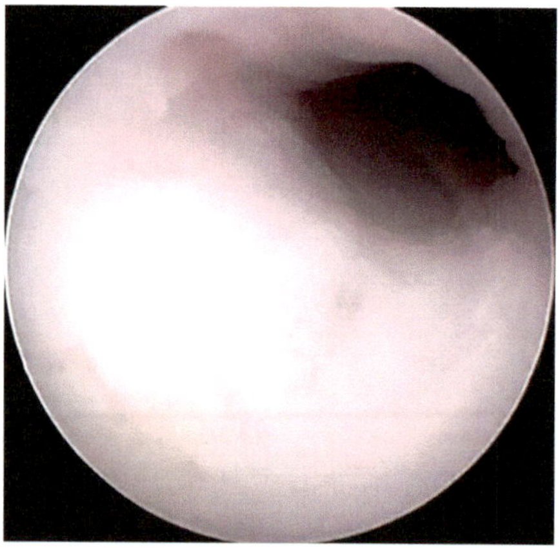

Fig. 5.95 Mega-aqueduct at pentagonal level

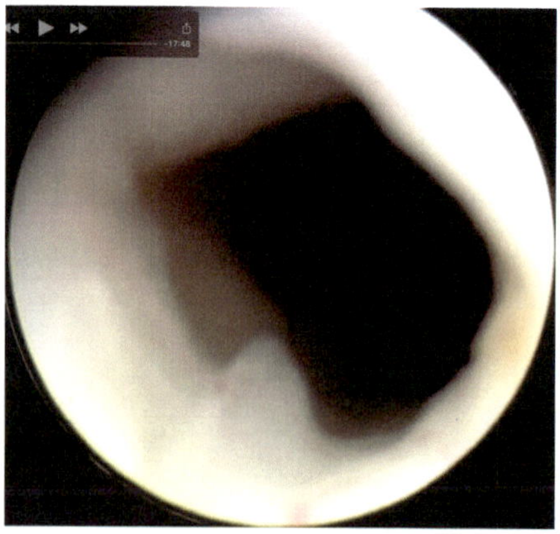

Fig. 5.96 Entrance of 4th ventricle

Once, the endoscope is deep enough and light-reflection from the white tissue will become weak, only then the light-automatism will allow to see the plexus of the fourth ventricle on the surface of the bulging inferior vermis (Fig. 5.97).

The anatomical features of the brain-stem are better seen on the next figure.

This is the view into the fourth ventricle of a non-fixed specimen. Ventral the typical landscape of the brain-stem with median sulcus, facial colliculi, and directly below the eminentia mediana on both sides. From above, coming from the aqueduct, the figure remembers to quadrigeminal lamina. But the landmark of plexus makes sure, where we are. The plexus runs through foramina Luschkae on both sides, just below the acoustic striac and vestibular area. The median sulcus leads to the depth of foramen Magendi and cisterna magna (Fig. 5.98).

The view of the same fourth ventricle shown from latero-dorsal and with a 45° left-rotation of the 0°-endoscope shows the features in a completely different

Fig. 5.97 Entering 4th ventricle

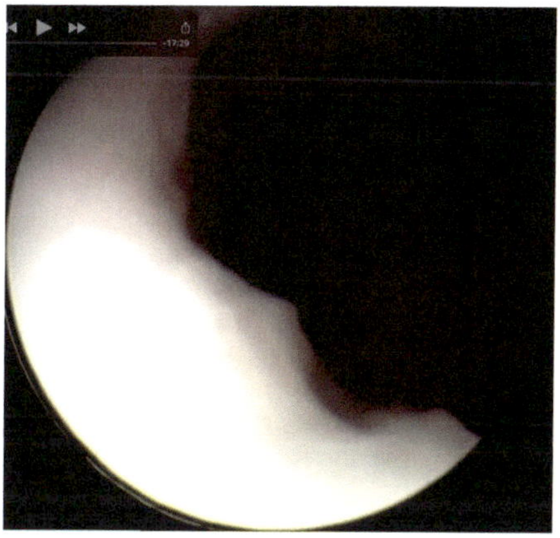

Fig. 5.98 Anatomical overview of 4th ventricle

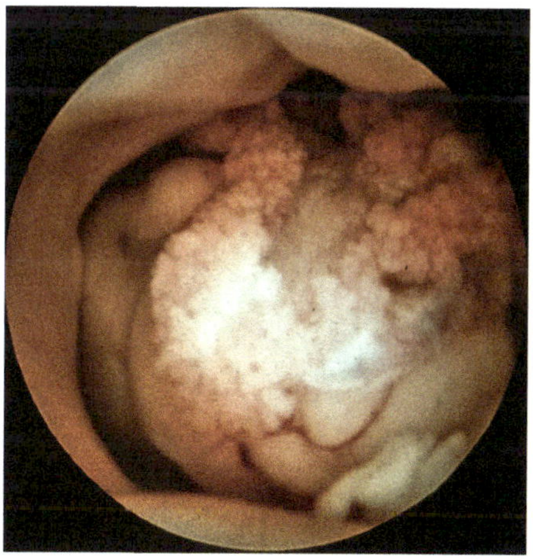

appearance. The view is dominated by the plexus, below of which the inferior fourth ventricle lies in the dark. There is also in the depth foramen Magendi (Fig. 5.99).

This aspect of the superior fourth ventricle of a baby shows an opaque liquor due to a tumor (O). The child came in deterioration and MR showed a sever internal hydrocephalus. A tumor at foramen Magendi level was the reason for this danger situation (Fig. 5.100).

In an emergency endoscopic procedure we met this opaque CSF and did first, after ICP release, a sample of liquor for cytology and a ventricular-lavage. As the tumor was very deep and well vascularized, biopsy seemed too risky.

In this case the aqueduct was very wide enabling a glance into the fourth ventricle. The tumor was hardly visible, but the locally displaced strong plexus gave a kind of contrast.

Rickham-reservoir was placed, and as it was a clear obstructive Hydrocephalus, an emergency ETV was done.

In a neuro-oncological pediatric unit the further non-operative therapy was taken over.

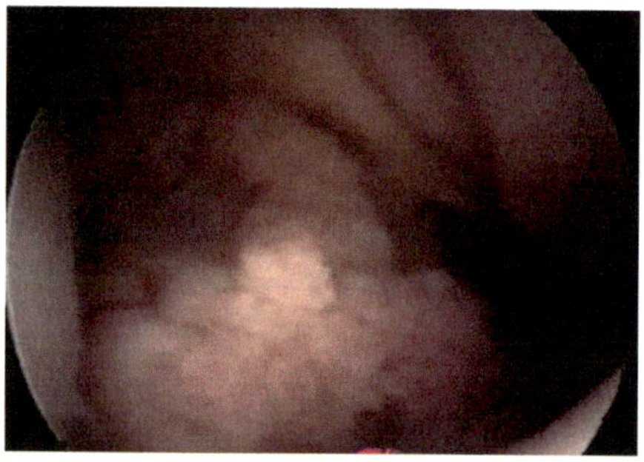

Fig. 5.99 Anatomical overview of inferior 4th ventricle

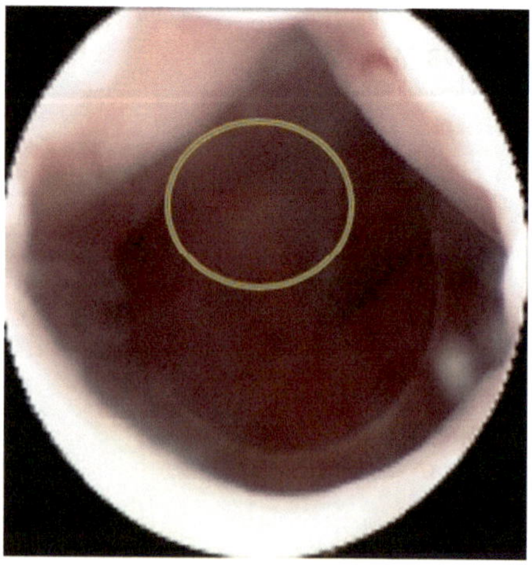

Fig. 5.100 Inferior 4th ventricle obstructed by a leftsided tumor

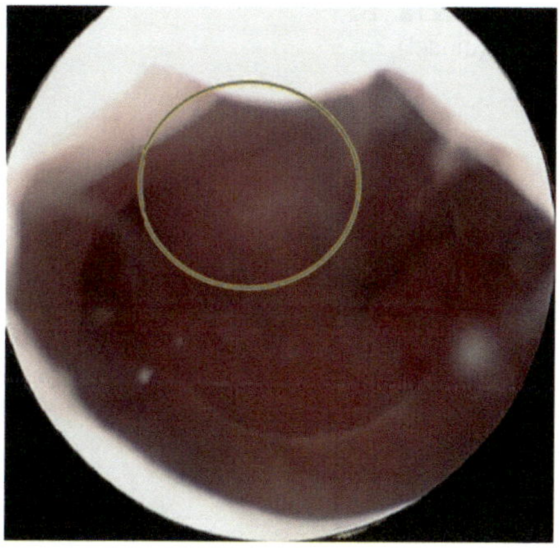

Fig. 5.101 Overview and obstruction of the 4th ventricle inferior-lateral left

The endoscopy was done along midline ("kings-route"), and only through corpus callosum, to preserve cerebral tissue. This should only be done by very experienced neuro-endoscopists! (Fig. 5.101).

5.5 Subarachnoidal Space

5.5.1 Retro-Clival Cisterns

The subarachnoid space is a complex system of compartments named cisterns (Graph 5.21a). There are 6 retro-clival cisterns (s. Graph 5.21b). These cisterns show a large variety, but the content is constant, representing embryological reasons. The membranes, separating the compartments, have openings for the vessels and are very variable in texture, thickness and mechanical properties. Since the routine use of endoscopy in neurosurgery, the understanding and imagination of the subarachnoid space has changed a lot. While the subarachnoid cisterns should be the roadmaps of micro-neurosurgery (Yasargil), in neuro-endoscopy they are the environment within the endoscopy is moving and viewing like a submarin.

The journeys through the anatomy of the retro-clival cisterns have been shown first time in the video "MININVSIA: A journey through the brain", 1993, during the (1) International Congress on Minimally Invasive Techniques in Neurosurgery (Wiesbaden/Germany), and on the X. WFNS Congress (Acapulco/Mexico) in the same year.

This space can present new membranes, causing obstructions for the CSF flow at lower levels, and causing obscure or not recognized failures of ETVs.

This perforation of a parenchymatous premamillary membrane during ETV in a step-wise fashion, preserved a rather fibrous Lilliquist membrane with a nicely visible texture. The endoscope is rotated 90° left and the basilar trunk is running beneath the arachnoid membrane. In such high basilar positions and a small space, perforation during ETV should always be done stepwise and the last step under

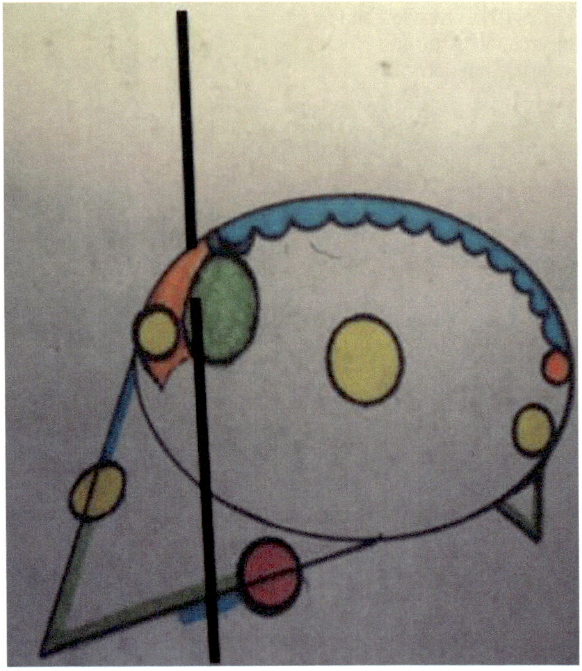

Graph 5.21a ETV route (model)

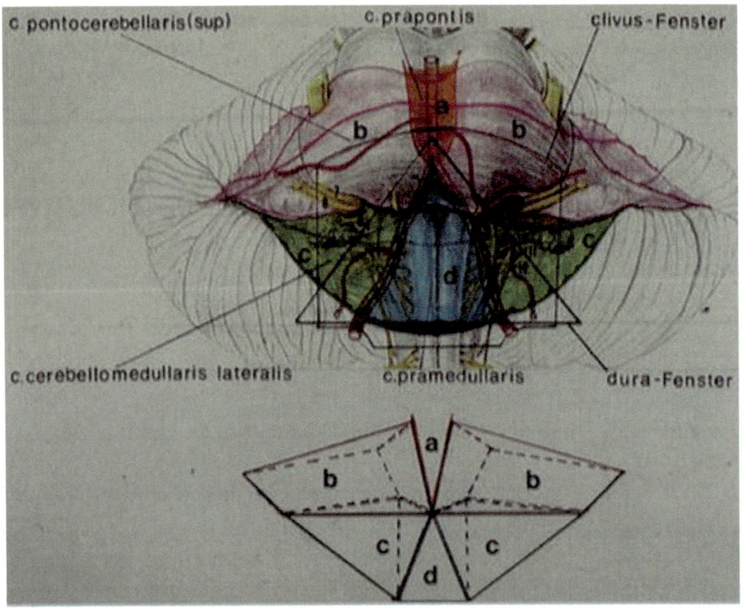

Graph 5.21b Model of ventral basal cisterns

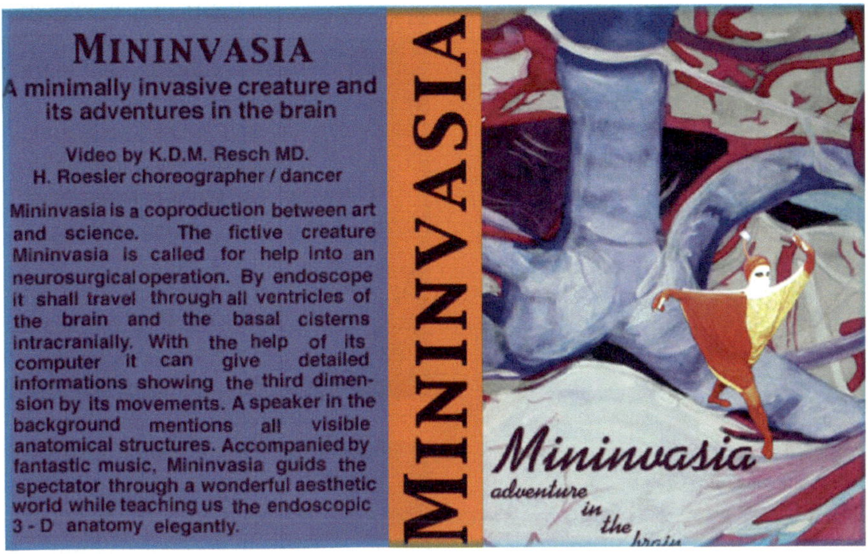

Graph 5.21c Premiere 15.–19.06.1993

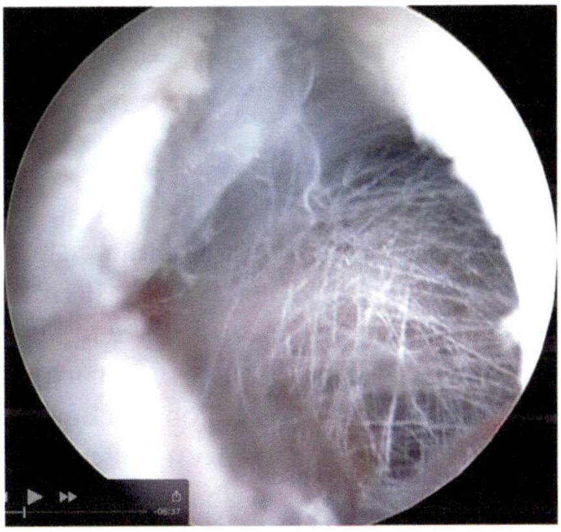

Fig. 5.102 Subarachnoid fibers of liliequist membrane

visual control. Such cases are also good indications for LASER application (Fig. 5.102).

After opening and preparing most of the fibers, a well-formed remnant of the Lilliquist membrane becomes visible. It is located at the typical level between the oculomotor nerves. The complete opening gives view on the long basilar trunk, and on both side the pontine arteries. Typically, the opposite side can be seen better than the ipsilateral, for the same reason, the opposite oculomotor nerve can be well seen (Fig. 5.103).

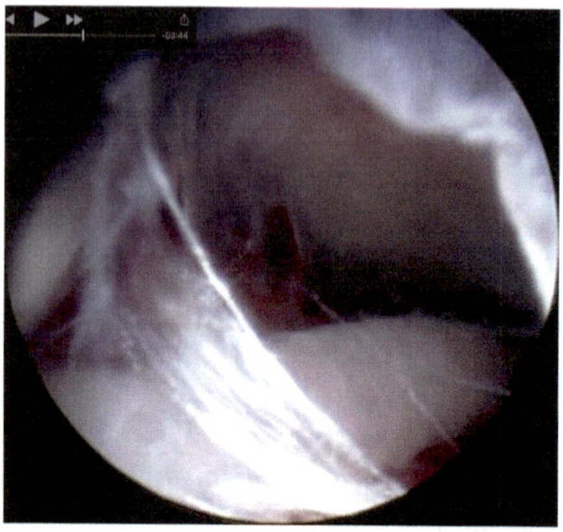

Fig. 5.103 Remnant of liliequist membrane

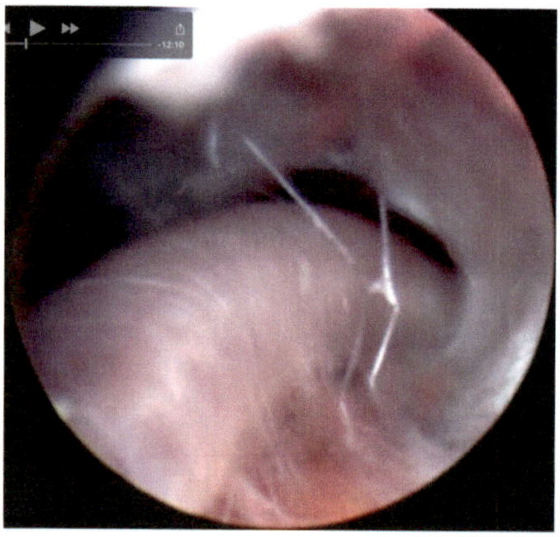

Fig. 5.104 Deep Second membrane arround basilar trunc

This view into the prepontine cistern, along a strong basilar trunk, shows the opening into the pre-medullary cistern. The separation by this opaque membrane is normal, but the in-transparency and tuff consistency are caused by a former infection. If the opening becomes too narrow, a resistance for the CSF flow may occur, causing the intra-cisternal obstruction, only known by endoscopic experience, and only presented as a theory until today. This is the morphological correlate. The membrane can develop a bulging caudally, causing a ventil-mechanism and increase the obstructive effect (Fig. 5.104).

This is the beautiful appearance of a basilar bifurcation with a system of perforators in a 6-month young baby, seen through an ETV. All structures look innocent without any changes seen usually in grownups. The interpeduncular cistern we are looking into is the only one that is bigger in babies than in adults. We have space to see and can recognize a small remnant of the caudal Lilliquist membrane. The area we are looking at is usually a "no-go area" for ETV containing all the perforators and the basilar bifurcation. The basilar system is very close to the clivus, causing a very limited space for ETV. The relatively big size of the cistern in cranial direction allowed to do the ETV anyway. But in such conditions, which must be analysed precisely on the tri-planar MR, the perforation must be done stepwise. After opening the in-transparent parenchymatous cinereum or pre-mamillary membrane, the arachnoid roof of the interpeduncular cistern, the Lilliquist membrane, must be opened under direct and close visual control. In this case it was then visible, that the basilar system is far below and perforation safe (Fig. 5.105).

In cases when the obstruction will predictable persist, a primary ETV postnatal in combined with a shunt, may be considered, and especially, when shunt is not possible for longer time.

This is a high basilar bifurcation, appearing rather dark because of a strong reflecting flat dorsum of sellae (Fig. 5.106).

There are no normal superior cerebellar arteries visible. Variations are very common there. The basilar artery is very close to the clivus.

Here we see a perfect basilar bifurcation, rather low, the dorsum sell is not visible being rather high. The course of the basilar trunk can be followed into the depth, as it is far away from the clivus in a big pre-pontine cistern. The left cerebellar superior artery is strong, the right one is rather slim (Fig. 5.107).

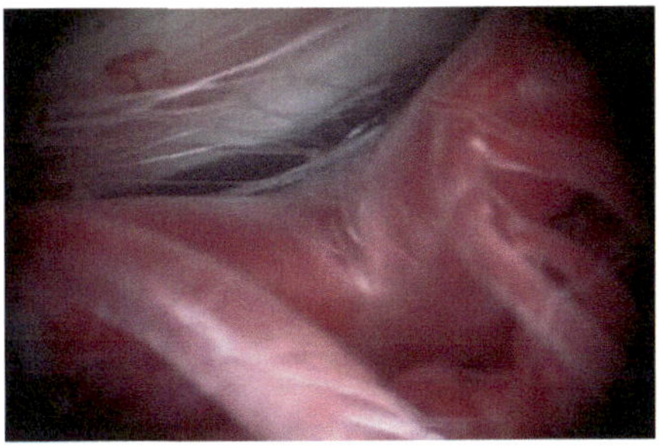

Fig. 5.105 Basilar Head through etv (6 month baby)

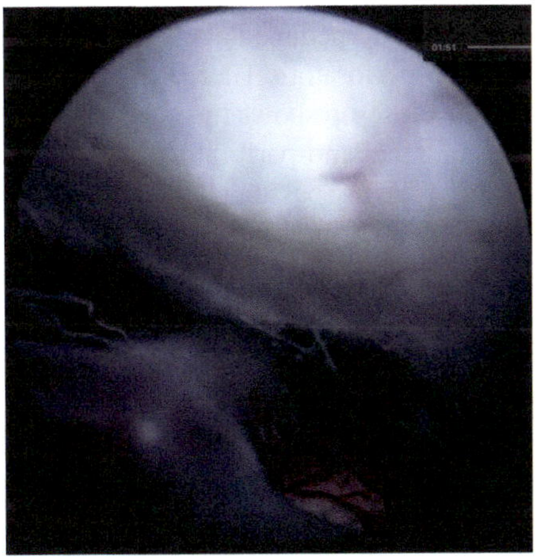

Fig. 5.106 Basilar head through ETV and strong dorsum of sellae

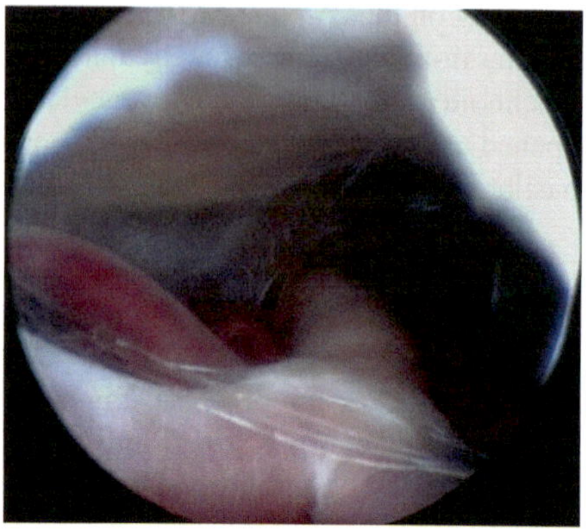

Fig. 5.107 Second membrane in the depth transgressed by basilar trunk (cisternal obstruction)

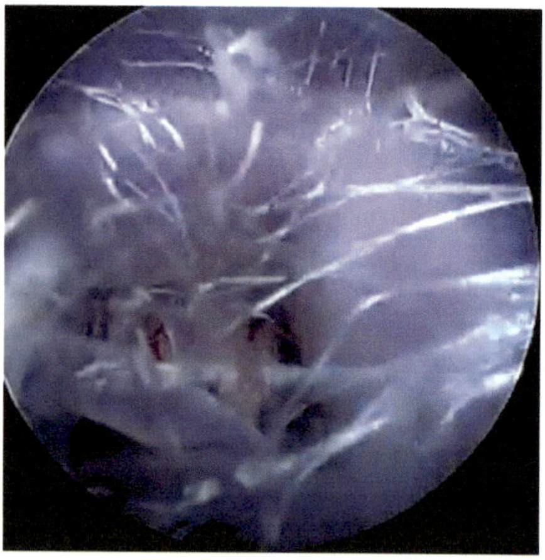

Fig. 5.108 Basilar division with richly subarachnoid trabecula

In this interpeduncular cistern we see a strong basilar bifurcation with also strong thalamo-perforators. There is especially a very rich trabecular system (Fig. 5.108).

The close-up shows the beautiful art-work of the trabecular network. The strong and long thalamo-perforators are well present (Fig. 5.109).

After having passed all the trabecular fibers, one can see the forest of perforating vessels to the midbrain in the interpeduncular fossa. This amount of vessel in a small field is more than usual (Fig. 5.110).

Following with the eye into the depth the course of the basilar trunk, one can see thin arachnoid wall. On the right side of the basilar trunk the area of foramen magnum can be recognize in a dark field (Fig. 5.111).

The pontine arteries are well visible left.

The view into a big interpeduncular cistern with a somehow freely floating basilar head is visible. The arachnoid membranes and cisternal walls are in-transparent. The left posterior cerebral artery presents a narrow whiteish part under which the oculomotor nerve is running (Fig. 5.112).

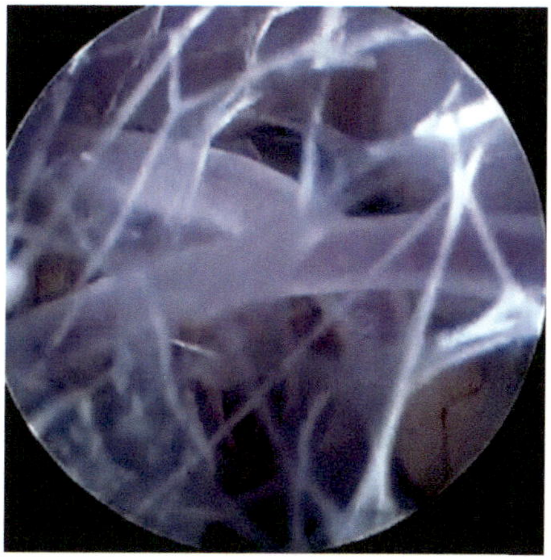

Fig. 5.109 Thalamo-perforators with subarachnoid trabecula

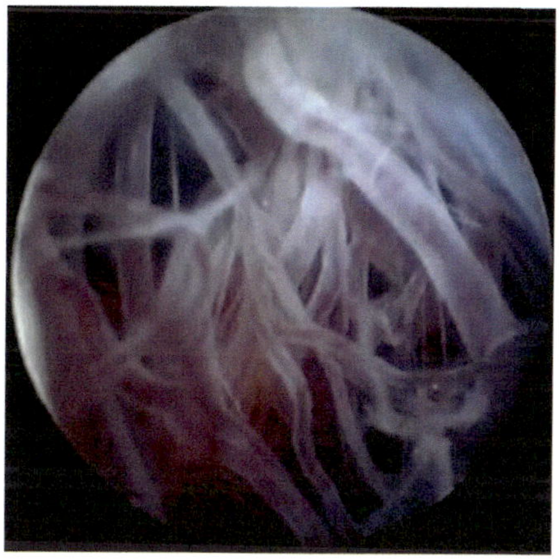

Fig. 5.110 Forest of thalamo-perforators (no mans land for ETV!)

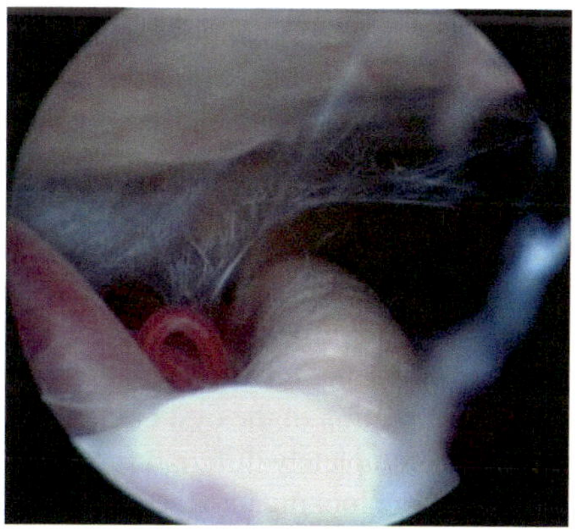

Fig. 5.111 Cisternal Wall of CPA Cisterns left, seen from Prepontine Cistern

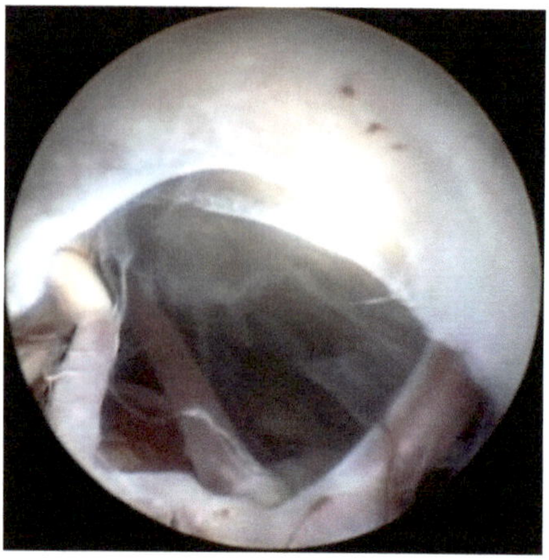

Fig. 5.112 Big interpeduncular cistern with riding P1 on CN III left

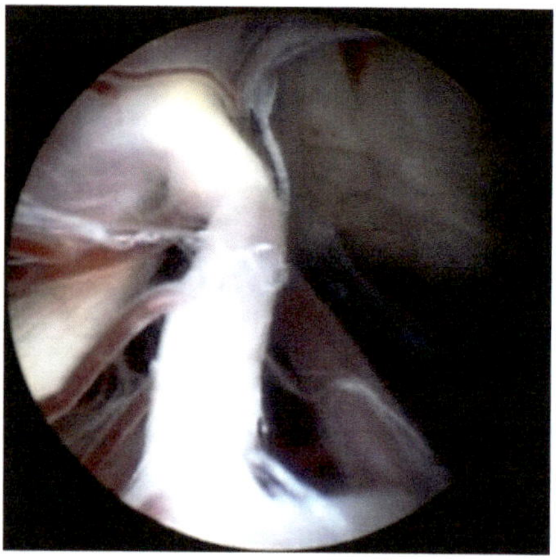

Fig. 5.113 Closeup 5.112; riding P1 on CN III left

In the close-up this narrow whiteish part came out to be entering into the left carotid artery, the so-called embryonal type. Just beneath this connection the strong oculomotor nerve is visible, slightly pushed caudal. In the endo-video, this connection was jumping on the nerve. There were, however, not deficits of the nerve. Such finding can be seen time by time (Fig. 5.113).

This case shows a view inside a suprasellar cyst in the third ventricle. In such cases the floor of the ventricle seams to be the cyst, however, the cranial side is usual parenchymatous and should not be resected too radical (Fig. 5.114).

Here we see the basilar bifurcation and the thalamo-perforators, covered by the inferior arachnoid layer of the cyst. This cysts cause a large space between the brainstem and the clivus, so the endoscopic view is perfect and perforation and partially resection of the cyst cranially not very difficult.

This is completely different for the perforation of the caudal wall for such cysts. Again, we are looking into the cyst to the lower boarder, still seeing the basilar head

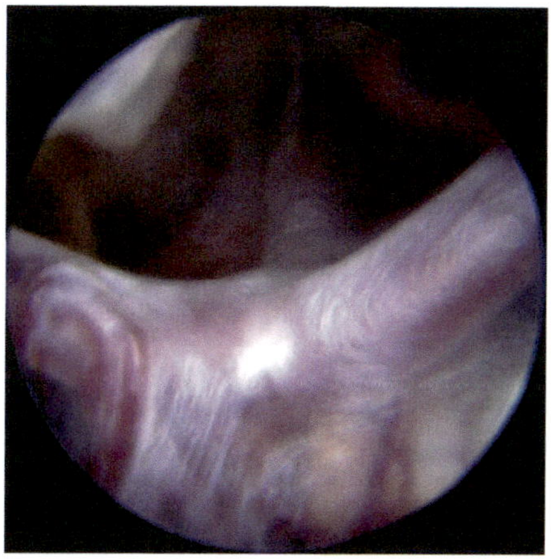

Fig. 5.114 View into prepontine cyst, caudal wall on basilar artery

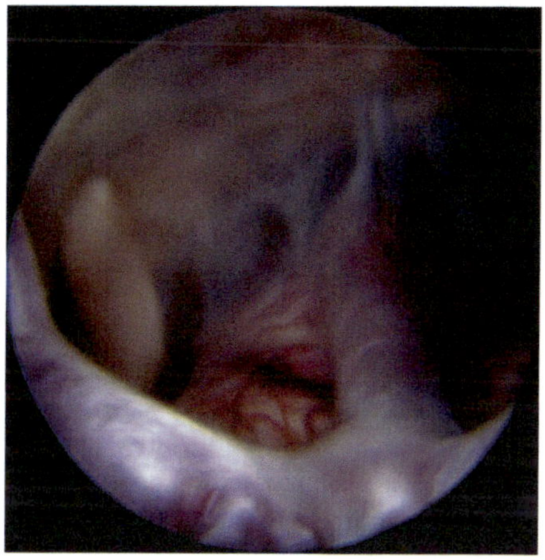

Fig. 5.115 Basal membrane of the prepontine cyst

and on the left side even the trigeminal nerve entering the scull base. The caudal cyst wall is very tuff and only few mm space is given for a perforation to get free flow of CSF (Fig. 5.115).

We see the basilar trunk getting close to the clivus, laying on the pons.

The cyst may leave regularly flow in the aqueduct, but once this compensation ends for what reason ever, the decompensation can occur fast. Symptoms are often overseen for years and typically have an undulating course. Ophthalmic symptoms and findings may predict the end of tolerance.

In many cases it is possible, according to imaging findings, to view into the depth of the basal cisterns (Fig. 5.116).

We see the basilar trunk on the better visible opposite side and some remnants of the trabecular system. Moreover, we recognize the sympathetic fibers on the walls of the big vessels. Some pontine arteries are floating freely in the cistern, indicating some atrophy of the brain-stem.

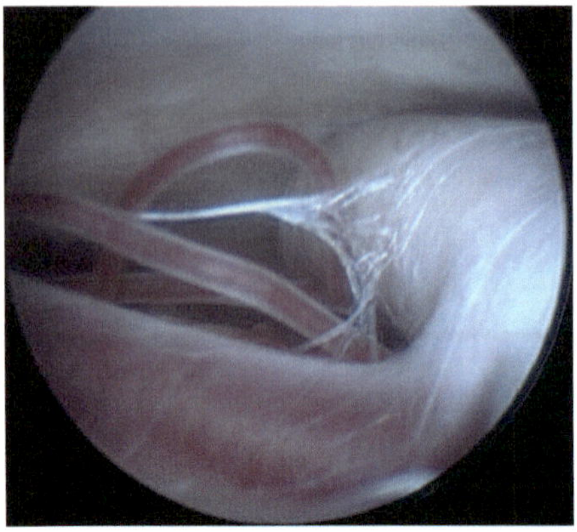

Fig. 5.116 Basilar trunk with free floating pontine arteries left side

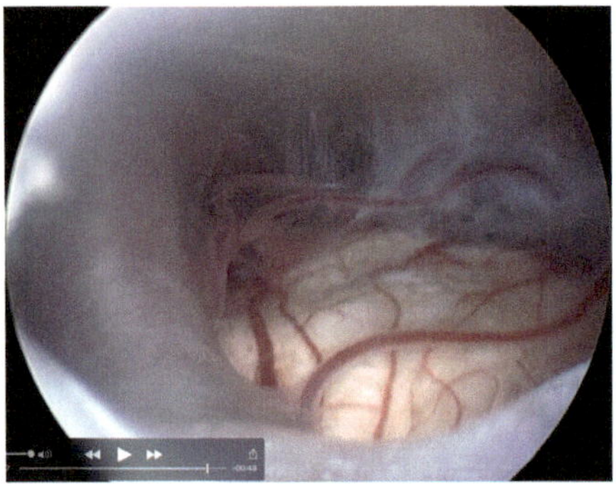

Fig. 5.117 Deep arachnoid membrane para-basilar right

On the ipsilateral right side, the difficult side for viewing, we see along the curved basilar trunk a strong developed arachnoid wall, causing some resistance for the CFS flow in the depth. It may primarily have been the lateral wall of the prepontine cistern (Fig. 5.117).

However, such membranes, if the natural openings are closed, can cause an intracisternal obstruction, and may cause an ETV failure in the depth. Perforation here need experience and meticulous analysis of imaging!

This view into the left pre-pontine cistern shows well visible arachnoid fibers lateral to the basilar trunk. It may be parts of an incomplete lateral wall of the prepontine cistern. Left cerebellar artery is shown and part of posterior cerebral artery. In the depth the left abducens nerve can be recognized. The caudal arachnoid wall of the pre-pontine cistern is present also left. Pons is slightly atrophic causing a good space for viewing (Fig. 5.118).

This view into pre-pontine cistern shows an occasional finding when the AICA runs through the abducens nerve. Such variations do not cause symptoms. The arachnoid fibers and membranes are visible, appearing in endless variations (Fig. 5.119).

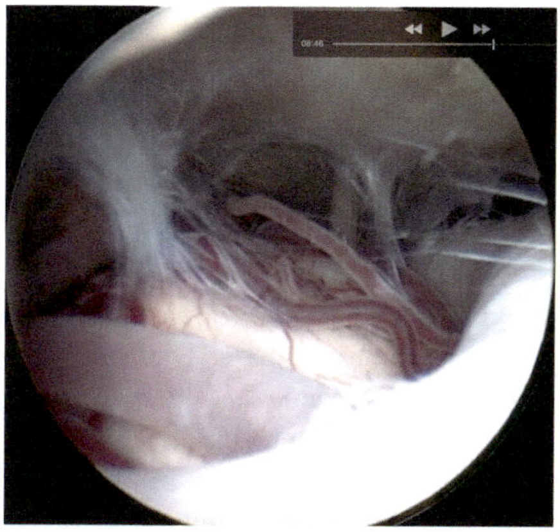

Fig. 5.118 Strong subarachnoidal fibers parabasilar left

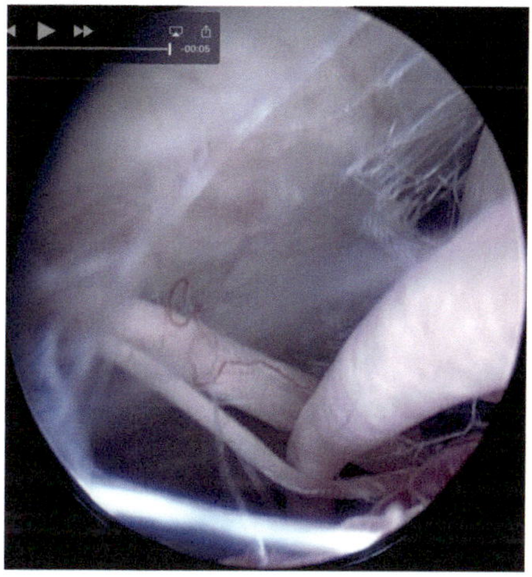

Fig. 5.119 Trangression of AICA through CN VI left

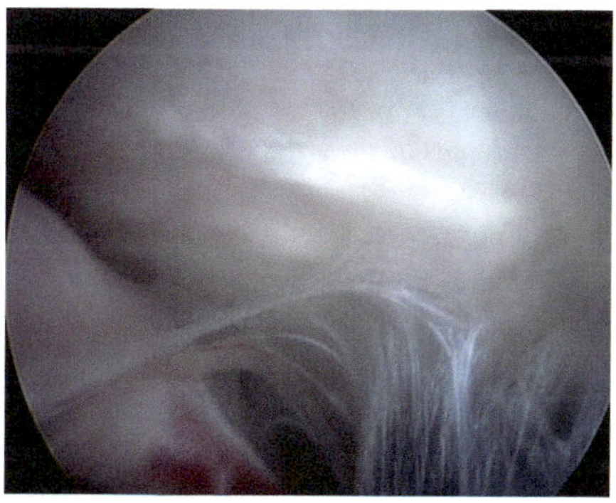

Fig. 5.120 Fixed artery at basal dura prepontine

This view shows a nice detail of trabecular system and an arachnoid fixation of basilar artery at the clival dura (Fig. 5.120).

Such kind of vessel fixation are rather frequent and are usually very tuff. Detaching or cutting such fixation is challenging and might result in a fatal bleeding. Therefore, it should only be done in very rare situations when the whole fixation is broad, causing an obstruction and failing an ETV.

The view into the right side of pre-pontine cistern, which is the difficult side for viewing in endoscopy from a right sided burr-hole. We see the strong basilar trunk and a group of pontine arteries. In this case there is a xanthochromia between the arteries recognizable and in the depth a well preserved arachnoid membrane, belonging to the latero-basal wall of the cistern (Fig. 5.121).

In this case, looking like from an patient over 80 years, with sever atrophy, causing a big pre-pontine cistern with the basilar artery far from the clivus, we see a result in a 31 years woman. She had to wait for regulation of the CSF flow disturbance 3 decades (Fig. 5.122).

Fig. 5.121 Strong basilar trunk and pontine arteries parabasilar right (see pontine xanthochomia)

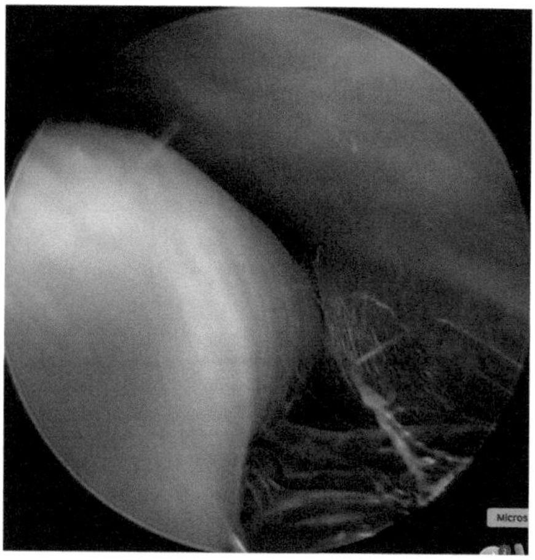

Fig. 5.122 Large prepontine cistern with strong atrophia of parenchyma ((80 Y)

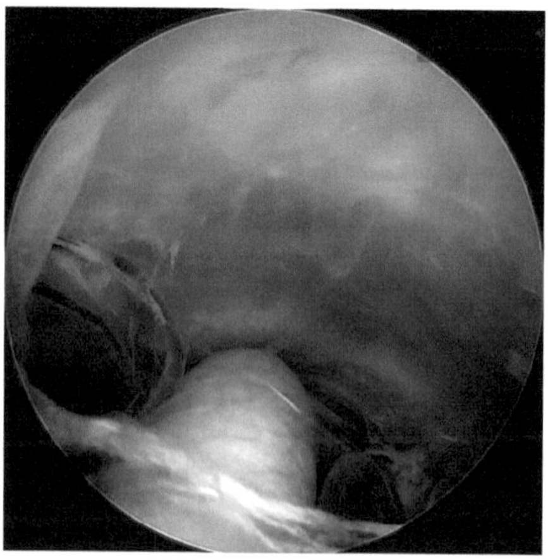

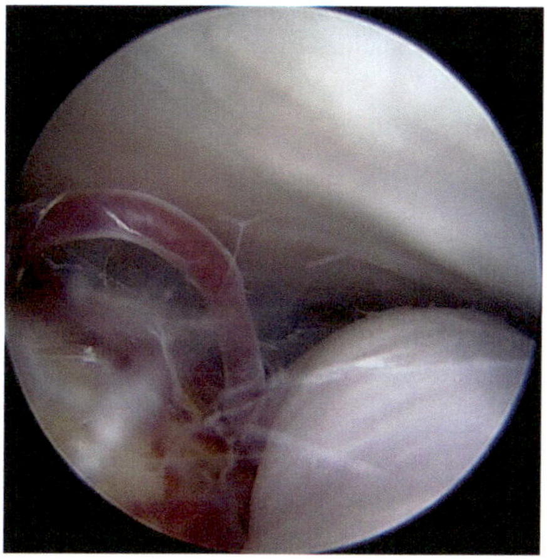

Fig. 5.123 Parapontine atrophia parabasilar left (xanthochromia!)

We see remnants of Lilliquist membrane and the pontine arteries floating in the cistern.

The close-up shows better the irregular course of a pontine artery and the trabecular system holding it in place (Fig. 5.123).

The pontine surface has a xanthochromia, indicating a former bleeding.

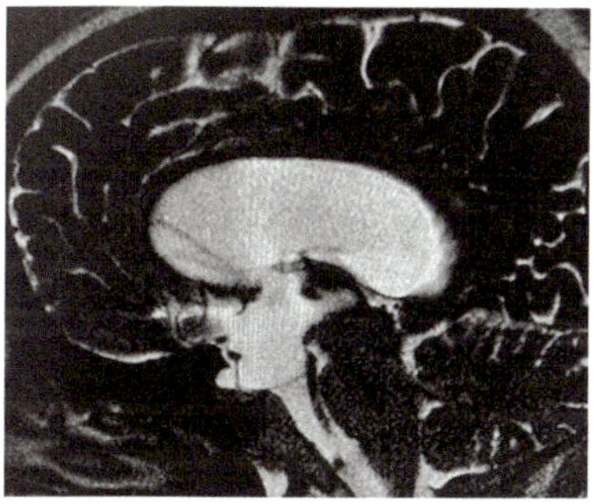

This primarily pre-mature case was therapied for obstructive hydrocephalus by shunting.

We see through the perforation of an emergency- ETV the basilar head and a small bleeding from the Fogarty catheter, indicating the vulnerability of the premature tissue. The vessels are not normal, and the basilar artery is paralytic after a herniation phase during emergency time (Fig. 5.124).

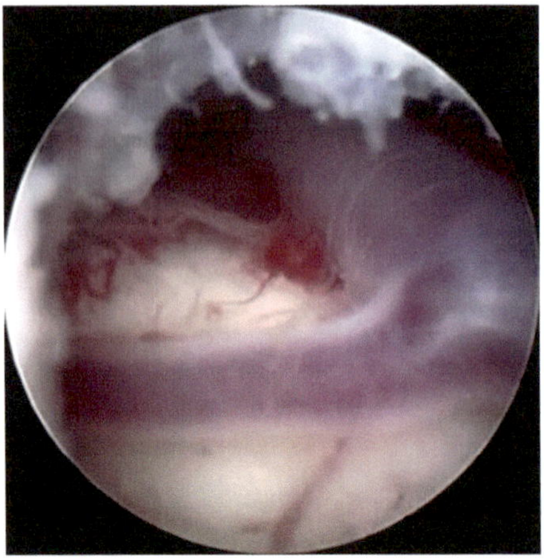

Fig. 5.124 Small bleeding parabasilar pontine during ETV

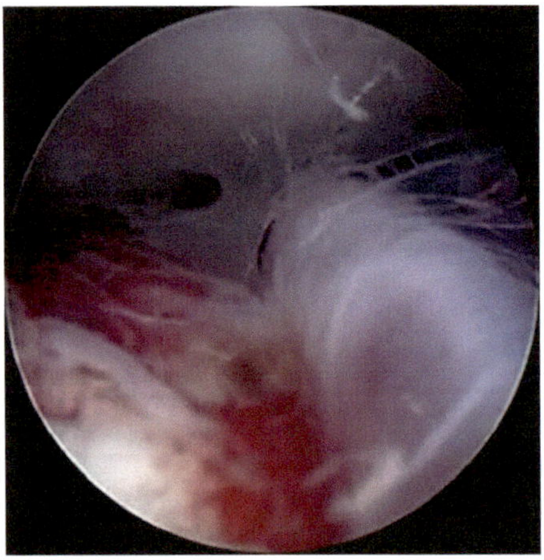

Fig. 5.125 Deep second arachnoidal membrane (post-inflamation)

The pontine arteries show a vasospasm and the pontine tissue is edematous.

The caudal arachnoid wall of the pre-pontine cistern is opaque and tuff (Fig. 5.125).

This kind of ETV was an individual decision and it is absolutely difficult, needing a lot of experience! Such situations should be avoided, and ultrasound aid is needed.

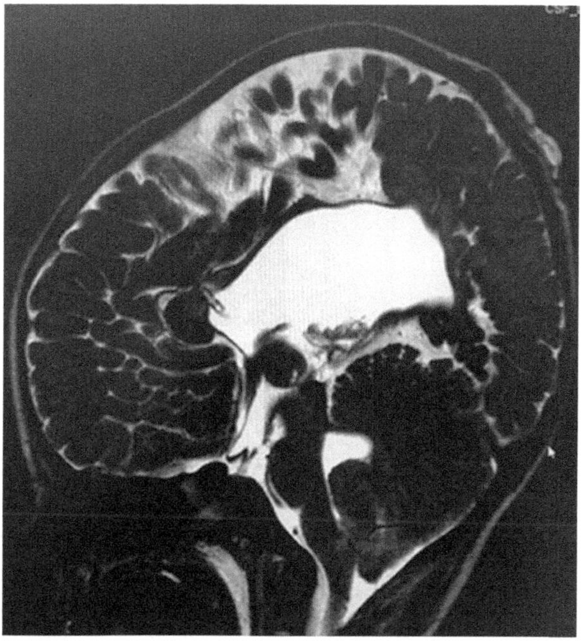

5.5.2 Caudal Basal Cisterns, Foramen Magnum

Endoscopy of ventral foramen magnum and the cisterns there are not well documented and described. Clinically, this is a "no man's land" and seldom approached. It needs a high-level experience and a large mental library with anatomy and video-clips. The equipment to approach there must be perfectly known and mastered. A talent for mental navigation and feeling for the special tissue characteristics must be trained over years in the laboratory. This is recommended and an ethical imperative.

The view in a specimen endoscopy reaches the foramen magnum ventral the medulla oblongata, where fibers of the C_1 and C_2 level are present on the right side. The impression of the dens axis can be seen and the right vertebral artery (Fig. 5.126).

In the specimen endoscopy C_1 fibres left and hypoglossal nerve are visible, but also fibers of denticulate ligament. The angle of view allows to see into cervical canal. The space for the medulla within foramen magnum is well recognized. There is pre-medullary cistern (Fig. 5.127).

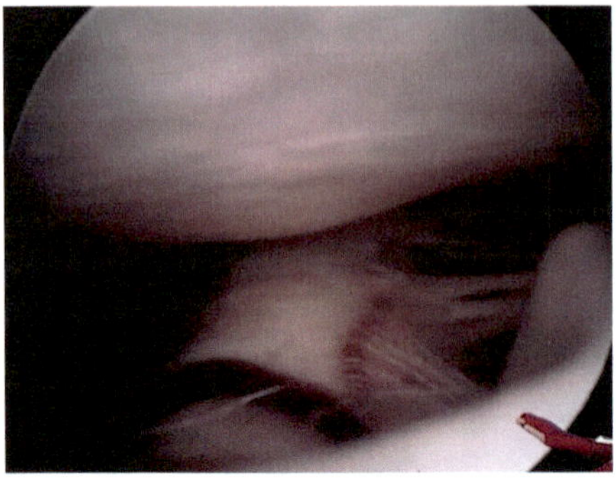

Fig. 5.126 Anatomica view premedullary to foramen magnum

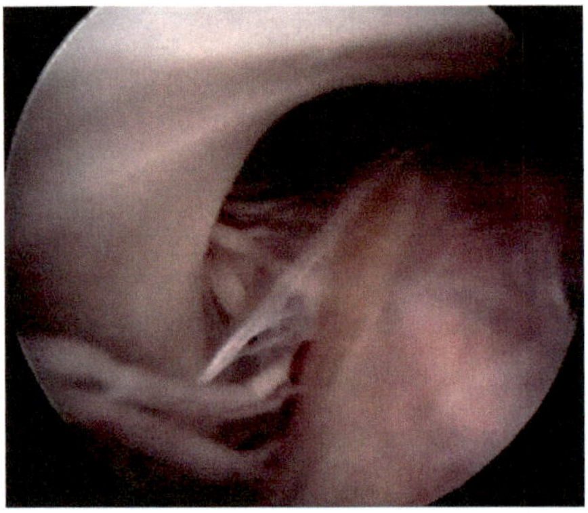

Fig. 5.127 Left Side view to foramen magnum premedullary

This view into the left CPA, the ponto-medullar cistern, gives sight internal acoustic meatus with CN 7/8 and AICA loop with the tiny labyrinth artery. One level deeper the jugular foramen with CN 9/10/11 can be seen. CN 11 runs over the jugular tubercle, and there in the dark area the lateral left side of foramen magnum can be found. Such a lateral view needs a 30° endoscope, rotated laterally 90° (Fig. 5.128).

This is a high-level endoscopy.

In this case we have a view into an enlarged pre-pontine and pre-medullary common cistern, as separating arachnoid walls do not exist. We see the foramen magnum level along the clivus and the CN 12 left side. The typical impression of dens axis is well developed here, marking best the C_1 level. Normally the deepest insight reaches until C_2 fibers, then the cervical medulla curves out of sight (Fig. 5.129).

An extra-ordinary view is given in this case of a non-treated chronic hydrocephalus for 5 decades. The atrophy of the brain stem tissue enables an unbelievable perspective, from the entrance of CN 6 left into the foramen magnum. The basilar trunk and a strong vertebral artery left with a deep originating PICA can be seen. Several pontine arteries left are present and in the foramen magnum the medulla is hardly recognizable (Fig. 5.130).

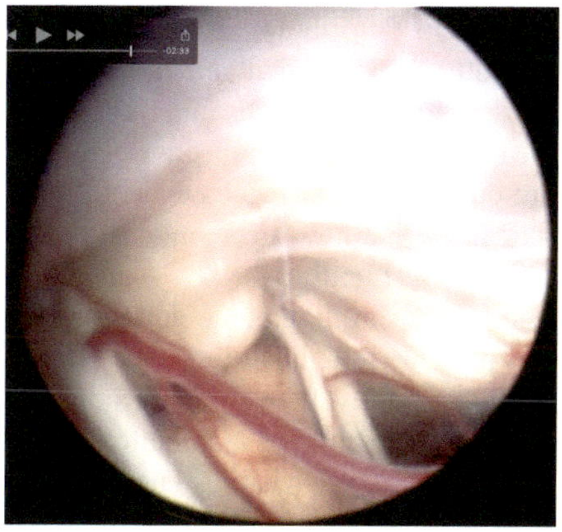

Fig. 5.128 Left CPA and jugular foramen ventrally

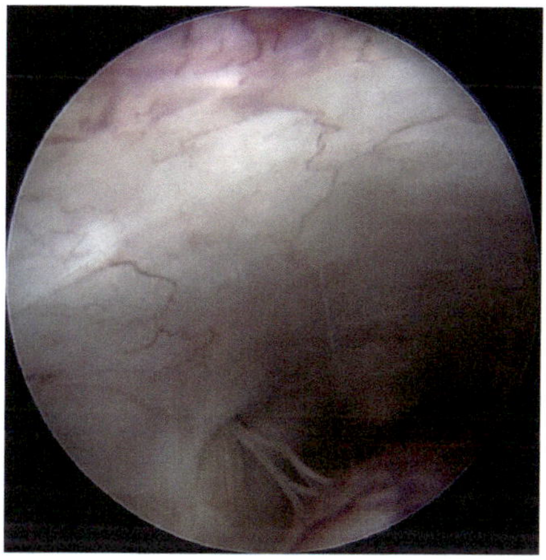

Fig. 5.129 Big prepontine cistern with view to foramen magnum

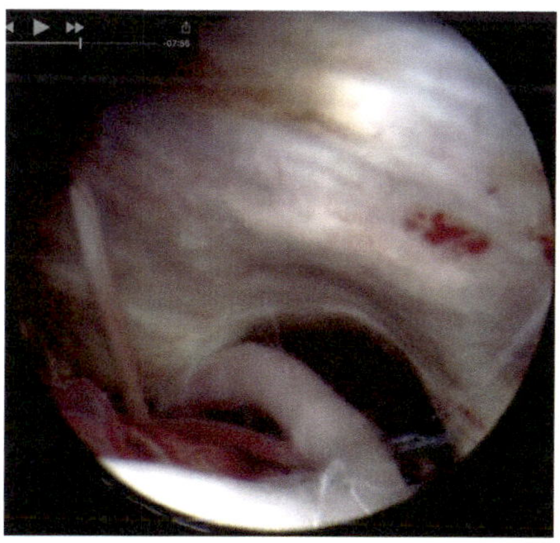

Fig. 5.130 Left vertebral artery and CN VI with view into F. Magnum

The close-up shows an unusual sight on the medulla with all the vascularization ventrally. A right sided vertebral artery is crossing freely through the enlarged cistern. The arteria spinalis anterior system is present with several branches entering the medulla. Such a perspective is rarely possible and can be planned in the imaging pre-op. But the existence of membranes cannot safely predicted, even in the CISS sequences (T2 flow). The only safe prove is the endoscopy all along the clivus and brain-stem (Fig. 5.131).

After perforating some adhesions and arachnoid membranes, finally the sight was possible into this foramen magnum region. The typical impression of dens axis marks the C_1 level, and on both sides the C_1 and C_2 fibers are present. A strong vertebral artery right runs at the inferior border of the view. The medulla and its vascularization are well visible. The arachnoid cover of the medulla can be recognized (Fig. 5.132).

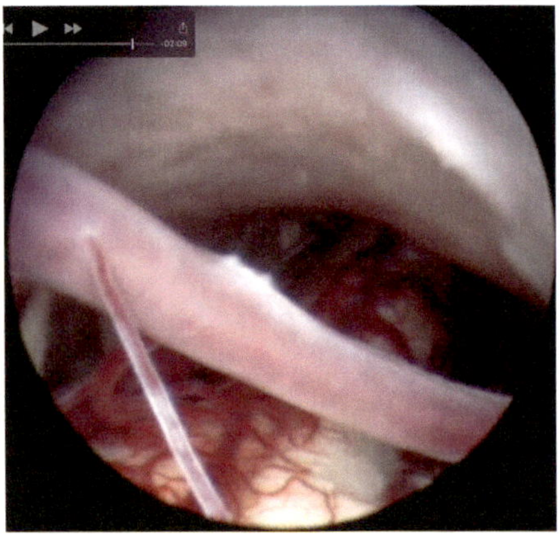

Fig. 5.131 Right vertebral artery and medulla oblongata/spinalis (C1+C2); F. Magnum

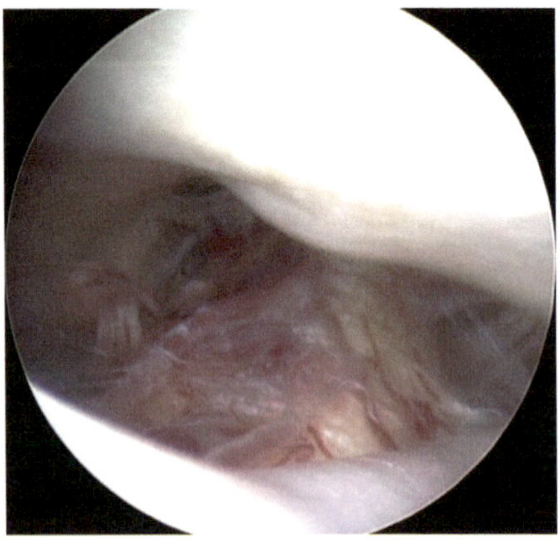

Fig. 5.132 Medulla oblongate/spinalis C1

The close-up of the left side in the foramen magnum level shows the typical C_1 root having mainly sensible dorsal fibers. The beginning of cranial C_2 fibers can be hardly recognized and a strong fiber of the denticulate ligament. The arachnoid membrane with stronger fibers at the origin of the cervical roots is impressive. The medulla spinalis looks a bit opaque and a rarefication of the vascularization is striking (Fig. 5.133).

In this case a symmetrical view was possible showing a weak impression of the dens axis. The space ventrally is slim and the view dark, but anyway the C_1 roots are well visible, but also on the right side a denticulate ligament in the dark depth. Arachnoid fibres are well developed, and vascularisation of the medulla is normal (Fig. 5.134).

At the entrance of left vertebral artery at C_1 level, the C_1 root and the fibers of the first denticulate ligament are meeting. In the depth we see also C_2 and even one fiber of $C_{3.}$ The impression of dens axis is not present in dis case, but the angle of view enables to see three cervical levels of the cervical medulla (Fig. 5.135).

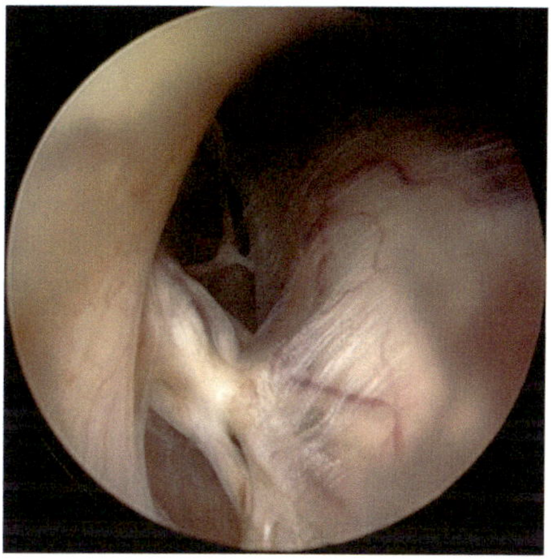

Fig. 5.133 Left side F. Magnum with C1 fibers

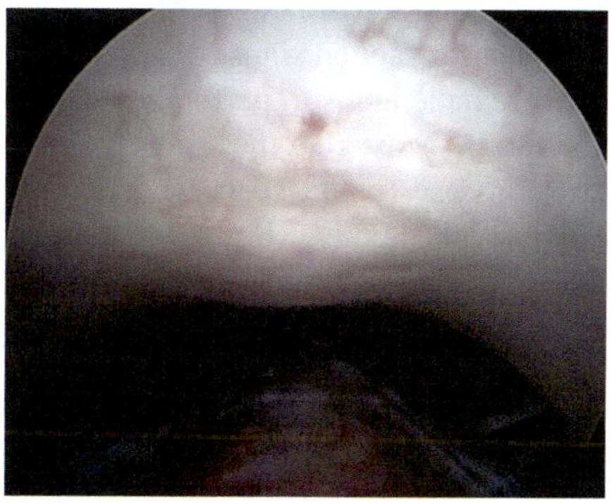

Fig. 5.134 Foramen magnum with slide impression of dens axis

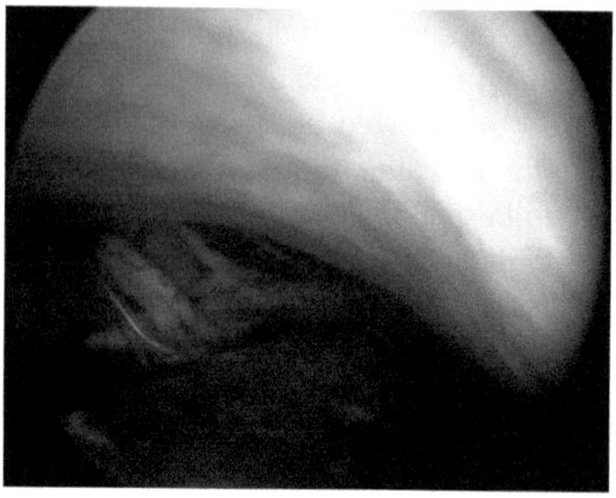

Fig. 5.135 Left F. Magnum with C1+C2 fibers and denticulate ligament

In summary, the visualisation of the ventral foramen magnum region, traveling into the depth, between brain-stem and clivus, is rarely necessary. The anatomical landmarks and landscape to know is a prerequisite, especially in such difficult conditions with the rare chance to get clinical experience or even routine.

But if the arachnoid presentation is striking after bleedings or infections, it is the only safe way to rule out an obstruction for CSF below the ETV level, pretending that ETV is not working.

However, this is a **high-level endoscopy** procedure and indication must be precise as well as planning by sufficient imaging. In such problems a high quality CISS sequence MRT (T_2 flow) is mandatory.

However, such cases are often desperate cases, with multiple shunt failures, and it should be an ethical demand, if necessary, to send the patient to an experienced neuro-endoscopist instead preserving always the same problem unsolved until the next complication!

5.6 Approach-Canal

This is an approach canal, using the "**kings' route**", which is the midline interhemispheric path, perforating only corpus callosum. But this is high-level endoscopy, because this approach passes many strong vessels epicortical, and frontal interhemispheric adhesion are quite common. The burr-hole must be placed directly beside the superior sagittal sinus, which must be planned in MRA and CTA. However, the mayor challenge is to perforate the corpus callosum as lateral as possible to spare the fornix. Mastering these tasks, one is able to enter even a very small third ventricle in cases of slit-ventricle syndrome. It is well known, that such cases are not solved with shunting (Fig. 5.136).

The corpus callosum canal is short and only white without a vascularized cortical level (Fig. 5.137).

A trans-callosal perforation via an interhemispheric approach in a case with an unusual strong cortical artery arising from the right pericallosal artery. The

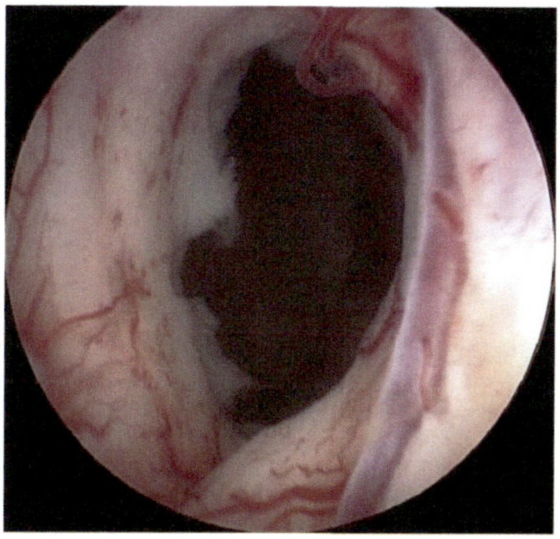

Fig. 5.136 Approach canal transcallosal (kings route)

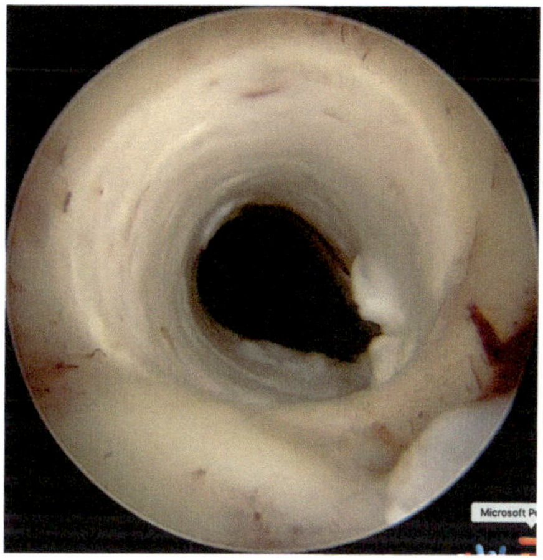

Fig. 5.137 Short, tapical transcallosal canal (no cortical level!)

perforation site is laterally from the pericallosal artery. The canal is only 5 mm long and presents a typical ependymal preserved layer replaced after the endoscope is retracted. It is intuitively recognizable that an extreme precision is needed, defining the correct entry-point (Fig. 5.138).

This endoscopic controlled visual precision is far beyond that of so-called neuronavigation sytems.

The approach canal shows three levels: in the depth the ependyma, followed by the long white matter part, and finally the cortical level. Before leaving the canal, it should be observed if there are bleedings. Mostly some irrigation for some minutes will stop bleedings, otherwise coagulation is needed (Fig. 5.139).

Usually, after retraction of the endoscope, the approach canal will collapse and nearly close.

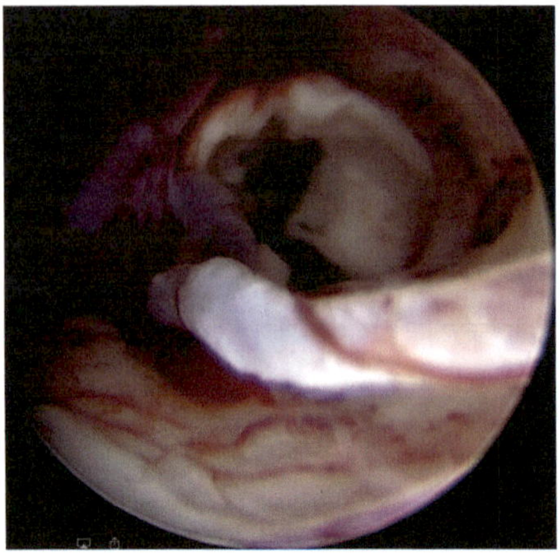

Fig. 5.138 Complaex difficult transcallosal canal with a strong pericallosal artery-system

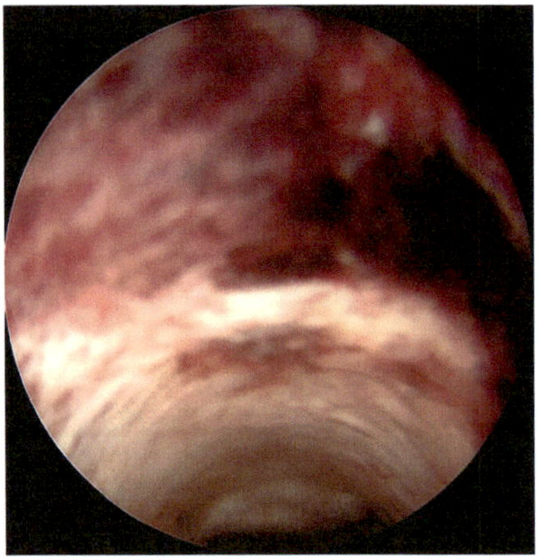

Fig. 5.139 Typical transcortical canal

5.7 Subarachnoidal Space, Basal and Hemispheral

Two main-indications in the epi-hemispheral space are convenient for endoscopy: the subarachnoid cysts and the chronic subdural hematomas. Here we see ab opening in the parietal arachnoid layer of an arachnoid cyst fronto-temporal right, viewing into the insular region. The arachnoid membranes are mostly unusual tuff and in-transparent, needing a sharp dissection and ultrasound-navigation. Mostly this type has a connection to the supra-sellar space (Fig. 5.140).

In such cases the stalk can be visualized endoscopically, leading to the final fenestration into the interpeduncular cyst. Lack of training in two key-techniques in MIN, endoscopy and ultrasound, is the main reason to operate this lesion microsurgically (Fig. 5.141).

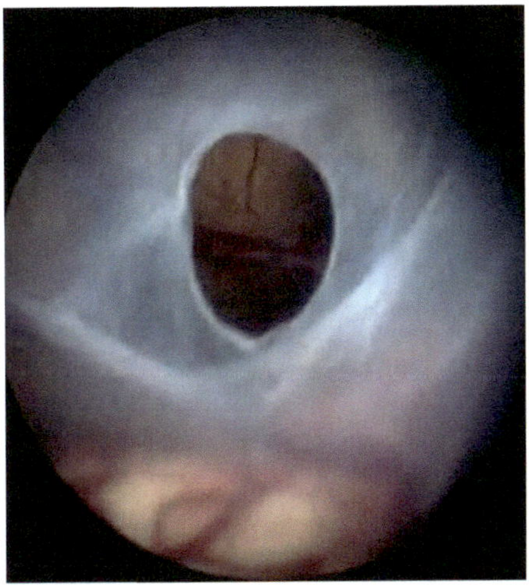

Fig. 5.140 Arachnoidal approach of subarachnoidal cyst

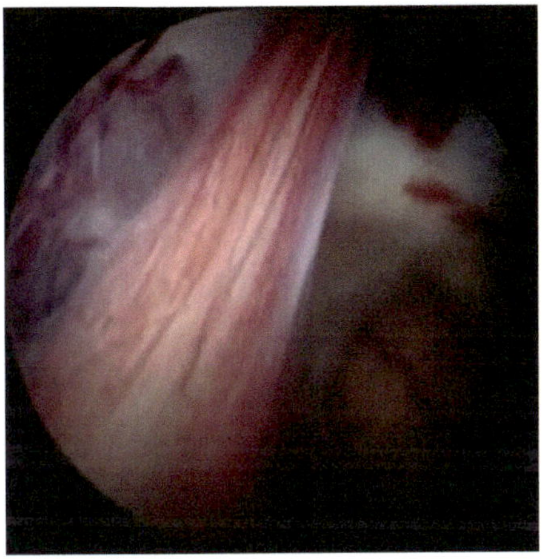

Fig. 5.141 Suprasellar subarachnoidal cyst with pituitary stalk

We see into an arachnoid cyst fronto-temporal on the right ICA bifurcation. A strong MCA is running under the frontal operculum into Sylvian cistern. The ACA runs free through the cyst, and few tiny perforators in a spiral fashion are visible. Orientation may be difficult regarding the stoma-point (Fig. 5.142).

A intra-operative sono-navigation might be necessary to image through opaque membranes.

After opening the parietal arachnoid membrane, we look into the insular region. A air collection in the superior third of the image shows the "mirrow artifact", pretending the oposit operculum. Indeed this "mirrow" shows a virtual operculum, where in real is air bubble in the cyst. Ultrasound would uncover this by presenting the air artifact (Fig. 5.143).

The right angle for reaching perforating-point needs to be planned.

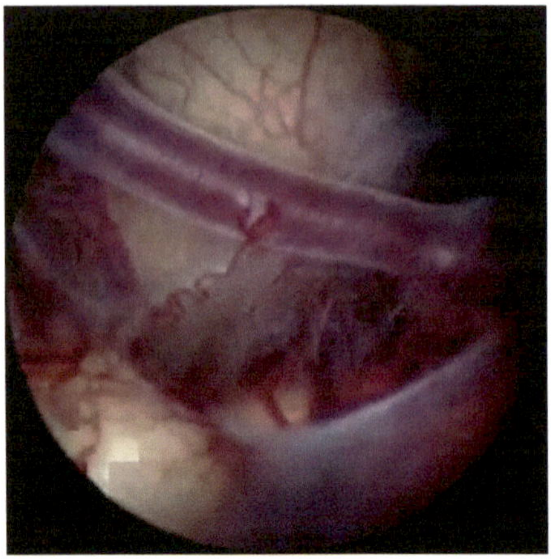

Fig. 5.142 Right frontomedial arachnoidal cyst with free ICA bifurcation

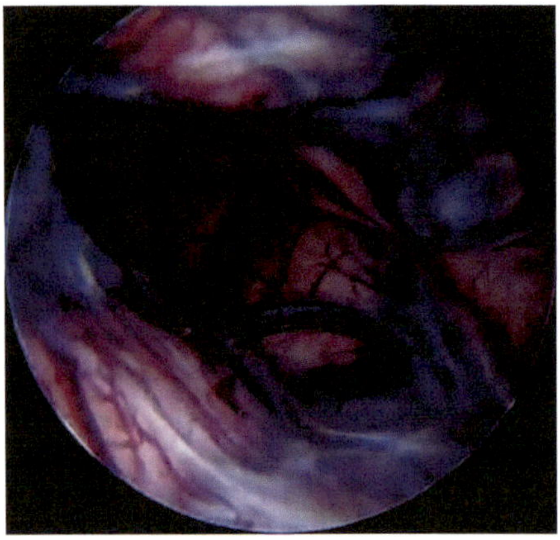

Fig. 5.143 Entrance into sylvian arachnoidal cyst

The close-up gives view on the insula and in the superior part of the image shows a "mirrow-artifact". The insular veins lead to the sylvian vein more in the depth than normally. Here an anatomical navigation to the target by following biological morphological coordinates, instead of mathematical once, as neuro-navigation systems do, is applied (Fig. 5.144).

Finally, the MCA, covered by normal arachnoid membrane, is reached (Fig. 5.145).

Form of surface and course of vessels can have markedly deviations, causing difficulties in orientation. This can be compensated by ultrasound in correlation with pre-op imaging like MRA.

Liquor-release may cause shifting, making neuro-navigation system helpless.

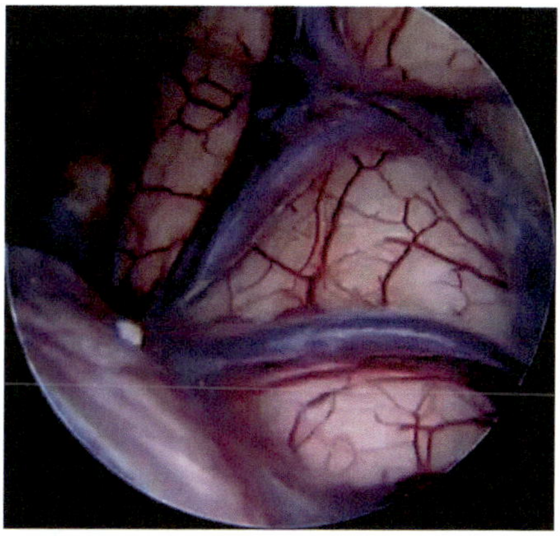

Fig. 5.144 Insular region in an sylvian arachnoidal cyst

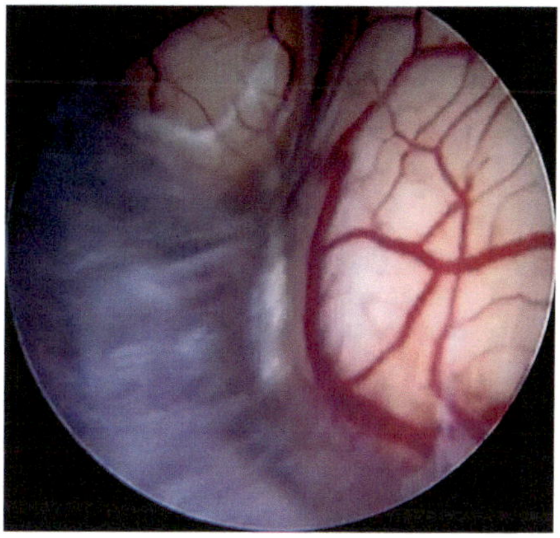

Fig. 5.145 Proximal basal right m1 within sylvian cyst

5.8 Supra-Orbital Key-Hole

The supra-orbital approach is 2x3 cm in size according to the anatomical conditions, enlargements cause risk of damage. So far this is an example when size matters. In the endoscopy design the burr-hole is lateral under the temporalis muscle. It is one of the most used key-hole for some of the most target regions like suprasellar, fronto-temporal basal and circle of Willis (Graph 5.21d).

From the burr-hole in the exoscopic use of the endoscope, supra-orbital right the fronto-orbital region and the orbital roof are visible. In the depth right optic nerve and ICA are present. Regard the faciform ligament covering entering zone of optic nerve and location of injury of the nerve (Fig. 5.146).

Graph 5.21D Supra-orbilal approach right

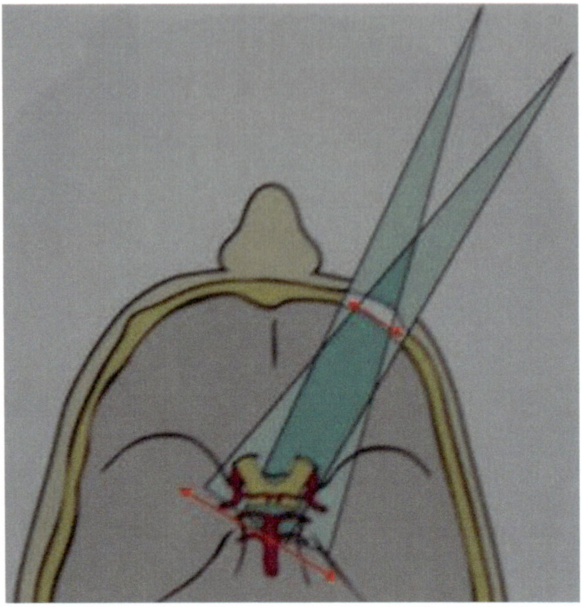

Fig. 5.146 Anatomical supra-orbital route to the supra-sellar region right

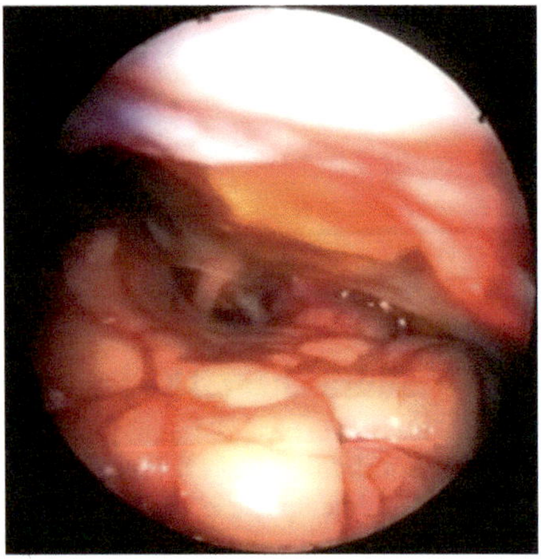

The optic range is large in contrast to the microscope and zooming is done by advance the endoscope, causing the risk of being intra-cranial or intra-cerebral. Here we see it in a non-fixed specimen during MIN-simulation lab training.

The close-up by traveling into the intracranial space, after liquor release, we see remnants of Liliequist membrane and the basilar Head with part of the oculomotor nerve. All main vessels are present, here with an unusual strong P. com. (embryonal type) and the pons with its vascularization. The P_1 right is in tough with the right mamillary body. The roof of the interpeduncular cistern (Liliequist m.) is fixed at the dorsum of sella and runs until the mamillary body line of the hypothalamus (Fig. 5.147).

Advancing the endoscop into the cistern and under the L. membrane, the basilar head is symmetrically seen and the opposite oculomotor nerve in a long course.

The illumination at that time was not as perfect as todays automatic light control do. There is an overgained area resulting. The endoscope is rather close to the brain and must be controlled perfectly, especially in the free-hand use (Fig. 5.148).

During clinical application, the visualization is very convincing, reaching each corner of the region. The right optic nerve looks swollen due to a sub-chiasmatic cyst of a craniopharyngioma remnant. ICA as well as the fronto-orbital basis covert by arachnoid membrane is visualized perfectly. The closer the endoscope advances, the more brilliant, bright and zoomed the image will be, causing the risk of injury for the tissue (Fig. 5.149).

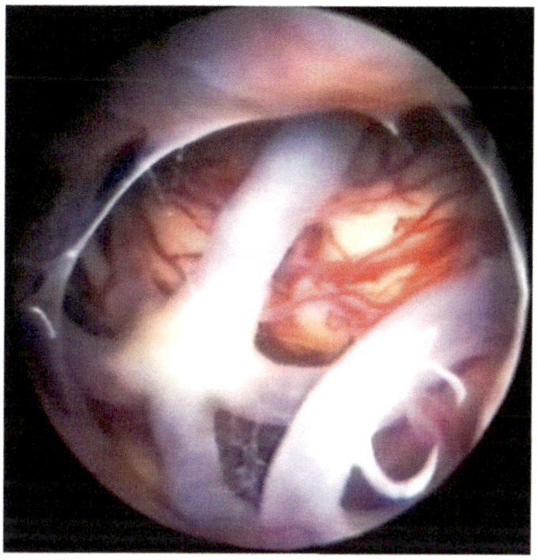

Fig. 5.147 Approaching interpeduncular cistern with remnant of liliequist membrane and basilar head

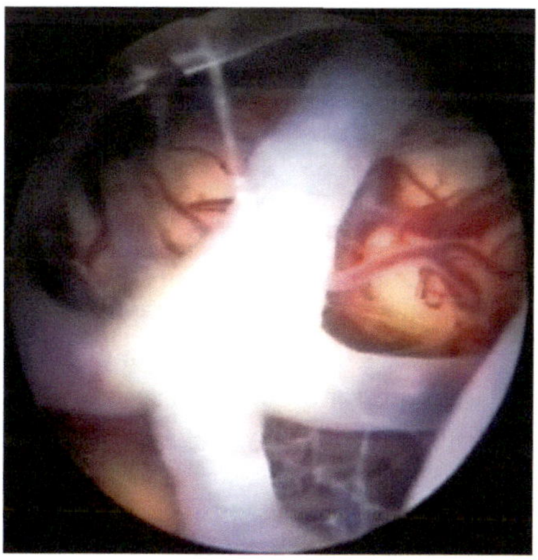

Fig. 5.148 Interpeduncular cistern and basilar heda with opposit (left) CN III

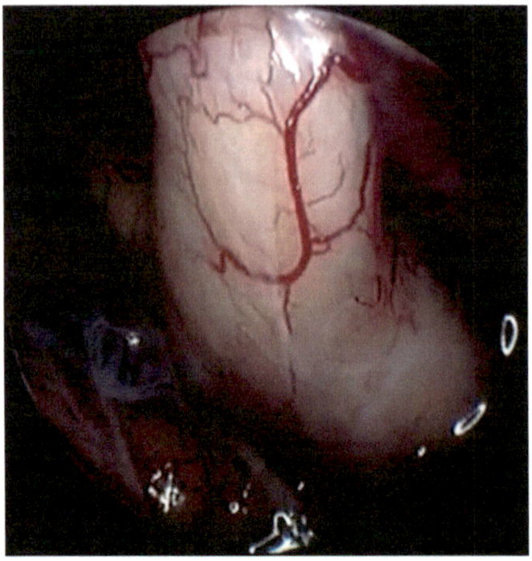

Fig. 5.149 Optic chiasm from supra-orbital

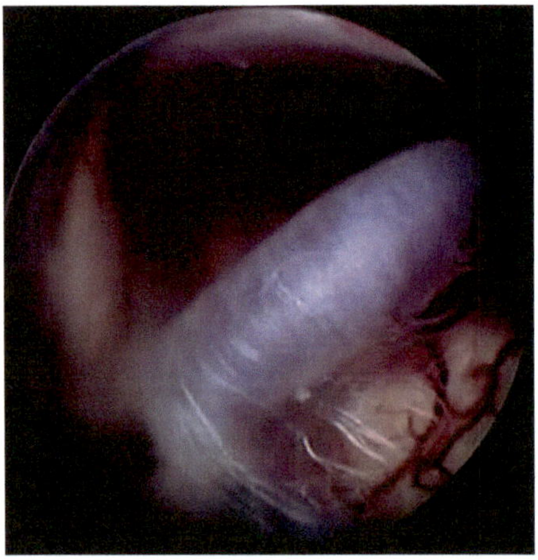

Fig. 5.150 Interpeduncular cistern with basilar artery and strong opposit CN III

Entering clinically the interpeduncular cistern, after perforation the Liliequist membrane, we see a strong basilar artery, and the basilar head within liquor. Sucking it away for better seeing, is a difficult maneuver para-endoscopically, especially in free-hand use (Fig. 5.150).

The oculomotor nerve can be see as well as part of the pons with vascularization. For entering the sucker parallel to the endoscope, there is very limited space of few mm. There is the sever problem of endoscopy instrument space competition, causing iritations for the surgeon and danger of injury to the involved structures. Its high level neuro-endoscopy!

5.9 CPA Exoscopic Surgical Anatomy

The cerebello-pontine-angle CPA is, after the target-cistern of ETV (interpeduncular cistern) the next frequent endoscopically approached region. Mostly this is used in an endoscopy-assisted microsurgical fashion to visualize the internal acoustic meatus (Graph 5.21e).

Graph 5.21e Endoscopic pathways of lateral posterior fossa

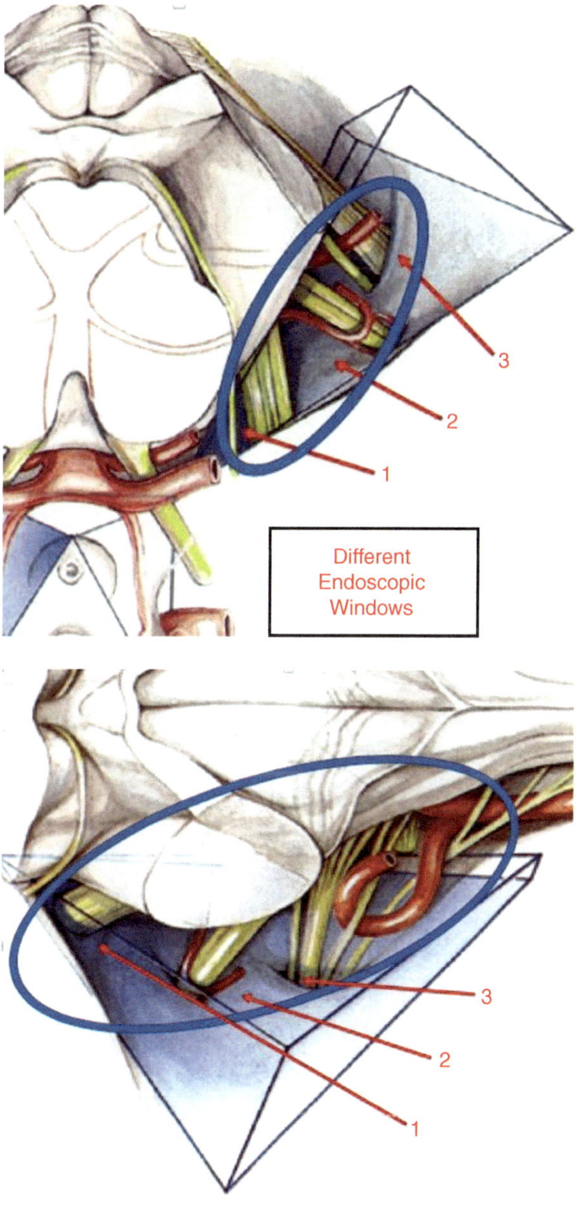

This is of great benefit, approaching each and deep corner of the involved space and different structures. The visualization angle can be varied in many ways, focused and zoomed by advancing the endoscope. The perspectives gained that way are impossible by microscopy alone. Endoscopically there are multiple windows and pathways to be used and target all structures of interest (Figs. 5.151, 5.152, and 5.153).

The CPA can be made comprehensive by differentiate it into endoscopic windows and pathways.

The windows at the surface and the pathways into the depth will provide the target route to all structures in very different perspectives, that can not be seen by the microscope. With angled lenses one can see below structures without moving

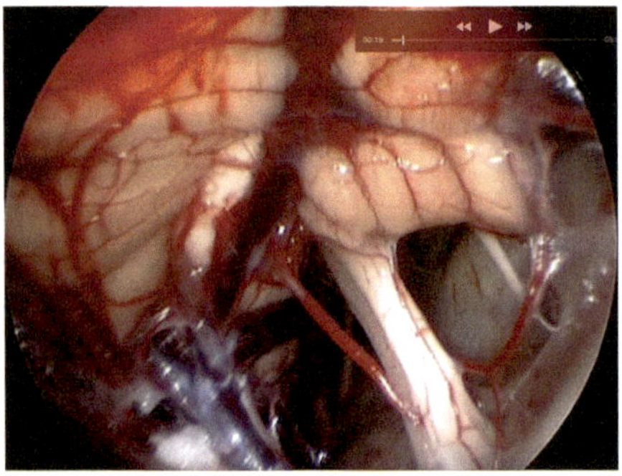

Fig. 5.151 CPA left : CN 7/8, dandy vein

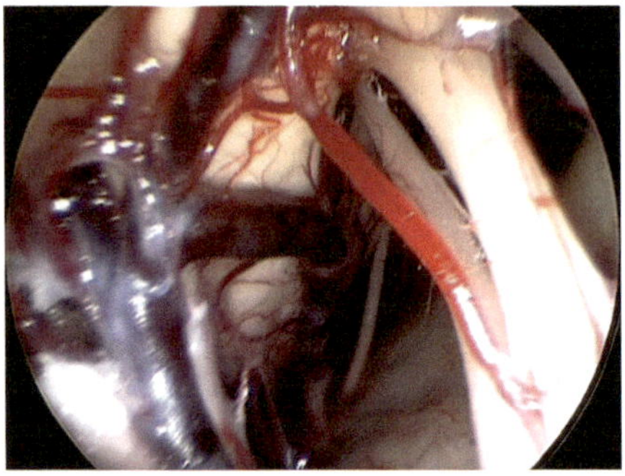

Fig. 5.152 CPA left : dandy vein CN V; CN 7/8; CN VI

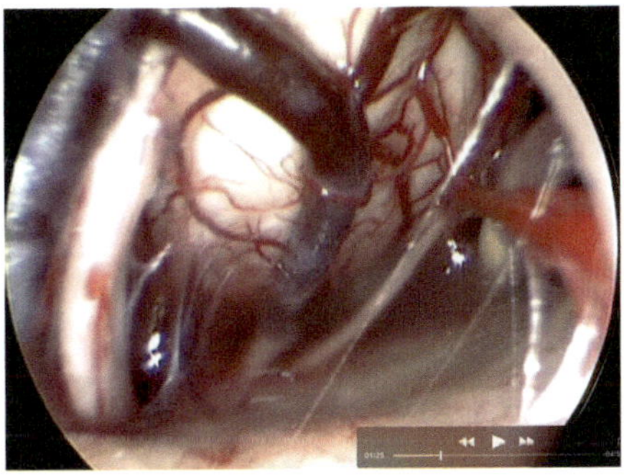

Fig. 5.153 CPA left dandy vein + CN V; lateral pontine vein; CN VI

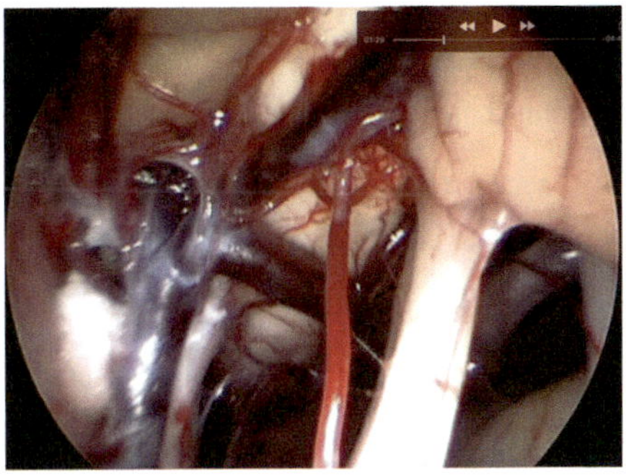

Fig. 5.154 CPA left: dandy vein + CN V; CN 7/8; basilar artery (in the depth)

them. Viewing a target structure through the different windows enable to see them from different sides.

In endoscopy assisted use one can provide difficult dissections with the best visual control. However, superficial structures are not visible anymore, when the endoscope has passed the window. This causes a typical danger in endoscopy assistance use. This is high-level endoscopy! (Figs. 5.154, 5.155, and 5.156).

5.10 Endoscopic Anatomy of Latero-Basal Key-Holes to the Posterior Fossa

5.10.1 Introduction

Endoscopy is becoming accepted in assisting micro-neurosurgery. The application of endoscopy for minimizing the dimension of an approach is not usual. According to the para-endoscopic preparation concept very small approaches will be usable when the microscope is substituted by the endoscope. This evolution makes a more

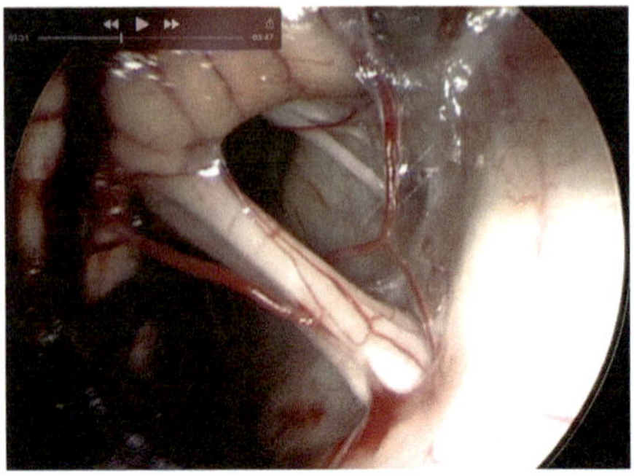

Fig. 5.155 CPA left: CN 7/8; jugular foramen witn CN IX

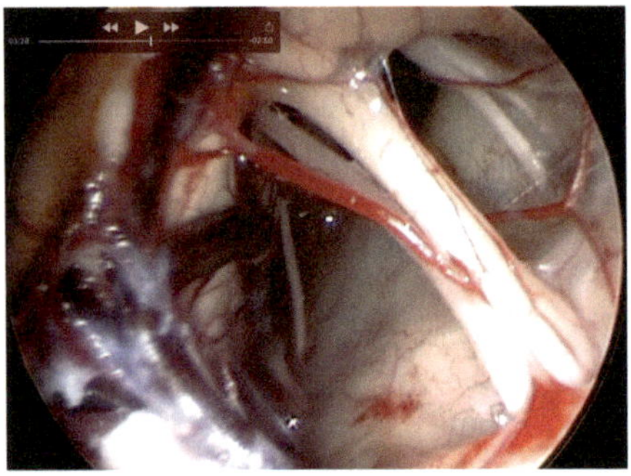

Fig. 5.156 Overview of 3 levels of posterior fossa endoscopy-routes: trigeminal; meatal; jugular

precise description of endoscopic approaches necessary. In this part four latero-basal key-hole approaches of the posterior fossa are differentiated and described in addition to and supplementing the first atlas-book from 1993 (Perneczky/ Tschabitscher/Resch: Endoscopic Anatomy for Neurosurgery, Thieme) (Graph 5.22).

There are two concepts to avoid sever trauma to the complex muscle apparatus suboccipital: the midline-approach causing large openings with a long incisions, or to place the approach peripheral of inferior nucheal line, making small approaches necessary. In the latter concept, the MIN concept of posterior fossa, a high precision of approach planning and approach design is mandatory. The differences in postop morbidity and clinical results of this MIN approaching to posterior fossa lesions is convincing. The content and size of posterior fossa, obscured by a mass of complex muscle system which is very sensitive for functional disturbance due to its task to balance the head on the most mobile part of the spine, causes a mismatch between size of standard-approaches and size of target region. For all these reasons the need of MIN concept regarding posterior fossa lesions is prominent.

In 12 specimen endoscopy-assisted (para-endoscopic) preparations were done. Laboratory setting was designed for surgical simulation conditions. Exclusively key-hole approaches were examined. Rigid endoscopes with 3mm and 5 mm diameter and 5° or 30° (R. Wolf, Knittlingen /Germany) were used. All instruments of

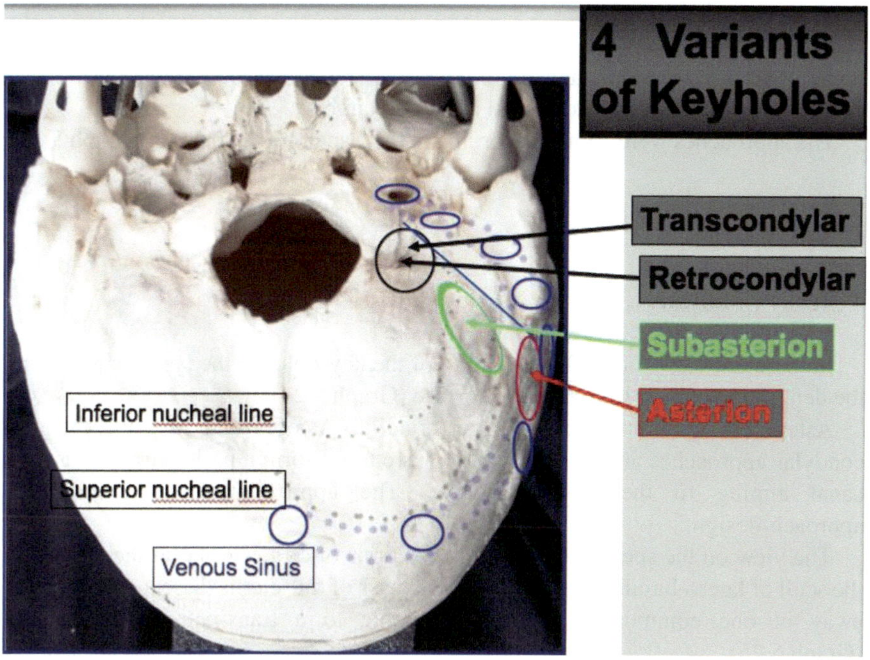

Graph 5.22 4 MIN-Approach Model Endoscopy Lateral Posterior Fossa

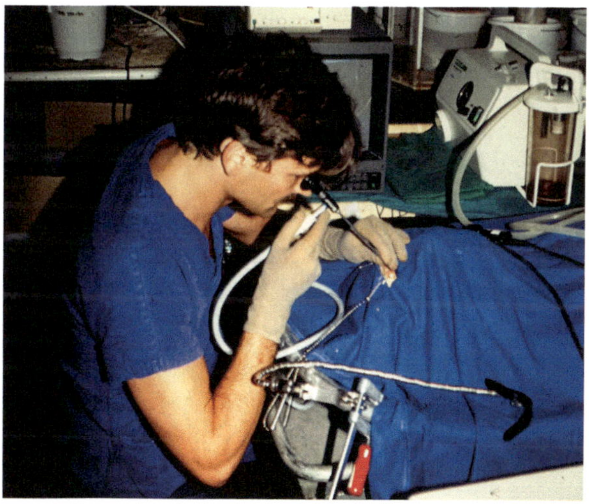

Fig. 5.157 Laboratory work

microsurgery were ad hand and clipping action has been done to show the spatial conditions in the depth. In each type of approach trans-endoscopic ultrasound equipment (ALOKA, Meerbusch/Germany) was applied, technique and imaging were described elsewhere. Results were documented by digital tape recording (Fig. 5.157).

5.10.2 Results

It was obvious by anatomy and ergonomics to differentiate four typical locations of key-holes enabling to expose special areas and targets.

The approaches were named according to their anatomical location (Graph 5.23).

- **Asterion**
- **Sub-Asterion**
- **Retro-Condylar**
- **Trans-Condylar**

Intracranial view shows the proximity of the key-holes to the parent sinuses and the depth of approach from the skin surface (Graph 5.24).

Asterion is superficial and easy to perform, Sub-Asterion and especially the peri-condylar approaches are dificult and one have to perform a perfect trans-muscular canal aiming to the target key-hole. The approach-analysis defines the approach-design.

The view on the specimen shows the positions of all key-holes at the surface of the scull of latero-basal posterior fossa. The mass of muscles is dissected and spread away in one common approach comparable to a trans-petrosal preparation (Graph 5.25).

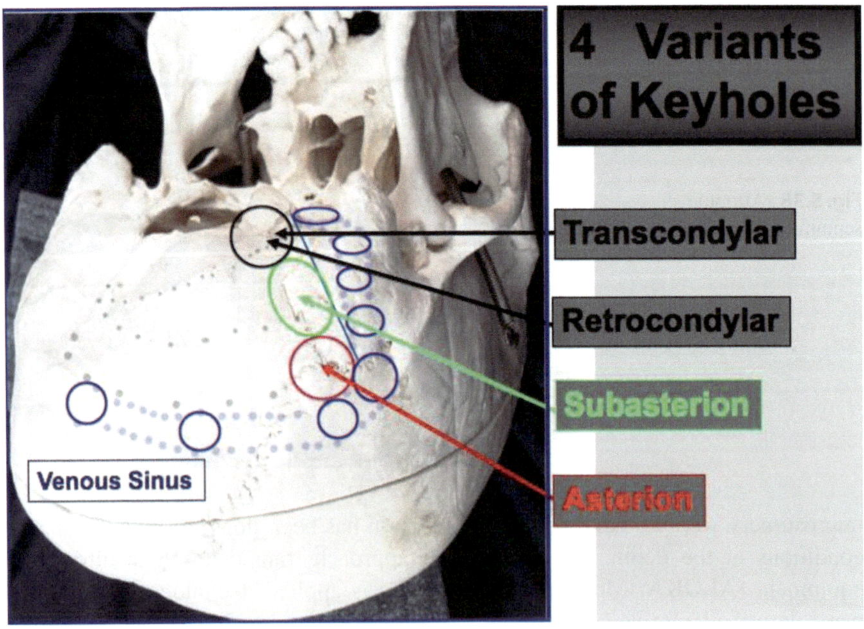

Graph 5.23 Topology of key-holes I

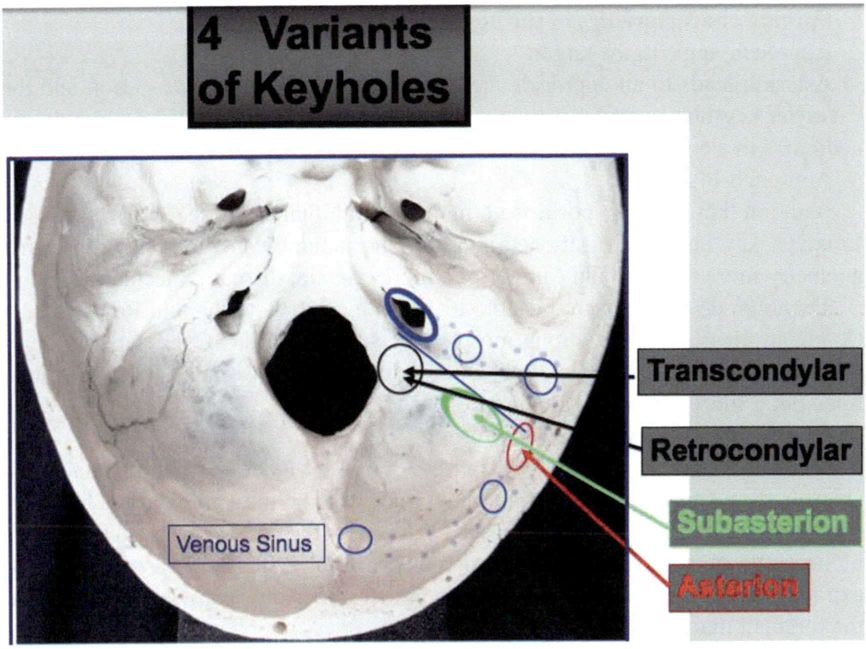

Graph 5.24 Topology of key-holes II

Graph 5.25 Topology of Key-holes III (right)

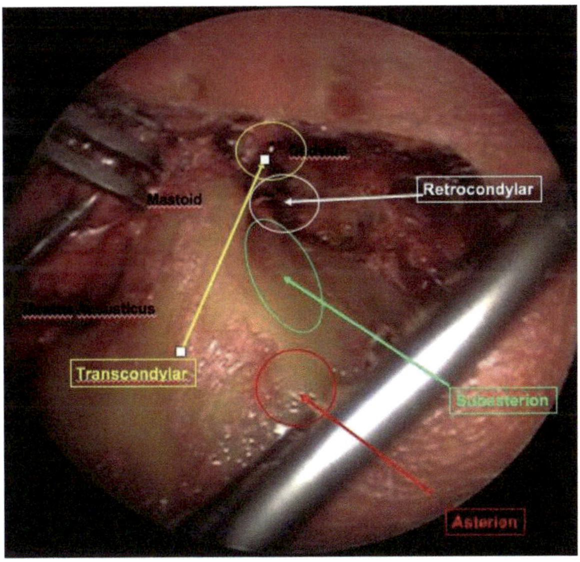

Specific characteristics of the key-holes:

Approach-angle to the target:

Asterion leads to an approach-angle more than 45°, while Subasterion and the Condylar key-holes present targets in a small angle of less than 45° presenting the neuroaxis in a co-axial view and causing very unusual aspects of imaging.

Approach-Plane:

Asterion key-hole is positioned in a vertical plane, while Subasterion and Condylar key-holes are positioned in a more horizontal plane, making preparation geometry more difficult and calling for a more co-axial dissection technique.

The three deep approaches caudal to Asterion, Sub-Asterion and Retro-/Trans-Condylar, are covered by strong muscle layers. Retro-/Trans-Condylar Key-holes are very deep, close to foramen magnum with a long trans-muscular extracranial route but a short subdural (Graphs 5.26 and 5.27).

Extra- intracranial route correlation:

From "Asterion key-hole" to "Transcondylar key-hole" the extracranial route changes from short to a long extra-cranial route and the intracranial route becomes shorter. Asterion is the easiest approach and enables the best overview to the targets of latero-basal posterior fossa. Retro-condylar and Trans-condylar need a deep extracranial preparation but enable a short intracranial pathway to the targets around the foramen magnum (Graph 5.28).

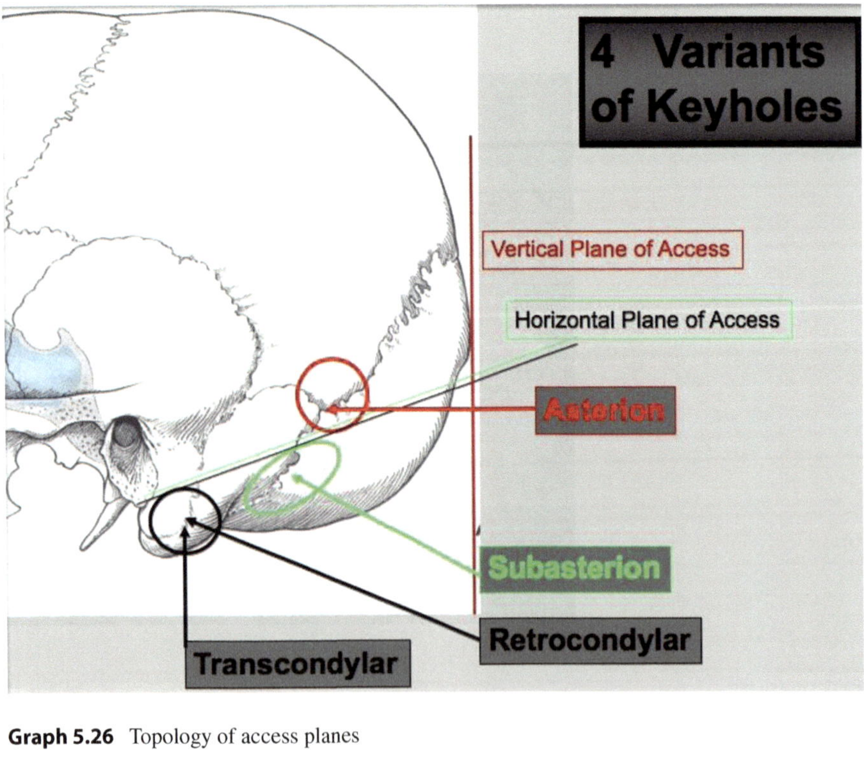

Graph 5.26 Topology of access planes

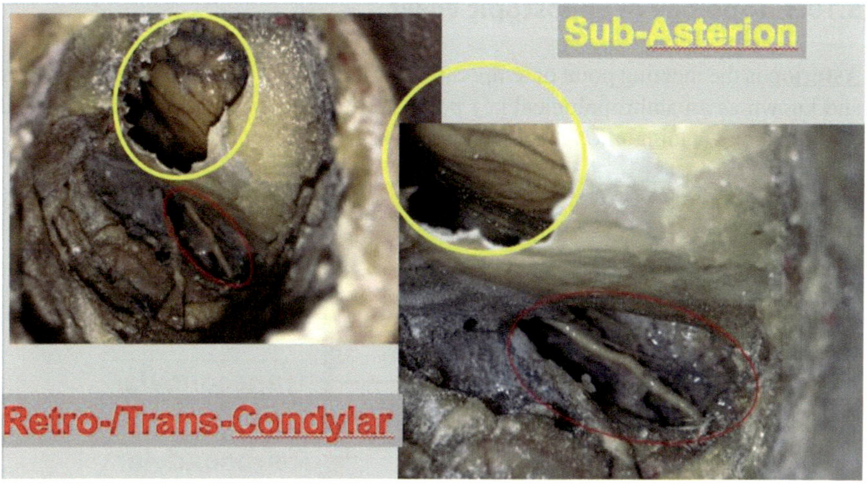

Graph 5.27 Anatomica preparation

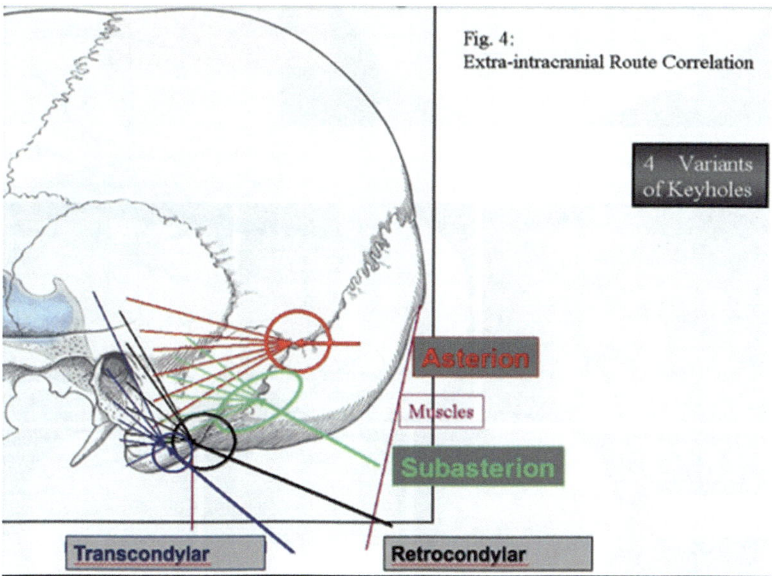

Graph 5.28 Topology of intra-extra route correlation

5.10.3 Topology of Endoscopic Views

<u>**Asterion**</u> is the meeting point of temporal, occipital and parietal bone, well described and known as an anthropological key point. It is within the knee of transverse and sigmoid sinus leading into lateral posterior fossa superiorly (Graph 5.29).

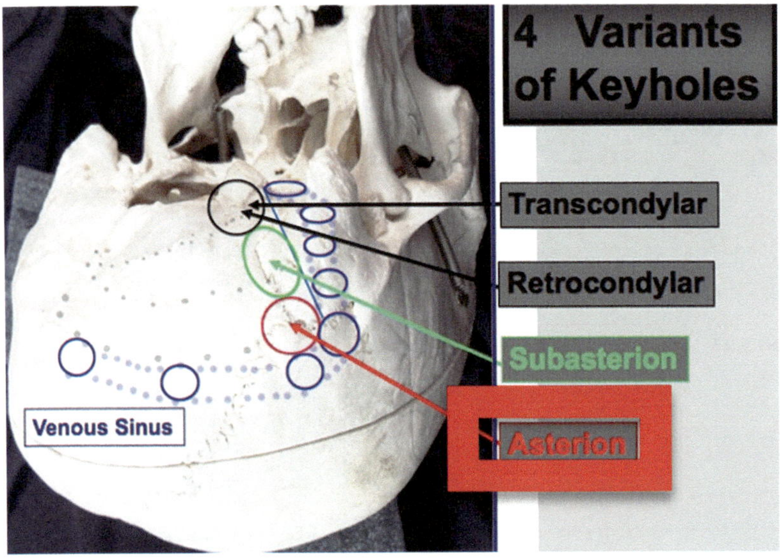

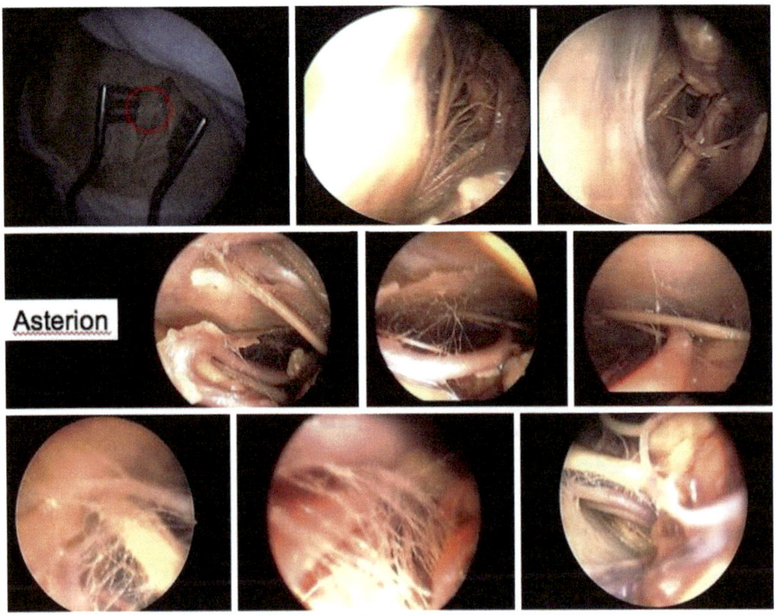

Graph 5.29 Anatomica preparations (few selected aspects)

Subasterion is defined in this study as the area medial to the mastoid process, between superior and inferior nuchal line. It leads to the CPA from a para-sigmoid route (Graph 5.30).

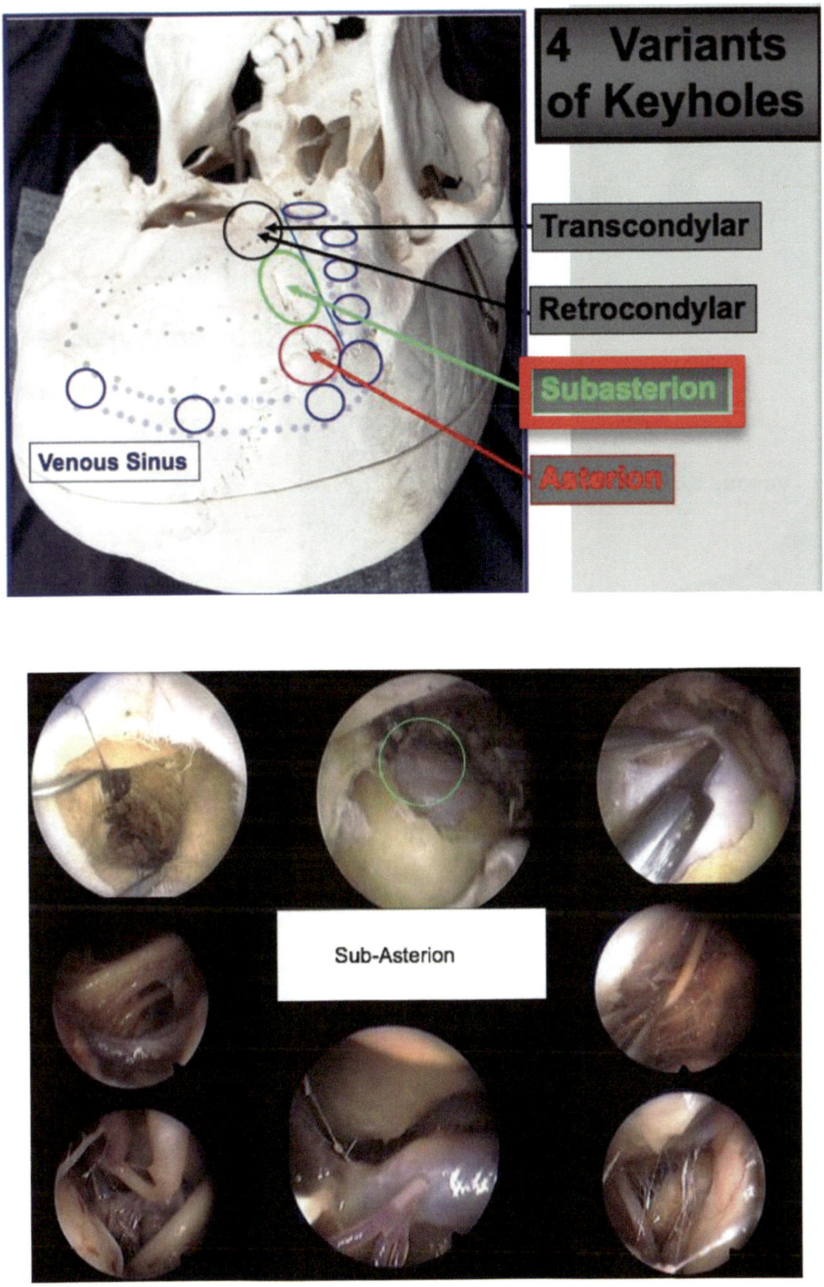

Graph 5.30 Anatomical preparations (few selected aspects)

Retrocondylar area is a well-known skull base approach located directly dorsal to the atlanto-occipital joint and close to the foramen magnum. It is deep seated and gives access into a short corridor intracranial at the level of the tonsils of cerebellum (Graph 5.31).

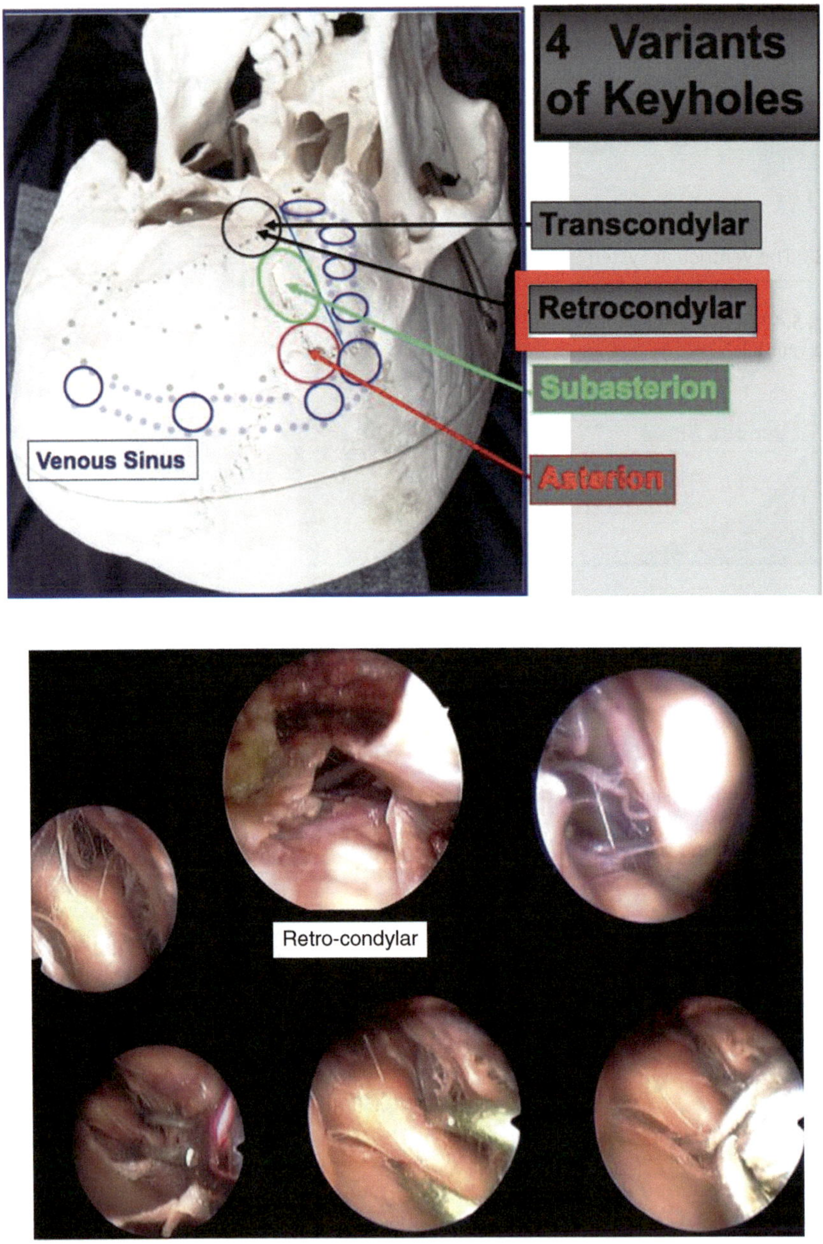

Graph 5.31 Anatomical preparations (few selected aspects)

<u>Transcondylar</u> route is also a well-known approach in skull base surgery. In this study it was designed from a far lateral angle by drilling posterior part of mastoid process and posterior third of the occipital condyle with opening the hypoglossal canal. It is also deep seated, approaching intracranial space at a sub-tonsillar level ventro-lateral of medulla oblongata (Graph 5.32).

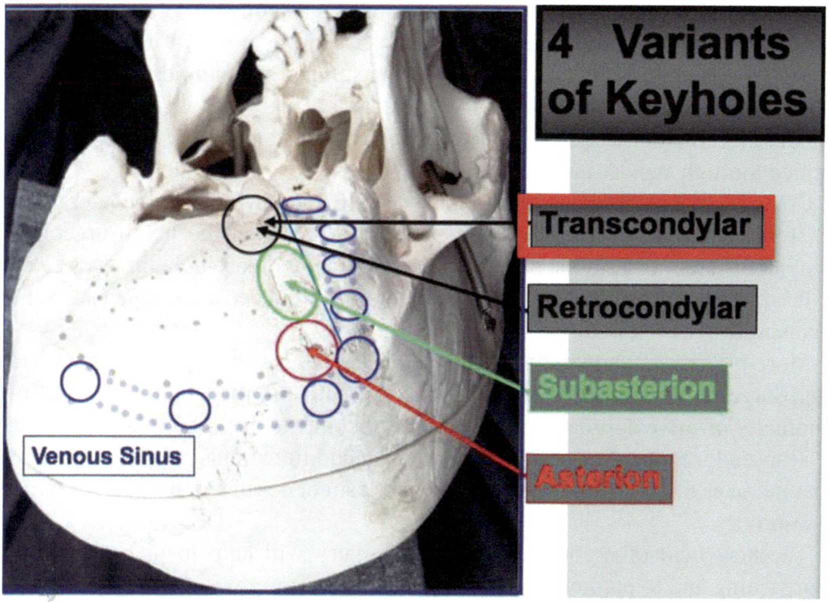

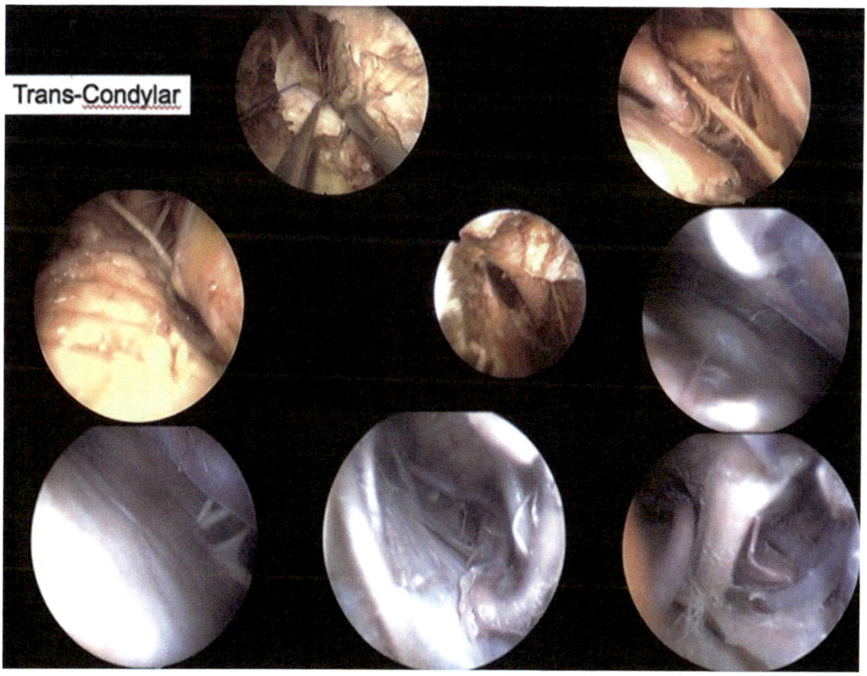

Graph 5.32 Anatomical preparations (few selected aspects)

5.10.4 Discussion

Standard approaches can be described as a sum of burr-holes. If we can plan which burr-holes will be the really necessary one to reach the target area directly we apply minimally invasive approaching. Endoscopic anatomy was described already, however, the posterior fossa presents very sophisticated anatomical details making election of precise key hole more important than supra-tentorial. This is due to the consequences of different key holes regarding surgical ergonomics, approach angle to the target area and depth of approach corridor. This was not shown in the atlas book and detailed description is lacking. Today endoscopy to the posterior fossa is applied through standard approaches.

Creating an approach corridor to the four different keyholes to the lateral posterior fossa, comparable to a per-nasal route, will confront the neurosurgeon with complet different conditions depending from the type of key-hole that was elected.

It has been shown, that approach corridors from the per-nasal type can be managed surgically even in difficult lesions.

More-over it was shown that cranial nerve decompression is an important field of neurosurgery. However, if it should be an alternative to medication we must offer minimally invasive approaching techniques to compete.

Transendoscopic ultrasound can provide an additional imaging for targeting and orientation as has been reported already. This cannot be substituted by conventional sono-probes.

The new field of ergonomics in neurosurgery will help to plan new kind of approaching and to predict possibilities and limits.

5.10.5 Conclusion

Four distinct key-hole approaches to the latero-basal posterior fossa can be differentiated. They are characterised by typical features regarding extra-intracranial route correlation, approach plane, target visualization, and angle of approaching. The results of this study make an approach analysis and approach planning more comprehensive.

5.11 Per-nasal Endoscopic Anatomy

Example 1: (1987)

Trans-nasal approach retro-sellar to the basilar tip (with base of an aneurysm) through the **microscope.** It is the view of anatomy through a narrow approach canal ending in a clivus window aiming at basilar head. The usual target to this time (1988) is the sella with pituitary gland, here with untoughed dura. The case belongs to the series of surgical simulation/ approach-analysis and approach-design for MIN (s. Vol. 2, Chap. 3).

In such non bleeding situation the anatomy Post-mortem is well preserved to examine the subarachnoid space (Fig. 5.158).

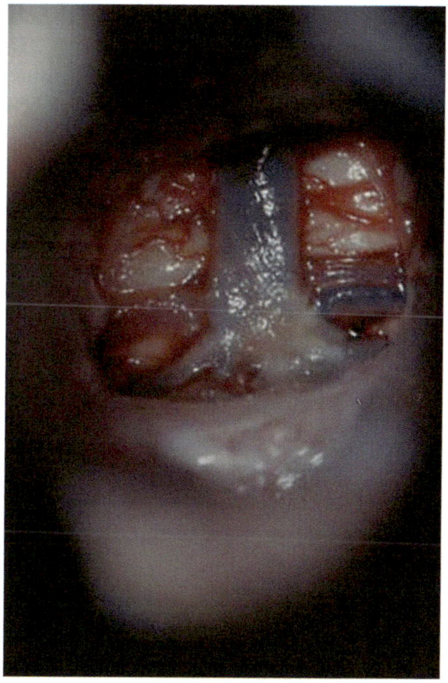

Fig. 5.158 Surgical simulation pernasal to the basilar tip aneurysm

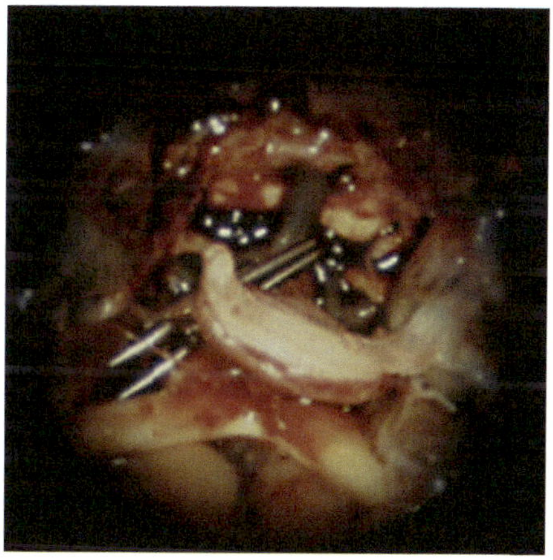

Fig. 5.159 Pernasal endoscopy-assisted simulation for clipping analysis of basilar tip aneurysm

Example 2: (1990)

Trans-nasal approach with the same target, but with all the peri-sellar region, post bleeding and post clipping, visualized through an **endoscope**.

The narrow approach canal became optically a wide field allowing dissection and viewing the parent areas to the retro-sellar target trans-clival. Pre-sellar area with optic chiasm and para-sellar with cavernous sinus are approached also (s. Vol. 2, Chap. 3).

5.11.1 Technique and Video-Chain

The endo-neurosurgeon must be absolutely firm in dealing with the video-chain and all the equipment of the video-tower. The surgery is complete dependent from the undisturbed function of the equipment. This cannot compensate by the abilities of the surgeon. Therefor he must be safe in trouble-shooting regarding problems with the technical system he uses. It is best to virtually follow the way of the light signal through the system to diagnose the location of the problem. Before starting the surgery, all parts need to be tested and the endo-cam must show a picture on the monitor before connected with the endoscope (Graph 5.33).

5.11.2 Adjustment of the Endoscope

After the endoscope and the video-chain is prepared producing an image on the monitor, the endoscope is still a danger tool and must be adjusted to become a surgical tool by adjustment of three parameters:

1. Convergence of coordinates of the endoscope and the monitor
2. Focus of the endoscope for average working distance
3. White balance to get real colors on the display (monitor)

These adjustments are made outside of the patients' head! They are mandatory if not to risk a disaster with mayor injury in the brain/cranium! (Graph 5.34)

5.11.3 Application-Fields of Neuro-Endoscopy

For 80 years endoscopy was used just with the concept of trans-endoscopic instrumentation. This is still the main way of use and was refined more and more by technical evolution and a big spectrum of instrumentation (Graph 5.35).

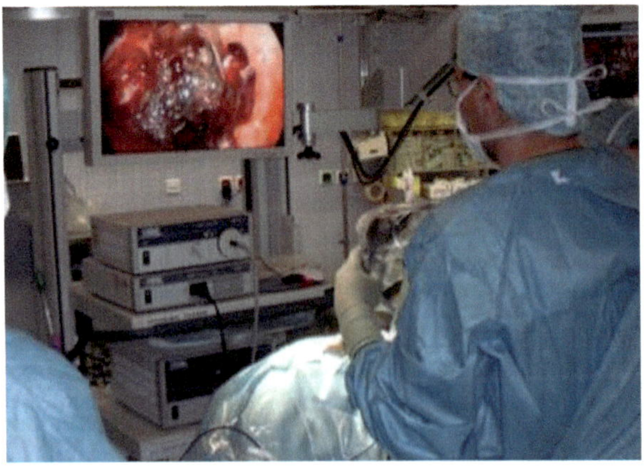

Graph 5.33 Equipment for pernasal endoscopic-assisted microneurosurgery

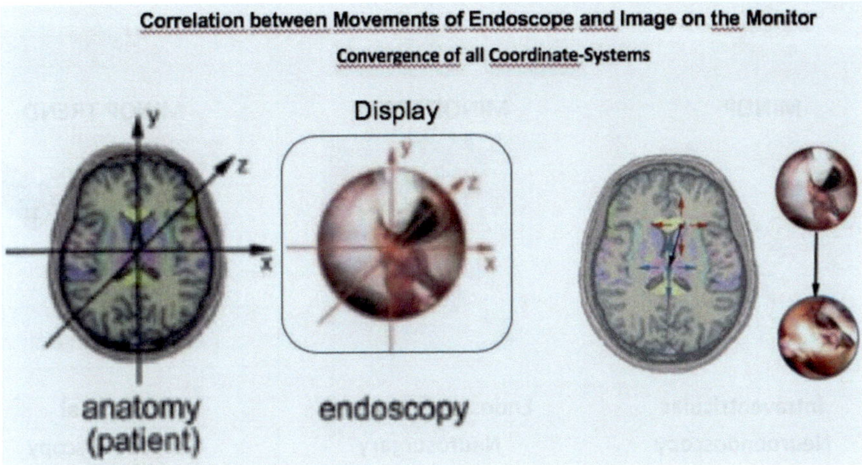

Graph 5.34 Convergence and synchronization

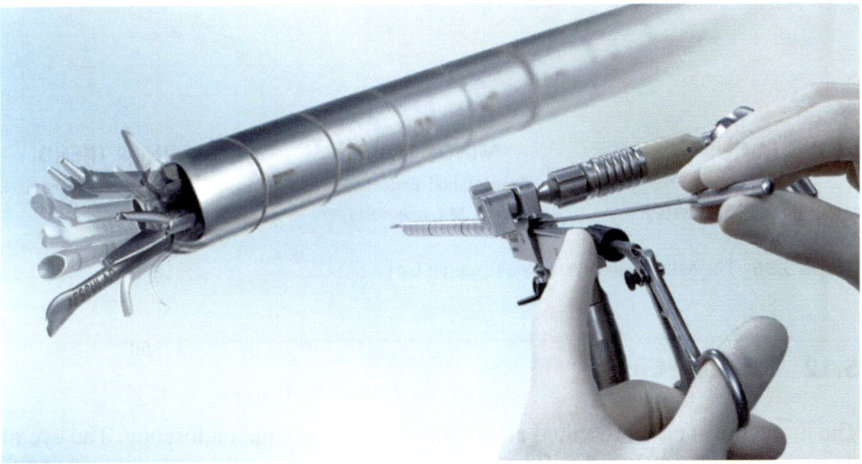

Graph 5.35 MINOP system (aesculap Co)

During this evolution different fields of application were developed and instruments adapted to this application. In the Perneczky-time the MINOP-system for ventricular endoscopy and the MINOP TEAM-system for endoscopy-assisted micro-neurosurgery was elaborated and introduced to the practice (s. Chaps. 2–4).

The instrument-family was supplied by the MINOP TREND-system for per-nasal neuro-endoscopy in sellar- and peri-sellar lesions by R. Reisch. This new field of per-nasal endoscopic scull base surgery was started by H. D. Jho (s. Graph 5.6).

Working with these systems need a training time and should be done in courses offered in several centers and the Aesculap Academy (Aesculap-Neuro.com) (s. Vol. 2).

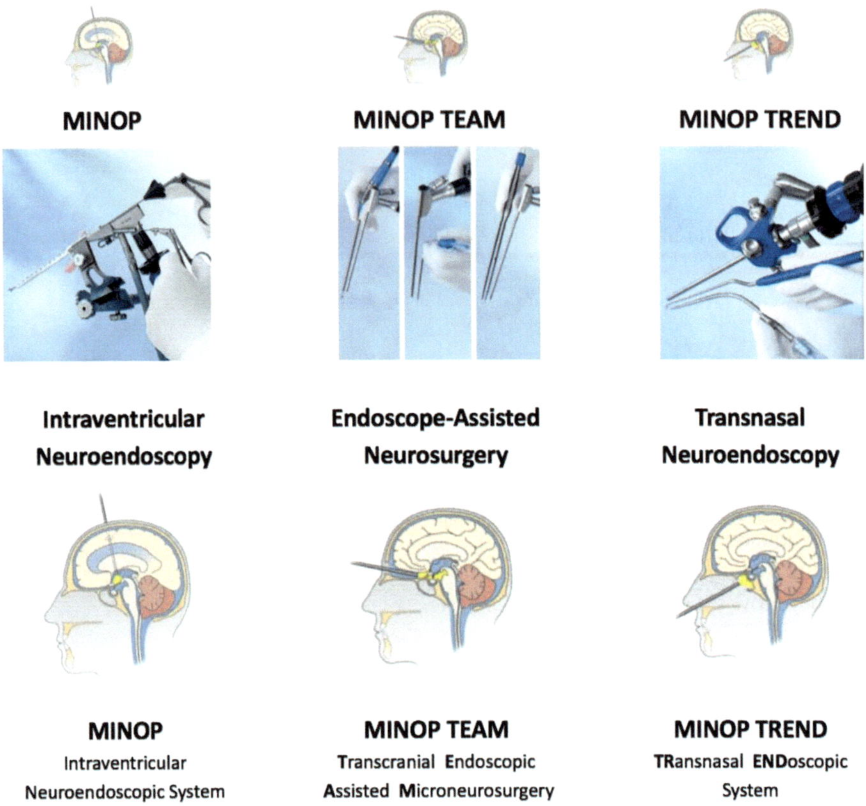

Graph 5.36 The MIN endos systems (Aesculap Co)

5.12 Clinical Cases

The first clinical case presented all conditions in ventricular endoscopy: The eye of the surgeon (KDMR) is fixed on the monitor, which is still in an uncomfortable lateral position. Camera cable, light cable, irrigation- and drainage-line need to be controlled and kept sterile. The heavy camera and too long endoscope could only be controlled safe by four hands. A good overview of the third ventricle was posible and no obstructive membrane found in this dysmorphic brain, and a shunt was implanted.

Before going into endoscopic neurosurgery, it is indispensable to know and follow some rules:

1. Always fix your eyes on the monitor, whatso ever may happen!
2. Do everything inside the brain very very slowly
3. Stop moving the endoscope if you do not understand the image on the monitor

4. In case of a bleeding never leave the intracranial space/ventricle
5. Advance the tip of the endoscope close to the bleeding origin
6. If the irrigation pressure does not stop the bleeding carefully press the lens on it and wait long enough and prepare the trans-endoscopic bipolar
7. Continuously control patency of irrigation and the free outflow
8. End of outflow line must be placed not below F. Monroi
9. One canal of the endoscope must always stay open
10. In bleeding accident cases leave a drainage at the end of the procedure

It was the anatomist S.T.v. Soemmerring, who described the cranial nerves, but also believed the soul of human to be located in the CSF. Regarding the clinical symptoms of hydrocephalus patients, this was a logical concept. It represents the effects of liquor dynamics to the peri-ventricular regions of the third ventricle, the location of human soul functions, according to recent results of neurobiology (G. Roth). Patients of this group are neglected usually regarding hydrocephalus therapy (Graph 5.37).

Case 1 (Graph 5.37).

Case 2 shows a 49y old woman with chronic headache and NSAR abuse presented with gait walking, stress incompliance and decline in performance. She could only sleep in semi-sitting position. Concentration and memory became worse (Graph 5.38).

Intermittent scotomas during last year; atrophy of optic nerve OS > OD without glaucomatous excavation; visual field defects left > right with new central scotoma, as well as central oculomotor disturbance were present. Preexistent congenital strabismus and nystagmus were additionally found.

MR showed giant hydrocephalus and AV-shunt dysfunction due to displacement of cardiac catheter. More-over a functional aqueduct-stenosis can be diagnosed.

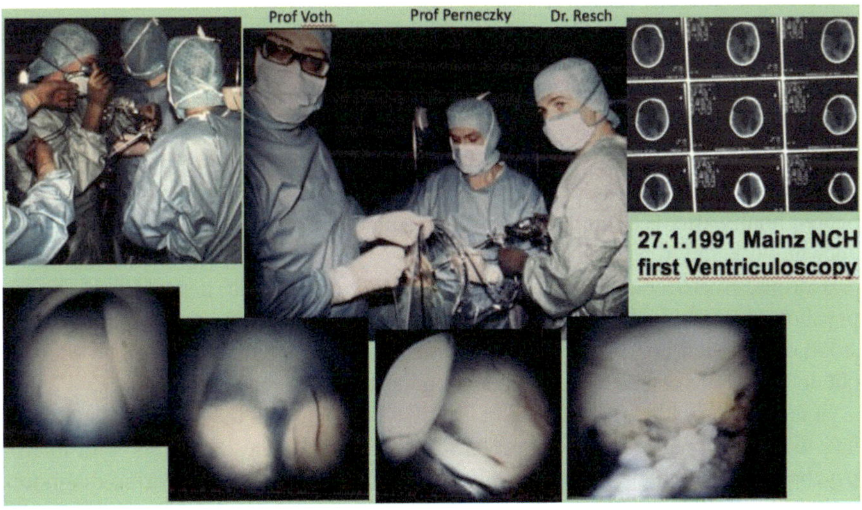

Graph 5.37 Case 1

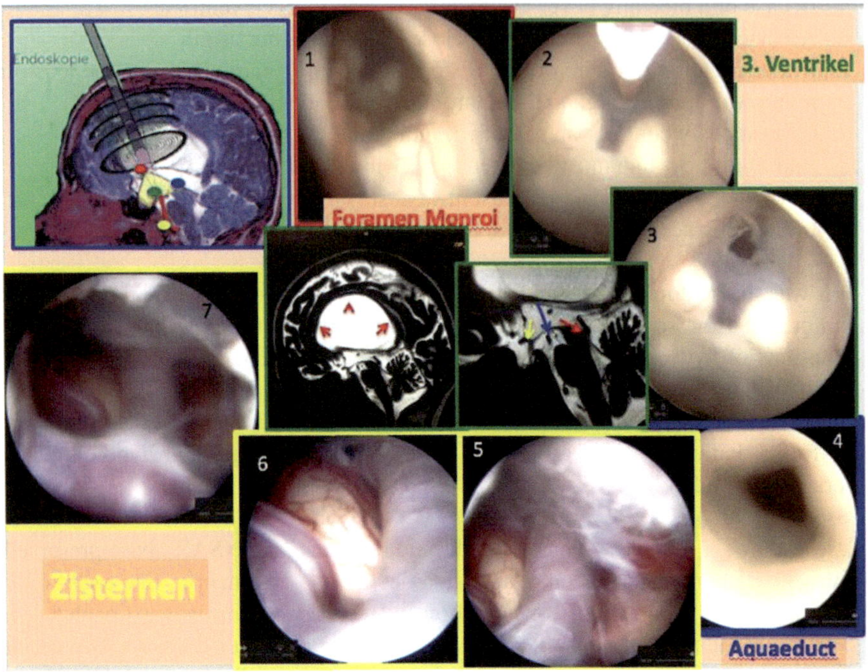

Graph 5.38 Case 2

Emergency lumbar punction (LP) to release ICP was done after a sever attack of raised ICP. According to the patients' prefer, an endoscopic repair of liquor pathway by ETV was performed.

During endoscopy (Graph 5.38) a caudal second arachnoid membrane was found (5) and a deep stomy (6) was performed. This should be done only by experienced endo-neurosurgeons. The view into the fourth ventricle through the aqueduct was obscured (4). Though the third ventricle seamed small in sagittal view on the MR, the distance of mammillary bodies (1–3) represent clearly the enlargement of the third ventricle.

Visual field and also RFNL recovered soon after LP. All other symptoms disappeared finally after ETV and she received good feedback for her new mental appearance at work and in privacy.

Chronic hydrocephalus causes multiple ophthalmological, neurological, neuropsychological and endocrinological signs and symptoms. Often these are misunderstood due to their complexity. Neuro-ophthalmological symptoms may be decisive but are mostly overseen and not understood as a hydrocephalus symptom.

It is necessary to recognize and believe the complains of the patient. This patient had to wait 10 years to get her problem solved. All of these patients with neuropsychological findings present an odyssey of misunderstandings, wrong diagnoses and abuse of medicaments, especially psycho-pharmacological drugs, causing sever side-effects. Regularly these side-effects are later misunderstood as part of the disease.

With meticulously detection and documentation in a closely coworking, difficult neuro-ophthalmological-neurosurgical cases can be solved.

In Case 3 we see comorbidity of chromosomal pathology (M. Perthes) and aqueduct-stenosis. Such conditions cause, that hydrocephalus symptoms were overseen or mixed up with the main disease.

Despite the diagnosis of aqueduct-stenosis it is common to implant a shunt-system and to ignore the "evidence". This will be followed in a course of shunt-revisions because of two main reasons. It is difficult to differentiate the symptoms of shunt-disease from those of the main disease. More difficult is the fact, that disabled patients and brains use to have a small compliance window to compensate the on-off effect of the shunt-system.

Even if there is no aqueduct-stenosis, it is wise to start liquor dynamic regulation by ETV and to add a shunt only in the minor number of cases, when ETV is not enough.

This patient had to wait 30 years, under control, to get the liquor dynamic regulated and to prevent further cerebral changes of the hydrocephalus. The deterioration were mainly neuro-psychological making integration and inclusion more and more impossible.

Post ETV the morphological changes (s MR, Graph 5.39) were extraordinary strong, but the symptoms changed very slowly, indicating that the disturbance of the liquor flow pathology lasted rather long, too long. However, she recovered within few months good enough, that she could change to a private and near independent life. Here endocrinological functions normalized and she even went through a late

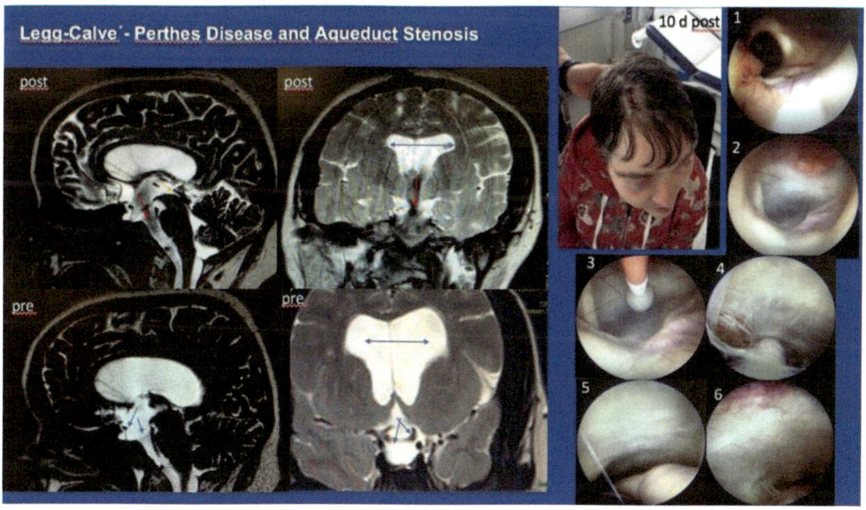

Graph 5.39 Case 3

puberty. All the communication functions improved, and she developed cognitively, being more alert and shows better self-control. This was also represented in her mimic appearance. In summary a complete spectrum of improvements, even in this disabled life, leading to near independency, in a chronic hydrocephalic pathology was possible by ETV.

The variations of symptoms in liquor dynamic pathology are many and may not fulfill the rules of medical book knowledge: "Nature does not read our books!" Especially psychological dominant symptom cases may present far away from the classical Hakim-trias of normal pressure hydrocephalus (NPH), and curiously complains and symptoms may dominate the clinical presentation. The patients may live a normal life during a long time and the symptoms estimated not in relation to the documented or unknown hydrocephalus. But seemingly suddenly the compensation mechanisms may break down and deterioration occurs (Graph 5.40).

Quite commonly the patients may have a long psycho-pathological history using psycho-pharmacological drugs for a long time, like case 4, it is nearly impossible to differentiate the primary symptoms from the drug side-effects and the secondary reactive course. Neither neuro-psychological test series nor lumbar drainage may

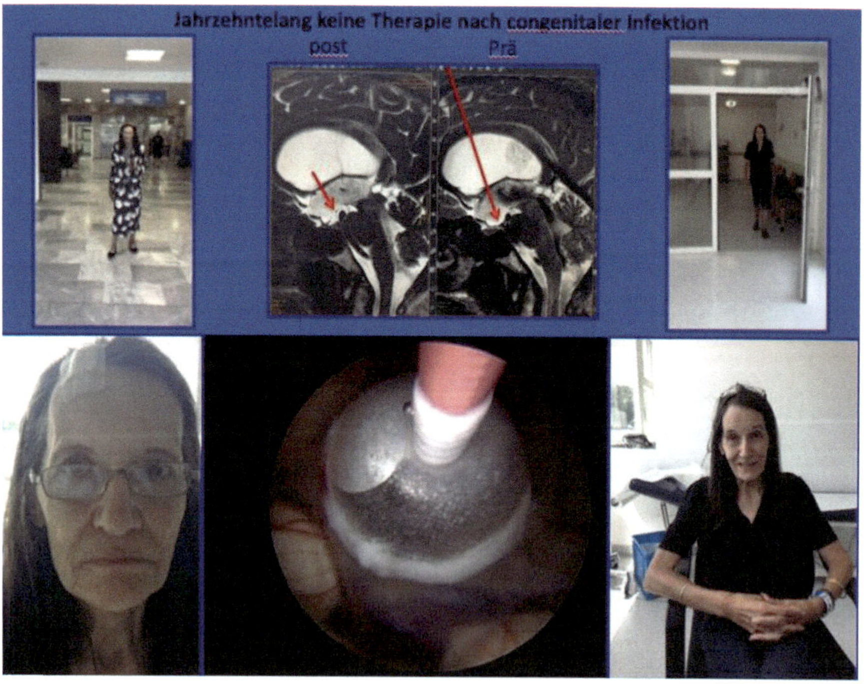

Graph 5.40 Case 4

show up with convincing results, but the patient is in a desperate situation. Only small beneficial results may keep them away from dependency or high cost therapeutic situation.

The clinical history often is unbelievable and very sad and intolerable, and whole lives and families are destroyed.

However, case 4 patient presented all symptoms of NPH in a low expression, but psychological symptoms were very dominant, and the relatives are completely decompensated. Though the diagnosis was quite clear by Hakim-trias and the typical MR, she was kept in a geronto-psychiatric unit under drugs causing a difficult pre-op time and handling.

Finally, the ETV was successfully done and she recovered fast from NPH symptoms and within weeks from depression changing her complete appearance in mimic, moving and behavior. The mayor difficulty was the reduction of psychopharmacological drugs. It is one of the most impressive clinical courses, which are presented by these patients psychologically and to free them from the pharmacological straight jacket (s. also Graph 5.3).

Population in industrial countries are becoming older and the multimorbid fraction of patient is growing fast. Endoscopy as a minimal invasive key-technique is becoming an important role, avoiding trauma and enabling procedures in old people, which otherwise would be contraindicated.

A 77 years old patient was first admitted by his ophthalmologist to clarify vertical diplopia with hypotropia of OD. cMR showed an **acoustic neurinoma** (AKN) at his right side. The patient decided for follow-up strategy. Fife month later he already appeared again with several new problems (Graph 5.41):

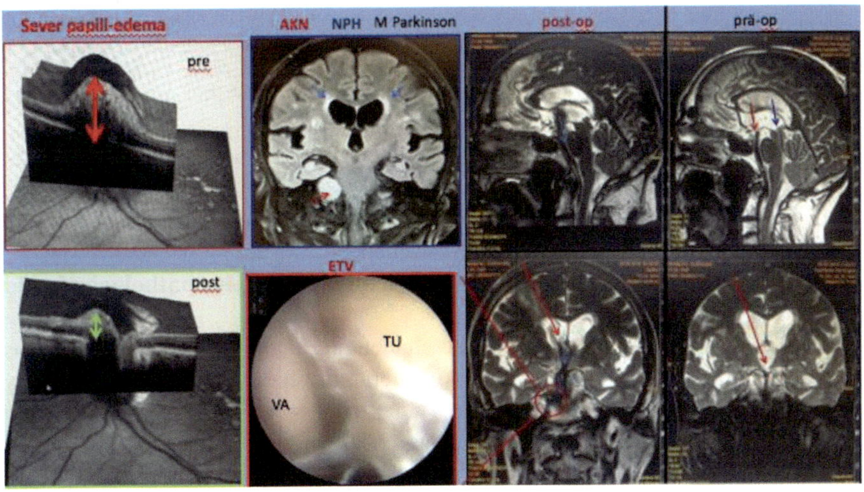

Graph 5.41 Case 5

Since several month the patient experienced changing visual field defects and actually, surprisingly showed **dramatic papilledema** on both sides combined with signs and symptoms of normal pressure hydrocephalus (**NPH**) with complete Hakim-Trias. Emergency lumbar punction (LP) to release ICP was performed.

Lumbar liquor drainage proved a high positive effect of recovery and shunt implantation indication. A standard liquor shunt procedure was not possible due to **very high albumin levels in the liquor**. A shunt procedure was contraindicated and an endoscopic repair of the liquor pathway (ETV) was performed. However, configuration of a supratentorial hydrocephalus indicated a functional aqueduct-stenosis. MR showed sudden stop of flow signal at entrance of third ventricle! Qualitative changes of flow signal in the aqueduct are often ignored.

During endoscopy (Graph 5.41) an opaque liquor was found and strong changes of arachnoid membranes. The cisternal route was followed until the view of foramen magnum showed free liquor pathways. View of fourth ventricle through aqueduct was obstructed.

LP was followed by a fast recovery in visual fields and NPH symptoms. ETV saved the results and improved them additionally without shunt-system implantation.

After radiation of the acoustic neurinoma (an operation was refused and also not convenient regarding overall conditions), the albumin level decreased to half and the papilledema decreased fast. NPH symptoms decreased markedly but only after ending the neuroleptic medication, waking became normal. All visual problems recover completely.

Shunting one year later resulted in complications and revisions and did not improve the post ETV and post-radiation results, but the patient lost his gross independency.

In summary:

Neuro-ophthalmological findings led to diagnosis and fast therapy of a **complex multimorbid case.**

Close cooperation of ophthalmology and neurosurgery prevented decompensation of vision and NPH.

In cases of functional aqueduct-stenosis and shunt contraindication, ETV can solve the problems of liquor dynamics disturbance.

Minimally invasive neurosurgery (MIN) can face complex pathologies in old patients.

Large cystic craniopharyngiomas are challenging multi-disciplinary as well as for neurosurgery. Even the best series in literature report 15% overall mortality (Graph 5.42).

A supra-sellar located tumor injures the visual pathway at its' most vulnerable site functionally: compression of the chiasm, less than a half of cubic centimeter, causes blindness! There is a narrow window of time for acting to save vision! This s also true for endocrinological and neurological preservation of function. The time

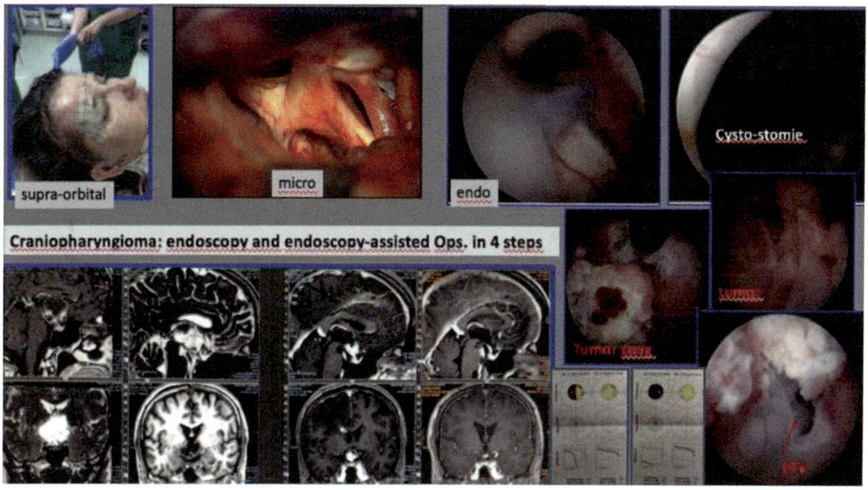

Graph 5.42 Case 6

is a mayor factor for preserving function. Time is not visible on imaging but in the clinical course.

Intense interdisciplinary interaction, and application of minimally invasive concepts and techniques, preserved vision and visual field, endocrinological and neurological symptoms in each of the four phases of therapy in this case. Finally, also the liquor pathway was kept intact.

In this case of a 53 years old female acute decrease of vision and visual field was diagnosed by ophthalmologist followed by emergency MRI and immediate sending to neurosurgery. Visual acuity and perimetry as well as MRI and endocrinological laboratory pre- and postop were analyzed retrospectively in all four phases of therapy.

Neurosurgical micro- and endo-surgical operations assisted by neuro-sonography and LASER were applied according to minimal invasive concept (Table 5.1).

Recovery of visual function after all four surgical intervention and finally complete resection of tumor was achieved. Return to normal vigilance without disturbance of liquor dynamics and minor hydrocortison substitution postop during 1 year could be accomplished.

In summary immediate, direct and systematically interaction between ophthalmology, endocrinology, neurology, neuroradiology, neuro-anästhesia and neurosurgery prevented visual loss in all four phases of therapy in this case of a large cystic craniopharyngioma of the third ventricle. Ophthalmology gained a trigger function to get neurosurgical intervention in time. Minimal invasive concept and staged therapy resulted in excellent outcome. In this cystic case the new concept of stepwise approaching, starting with cyst control by endoscopy and advancing with endoscopy

Recovery of visual function after all four surgical intervention and finally complete resection of tumor was achieved. Return to normal vigilance without

Table 5.1 Ophthalmological results

Results of: **OP***					
Visual ac.: pre * postop	R 0,4 0,25*0,6	0,1*0,6	0,5 0,25 0,3*0,3 0,4	0,3*0,25	0,4
	L 0,4 0,5*0,8	0,4*0,6	0,5 0,4 0,3*0,4 0,6	0,5*0,4	0,6
Exam Dates	11/2013	3/2014	5/2014	8/2014	4/2015
Visus-Amplitude:	**OD** 0,1 ➔ **0.4** **OS** 0,3 ➔ **0.6**				
	During the four Phases good visual acuity again after all operations				
Vigilance:	Complete recovery (**risk**: hypothalamic/thalamic coma vigile!)				
Perimetry:	Recovery of both eye (MD (dB): **OD** 24,3 ➔ 5,1 **OS** 14,4 ➔ 5,6)				
Endocrinological:	Intact with finally no hydrocortison substitution 1 year				
Tumor status:	Complete resection without recurrency since 6 years				
Status of hydrocephalus:	No shunt necessary				
Radiation:	Non				

disturbance of liquor dynamics and minor hydrocortison substitution postop during 1 year could be accomplished.

In summary immediate, direct and systematically interaction between ophthalmology, endocrinology, neurology, neuroradiology, neuro-anästhesia and neurosurgery prevented visual loss in all four phases of therapy in this case of a large cystic craniopharyngioma of the third ventricle. Ophthalmology gained a trigger function to get neurosurgical intervention in time. Minimal invasive concept and staged therapy resulted in excellent outcome. In this cystic case the new concept of stepwise approaching, starting with cyst control by endoscopy and advancing with endoscopy assisted microsurgery. This MIN concept allows the extremely sensitive and disabeled tissue to recover before the next step is done. Especially normalization of liquor pathway and release of ICP for better perfusion and decompression is essential. As shown by Yasargil attention must be drawn to the anterior circle of Willisii obstructing the visual pathways from superior during the growing from inferior.

Arachnoid cysts are benign congenital cystic lesions, a duplication of arachnoid membranes. Most frequent they appear in the Sylvian fissure (15%) but also in the pineal recess (10%), CPA (10%), cerebellar vermis (10%) and inter-hemispheric (5%). Lesions of supra-sellar location do protrude into the third ventricle, causing neurological, neuro-psychological, ophthalmological and endocrinological symptoms. Complexity of clinical appearance may lead to misinterpretation of single symptoms, and diagnostic procedure may be delayed. Sometimes the diagnosis is made accidently, but when it comes to liquor dynamic problems, deterioration may occur and emergency situations can happen, sometimes with sever outcome.

Ophthalmological dysfunctions can lead to the diagnostic actions finally, or just in time.

In case 7 (Graph 5.43), the 16 years pupil experienced a two-hour speech arrest, paresthesia at the tongue, hypesthesia at the right hand, sever cephalgia with vomiting attacks during fife (!) years. Visual disturbance with diplopia, and arterial

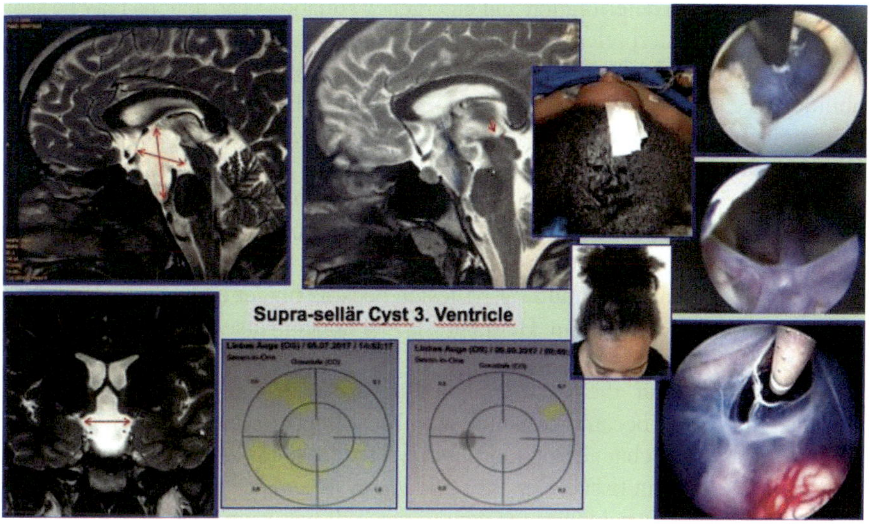

Graph 5.43 Case 7

hypertension were present, and she developed adipositas. Finally, problems at school with difficulties to concentrate and memory deficits led to presenting her to an ophthalmologist, to neuro-imaging and to neurosurgery.

Intermittent visual field deficits appeared, and MR showed a big suprasellar cyst of the third ventricle and prepontine extension, compressing trigeminal nerves, optic pathways, and obstruction of aqueduct, causing tree-ventricular hydrocephalus.

Finally, an ETV was successfully performed on superior and inferior level of cyst membranes, and a subtotal resection of the cyst.

The neuro-psychological and ophthalmological problems disappeared within weeks and she could finish her college in time. However, a reduction of peri-papillary retinal fiber layer (RFNL) nasal inferior (left > right) was now present as a remnant of the chronic hypertension of liquor.

During endoscopy, additional to the cyst she had a membranous stenosis with a small hole of the aqueduct, explaining the typical extraordinary "jet-flow" in such cases, and usual misdiagnosed as normal aqueduct because of flow signal. This membranous stenosis of aqueduct can only be seen in CISS sequences (T2 flow) sagittal. After ETV, the "jet-flow" disappeared and an very strong flow signal through the stomies is visible. Bulging of the cyst is not present anymore and cyst wall in the third ventricle is not visible anymore.

In Summary, the case shows a chronic and complex symptomatology due to a supra-sellar arachnoid cyst of the third ventricle.

The membranous stenosis of the aqueduct was hidden radiological by the huge finding of the cyst, but only indirectly announced by the typical "jet-flow" in this type of stenosis with a small hole in the membrane. This compensates peaks mostly, leading to a very slow chronic course, but with the possibility to suddenly decompensate with deterioration and emergency situation. The deficits may be seen very late and misinterpreted.

This woman (case 8), 69 years at admission, is a complex case, representing the abilities of endoscopy to anyway realize a good result, after a series of traumatic operations in two university-departments of neurosurgery, with complications followed by functional deficits.

In Graph 5.44 the skull shows a battlefield of approaches. After fife procedures, the sixth one by ultrasound assisted endoscopy through a burr-hole, stopped the progress of neurological deficits and seizures.

Now, after 6 years, the multi-cystic lesion became bigger again and some of the old symptoms appeared again. Hemiparesis and nonconvulsive local seizures were in progress.

MR showed, that the endoscopic cysto-ventriculostomy has closed again. Approach-planning became essential with regarding of all the former results of operations and the changes of morphology. To use the given condition without transgression of brain tissue, but only through cysts and scarfs, and connecting all volumes into the ventricle, was the goal to reach. It was soon clear, that endoscopy alone would be dangerous, and computer-neuro-navigation cannot serve with real-time capacity.

The only real-time imaging is ultrasound with a true navigation capacity and without bias by shift or loss of liquor. Moreover, ultrasound is applied through the surgical approach with the perspective of the surgeon, an fitting easily into the workflow (ergonomics) of the procedure.

Through a small precise approach () (Graph 5.44), leaving all bone-flaps in place, the endoscope was introduced through inter-flap scarf tissue, and guide from

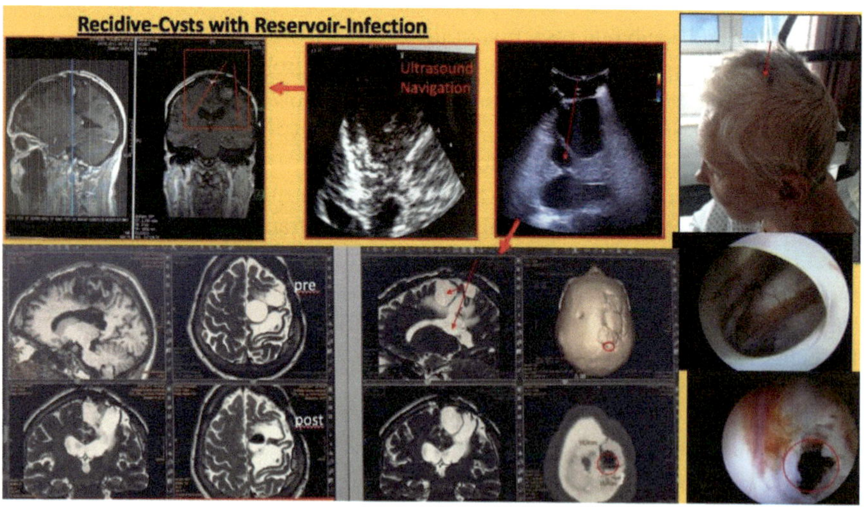

Graph 5.44 Case 8

one cyst to the other by perforating the membranes at the needed point by trans-endoscopic bipolar. Finally, the connection was done until the lateral ventricle.

Post-op the hemiparesis improved fast and the focal seizures within 1 month and she was full mobilized.

The post-op imaging showed air in the cysts, marking the endoscopy route, the cyst decreased and a flow signal into the ventricle was documented.

To avoid a shunt-system the patient had to travel outside her city and her country, in Europe! Six of the seven procedures were done after 1996, that means after the international congress of MIN in 1993 and after the mayor publications were done. This problem has not changed jet and paticnts have to search long and far to find a well-trained endo-neurosurgeon. The manual difficulties are of medium level, but the concept and the planning are high level. Judgement of level should be done according to difficulty managing intra-operative and peri-operative complication. Complications do not disclose endoscopy rather than the typical complication need to be known and managed. In cystic lesions endoscopy is always an option.

Liquor dynamic problems in complex constellations should not be answered by standard procedures.

The 34 years man of case 9 (Graph 5.45) suffers from a battered child syndrome with ICH and hydrocephalus as well as optic nerve injury (right amaurosis, left decreased visual acuity). He got a ventricular and subdural shunt with complications and revisions.

Actually, he was admitted with visual acuity of 0.1 and intermittent amaurosis fugax attacks at his last eye. Retinal sever atrophies of capillaries and optic nerves were present. One week before additional he experienced hypakusis and NPH

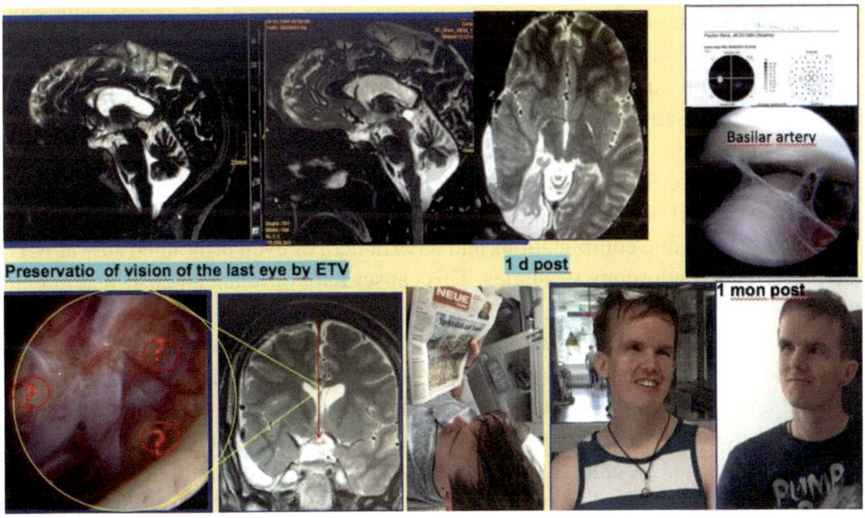

Graph 5.45 Case 9

syndrome (Hakim) and need to stop working. He showed a progredient change of behavior and parkinsonoid signs left side.

A lumbal drainage was applied (danger!) with minus pressure syndrome forcing him to bed, but the visual acuity recovered to 0.4, which was not a benefit in that situation.

MR showed an aqueduct-stenosis and a chronic slit ventricle syndrome due to long time over-drainage.

After worsening of lumbar over-drainage and visual acuity to 0.1, an emergency ETV was performed.

Due to the slit-ventricles and variations of strong midline vessels endoscopy through the midline route transcallosal was a master task, not recommended for unexperienced endo-neurosurgeons. The very short route through corpus callosum is the best way in a handy caped brain functionally.

First view hours the patient did not see anything. Within 1 day he recovered to 0.6 visual acuity and his vision normalized for the patient on his last eye. All NPH symptoms disappeared fast and performance improved markedly and visible. His mimic and personal expression changed positively and back at his working place he was admired.

The most difficult change was the slow exit from his (contra-indicated and mis-used) neuroleptic medication. With the help of his fabulous grandmother he mastered this, and left side parkinson symptom disappeared and showed only remnants in very stressful situations. The newly gained vigilance and self-awareness led to irritating observation on his disabled short memory. The critical self judgement of remnants must not be misinterpreted as failure of ETV rather than expectable course of inhomogeneous recovery. However, the strong recovery of visual function and NPH symptom as well as better overall life quality made ETV successful.

Such success needs a close interdisciplinary cooperation and the understanding that complex pathology do need an individual solution and not a standard procedures. Disabled brains cannot compensate shunt-disease!

Case 10 represents a mayor conceptual problem regarding indication of endoscopy, because it is the combination of aqueduct-stenosis and ICH. The ideology, implemented since STICH-tails, caused a prohibitive environment for ICH surgery, and in case of IVH or additional liquor involvement, drainage strategy with series of surgeries, predictable complications and days in bed, or even prolonged stay at ICU, are preferred to endoscopy strategies. This causes a lot of financial costs and human tragedies.

This 57 years patient was admitted to ICU with stroke symptoms of the brainstem. She was somnolent, complained headache and visual problems. Speaking was difficult and slow and the right side sowed dysmetria. Ophthalmological follow-up diagnosed incomplete Parinaud-syndrome and paresis of abducens and trochlear nerves. Dysarthria, dysphagia and hemi-dysmetria were found initially.

Emergency CT showed an obstructive hydrocephalus and MR presented an ICH, dorsal to the upper brain stem with compression of aqueduct and superior medullary velum.

As usual, a ventricular drainage was inserted and changed one times, when liquor became signs of infection. Moreover, it became clear by control imaging, that the blood clot will not be solved soon, and by laboratory results of liquor, that shunting will also not be able for a unpredictable and long time.

Under this condition an indication for EVD could be pushed through, according additional with the authority of the wish from the relatives. This strategy allows to shorten the drainage (when already placed), to shorten the ICU time and the bed-rest time, last but not least, it shortens hospital stay and costs of the therapy financially and ethically.

The ETV was successful and she could be mobilized soon there-after and sent to rehabilitation earlier as usual.

During endoscopy (Graph 5.46) an injury from the ventricular catheter was seen close to the fornix (1), as already visible in the coronary MR (→) post-drainage. There was an extraordinary strong choroid plexus (2), narrowing the foramen of Monroi. The anatomy of the floor of third ventricle was normal and well developed (3) with a narrow infundibular recess, a well resent impression of optic chiasm and lamina terminalis also visible in the 30° view.

The stomy was placed rather ventral (4) close to the infundibulum, according to the planning at the sagittal MR, where the interpeduncular cistern was small due to the dorsal ICH volume. The gap for perforation was small and close to the dorsum of sellae.

The arachnoid membrane of the prepontine cisterns showed already xanthochromia (4). In all of these cases an early ETV is recommended and drainage should be avoided to get the patient out of the bed and to avoid infections, as well as multiple surgeries.

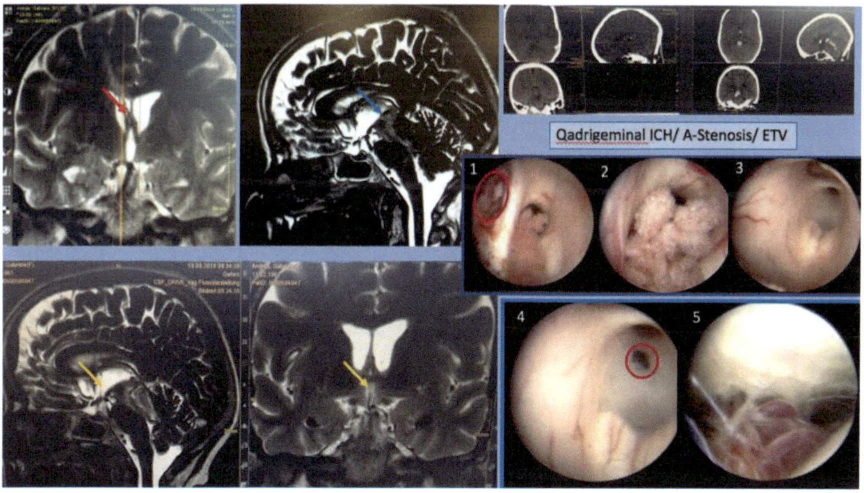

Graph 5.46 Case 10

Minimally invasive evacuation of the bleeding was not necessary, as she recovered after ETV fast, and was mobilized. Ophthalmological and neurological symptoms recovered, and CT showed release of the blood clot by the liquor transportation rather early.

An optimistic indication for ETV is rectified in such cases from the viewpoint of pathophysiology, but also from view of psychology and humanism (ethical). Of course equipment and trained surgeons need to be on side.

The case 11 represents aqueduct-stenosis by tumor with the indication of ETV at the ventral third ventricle and biopsy at the posterior third ventricle. Usual this is done, if endoscopically, by planning and performing two surgeries with two different trajectories, causing a lot of trauma and surgical-, and logistical, and financial efforts (Graph 5.47).

At the sagittal MR it can be calculated and planned (Graph 5.47), in relation to the diameter of the foramen Monroi and the diameter of the endoscope, if both procedures can be done with one trajectory only. However, this concept is theoretical, because there are more parameters involved like the size of the ventricles (especially the third) and the rigidity of the brain. The anatomical variation of plexus and veins of the foramen Monroi (1) also plays a critical role.

For one year, she had head-ache attacks, some-times with vomiting and even with visual symptoms. These attacks and waveform symptoms are often underestimated, and experienced surgeons remember tragic cases for example in colloid cyst cases.

During endoscopy foramen Monroi showed well present veins (1), and the vein most underestimated and without a name, and the main bleeding cause (not the plexus as always believed) at the foramen Monroi during endoscopies. This vein may be hidden by ependyma and by the fibers of stria terminalis of lateral ventricle

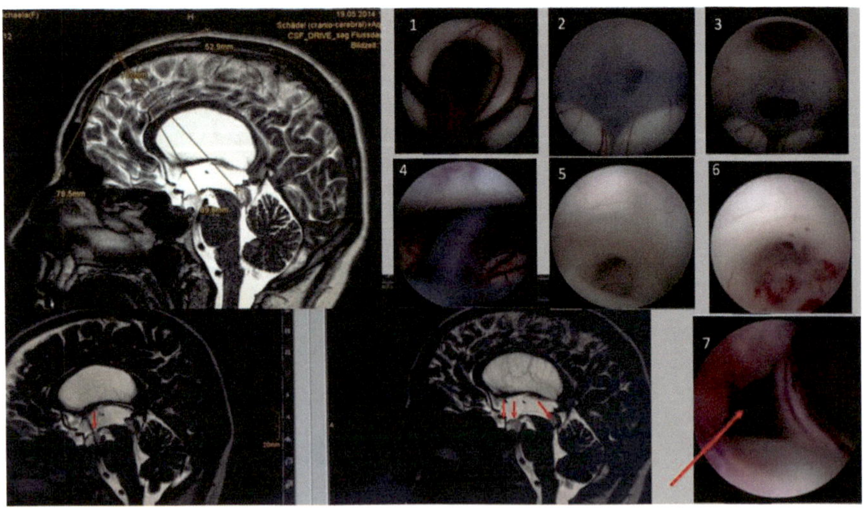

Graph 5.47 Case 11

(part of the limbic system), may cause bleedings, which are difficult to coagulate, being in an unfavorable angle to the working canal of the endoscope. Especially in the 30° endoscope it can be in the blind angle, when the lens is directed anteriorly. Turned 180° dorsally, which is difficult to navigate free-handed, this area is well visible and the favorable way to coagulate the vein, but also to view into the posterior par of third ventricle. Mostly, in such circumstances, a very slight compression with the endoscope- shaft may serve as first aid, is often sufficient. However, this is a holy area, as the paraventricular nucleus is nearby, which is the "switch of awareness"!

The cinereum, entrance to the infundibular reccss and pre-mammillary located, was still parenchymatous (2) which tends to have vessels for bleedings, but the stoma (3) could be placed in the usual typical place.

A highly placed head of the basilar artery was not visible (predictable by sagittal MR), but the strong basilar trunc just below the origin of superior cerebellar artery is well seen (4). Turning the 30° endoscope 180° posterior with a light carefully minimal angling of the endoscope-shaft gives view to the posterior part of third ventricle and the entrance of aqueduct (5 + 6). One has to decide to go anterior to the inter-thalamic adhesion or posterior. The latter route (7) is dangerous and should only be used after careful proving the anatomy regarding veins and plexus. An injury of the angle of the veins, mostly hidden by the plexus, can end-up in a disaster.

Biopsy (6) was uneventful and astrocyte tissue (no tumor) was found. Pre- and post-op MR showed the lifting of floor (→) of the third ventricle to normal level, increasing interpeduncular cistern, and the change of the chiasma-angle (most unaware), in this case only minimally (↗). A typical flow void through the stomy became visible but is not necessary to prove the patency of the stoma, only in the positive case it is predictive.

Case 12 represents the always present task in surgery to control and preserve coagulation on a level allowing surgery and safety regarding bleeding on one side and thrombosis on the other side. To balance this parameter is a mayor issue and became very difficult due to multi-morbidity and complexity of recent thrombosis prophylaxis with new drugs coming up (Graph 5.48).

In this case of a mechanic heart valve anticoagulation is indispensable and surgery becomes a balancing act. The trias of Virchow seams not to be aware anymore and drugs, anti-drugs and laboratory values rule over the surgical field.

The 88 years patient had already two operations on both sides (1). During rehab-therapy she deteriorated, and CT presented a subdural rebleeding left (2) with sever midline-shift and somnolence.

An emergency-operation was necessary under unfavorable coagulation conditions. Under these circumstances a usual blind drainage procedure with a rather insufficient irrigation effect was not recommended rather than to control subdural procedure endoscopically and to avoid trauma. Also, it seemed wise to avoid a persistent catheter and, to control also endoscopically the completeness of evacuation.

In such a case planning of burr-hole position is totally different to usual strategy. In MIN any intracranially volume should be approached from the small side if possible (3).

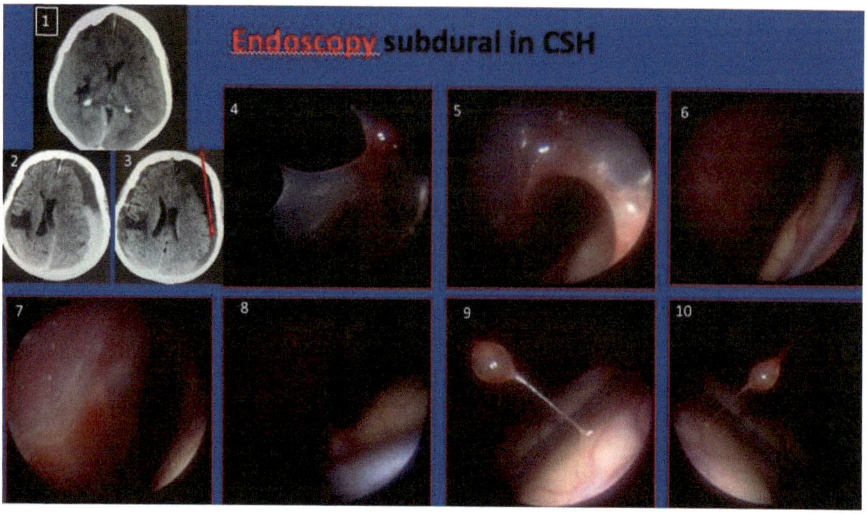

Graph 5.48 Case12

In the hype-time of endoscopy, subdural hematomas were evacuated frequently under endoscopic control. During this phase experiences were possible about the visual appearance of the subdural space under CSH-conditions. We have learned, what curiosities Jackson-Pratt catheters can show, how they can injure the cortex and bend or tear bridging veins. The change of color of the meninges and the kind of compartmentation by membranes could be studied.

During endoscopy in this case something new was experienced. Additionally, different changes and membranes (4 + 5), and the appearance of the capsule (6), very tiny "bridging veins" at unusual sites were encountered and, moreover, venous aneurysms were visible (8–10). Remnants of blood-clot (7) could be made visible in deep regions and a visual controlled decision made, if to evacuate radically or not. The sever vulnerability regarding bleeding could be seen, supporting the strategy not to leave a drainage in such circumstances, to control completeness visually and by decision making, and to avoid unnecessary trauma like craniotomy. It became clear, that the balance between heart and brain was critical in this case, and a bit in favor for the heart, because of the danger to get a sudden death. Rebleeding, in contrast, would cause another operation.

Finally, in this case, after two good weeks, another rebleeding occur, but without a server symptom. Prognosis depends from the, above mentioned, balance, and the multi-morbidity in the old patient, however, craniotomy would be far too dangerous, and MIN strategies is the only way to go, or to give up. In conclusion, endoscopy is especially helpful in the elderly patient group.

Case 13 represents the situation to do an emergency endoscopy procedure. In addition, it teaches again, that deterioration in posterior fossa pathology can occur unpredictable and without the symptom-cascade of supra-tentorial lesions! However, we have learned also, that neuro-ophthalmological signs and symptom,

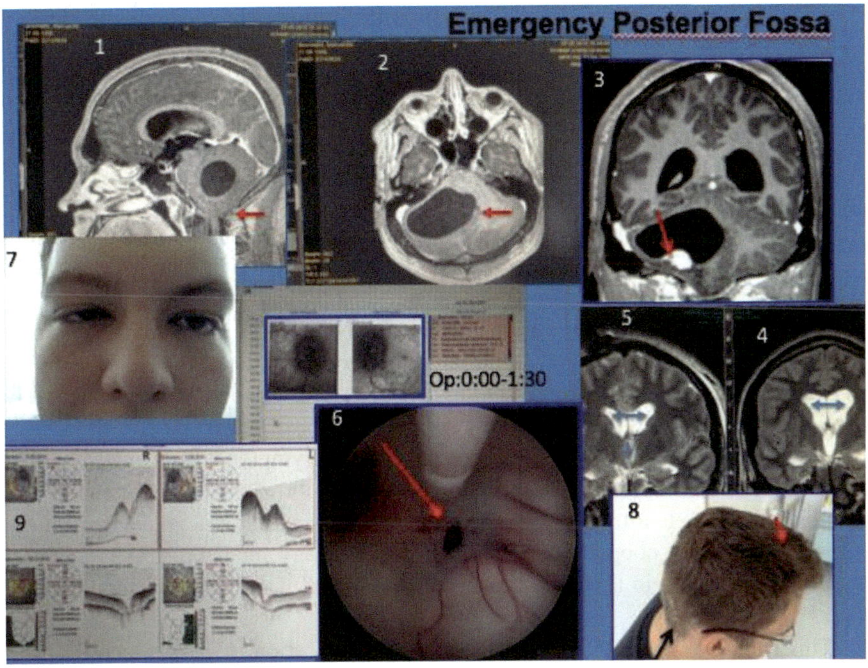

Graph 5.49 Case 13

but also such complains of the patient, are often the only and last warning sign before sudden emergency happens. A crucial balance, it might be for a long time, should never be the reason to neglect the life-tethering danger (Graph 5.49).

In this case, the danger was visible by warning signs in the MR: cerebellar were already herniated deep into foramen magnum! The brainstem was severely compressed and shifted. Several cranial nerves showed symptoms, and this was present since some weeks, evoking standstill instead of surgical alarming. This type of gaming-behavior can be observed more and more. The patient even was allowed to have meal at the evening. But neurological deterioration happens suddenly and unpredictable. Ophthalmological signs and symptoms are often decisive to predict developing of emergency.

A 17 years boy suffer from intermittent headache, dizziness and ataxia for about 2 weeks, with diplopia and vomiting he was admitted.

Ophthalmological findings: Visual acuity OD 0,9 OS 1,0 without RAPD, IOD 17/14 mmHg. Edema of the optic nerve OU with radial hemorrhage OD. Ptosis OD, palsy of abducens nerve OU, trochlear nerve OS, pathological saccades, up gaze nystagmus and nystagmus to both sides. Anterior eye segment and ocular fundus moreover normal OU.

Imaging findings: Big cystic tumor of the cerebellum, compressing fourth ventricle causing obstructive hydrocephalus. Herniation of tonsils and midline shift with compression of the prepontine cistern were visible. Torsion and dislocation of optic chiasm with compression of the stalk at the dorsum sellae.

To decrease the raised ICP in this case of obstructive hydrocephalus an endoscopic emergency procedure was done. Perforation of the floor of third ventricle and the roof of the interpeduncular cistern created a new pathway for the liquor to flow, releasing the raised ICP to normal and decreasing the risk of hydrocephalus post tumor operation to 50%.

Vomiting and headache disappeared immediate and diplopia resolved within days. The overall condition was normalized without external drainage having 50% less risk for postop shunt dependency.

After a sudden increase of diplopia, followed by headache and vomiting, at midnight, an emergency ETV had to be done. The procedure was uneventful, but it lasted longer than usually during day-surgery does. The liquor ICP intra-operatively was 30 cm H_2O. The tumor histology was cystic astrocytoma WHO I.

This case shows impressively, ophthalmological symptoms as typical warning signs before neurological deterioration in case of tumor of the posterior fossa. The ophthalmology gets a trigger function before neurosurgical emergency operation. Close and routinely cooperation between ophthalmology and neurosurgery can best manage these cases in time and the ophthalmological warning-signs should never ever be neglected.

This case is representative for planned ETV indication to resolve acute symptomatic in a first step, before going into the main surgery of a big tumor causing an obstructive hydrocephalus. This strategy, avoiding a ventricular drainage, enables to do the main surgery at an optimal timing, without keeping the patient in bed and having the infection risk additional. The patient and the tissue can recover under relaxed ICP (Graph 5.50).

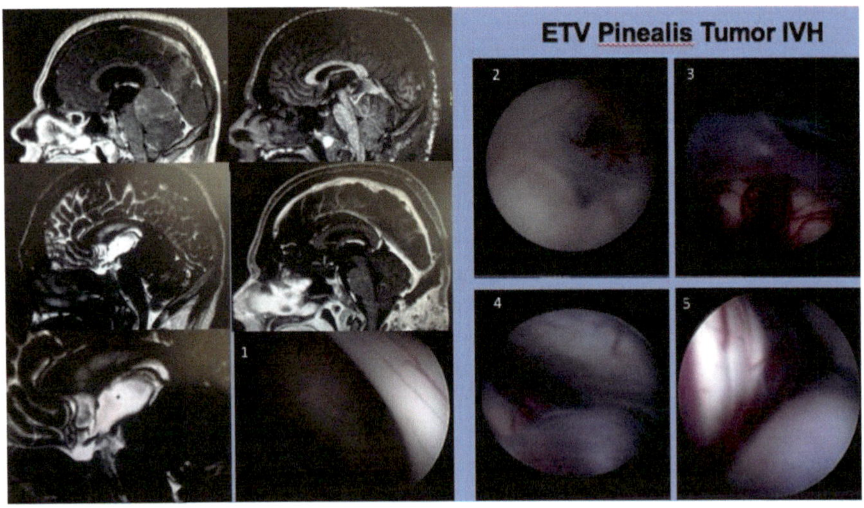

Graph 5.50 Case 14

Endoscopy was not of average difficulty, as the brain-stem and basilar artery were shifted ventrally. The tumor was visible in the posterior part of the third ventricle because the space between venous angle and inter-thalamic adhesion was large (1), and the tumor tended to bleed. The perforation was done rather ventrally (2) and due to the shift, all the thalamo-perforators were stretched (3) The way along the clivus and pons showed the bulging tissue of pons on both side of basilar artery (4) and compression of the pontine subarachnoid veins (5).

The patient went well recovered into the main surgery 14 days later, which was successful. After radiation, this pinealoblastom WHO 4° disappeared completely without recurrency and no hydrocephalus thereafter.

The step-wise strategy by MIN in lesions, in and around the brainstem and third ventricle, is well tolerated and prepares the patient for the main surgery without acute symptoms. After this experience the patient will have a strong trust to the following therapeutic steps.

This case represents, in contrast to case 6, an adjuvant indication, having a sever clinical course for one year after primary surgery of a craniopharyngioma (Graph 5.51).

Also, it represents the extremely anatomical- and tissue changes in such a case, making orientation and identification difficult.

Finally, it is an indication of ventriculoscopy and lavage of IVH in a situation, when shunting is contraindicated for an unpredictable time.

MIN should be applied in desperate cases rather than a complication promising shunt procedure.

This patient had, at the end, a list of 16 diagnoses with several sub-diagnoses in a multimorbid and minimal consciousness state, after two operations of a craniopharyngioma. Eight months before the second operation an IVH has occurred

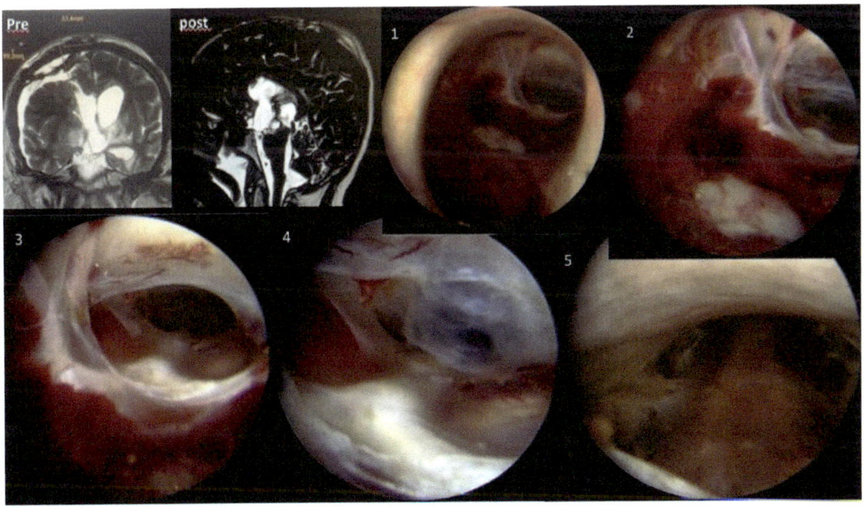

Graph 5.51 Case 15

causing an incomplete obstructive hydrocephalus. After growing of the remnant, visible in the MR, the second operation was indicated, the patient was in an absolute dependent status. The neuro-endocrinological deterioration did not recover and led to the above described situation.

During the ventriculoscopy for irrigation of the ventricles, and to prove the communication of the third ventricle into basal cisterns, until he level of $C_{2,}$ all the sever changes of arachnoid membranes, bleeding remnants and tumoral changes of the ependymal tissue were visible. The wall of prepontine cistern was, in addition, fenestrated into subdural (epi-arachnoidal) space, to further promote free flow into the spinal canal for compensating liquor peaks. During ICU stay and the many problems coming up, the concept of solving the liquor dynamics by MIN in a desperate situation, was not followed and a shunt was placed before he died. Here we see finally an ethical question, which can be answered in different ways.

The author would always prefer an individual decision over a standard decision in such a situation, recommending only MIN strategies in such a late phase, if not to stop therapy at all.

This case of a pre-term (24 SSW/700 g) baby represents a very rare compensation mechanism of an obstructive hydrocephalus due to obliteration of the aqueduct, and within peri-natal infection and bleeding. Moreover, it represents the need to analyze in each shunt-revision situation the possibility to substitute the shunt by an endoscopic procedure or to combine both concepts (Graph 5.52).

A complex dysplasia of the brain, compromising corpus callosum and cerebellum showed a cystic lesion supra-cerebellar (B)

In only 15 month of age the baby experienced already several operations with reservoir implantation, multiple shunts and shunt-revisions, and several drainages. The baby was born 2016, that means, more than 23 years after the first MIN

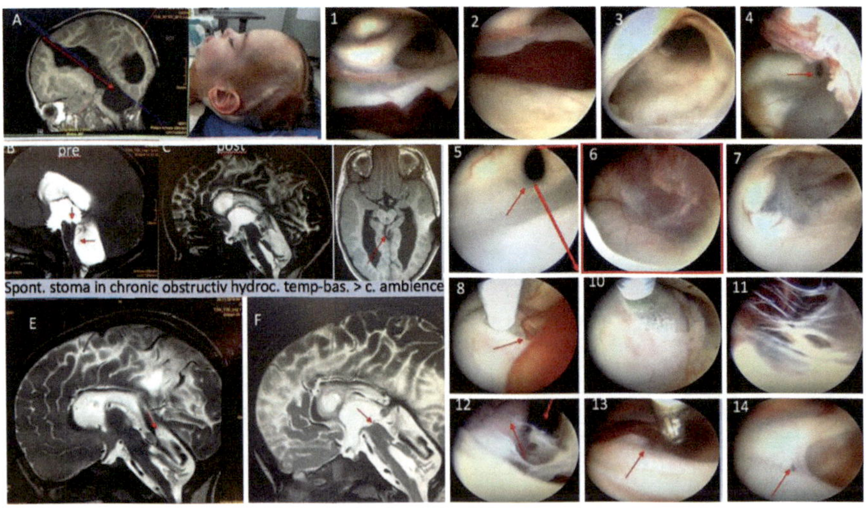

Graph 5.52 Case 16

Congress in Wiesbaden 1993 and appearance of the mayor papers of MIN, also 15 years after the founding of IFNE, primarily as a research worldwide net for neuro-endoscopy.

It needed 2 years after the first shunt to indicate an endoscopy through an already given approach in this difficult case, at least, to verify the unusual findings in the imaging: from the ventricular trigone left side an invagination into the posterior fossa has developed during the first year of life (axial MR), indicating a chronic obstructive hydrocephalus. It would be difficult to explain this in a communicating hydrocephalus. Even the pre-existence of an arachnoid cyst supra-cerebellar would not support communication in the aqueduct, moreover, the aqueduct is obliterated completely (B, C, F).

Even in CT, the trajectory to the natural developed stoma was given from the existing burr-hole of former operations (A). The pressure and shift in the posterior fossa was well visible in the pre-op MR for the endoscopy procedure (B). Post-op MR (E, F) showed release of the suprasellar arachnoid cyst and invagination of the left trigone of lateral ventricle. The fourth ventricle is visible again and the pre-pontine cisterns are present normally. The ETV flow is seen (C) but also the flow through the invagination pathway to the posterior fossa (E, F).

The insufficient shunt was substituted by a new one during the same operation to meet, additional to the obstruction (ETV), the post-bleeding and post-infection situation.

During endoscopy one first came into a mono-ventricle giving view to the fornix on both side and the left enlarged foramen Monroi. The right one was not visible due to a blood-clot recently set by a drainage catheter (1, 2). Directing the endoscope towards the trigon of the left ventricle, in the depth the natural developed stoma was visible soon (4). Approaching closer (5) it could be seen, that it is not artificial, and entering gave a view on the dorsal ambient cistern (6) with arachnoid covered pineal gland and the superior colliculi of quadrigeminal lamina.

Back in the frontal horn and passing wide open foramen Monroi on right side (3), the broad unusual overview of the third ventricle (7) with immature hypothalamic walls and typical signs of former bleedings in the ependyma is visible. Mammillary bodies are well in size, but ventricle width and floor was five-times too broad.

During the first perforation, a vein of the cinereum (pre-mammillary membrane) started to bleed (8) and could be easy be stopped by blowing up the balloon again (10). There-after the arachnoid trabecula of the Lilliquist membrane (roof of the interpeduncular cistern) became visible (11). After complete opening, the view goes along the basilar artery with remnants of the Lilliquist membrane (12). Finally, after taking away the blood clot with the biopsy forceps (13), the origin of that bleeding and primary catheter injury became visible. It is quite typical, that even soft catheters cause, some- times sever, injuries in the premature tissue.

At the end, the ventricular system, with obstructive and infect and post bleeding changes, has three pathways to compensate:—the naturally developed into posterior fossa through an absolute rare stoma,—the ETV and—the shunt system. According to the recent imaging controls, the stoma is strongly, the ETV less used, the shunt is at 130 cm H_2O adjusted.

Function of cerebellum and visual function will most sensitive show if the system is balanced well. Anyway, each case of surgical revision of the shunt system must be used to re-evaluate, if the shunt is necessary, or even, if the stoma must be closed, of course endoscopically.

The traumatized head of the baby tells us that we do not recognize MIN enough, even not in our youngest patients. But they need it most!

This recent case 17 (2/2019) represents together with case 1, a period of 32 years endoscopic procedures. It is, at the end of this series with few examples (what can I do with endoscopy?), an easy and benign case, the usual case one can meet at congresses and in industry launched brochures. It is the case often done with 6.5 mm endoscopes with nice, wonderful images, done with fixed endoscopes, stereotactically or by any kind of rigid holding-device, a technique with which you can do such cases, but not that once seen above.

However, even such cases are not free of long time of missing diagnosis and therapy, mostly and everywhere solved by implantation of shunt systems. By accident, and because the patient refused constantly a shunt implantation, in case 17 of the series, it was rather easy to get the indication brought through and to do the procedure within 40 minutes. Even in the convincing changes of the third ventricle (Graph 5.53), and the typical three-ventricular constellation of the ventricular system, the jet-flow of the aqueduct caused the primary advice for a shunt-procedure. (Aqueduct is believed to be patent according to flow void signal!)

During endoscopy, we look into a ventricular system of a 22-year young lady, that looks like that of an 80-year old man after a life in alcohol miss-use (1, 2). All the peri-ventricular tissue is atrophic and the floor of the third ventricle is only a

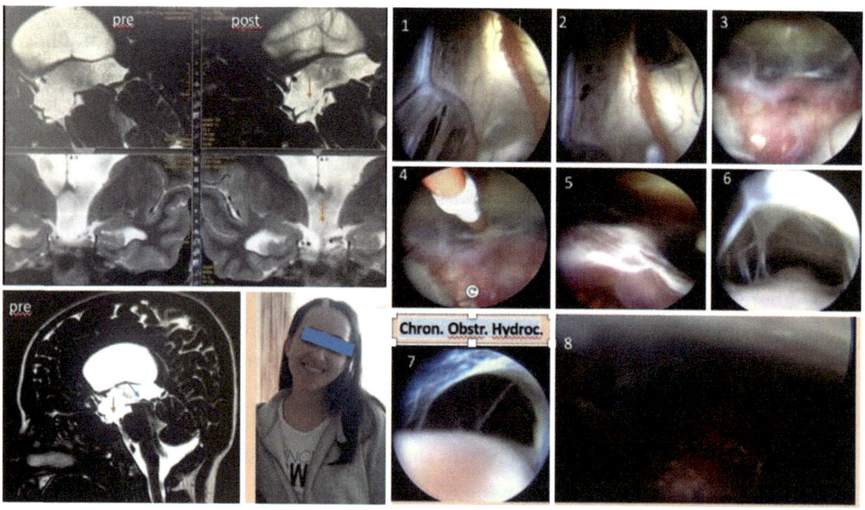

Graph 5.53 Case 17

membrane (3, 4). The complete posterior circle of Willis is visible through the slightly opaque membrane pulsating strongly. The transparency is surgically of benefit because one can choose the perfect place to perforate under visual control, well knowing that this kind of membrane may be very tuff, and it's easy to run into the pons and hurt it. Here it was surprisingly easy to perforate the membrane (4). However, with the correct angle and a given pathway from basilar tip (5) along the basilar trunc (6) and through an enlarged cisternal hiatus (7), a visual control until foramen magnum (8) could be done. This can guaranty, that no second or third membrane is overseen and obstructs the liquor flow at a deep caudal level.

The subarachnoid space always has a complex compartmentation with openings to let the mayor vessels pass through different compartments. These cisterns have therefor hiatus construction, and the most known is that of basilar artery, close to the meeting point of all 6 ventral cisterns of the brain stem (Graph 5.54).

However, this control tour through the basal cisterns along the basilar artery should only be done with good reasons, and by very experienced endoscopic neurosurgeons (with laboratory-training, and over 100 cases)!

Nobody can know which destructions, acquired for 22 years, this brain of our patient case 17, has to compensate, so that the she looks and functions normal during daily life.

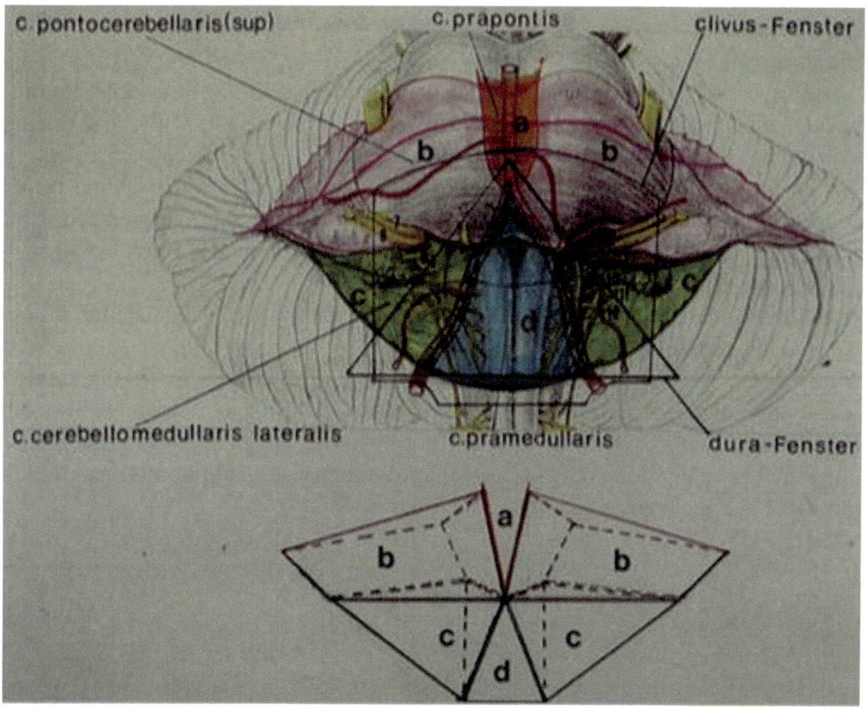

Graph 5.54 Compartmentation of ventral cisterns posterior fossa

The tiny flow through ETV in the postop imaging is no problem as we know from the endoscopy journey that the liquor-pathway is open. The signs and symptoms disappeared all soon and further recoveries in neuro-psychological field might be encountered in the near future. Such changes were observed regularly after ETVs.

In summary, in MIN the endoscope is not just a tool but a corner stone of the new concept and it is a **key-technique**. This technique must be learned in the lab and courses. Actually, though many advances of technique and trials were able, endoscopy is not used according to its unique possibilities to realize MIN with a broad spectrum of indications. The lag of trained endo-neurosurgeons, limits most the application actually. This is not acceptable for the patients, and for developing neurosurgery of the future. The actual model in neurosurgery is simply mechanistic and static, neglecting dynamic parameters and the complex CSF homeostasis. Modern, expensive shunt evolution try to compensate problems created by shunting. Endoscopy has therefor been established to have an alternative, if making sense, not disturbing the physiology of CSF and the brain. But all the efforts and advances are ineffective, if neurosurgery refuse to apply this MIN key-technique and to train the next generations.

The main contribution on the increasing of applications and indications in endoscopy came from the **Perneczky-school** by introducing **endoscopy-assisted microneurosurgery.** If the new industrial course of "exoscopy" will culminate in a benefit for the patients, or only support the industry and neurosurgeons' comfort, is not clear.

However, according to the core-philosophy of MIN, endoscopy needs to be combined with other key-techniques, like ultrasound (Chap. 4), LASER and sealing techniques (s. Vol. 2; Chap. 1 and 2) to provide the best MIN effects and results.

Finally, and according to another core-philosophy of MIN, this is an individual decision in each patient, and it needs an ethical education and open-minded habit rather than following statistical rules and not falsified recommendations by not clinical working professionals.

All the research results, and influences of CSF flow, cannot regulate the disturbance and secondary pathogenetic effects, and do not restore liquor pathways, if neurosurgery does not take over the responsibility. Neurosurgery is still the stenosis in providing the benefit of science for the patients, suffering of hydrocephalus. This is also true for other pathologies treatable by MIN strategy and endoscopic procedures. Other scientific results beside liquor-dynamics, actually do not lead to any therapy, but may be the future (Graph 5.55).

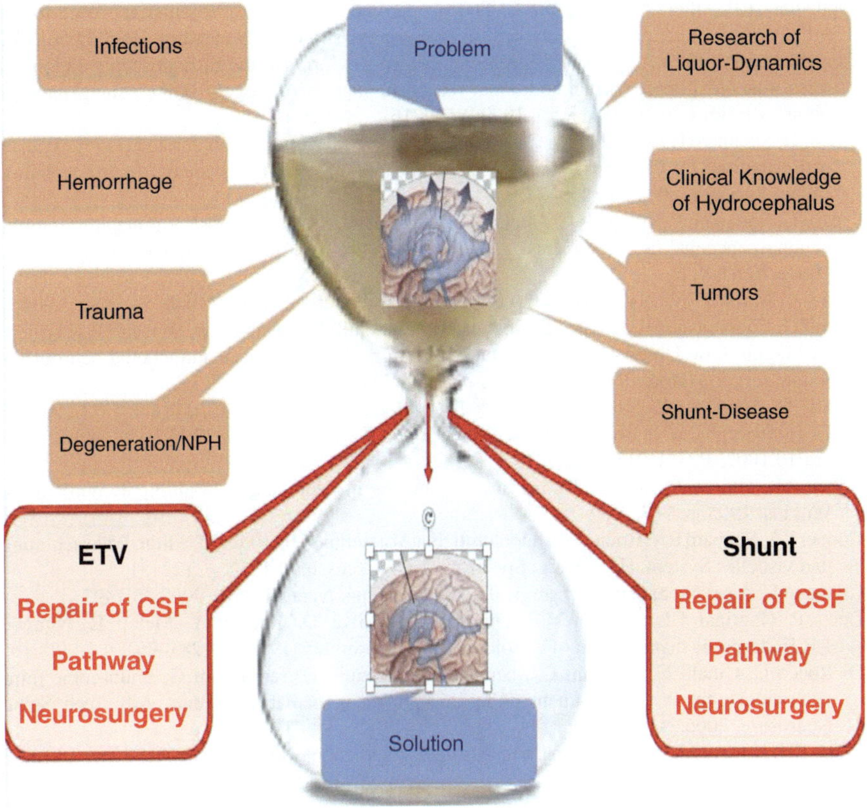

Graph 5.55 The therapeutic stenosis

Suggested Reading

Amberg R, Lemmen KD, Pollak S, Strutz I, Unsold R. Postmortale endoskopische Diagnostik. Beitr-Gerichtl-Med. 1992;50:241–50.

Auth DC. Fundamentals of lasers for endoscopy and la-ser tissue interactions. In: Jensen DM, Brunetaud J-M, editors. Medical laser endoscopy. Dordrecht: Kluwer; 1990. p. 1–16.

Bauer BL, Hellwig D. Minimally invasive neurosurgery I. Acta Neurochirurgia Supplement, vol. 54. Wien, New York: Springer; 1991.

Bauer BL, Hellwig D. Minimally invasive neurosurgery II. Acta Neurochirurgia Supplement, vol. 61. Wien, New York: Springer; 1994.

Caemaert J, Abdullah J. Endoscopic management of colloid cysts. Tech Neurosurg. 1996;1:185–200.

Caemaert J, Abdullah J, Calliauw L, Carton D, Dhooge C, Coster VR. Endoscopic treatment of suprasellar arachnoid cysts. Acta Neurochir. 1992;119:68–73.

Cappabianca P, Alfieri A, de Divitiis E. Endoscopic endonasal transsphenoidal approach to the sella: towards functional endoscopic pituitary surgery. Min Invas Neurosurg. 1998;41:66–73s.
Cappabianca P, Alfieri A, Colao A, Ferone D, Lombardi G, de Divitis E. Endoscopic endonasal transsphenoidal approach: an additional reason in support of surgery in the management of pituitary lesions. Skull Base Surg. 1999;9:109–16.
Cinalli G, Spennato P, Del Basso De Caro ML, Buonocore MC. Hydrocephalus and Dandy–Walker malformation. In: Cinalli C, Maixner WJ, Sainte-Rose C, editors. Pediatric hydro-cephalus. Milan: Springer; 2004a. p. 259–77.
Cinalli G, Spennato P, Ruggiero C, Aliberti F, Maggi G. Aqueductal stenosis 9 years after mumps meningo-encephalitis. Treatment by endoscopic third ventriculostomy. Childs Nerv Syst. 2004b;20:61–4.
Cinalli C, Maixner WJ, Sainte-Rose C, editors. Pediatric hydrocephalus. Milan: Springer; 2004c.
Cinalli G, Cappabianca P, de Falco R, Spennato P, Cianciulli E, Cavallo LM, Esposito F, Ruggiero C, Maggi G, de Divitiis E. Current state and future development of intracranial neuroendoscopic surgery. Expert Rev Med Devices. 2005;2:351–73.
Cinalli G, Spennato P, Nastro A, Aliberti F, Trischitta V, Ruggiero C, Mirone G, Cianciulli E. Hydrocephalus in aqueductal stenosis. Childs Nerv Syst. 2011;27:1621–42. https://doi.org/10.1007/s00381-011-1546-2.
Cohen AR, Haines SJ. Minimally invasive techniques in neurosurgery. Baltimore: Williams & Wilkins; 1995. p. 1–72.
Cooper LA, Shepard RN. Rotation in der räumlichen Vorstellung. In: Ritter M, editor. Wahrnehmung und visuelles System. Heidelberg: Spektrum der Wissenschaft; 1987. p. 122–31.
Cristante L. A set of coaxial microsurgical instruments. p4. Neurosurgery. 2000;45:1492–4.
Decq P, Guerinal CL, Brugieres P, Djindjian M, Silva D, Keravel Y, Melon E, Nguyen J-P. Endoscopic management of colloid cysts. Neurosurgery. 1998;42:1288–96.
Di Rocco C, Cinalli G, Massimi L, Spennato P, Cianciulli E, Tamburrini G. Endoscopic third ventriculostomy in the treatment of hydrocephalus in pediatric patients. Adv Tech Stand Neurosurg. 2006;31:119–219.
Frank EH, Horgan M. An endoscopic aneurysma clip applicator: preliminary development. Minim Invas Neurosurg. 1999;42:89–91.
Fries G, Perneczky A. Endoscope-assisted brain surgery: Part 2. Analysis of 380 procedures. Neurosurgery. 1998;42:226–32.
Fukuhara T, Vorster SJ, Luciano MG. Risk factors for failure of endoscopic third ventriculostomy for obstructive hydrocephalus. Neurosurgery. 2000;46:1100–11.
Gaab MR, Schroeder WS. Neuroendoscopic approach to intraventricular lesions. J Neurosurg. 1998;88:496–505.
Gillam B. Geometrisch-optische Täuschungen. In: Ritter M, editor. Wahrnehmung und visuelles system. Heidelberg: Spektrum der Wissenschaft; 1987. p. 104–13.
Griffith HB. Endoneurosurgery: endoscopic intracranial surgery. In: Symon L, editor. Advances and technical standard in neurosurgery, vol. 14. Heidelberg/Berlin/New York: Springer; 1986. p. 3–24.
Grotenhuis JA, Bartels RHMA, Tacl S. Intraoperative dislokation of the distal lens of a neuroendoscope: a very rare complication: technical case report. Neurosurgery. 1997;41:698–700.
Grunert P. From the idea to its realization: the evolution of minimally invasive techniques in neurosurgery. Minim Invasive Surg. 2013;2013:171369. PMCID: PMC3877623.
Grunert P, Perneczky A, Resch KDM. Endoscopic procedures through the foramen interventriculare of Monro under stereotactic conditions. Minim Invas Neurosurg. 1995;38:2–8.
Grunert P, Gaab MR, Hellwig D, Oertel JMK. German neuroendoscopy above the skull base. Neurosurg Focus. 2009;27(3):E7.
Guber AE, Wieneke P, Doczi T, Fries G, Gabbert U, Hüwel N, Lauschke U, Menz W, Neuhäuser U, Perneczky A, Reindl M, Schmidt F, Vogler K. Miniaturisiertes neuroendoskopisches Operations system. Minim Invas Med. 1996;7:23–31.
Haase J. Image-guided neurosurgery/neuronavigation/surgiscope—reflections on a theme. Minim Invas Neurosurg. 1999;42:53–9.

Heilman CB, Shucart WA, Rebeiz EE. Endoscopic spenoidotomy approach to the sella. Neurosurgery. 1997;41:602–7.
Heilmann CB, Cohen AR. Endoscopic ventricular fenestration using a "saline torch". J Neurosurg. 1991;74:224–9.
Hellwig D, Bauer BL. Minimally invasive techniques in neurosurgery. Heidelberg, Berlin, New York: Springer; 1998.
Hellwig D, Bauer BL, List-Hellwig E. Stereotactic endoscopic interventions in cystic brain lesions. In Meyerson BA, Ostertag C (eds) Advances in stereotactic and functional neurosurgery II. Acta Neurochir Suppl. 1995;64:59–63.
Hellwig D, Grotenhuis JA, Tirakotai W, Riegel T, Schulte DM, Bauer BL, Bertalanffy H. Endoscopic third ventriculostomy for obstructive hydrocephalus. Neurosurg Rev. 2005;28:1–34.
Hopf N, Perneczky A. Endoscopic neurosurgery and endoscope-assisted microsurgery for the treatment of intracranial cysts. Neurosurgery. 1998;43:1330–7.
Hopf NJ, Grunert P, Fries G, Resch KDM, Perneczky A. Endoscopic third ventriculoscopy. Neurosurgery. 1999;44:795–806.
Jendrysiak U, Resch KDM. Ergebnisse der klinischen Erprobung der Operationszugangsplanung mit NeurOPS. In: Bildverarbeitung für die Medizin, editor. Algorithmen-Systeme-An- wender. Proceedings-band. Heidelberg: Springer; 1999. p. 187.
Jho H-D, Carrau RL. Endoscopic endonasal transsphenoidal surgery: experience with 50 patients. J Neurosurg. 1997;87:44–51.
Jho H-D, Carrau RL, Ko Y, Daly MA. Endoscopic pituitary surgery: an early experience. Surg Neurol. 1996a;47:213–23.
Jho H-D, Carrau RL, Ko Y, Daly MA. Endoscopic pituitary surgery: an early experience. Surg Neurol. 1996b;47:213–23, 31.
Knoll M. Microsensors. In Wickham J (ed) Minimally invasive therapy 3. Medtech. 1994;94(Suppl 1):16.
Knosp E, Perneczky A, Resch KDM, Wild A (1993) Endoscopic assisted microneurosurgery. 1. International congress on minimally invasive techniques in neurosurgery. Abstractbook. p. 22.
Kockro RA, Serra L, Tseng-Tsai Y et al. Neurosurgical planning and training in a virtual reality environment. In Eleventh European congress of neurological surgery abstract book. 1999. Published online 17 Dec 2013. https://doi.org/10.1155/2013/171369.
Kockro RA, Serra L, Tseng-Tsai Y, et al. Planning and simulation of neurosurgery in a virtual reality environment. Neurosurgery. 2000;46:118–37.
Kriz W. Die Bedeutung der Anatomie in Forschung und Lehre. In: Bauer A, editor. Theorie der Medizin. Dialog zwischen Grundlagenfächern und Klinik. Heidelberg: Barth; 1995. p. 60–9.
Lang J. Klinische Anatomie des Kopfes. Berlin, Heidelberg, New York: Springer; 1981.
Limbrick DD Jr, Leonard JR, editors. Cerebrospinal fluid disorders. Switzerland: Springer; 2019.
Linke DB. Cognitive neuroscience foundations for a theory of neuronavigation. Comput Aided Surg. 1997;97:3–16.
Manwaring KH, Crone KR. Neuroendoscopy, vol. I. New York: Mary Ann Liebert Inc.; 1992. p. 1–2.
Marketresearchfuture: Neuroendoscopy market research report—forecast to 2023. ID: MRFR/MED/4380-HCRR | May, 2019 | Region: Global | 100 pages | Half-cooked research reports. https://www.marketresearchfuture.com/reports/neuroendoscopy-market-5836.
Matula C, Tschabitscher M, Reinbrecht A, Koos WT. Endoscopically assisted microneurosurgery. Acta Neurochir. 1995;134:190–5.
Misra M, Dujovny M, Alp MS. Endoscopic instruments. Surg Neurol. 1997;48:140–2.
Oka K, Go Y, Yamamoto M, Kumate S, Tomonaga M. Experience with an ultrasonic aspirator in neuroendoscopy. Minim Invas Neurosurg. 1999;42:32–4.
Perneczky A. Planning strategies for the suprasellar region—philosophy of approaches. Neurosurgeons. 1992;11:343–8.
Perneczky A, Fries G. Use of a new aneurysm clip with an inverted-spring mechanism to facilitate visual control during clip application. J Neurosurg. 1995;82:898–9.

Perneczky A, Fries G. Endoscope-assisted brain surgery: evolution, basic concepts, and current technique. Neurosurgery. 1998;42:219–25.
Perneczky A, Tschabitscher M, KDM R. Endoscopic anatomy for neurosurgery. Stuttgart: Thieme; 1993.
Perneczky A, Müller-Forell W, van Lindert E, et al. Keyhole concept in neurosurgery. Stuttgart: Thieme; 1999.
Poggio T. Wie computer und Menschen sehen. In: Ritter M, editor. Wahrnehmung und visuelles system. Heidelberg: Spektrum der Wissenschaft; 1987. p. 78–89.
Regan D, Beverly K, Cynader M. Die Wahrnehmung von Bewegungen im Raum. In: Ritter M, editor. Wahrnehmung und visuelles system. Heidelberg: Spektrum der Wissenschaft; 1987. p. 90–103.
Reisch R. Aesculap neurosurgery MINOP TREND: the endonasal transsphenoidal biportal binostril approach. Germany: Aesculap Co./B Braun Co.
Resch KDM. Beitrag zur Zugangsanalyse und zum Zugangsdesign des transoral-transpharyngealen Weges zum Hirnstamm. Dissertation: University of Heidelberg; 1991.
Resch KDM. MIN: transoral transpharyngeal approach to the brain. Neurosurg Rev. 1999;22:2–25.
Resch KDM. Endo-neuro-sonography: I. Technique, methods and anatomy. Neurosurg Rev. 2001a.
Resch KDM. Endo-neuro-sonography: II. First clinical series (52 cases). Neurosurg Rev. 2001b.
Resch KDM, Perneczky A. Endoscopic approaches to the suprasellar region: anatomy and current clinical applications. In: Bauer LB, Brock M, Klingler M, editors. Advances in neurosurgery, vol. 22. Heidelberg: Springer; 1993. p. 126–33.
Resch KDM, Perneczky A. Endoneurosurgery: anatomical basics. In: Samii M, editor. Skull base surgery. Basel: Karger; 1994. p. 78–80.
Resch KDM, Perneczky A. Mikro- und endoskopische Anatomie des supraorbitalen Zugangs. Beispiel für das Konzept minimalinvasiver Technik in der Neurochirurgie. In Schmelzle R (Hrsg.) Schädelbasischirurgie. 6. Jahrestagung der Deutschen Gesellschaft für Schädelbasischirurgie e. V. 2000. S. 230–6.
Resch KDM, Bohl J, Perneczky A. Postmortal inspection, a new pathoanatomic method. Clin Neuropathol. 1992;11:191.
Resch KDM, Bohl J, Perneczky A. Endoscopy to the basilar artery: anatomical and pathological aspects. Clin Neuropathol. 1993;12:263.
Resch KDM, Perneczky A, Tschabitscher M, Kindel S. Endoscopic anatomy of the ventricles. In: Bauer BL, Hellwig D, editors. Minimally invasive neurosurgery II, Acta Neurochirurgieca Supplement, vol. 61. Wien: Springer; 1994. p. 57–61.
Resch KDM, Reisch R, Hertel F, Perneczky A. Endo-Neuro-Sonographie: eine neue Bildgebung in der Neurochirurgie. Endoskopie Heute. 1996;76(12):152–8.
Resch KDM, Mazánek M, Perneczky A, Stoeter P. Grenzen der 3D-CT Planung in der Endoneurochirurgie. Endoskopie Heute. 1997;3:14–20.
Reulen HJ, Steiger HJ. Training in neurosurgery. Acta Neurochir (Wien) Suppl. 1997;69:58–82.
Riegel T, Hellwig D, Bauer BL, Mennel HD. Endoscopic anatomy of the third ventricle. In: Bauer BL, Hellwig D, editors. Minimally invasive neurosurgery II, Acta Neurochirurgia, Supplement, vol. 61. Wien, New York: Springer; 1994. p. 55.
Sainte-Rose C. Third ventriculostomy. In: Manwaring KH, Crone KR, editors. Neuroendoscopy. New York: Mary Ann Liebert; 1992. p. 47–62.
Sampath P, Long DM, Brem H. The Hunterian neurosurgical laboratory: the first 100 years of neurosurgical research. Neurosurgery. 2000;46:184–95.
Schmidt RH. Use of a microvascular Doppler probe to avoid basilar artery injury during endoscopic third ventriculos- tomy. J Neurosurg. 1999;90:156–9.
Schroeder HWS, Warzok RW, Assaf JA, Gaab MR. Fatal subarachnoidal hemorrhage after endoscopic third ventriculostomy. J Neurosurg. 1999;90:153–5.
Seeger W. Atlas of topographical anatomy of the brain and surrounding structures. Vienna, New York: Springer; 1978.
Seeger W. Planning strategies of intracranial microsurgery. Vienna, New York: Springer; 1986.
Tandler J. Lehrbuch der systematischen Anatomie, Band 4. Leipzig: Vogel; 1929.

Taniguchi M, Takimoto H, Yoshimine T, Shimada N, Miyao Y, Hirata M, Maruno M, Kato A, Kohmura E, Hayakawa T. Application of rigid endoscope to the microsurgical management of 54 cerebral aneurysms: results in 48 patients. J Neurosurg. 1999;91:231–7.
Teo C. Third ventriculostomy in the treatment of hydrocephalus: experience with more than 120 cases. In: Hellwig D, Bauer B, editors. Minimally invasive techniques for neurosurgery. Berlin: Springer; 1998. p. 73–6.
Teo C, Jones R. Management of hydrocephalus by endoscopic third ventriculostomy in patients with myelomeningocele. Pediatr Neurosurg. 1996;25:57–63.
Von Weizäcker V. Der Gestaltkreis. Stuttgart: Thieme; 1950.
Wallach H. Wahrgenommene Stabilität der Umgebung und Eigenbewegung. In: Ritter M, editor. Wahrnehmung und visuelles system. Heidelberg: Spektrum der Wissenschaft; 1987. p. 114–21.
Weinberg R. Future direction in neurosurgery visualization. In: Apuzzo MLJ, editor. Neurosurgery for the third millennium. USA: AANS; 1992. p. 47–64.
Witt H, Kozianka J, Waleczek H, et al. Das Erlernen und Optimieren minimal-invasiver Operations verfahren am menschlichen Leichnam. Chirurg. 1999;70:923–8.
Yasargil MG. A legacy of microneurosurgery: memoirs, lessons, and axioms. Neurosurgery. 1999;45:1025–91.

Scientific Conditions for MIN

6

If a surgeon is doing a bad job due to lack of education, knowledge and experience, he will produce harm to some hundred or even thousand victims. However, if a doctor or scientist produces wrong theories and misleading statistics he will damage through a long time period and wide spread geography, causing harm to all patients of perhaps a whole generation of surgeons and physicians (Graph 6.1).

Developing the future of medicine therefore needs the **balance between technical and intellectual infrastructure (Resch 1991)**. The brain and intellect of the surgeon is the most important "surgical" instrument. It needs an education in science theory and history, not only in bio-mathematics, to accomplish the mission.

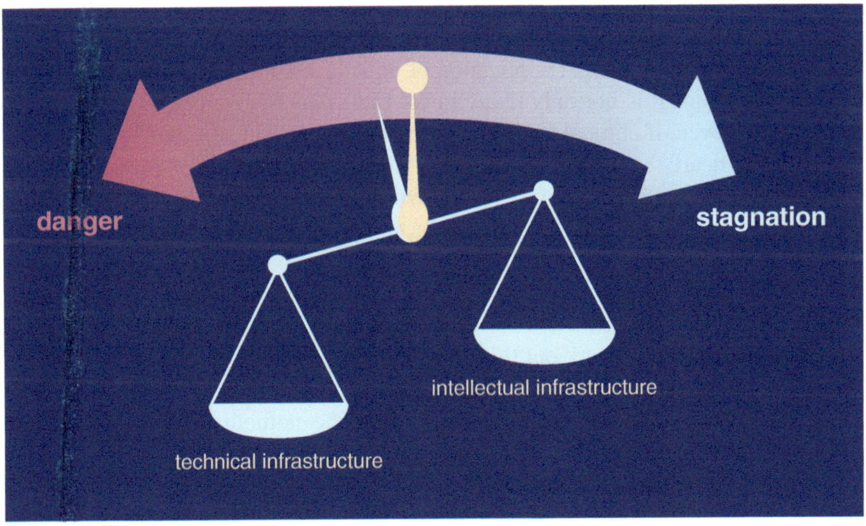

Graph 6.1 Ergonomics libra

K. D. M. Resch, *Key Concepts in MIN - Intracerebral Hemorrhage Evacuation*, Key-Concepts in MIN 1, https://doi.org/10.1007/978-3-030-46513-1_6

6.1 The Limits of Clinical Trials Within the MIN-Key Concept

STICH I and II did not take any effective regard of MIN. The mayor concepts, techniques and strategies of MIN had been already worked out before the beginning of STICH. The first international congress of minimally invasive techniques in neurosurgery took place already in 1993 (Germany) and the major paper by Perneczky et al. appeared in Neurosurgery 1998 Feb. 42(2): 219–24.

From an ethical, biological and logical standpoint of view it was critical to ignore MIN by STICH trials and to deliver a curiously amount of data to the neurosurgical community, data, which were not up to date with the best of surgical technique and concept available already since 1990. But why should patients in Europe be treated with a standard of poorly developed countries? However, neither Yasargil nor Perneczky ever applied systematically their surgery in intra-cerebral haemorrhages.

Actually, after the majority of patients are left back at the stroke units and ignored by neurosurgeons, at least in Europe, this lack of the STICH-era regarding MIN, is being tried to work up by new trials. However, all actual trials on MIN, like MISTIE, CLEAR or ENDO-STICH misunderstood the absolute core of the MIN concept, elaborated and proved by Perneczky and others:

MIN means to master a spectrum of techniques with minimally invasive characteristics, like high-end microsurgery, endoscopy, ultrasound or LASER and sealing techniques. In each individual patient the MIN-surgeon must decide which combination of techniques should be applied. The designs of all actual trials on MIN techniques, however, examine only one technique alone. By that, the virtual ineffectivity of MIN-techniques will result predictably due to methodological reasons.

By nature, clinical trials (RCTs) can only answer simple questions with monotechniques, regarding surgical methods. However, the superiority of MIN comes from the precise analysis and application of, mostly, combined techniques.

Regarding this, it is not only that clinical trials may be overemphasized in general, but also even, if clinical trials are of any benefit in evaluation of MIN.

It became nearly common and necessary to re-evaluate results of clinical trials by "real world data" and to confront efficacy with effectiveness. MIN should be evaluated only by real world data, because of the complexity, especially during the start-up phase. In such a phase, the meaningful equipment to evaluate the new methods are a matter of development in parallel. While it may be un-doubtful in daily practical reality, seeing a clear benefit for the patients, the scientific result measurement might be still unable to work out this benefit. This "**scientific gap**" can be analysed regularly during evolutions and innovation-times and moreover, it is an artefact of exclusively statistic-driven science, stripped of the nomothetic and ideographic background.

Such methodological handy-cap causes, that somehow scientists sometimes seem to be the blind group in the community. To the senior authors own experience through the STICH I recruitment time, even in well developed participating countries, the correct collection and handling of data became a critical issue. The high ranking of RCTs is not justified if nobody is able to and does control the confidence of the individual data. Usual such data input is already a multiple selection of the real cases that were treated in reality. These uncertainties at the very beginning of the trial bear an incalculable and unpredictable bias, no matter how big the amount of cases may be in such trials. The reliability in STICH I was doubtful from the very beginning, however, once the messages were established, these errors stay for a decade or even much longer, having a kind of autonomous life in the following papers and minds of most professionals. Such effects and side effects might have urged the authors of STICH I to a comment (in STICH update 2007) remembering the surgeons to go on with decision making and not just to hide behind the statistics of STICH. "**An unfortunate outcome of STICH I has been that many people have misinterpreted the results to argue that there is no need to operate on patients with ICH at all. To leave patients with lesions that should be removed (an unfortunate misinterpretation of STICH) would condemn such patients to non-operative treatment perhaps for evermore.**" This sounds like a weak excuse, but there is in real not a misinterpretation but a systematic failure of a type to practise science exclusively by RCTs and believing in them like in a religion.

"Statistic information from an RCT is virtually not interpretable and meaningless if stripped away from the backdrop of basic understanding of physiology and biochemistry."

6.2 Missconception of "EBM"

The critical question is: **What is the evidence of evidence?** And, what is the evidence of EBM?

If things are unclear or probability is completely contradictory to what we see in reality, it is wise to go back to history and one will mostly find answers. Future is a vector coming from the past. Philosophically to rise this question is enough, but not for scientific working and never for medical practise. The reason: the task for philosophy is to create an immense consciousness but, for science to create reliable theories and knowledge. However, for medicine the goal of learning is experience, dexterity and empathic emotional caring.

Since about **3000 years**, it was not possible for **philosophy** to create a general accepted concept of evidence, that could be the basis for scientific use. Scientifically, **Carl Popper** finally even had to try to preserve rationalism for science by the **theory of falsification**. Do not try to prove a theory by evidence rather than to try showing that it is false. If you cannot, the questioned theory is preliminary the gold standard.

Since **Kurt Gödel**, the belief in any evidence, even in **mathematics**, became nonsense! At least we have to accept now, that "evidence" philosophically, scientifically and mathematically is a crucial matter, and that we should be aware about these axiomatic conditions before we believe, that evidence is a safe procedure in science. We may see that something is true but we should be aware, that we cannot prove it and that "truth" and "evidence" is not the same! **"Evidence Based Medicine" is a label**, but it is not true. The term EBM provokes misunderstandings and is easily victim of dogmatism:

The term "evidence" is misused in EBM to rectify therapeutic procedures. But "evidence" is not a rectifying category! In EBM "evidence" is identical with probability (r) in the system: hypothesis (h) proven by experience/data (e) is probable (r). The conditions of this sentence are extremely complex and some how axiomatic. In reality, nobody can guarantee in a trial design these conditions. Neither any adequate hypothesis nor the experience data definition will be enabled according to the above standard. But once this is frozen into a formalism of bio-mathematic and statistic-rules the system can hardly be proven whether it generates truth. Few people will be able to prove the correlating grammar in scales and mathematical terms. Does the term and formalism represent and reliably generate true results? Reliability of this system cannot be controlled by any user. He has to believe it or not and if frozen into a rule or dogma it becomes the characteristics of a misusing religion. Can science only mean to follow rules? Actually rising use of "real word data" controls of RCTs is not just a repair of EBM but the result of the above described conditions and reasons for the lack of reliability of EBM. So, what is the evidence of EBM?

However, we can learn from scientific disciplines, which are used to work with high dynamic open processes in multi-level designs. In "Psycho-Neuro-Immunology" (PNI) this is normality. Their "**multi-level models**" were created to better understand and control scientific research of **complex systems**. They have very early learned throughout the last 5 decades, that **trial design can produce artefacts**, which have nothing to do with reality or experience. All levels A; B; C; D (Graph 6.2) must be accepted and applied in medicine. Actually, we have a mono-RTCs ideology that cuts medical science from multi-perspective and multi-methodology, major roots of science.

PNI has shown, that correlations to explain complex and non-linear processes, do not present in easy linear mono-correlations detectable by usual RCTs, but rather appearing in patterns, presented in multi-level and time-course research. Clinical processes do follow the rules of open non-linear and chaotic processes and need a multi-level design of research. RCTs do play a minor part in such a scientific environment.

Thomas Kuhn developed the perspective that science, from a historical point of view, does not function according to rationalism but is a **social process**. He claimed that there is an environment of beliefs by the scientific community, namely "**paradigms**", that change when new findings cannot be in concordance with the actually believes, causing a crisis that culminates in a "**paradigm shift**".

The great founder of quantum physics, **Max Plank**, even stated in resignation at the end of his life, that the improvement of science does not come by learning, but rather by the **change of a generation**. A vast amount of literature exists within this environment of high value to keep the perspective multiple and the awareness high for the on-going process of science.

In the **nineties** of the last century, a new believe grew within a very small part of the scientific community, coming from **bio-statistics** and **epidemiology**. This origin, far from true clinical work, must be kept in mind, when we analyse the kind and

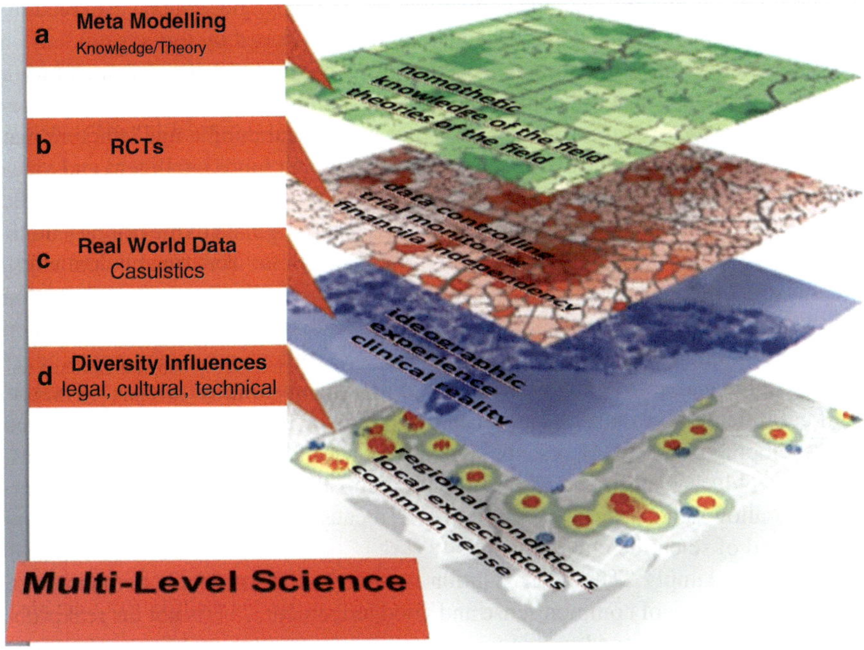

Graph 6.2 Multi-level research/root of science

way of "acceptance" career of EBM. It is best described by the term: "**hype**". **There was never a conclusive prove or better, a falsification of EBM**. What we saw was a suppression of poly-perspective and multiplicity that was not been described so far, even by Kuhn, as it remembers more to a political—than to a scientific—processes, becoming a strict dogmatism in a very short time. Especially Europe and North-America were place of this explosive event that desperately remembers characteristics of malignant growing. Not difficult to recognize, this could only spread within a generation of digital natives during an immense economic pressure and a change from medicine to medical industry.

Unbelievable the story, that in Germany, this started up with an epistemological accident, when the first authors took over the term "evidence" directly in their publications, which but has the meaning of obviousness in German language. This seemed to become the intellectual fate of the process in Germany. Until today the effects of this accident can be found in the minds of young doctors but reaching now already the higher levels of professionals.

At a first glance, the structure of EBM system seems simple, but by the time it became confusing complex and impracticable. Even specialists can have troubles in applying it correctly. To overcome this pitfall, a secondary simplified system was derived: "grade of recommendation". This is a rarefication method and the cut off criteria are quite inconclusive. However, the system offers a nice structure, very attractive for people, who like to just follow rules. I never experienced the situation that a medical professional analysed the exact level of evidence or at least the grade of recommendation. But if so, the temptation to follow a single scale instead reading the paper and analysing the data became now widely opened by this system.

All "benefits" of EBM however, become forgettable when it got frozen into a hierarchic system with a dogmatic character and a manipulative distribution of values. The hierarchic-pyramid as a mental order is a typical middle-age structure that causes until today many errors in science. Several much more intelligent and valuable intellectual structures have already been developed in different scientific fields. This rigid structure of a hierarchic-pyramid carries the stigmata of the primary motivation by which it seems to be driven and from its origin far from true clinical work. The poison which it implemented since then into the academic community can be experienced in each congress and the academic discussions, but even deeper, when dividing, into: political correct professionals and the others, who seem not to be up to date. In the end, maybe, intended or not, it has an anti-innovative effect. The scientific publication—"industry" produced all kinds of the problems that provoked the developing of the EBM system, so it simply tries to solve self-made problems. The new "industry" of trials causes more problems than it can solve. The clear tendency to allow innovation only through RCTs causes a very slow, expensive and boring kind of scientific environment.

But finally simply "The other solution is to insist that the levels be interpreted with a healthy dose of common sense and good judgement, …", is not the real problem. The above-mentioned problems are working by institutional force. Once you have established a certain ideology, you cannot avoid that it is miss-used or making continuing collateral side effects. These are typical mechanisms of steering function to bring evolution politically to a certain direction. It works like a very silent and calm corruption and destroys the illuminating spirit of science.

EBM invented a virtual world that has few to do with reality. It creates virtual patients and diseases, fitting into the formats of RCTs. It developed virtual doctors and a busy "scientific industry" that has nothing to do with clinical reality and real problems. And finally it causes virtual problems trying to solve them hardly with an enormous amount of efforts, which would be urgently needed in clinical real world.

The fatal fate of ICH-evacuation surgery is a paradigmatic example how a statistic powerful trial, according perfectly to all recommendations of EBM, can obstruct the evolution of a surgical discipline. Nearly a whole generation of neurosurgeons forgot to train and improve the art and techniques and even the concepts of ICH-evacuation. A complete technical evolution step, namely MIN, was and is neglected. Only few pathophysiological or biological reflexions could be seen anymore in the discussions and publications displaced by a digital coma of "data tsunamis".

Taking 4 Mill. ICH-patients worldwide, one can imagine how many patients with a reliable good chance of favourable outcomes suffer in the stroke units or elsewhere, obstructed from effective and fast recognition and therapy. Who has the right to statistically decide the worth of an individual life?

6.2.1 The Hippocratic Imperative

What was the first line imperative of our founding father Hippocrates during the axis-time of history? "At least do not harm!" But why did he demand on such a low goal from us, nearly the lowest aim one should expect? One degree lower, he would have to urge us, please do not kill them! Disappointed, one has to consider, however, that even this low imperative is the most hurt rule in medicine. MIN is dedicated to over-come this, in therapy and in science.

6.3 The MIN-Key Concept Between K. Popper and Th. Kuhn

The practical and conceptual basics of the MIN-Key Concept are described in Chaps. 3 and 4. The theoretical basics are now summarized in leaving the scientific methods of EBM behind in the dark past, where it comes from. The elaboration of the MIN concept was purely driven by real world data analysis and design. The evolution of the Key-hole Concept to the MIN-Key Concept is also described above. The scientific environment does function as an "Axiomatic Field" that can be placed in-between the two symbolic poles C. Popper and Th. Kuhn. Additional the "Gestalt theory", "neuropsychology for neurosurgery" and "fuzzy logic principles" belongs to the poly-perspective and multiplicity of the intellectual equipment of this axiomatic web environment.

The starting point of MIN-history may be seen in G.I. Rossolimo: "Despite success in surgical technique permitting to perform hemi-craniotomy for removing a small intracranial lesion, one cannot but admit, both a priori and from clinical experience, the necessity to minimize surgical injury and approve all methods of precise localization of a cerebral lesion." "GI Rossolimo, J Neuropath Psych Korsakow **1907**; 7: p 640."

"His brain…the soul's frail dwelling house" (W. Shakespeare) is the title of J. Thorwald's book of the History of Neurosurgery, reaching from the discovery of the Edwin Smith papyrus in 1862 until the death of W. Penfield in 1976. But the power of evolution towards MIN came from M.G. Yasargil in the 1970th and culminated with A. Perneczky, worldwide recognized during the "First International Congress on Minimal Invasive Techniques in Neurosurgery"; June 13–15, 1993; Wiesbaden/Germany (see "Recent roots of MIN"—Chap. 3).

The unique exoteric philosophy of A. Perneczky was, that the overall trauma needs to be kept as small as possible but in the same time reaching the surgical goal not less or even better. The ethical view was, that we have to involve all modern techniques available and integrate it early into the surgical environment. The new techniques and approaches were tested in a near surgical simulation laboratory setting regularly (Vol. 2). Feasibility, probability and workflow of ergonomics were tested and documented. Translational approach and direct knowledge transfer were part of the philosophy as well. This was a true ethic, not asking first, if it was commercial correct, but recommending the same rights for medicine as for military. The whole project was completely experience-based causing the fastest and meticulously controlled progress with emphatic discipline and a strong corporate identity of the team involved. This tightly monitored project reached a reliability and safety of all data, which the high ranked methods of EBM do not and cannot reach.

In the strict close to surgery simulation model (see Vol. 2 “Laboratory Simulation and Training for MIN”) proof and falsification became the same, and the social process was part of the project itself. This creative model of scientific clinical working, surgery and laboratory in parallel, would rank in EBM very low without any chance in competition with even the most stupid and unreliable data input RCT (who ever controlled input data on side in over 80 participating countries of STICH?). One must ask, if EBM can have any benefit for surgical procedure outcome measurement, when it is stripped of the complex effectors that make the surgical process. Surgical procedures outcome measurement can only be evaluated within its complete context and this is best done by real world data. The naive believe to find any “evidence” is in contradiction with the falsification theory of Popper and misuses the concept of Kuhn’s paradigm shift, as EBM authors themself believed to be a paradigm shift, but without underwent any falsification. But what we see real is only a clear communication-shift.

This book is an **Experience Based Medicine** (ExBM) contribution. It belongs to a scientific axiomatic web together with EBM, the falsification theory of C. Popper, the paradigm-shift concept of Th. Kuhn and many theories and concepts involved in MIN. The shortly described **scientific axiomatic web** is the theoretical basic of this book.

6.4 Misconception of “Image Guided Therapy” in Neurosurgery

Modern imaging techniques were strongly involved in advancing neurosurgery:

Roentgen	1895	Radiography
Dandy	1918	Encephalography
Moniz	1927	Angiography
Ambrose/Hounsfield	1972	CT
Hawkes	1980	MR
Since	1980	CTA, MRA, PET, SPECT

This speedy development now led partially to the field of “imaging-guided” neurosurgery which is believed to being the future. But this concept was also viewed critically, as the old paradigm of Da Vinci: “to see is to understand” is not true anymore. We seem to be far away from understanding all, that can be imaged today. The fast evolution has left behind critical understanding of limits in competence of imaging guidance.

For neuroradiology it is hardly possible to imaging needs of operation planning data for minimally invasive approaches or to interpret postoperative changes. Evolution from invasive to non-invasive imaging as well as from analogue to digital techniques has the important drawback of decrease of optical solution. At the same time technical abilities of neurosurgery allows an accuracy far superior to that of pixels or voxels. The actual course of development of computer assisted imaging is only safe in a close team-work management.

Neuro-radiological images have a different optical grammar being a result of mathematical and technical operations, compared to natural images as a result of bio-evolution and experience. To get a training effect, the artificial neuro-radiological image must be directly compared with the original anatomy.

There is a difference in visualisation by optical means and imaging by digital technique. V.v. Weizäcker already described the visual process in his book "Der Gestaltkreis": "The optical apparatus and the ZNS do not have a mathematical image of the space, indeed it forms the reality of space on-line". Digital media separate description of task and their hardware but the brain does it all together, there is a bio-evolution context.

If one try to get a computer to see it will be recognized that viewing is a complex process of information processing. From neurobiological point of view visualisation is a learning process of recognition. Digital imaging is the result of mathematical processing and ends in a pixel- and voxel-world. In a surgical operation visualisation takes place by using "biological-coordinates":

- 3D-system of the vessel tree
- Sulco-gyral system
- Cisternal space system
- Ventricular cavity system
- Dural cover system
- Bio-pattern "finger-print" system

The usual navigation systems cannot "understand" biological coordinates. It is one of the reasons why they choose mostly trans-cerebral routes instead of MIN corridors of the subarachnoid pathways. They do not calculate the biological values of the biological coordinates. The only MIN contribution of navigation systems (image guided) is the positioning of the craniotomy. After opening the dura and arachnoid space the data are not reliable, but even miss-leading. However, more important is that image-guiding concept of these systems leads to a non-MIN strategy. The lack of all the physiological and pathophysiological knowledge in the planning of the image-guided concept ends in a regression compared to M.G. Yasargils' micro-neurosurgery and to MIN. From the perspective of MIN image-guided therapy is a miss-concept. For MIN the concept-guided therapy is absolutely indispensable, using not only data but all kind of involved and comprehensive knowledge.

6.5 System Medicine and Big Data Medicine

We are hardly able to control bias of concepts bearing fundamental errors. Near future confronts us with the next tsunami of data challenge: Big Data Medicine and System Medicine. Why should we expect that medicine is immune for the pitfalls we saw in politics, secret services and digital companies like Facebook, Google and others: since Snowden, we cannot relay anymore, trust to anything in the digital world. Automatic of data-mining and deep learning provides us with complete new and in-transparent kinds of bias. Managing of manipulated bias will be a major task in future. However it remains unclear if the concept of big data medicine is valid or if it leads into a digital dementia of science. We have to be aware, if all this may even be part of cyber-illness, as M. Spitzer has shown for social life and society. One of the most reliable and transparent methods is the use of real-world data to control in parallel digitalized procedures. As shown above, this is the exotheric philosophy of this book regarding the MIN concepts. It is of immense importance for science to follow-up this revolution and to remember always that it is impossible to prevent the temptation for international business to abuse the opportunities of big data. The publication-industry itself has shown this in the recent past and dramatic events so far.

6.6 The AI-Systems

Neurosurgeons belong to the neuro-family, but have the extraordinary task at the heterochronic point, where function, structure and meaning is absolutely irreversible connected, giving neurosurgeons an additional obligation, as they are the only individuals in the family, that experience the brain in its' bio-phenomenological being, its' bio-physical fragility and vulnerability.

From this situation, we have this unique task to preserve not only the individual brains of our patients, but also the unique responsibility to preserve the incomparable meaning of the brain for mankind.

Within this context, we need to be aware of technical simulations and pretendings of brain abilities and to give testimony about the differences to biological reality. One of the most obscure activities of IT is to integrate and miss-use and transduce ideas, concepts and terms from the analogous world into the IT world, and to take over all the semantics, semiotics and meanings for procedures of incomparable lower technical abilities and functions. They do label fraud! This was allready described by Roger Penrose: "Shadow of the Mind"/ 1994 New Yourk.

If our human languages and thoughts and understanding of meanings is the mirror in which we see and learn the world, it could be self-murdering for mankind to allow some people this miss-use, just for a secondary of fast satisfaction and money-making. It would at least be the end of science, to destroy or manipulate the mental basis and framework of homo sapiens-sapiens.

The sparely controlled speedy revolution of so-called AI-systems has taken over the label of "human intelligence" from those who believe to be the winners of this trip, but they contribute to the "climate-catastrophe" of science.

AI-systems are only dominant symptoms how IT is working and functioning contemporary and against human brains. The neuro-community should be alarmed and take over the responsibility and task to avoid miss-use and protect the human brain at this key-level.

6.7 The Role of Philosophy in MIN

It has become usual to mix up "concept" with "philosophy". Contemporary philosophers claimed the term „exoteric philosophy "for this kind to express an overall concept, distinguishing it from esoteric philosophy. Philosophy in neurosurgery was mostly used exoteric and some times used only to glorify the profession.

However, there is an indispensable role for (esoteric) philosophy in neurosurgery:

First: "Medici sumus, homines simus" (L. Krehl; We are physicians, let us also be humans.) For to be a human, philosophy is a constitutional reality. Starting with the questions of children and of suffering patients, asking for the meaning of their suffering. Philosophy is allways present where humans are.

Second: it seems to be impossible to emphatic practice neurosurgery without philosophy, dealing with suffering patients in general and with the organ, that interacts with "spirit "creating consciousness, self- awareness and intentionality specifically.

Third: the reason to philosophize is not to gain specific knowledge on something, rather than to illuminate the own cognition and to raise the consciousness and self-awareness.

Fourth: to understand science and its role and meaning, but also its limits in medicine, we need to know about science theory which is primarily a subject of philosophy. The recent example is the appearance of "quantum philosophy" within application in brain research.

Fifth: neurosurgeons are the unique elite within the "neuro-community" which has the privilege to see and experience the living human brain directly. To know "…the soul's frail dwelling house", while standing at the crises of life, learning what might be once own future, is a irritating reality that needs a strong orientation and stand point. This can be created by philosophy.

Sixth: MIN is still in its childhood confronted with many obstacles and it needs a strong exoteric and esoteric philosophy being innovative within a very conservative and some-times, to MIN, hostile profession.

Seventh: philosophy represents best the phylogenesis of consciousness and anatomy of mind, especially in the perspective of evolution of culture, brain and mind.

6.8 The Meaning of Beauty and Aesthetics of the Brain in MIN

The decision for neurosurgery as profession can be promoted not only by ethics but also by aesthetics. The first glance on a living human brain can be an overwhelming experience due to its indescribable beauty. However, the real life will show through the daily work, that beauty is recognized through meaning, like all of recognition. Who cannot recognize it, he cannot see it. To recognize beauty of the brain is to accept and understand its unique meaning directly through the visual pathway. Opening the dura means to enter a complete different and most wonderful world, unique in its appearance. It means to enter a holy cathedral of incomprehensive complexity, aesthetics and vulnerability. Once you have seen the brain, it will not be the same anymore. Self-security and even the remaining self-secret of the brain are lost and something fundamental unusual happens: a brain (surgeon) is seeing a brain (patient). This is an evolution step. It has and needs mental consequences.

Plato, standing on the shoulders of Socrates, proclaimed that beauty leads to understanding by loving the object of understanding. In f-MRI we can image the neuropsychological experience, that things we love and in the status of love we can learn much faster and remember much safer. Beauty gives the meaning of an in-touchable status. The "**in-touchable of the brain**" is the ideal ethical and aesthetical grounding for neurosurgery and leads to a bridge between emotion and cognition. The object, here the brain, gets the highest meaning for the surgeon, bringing his brain into the experience of flow. This results in a rising level of sensitivity and the status of highest readiness, fitness and creativity. Recognition of beauty of the brain will result in a high value for the brain. Ethics and aesthetic are congruent in this process.

Suggested Reading

Baker M. Survey sheds light on the 'crisis' rocking research: 1,500 scientists lift the lid on reproducibility. Nature. 2016;533(7604). http://www.nature.com/news/1-500-scientists-lift-the-lid-on-reproducibility-1.19970.

Bauer AW. Der Hippokratische Eid. Medizinhistorische Neuinterpretation eines (un)bekannten Textes im Kontext der Professionalisierung des griechischen Arztes. In Zeitschrift für medizinische Ethik. Nr. 41. Schwabenverlag. 1995. p. 141–8. ISSN: 0944-7652.

Broderick JP. The STICH trial—what does it tell us and where do we go from here? Stroke. 2005;36(7):1619–20. https://doi.org/10.1161/01.STR.0000170714.43167.34. http://stroke.ahajournals.org/content/36/7/1619.full.

Button KS, Ioannidis JPA, Mokrysz C, Nosek BA, Flint J, Robinson ESJ, Munafò MR. Power failure: why small sample size undermines the reliability of neuroscience. Nat Rev Neurosci. 2013;14:365–76.

CEBM Working Group. Explanation of the 2011 Oxford Centre for Evidence-Based Medicine (OCEBM) levels of evidence. 2016. http://www.cebm.net/index.aspx?o=5653.
Clay RA. The pitfalls of randomized controlled trials: more than one way to measure. Randomized clinical trials have their place, but critics argue that researchers would get better results if they also embraced other methodologies. 2010. http://www.apa.org/monitor/2010/09/trials.aspx.
Edelmann GM. Göttliche Luft, vernichtendes Feuer. Wie der Geist im Gehirn entsteht. Munich: Piper München; 1992. p. 19–359.
Elliott JH, Grimshaw J, Altman R, Bero L, Goodman SN, Henry D, Macleod M, Tovey D, Tugwell P, White H, Sim I. Informatics: make sense of health data. Nature. 2015;527(7576):31–2. http://www.nature.com/news/informatics-make-sense-of-health-data-1.18691.
Feilden T. Most scientists 'can't replicate studies by their peers'. Science & Environment. 2017. http://www.bbc.com/news/science-environment-39054778.
Fiorella D, Zuckerman SL, Khan IS, Ganesh NK, Mocco J. Intracerebral hemorrhage: a common and devastating disease in need of better treatment. World Neurosurg. 2015;84(4):1136–41.
Gödel K. Über formal unentscheidbare Sätze der Principia Mathematica und verwandter Systeme I. Monatsh Math Physik. 1931;38:173–98. https://doi.org/10.1007/BF01700692. Zentralblatt MATH.
Gregson BA, Broderick JP, Auer LM, Batjer H, Chen X-C, Juvela S, Morgenstern LB, Pantazis GC, Teernstra OPM, Wang W-Z, Zuccarello M, Mendelow AD. Individual patient data subgroup meta-analysis of surgery for Spontaneous Supratentorial Intracerebral Haemorrhage. Stroke. 2012;43(6):1496–504. https://doi.org/10.1161/STROKEAHA.111.640284. http://www.ncbi.nlm.nih.gov/pubmed/22511006.
Grunert P. From the idea to its realization: the evolution of minimally invasive techniques in neurosurgery. Minim Invasive Surg. 2013;2013:171369. PMCID: PMC3877623.
Grunert P, Gaab MR, Hellwig D, Oertel JMK. German neuroendoscopy above the skull base. Neurosurg Focus. 2009;27(3):E7.
Hanley DF, Thompson RE, Rosenblum M, et al. Efficacy and safety of minimally invasive surgery with thrombolysis in intracerebral haemorrhage evacuation (MISTIE III): a randomised, controlled, open-label, blinded endpoint phase 3 trial. Lancet. 2019;393(10175):1021–32. https://doi.org/10.1016/S0140-6736(19)30195-3. Epub 2019 Feb 7.
Haug CJ. Peer-review fraud—hacking the Scientific Publication process. N Engl J Med. 2015;373:2393–5. https://doi.org/10.1056/NEJMp1512330.
Ioannidis JP. Why most published research findings are false. PLoS Med. 2005;2(8):e124. Epub 2005 Aug 30. http://www.ncbi.nlm.nih.gov/pubmed/16060722?dopt=Abstract.
Jaspers K. Kleine Shule des philosohischen Denkens. Müchen: Piper GmbH & Co. KG; 1965.
Knosp E, Perneczky A, Resch K D M, Wild A. Endoscopic assisted microneurosurgery. 1. International Congress on minimally invasive techniques in neurosurgery. Abstractbook. 1993. p. 22.
Kuhn T. Structure of scientific revolution. Chicago, IL: University of Chicago Press; 1962.
Langanke M. Medizinische Versorgung: Systemmedizin—Herausforderungen eines aktuellen Ansatzes. Dtsch Arztebl. 2015;112(31):1330–4.
Levy ML. Virtual endoscopic simulations in neurosurgery: technical considerations and methodology. Special Editorial Comment. Neurosurgery. 1998;43:538–48.
Linke DB. Hirnverpflanzung. Die erste Unsterblichkeit auf Erden. Reinbek bei Hamburg: Rowohlt; 1993. p. 11–309.
Linke DB. Kunst und Gehirn. Die Eroberung des Unsichtbaren. Reinbek bei Hamburg: Rowohlt; 2001.
Lisa Rosenbaum MD. Transitional chaos or enduring harm? The EHR and the disruption of medicine. N Engl J Med. 2015;373:1585–8. https://doi.org/10.1056/NEJMp1509961.
Maciunas RJ. Interactive image-guided neurosurgery. USA: AANS Publication Committee; 1994. p. 3–8.
McNutt M. Journals unite for reproducibility. Science. 2014;346(6210):679. https://doi.org/10.1126/science.aaa1724, originally published online 5 Nov 2014.

Mendelow AD, Unterberg A. Surgical treatment of intracerebral haemorrhage. Curr Opin Crit Care. 2007;13(2):169–74.

Mendelow AD, Gregson BA, Fernandes HM, Murray GD, Teasdale GM, Hope DT, Karimi A, Shaw MD, Barer DH, STICH investigators. Early surgery versus initial conservative treatment in patients with spontaneous supratentorial intracerebral haematomas in the International Surgical Trial in Intracerebral Haemorrhage (STICH): a randomised trial. Lancet. 2005;365(9457):387–97.

Oxford Centre for Evidence-based Medicine—Levels of evidence. Mar 2009. www.cebm.net.

Pancioli AM, Broderick J, Brott T, Tomsick T, Khoury J, Bean J, del Zoppo G, Kleindorfer D, Woo D, Khatri P, Castaldo J, Frey J, Gebel J Jr, Kasner S, Kidwell C, Kwiatkowski T, Libman R, Mackenzie R, Scott P, Starkman S, Thurman RJ, CLEAR Trial Investigators. The combined approach to lysis utilizing eptifibatide and RT-PA in acute ischemic stroke: the CLEAR stroke trial. Stroke. 2008;39(12):3268–76. https://doi.org/10.1161/STROKEAHA.108.517656. Epub 2008 Sept 4.

Perneczky A, Müller-Forell W, van Lindert E, et al. Keyhole concept in neurosurgery. New York: Thieme Stuttgart; 1999.

Penrose, Roger: Schatten des Geistes (Shadow of Mind Oxf. Univ. Press 1994) Spectrum Akad. Verl. Heidelberg/ Berlin/ Oxford. 1995.

Platon. Symposion: Rede des Socrates. In: Hildebrandt K, editor. Platon: Das Gastmahl oder Von der Liebe. Stuttgart: Reclam; 1979. ISBN: 3-15-000927-8.

Popper KR. Falsifizierbarkeit, zwei Bedeutungen von. In: Seiffert H, Radnitzky G, editors. Handlexikon zur Wissenschaftstheorie. München: Ehrenwirth; 1989. p. 82–5.

Prasad K, Mendelow AD, Gregson B. Surgery for primary supratentorial intracerebral haemorrhage. Cochrane Database Syst Rev. 2008;(4):CD000200. https://doi.org/10.1002/14651858.CD000200.pub2.

Prinz F, Schlange T, Asadullah K. Believe it or not: how much can we rely on published data on potential drug targets? Nat Rev Drug Discov. 2011;10:712. https://www.nature.com/articles/nrd3439-c1#top.

Resch KDM. Future concepts for minimally invasive techniques in neurosurgery. In: KDM R, editor. Transendoscopic ultrasound for neurosurgery. Heidelberg: Springer; 2005.

Resch KDM. ICH evacuation by MIN techniques. Stroke. 2013;333(Suppl 1):E183–4. https://doi.org/10.1016/j.jns.2013.07.748.

Rossolimo GI. Mozgovoi topograf. J Neuropath Psych Korsakow. 1907;7:640. In Grunert P. From the idea to its realization: the evolution of minimally invasive techniques in neurosurgery. Minim Invasive Surg. 2013;2013:171369. PMCID: PMC3877623.

Sano K, Asano T, Tamura A. Acute aneurysm surgery. Pathophysiology and management. Wien, New York: Springer; 1987.

Schipperges H. Ärztliche Bildung: Tradition—Situation—Projektion. In: Schipperges H (Hrsg.) Ausbildung zum Arzt von morgen. S. 8–22. 1971.

Schmucker C, Nothacker M, Rücker G, Muche-Borowski C, Kopp I, Meerpohl JJ. Bewertung des Biasrisikos (Risiko systematischer Fehler) in klinischen Studien: ein Manual für die Leitlinienerstellung Version 1.0 vom 04.05.2016 Cochrane Deutschland, Arbeitsgemeinschaft der Wissenschaftlichen Medizinischen Fachgesellschaften - Institut für Medizinisches Wissensmanagement. "Bewertung des Biasrisikos (Risiko systematischer Fehler) in klinischen Studien: ein Manual für die Leitlinienerstellung". 1. Auflage 2016. Verfügbar: Cochrane Deutschland: http://www.cochrane.de/de/rob-manual; AWMF: http://www.awmf.org/leitlinien/awmf-regelwerk/ll-entwicklung.html.

Schubert C. Psychoneuroimmunologie und Psychotherapie. Germany: Schattauer Verlag; 2011, ISBN: 978-3-7945-2700-7.

Sehon SR, Stanley DE. A philosophical analysis of the evidence-based medicine debate. BMC Health Serv Res. 2003;3:14. http://www.biomedcentral.com/1472-6963/3/14.

Spitzer M. Digitale Demenz. Berlin: Drömerverlag München; 2012.

Stegenga J. Down with the hierarchies. Nov 2013. https://doi.org/10.1007/s11245-013-9189-4. https://www.academia.edu/27276221/Down_with_the_Hierarchies.

Stegmüller W. Hauptströmungen der Gegenwarts-Philosophie Bnd. I:—Philosophie der Evidenz.—Methodische Phänomenologie.—Angewandte Phänomenologie.—Kritischer Realismus.—Moderner Empirismus.—Grundlagenforschung und Analytische Philosophie der Gegenwart. Kröner-Verlag Stuttgart. 1989.

The Lancet. Research integrity—have we made progress? Lancet. 2017;389(10081):1771. https://doi.org/10.1016/S0140-6736(17)31201-1. http://www.thelancet.com/journals/lancet/article/PIIS0140-6736(17)31201-1/fulltext?elsca1=etoc.

Thorwald J. Im zerbrechlichen Haus der Seele. Drömer-Knauer: Die große Odyssee der Gehirnchirurgie; 1986.

Weinberg R. Future direction in neurosurgery visualization. In: Apuzzo MLJ, editor. Neurosurgery for the third millennium. USA: AANS Publication; 1992. p. 47–64.

von Weizäcker V. Der Gestaltkreis. New York: Thieme Stuttgart; 1950.

Wikipedia: Hierarchy of Evidence. https://en.wikipedia.org/w/index.php?title=Hierarchy_of_evidence&oldid=811593079.

Wikipedia: Philosophy of Science. https://en.wikipedia.org/wiki/Philosophy_of_science.

Yasargil MG. A legacy of microneurosurgery: memoirs, lessons, and axioms. Neurosurgery. 1999;45:1025–91.

Xinyu Zhou, Jianjun Chen, Qi Li, Gaoping Ren, Guoen Yao, Ming Liu, Qiang Dong, Jìng Guo, Leilei Li, Jing Guo, Peng Xie. Minimally invasive surgery for spontaneous supratentorial intracerebral hemorrhage. a meta-analysis of randomized controlled trials. Stroke. 2012;43:2923-2930. http://stroke.ahajournals.org/content/43/11/2923.long.

GPSR Compliance

The European Union's (EU) General Product Safety Regulation (GPSR) is a set of rules that requires consumer products to be safe and our obligations to ensure this.

If you have any concerns about our products, you can contact us on ProductSafety@springernature.com

In case Publisher is established outside the EU, the EU authorized representative is:

Springer Nature Customer Service Center GmbH
Europaplatz 3
69115 Heidelberg, Germany

Batch number: 10732031

Printed by Printforce, the Netherlands